POLYCENTRIC GAMES
AND INSTITUTIONS

INSTITUTIONAL ANALYSIS

Editors, Michael D. McGinnis and Elinor Ostrom

Institutions link political, economic, social, and biophysical processes. *Institutional Analysis* was established to encourage multi-disciplinary, multi-level, and multi-method research on institutions for collective action, public services, resource management, development, and governance. Each book in this series investigates the origins and operation of institutions in particular empirical contexts or their broader roles in the constitution of order in human societies.

Titles in the Series:

Polycentric Governance and Development:
Readings from the Workshop in Political Theory and Policy Analysis
Michael D. McGinnis, editor

Polycentricity and Local Public Economies:
Readings from the Workshop in Political Theory and Policy Analysis
Michael D. McGinnis, editor

Polycentric Games and Institutions:
Readings from the Workshop in Political Theory and Policy Analysis
Michael D. McGinnis, editor

Related Titles on the Analysis of Institutions:

The Meaning of Democracy and the Vulnerability of Democracies:
A Response to Tocqueville's Challenge
Vincent Ostrom

Rules, Games, and Common-Pool Resources
Elinor Ostrom, Roy Gardner, and James Walker,
with Arun Agrawal, William Blomquist, Edella Schlager, and Shui-Yan Tang

Trust, Ethnicity, and Identity: Beyond the New Institutional Economics
of Ethnic Trading Networks, Contract Law, and Gift-Exchange
Janet Tai Landa

Laboratory Research in Political Economy
Thomas R. Palfrey, editor

POLYCENTRIC GAMES AND INSTITUTIONS

Readings from the Workshop in Political Theory and Policy Analysis

Michael D. McGinnis, Editor

Ann Arbor

THE UNIVERSITY OF MICHIGAN PRESS

2003   2002   2001   2000   4   3   2   1

A CIP catalog record for this book is available from the British Library.

Library of Congress Cataloging-in-Publication Data

Polycentric games and institutions : readings from the Workshop in Political
Theory and Policy Analysis / Michael D. McGinnis, editor.
    p.   cm. — (Institutional analysis)
    Includes bibliographical references and index.
    ISBN 0-472-09714-8 (cloth : alk. paper) — ISBN 0-472-06714-1 (pbk. : alk.
paper)
    1. Political science—Congresses. 2. Policy sciences—Congresses.
I. McGinnis, Michael D. (Michael Dean). II. Indiana University,
Bloomington. Workshop in Political Theory and Policy Analysis. III. Series.

JA35.5 .P65   2000
320'.01'13—dc21   00-055961

Contents

Figures

Tables

Series Foreword

Michael D. McGinnis

From its current location in a few scattered office buildings on the Bloom-ington campus of Indiana University, the Workshop in Political Theory and Policy Analysis lies at the heart of a worldwide network of scholars who use institutional analysis to understand and to strengthen the foundations of self-governance. Over the past twenty-five years the political scientists, policy analysts, economists, lawyers, anthropologists, sociologists, psychologists, bi-ologists, ecologists, and policymakers associated with the Workshop have in-vestigated diverse research topics. Results of these research programs have been published in books and journals from several disciplines. A portion of this work has been gathered in this volume; two related volumes are sched-uled to be published at approximately the same time.

Each of these edited volumes exemplifies what is special and distinctive about *institutional analysis* as it has been developed and practiced by Work-shop scholars. Institutions are ubiquitous in contemporary society, and the fields of political science and economics have experienced a recent renais-sance in the study of institutions. The Workshop approach is uniquely multidis-ciplinary, drawing on the complementary strengths of a wide range of social science methodologies: field studies, laboratory experiments, formal models, comparative case studies, opinion surveys, archival research, philosophical investigations, physical measurements, computer simulations, and, most re-cently, satellite imagery. Institutions affect all aspects of social life. Major Workshop research programs have focused on (1) police services in metro-politan centers in the United States; (2) the management of fisheries, irriga-tion systems, forests, and other common-pool resources from California to Nepal (and many places in between); and (3) the macro structure of constitu-tional order from imperial China to the contemporary international system, with particular emphasis given to the nature of American democracy.

Beneath this bewildering variety lies a core message, buttressed by rein-forcing methodological and political foundations. Politically, the goal is to establish and sustain capacities for *self-governance,* by which is meant the structured ways communities organize themselves to solve collective prob-lems, achieve common aspirations, and resolve conflicts. Methodologically, the goal is to understand the institutional foundations of self-governance, that is, to determine which conditions strengthen and which conditions undermine community capacities for self-governance.

In practice these goals have inspired a series of careful, detailed studies of narrow ranges of empirical phenomena. Yet, since each study draws on a single framework of analysis, the overall product has import far beyond the confines of these particular settings. The aggregate lesson of these empirical analyses is clear: many, many self-governing communities thrive, in all parts of the world.

By focusing on community efforts to resolve local problems, the writings of Workshop scholars are sometimes misinterpreted as lending credence to the "small is beautiful" slogan. For many public purposes local community action will be effective, but other circumstances require coordinated policies at the regional, national, or international levels. It is important to remember that public officials at all levels of aggregation have important roles to play in helping communities provide for their own needs.

Shouting slogans about the desirability of decentralization or civil society contributes little toward the crucially important task of sustaining capacities for self-governance. The challenge of institutional analysis lies in producing solid research findings, based on rigorous empirical tests of hypotheses grounded in carefully articulated theories and models. Institutional analysts have a responsibility to combine policy relevance and scientific rigor.

A basic tenet of institutional analysis is that multiple arenas, or centers, of interaction and participation need to be considered simultaneously. Self-governance works best if the overall governance structure is *polycentric*. The word itself may be awkward, but it encapsulates a way of approaching the study of politics and policy analysis that stands in sharp contrast to standard modes of thought. Governance does not require a single center of power, and governments should not claim an exclusive responsibility for resolving political issues. Instead, politics should be envisioned as an activity that goes on in many arenas simultaneously, at many scales of aggregation. Implications of polycentric governance for particular empirical and theoretical contexts are detailed in the readings included in these volumes.

To illustrate the *coherence* of the theoretical approach that underlies applications to a wide array of empirical domains, a selection of previously published articles and book chapters has been collected into three books with similar titles: *Polycentric Governance and Development, Polycentricity and Local Public Economies,* and *Polycentric Games and Institutions.* Each book addresses a separate audience of scholars and policy analysts, but all of them will be of interest to anyone seeking to understand the institutional foundations of self-governance.

The essays in *Polycentric Governance and Development* demonstrate that empirical analyses of the management of irrigation systems, fisheries, groundwater basins, and other common-pool resources have important implications for development policy. Long before "sustainable development" became an over-used slogan, scholars associated with the Workshop were trying to understand the myriad ways self-governing communities had already achieved that goal in practice.

After an initial section on the general conceptual framework that has influenced research on the full array of Workshop research topics, *Polycentricity and Local Public Economies* presents essays published from the first major empirical project associated with the Workshop, a comparative study of the performance of police agencies in metropolitan areas of the United States. Although most of the research results included in this volume date from over a decade ago, they remain relevant today. Recent trends toward community policing, for example, reflect the continuing influence of factors identified in this research program.

In *Polycentric Games and Institutions* the general concepts that guided these empirical analyses themselves become the focus of analysis. Workshop scholars use game theory and laboratory experiments to understand how individuals behave in the context of diverse political and economic institutions. Results from laboratory experiments and field settings show that individuals draw upon an extensive repertoire of rules or strategies from which they select different strategies, given their understanding of the nature of the situation at hand.

By collecting readings on similar topics that were originally published in scattered outlets, we hope to highlight the contribution these research programs have made to their respective fields of study. Any evaluation of the scholarly contribution of *institutional analysis* as a whole, however, must be partial and incomplete, for the Workshop remains an active place. Each of these research themes is being pursued by scholars who have long been associated with the Workshop and by a new generation of scholars.

Each article or book chapter is reprinted without changes, except for a few minor corrections to the published versions. To avoid duplication of material and improve the flow of this presentation, textual deletions have been made in a few of the selections. Citations to forthcoming books and articles have been updated, and cross-references to essays included in the other volumes have been added. Otherwise, reference and footnote conventions used in the original sources have been left intact.

Selection of an appropriate set of readings was a daunting task, for the list of publications is long and diverse. I enjoyed digging through the extensive files of reprints, and I wish we could have included many more readings, but that would have defeated the purpose of compiling accessible surveys of selected Workshop research programs. I tried to minimize overlap with the most influential and widely available books that have emerged from these research programs. Each edited volume includes an integrative introductory essay, in which frequent references are made to the many other books and journals in which the results of these diverse research projects are reported. Each book also includes an annotated list of suggested readings.

One final caveat is in order. Elinor Ostrom and Vincent Ostrom are authors or coauthors of many of the readings in all three books. Both have served jointly as co-directors since its establishment in 1973. Without doubt, these two individuals have been absolutely crucial to the success of the Workshop. Even so, they would be the first to insist that they have *not* been the only reason for its success. Collaboration has always been a hallmark of the Workshop. Many individuals have made essential contributions, as will be apparent throughout the readings in these books. Yet it is impossible to imagine how the Workshop could have been established or sustained without the tireless efforts of Elinor and Vincent Ostrom. Their influence will continue to shape the future direction of the Workshop for years to come.

Acknowledgments

In an edited volume composed of papers written over a span of thirty years, the assistance of many, many people and institutions deserves to be acknowledged. Fortunately, the contributors have already thanked those who assisted in the original preparation of each of the journal articles or book chapters reprinted here.

As editor, I am going to start my acknowledgments with Patty Dalecki. Her assistance has been invaluable for the preparation of all three of the volumes. Patty kept track of all the essays included in these volumes, arranging for copyright permissions and overseeing the process of translating them all into a single format. For this particular volume, however, most of the work was done by Anne Leinenbach, a relatively new addition to the Workshop staff. Anne ably guided this volume through several versions and typeset the final manuscript. It's clear Anne is off to a roaring start at the Workshop, and we hope she hangs around for many years to come.

I also want to thank all of the staff members who helped Patty, Anne, and me at various stages in this process. Amber Cleveland, Sara Colburn, Ray Eliason, Bob Lezotte, David Wilson, and Cynthia Yaudes all helped in scanning and proofreading the manuscripts or in posting drafts of the introduction and contents list on the Web. One of the strengths of the Workshop has always been our competent, hardworking, and friendly staff. Students also contribute in important ways. For example, T. K. Ahn prepared the comprehensive index for this volume.

At the other end of the publication process, I am deeply appreciative of the support and guidance offered by Colin Day, director of the University of Michigan Press. His comments and suggestions were crucial, all the way from initial planning of the volumes to preparation of the final manuscript. I would also like to thank two anonymous reviewers for their helpful comments and suggestions on drafts of the introduction and on the overall organization of the volume.

The research results published here emerged from a long series of overlapping research projects. Appropriate funding agencies are acknowledged in their respective essays. We also want to thank the Bradley Foundation for providing financial support for the preparation of this volume.

None of this activity would have been possible if the contributors to this volume had not written such excellent research reports in the first place. I thank each of the contributors for providing the foundational material out of which this book was constructed. They all returned corrections to the scanned versions in short order, helping us complete this book in a remarkably short period of time.

I also want to say a few words to those scholars associated with the Workshop whose work I was not able to include in this volume. I wish the book could have been twice as long, and I repeatedly found it necessary to restrain my enthusiasm for including more papers. I especially want to express my appreciation to those authors who were willing to proof papers that, for whatever reason, had to be cut from the final list.

Two of the contributors play uniquely pivotal roles. Elinor and Vincent Ostrom have inspired, encouraged, and supported all of the research included in this volume. Each has had a major impact on my own career, helping to broaden my interests and to sharpen my analytical skills. I can't thank them enough.

Introduction

Michael D. McGinnis

Since people establish and use institutions to help them achieve a wide array of individual and collective goals, a fuller understanding of institutions is crucial to any social science. In recent years, political scientists, policy analysts, economists, and other scholars interested in the analysis of institutions have increasingly used game theory, other formal models, and experimental research to study the complex decision problems confronting individuals and groups. Game models are especially useful for this purpose because they focus attention directly onto the underlying configuration of common, conflicting, and complementary interests facing participants in different settings.

Simplicity in representation is key to the success of game theory and laboratory experiments, but simple models are necessarily too abstract to capture all of the important aspects of inherently complex institutional arrangements (McGinnis 1991). Several scholars affiliated with the Workshop in Political Theory and Policy Analysis at Indiana University have made important contributions toward bringing a richer empirical content to formal models and experimental analyses of institutions. The readings collected in this volume illustrate several varieties of *institutional analysis* as exemplified in the research of Workshop scholars. Each reading uses different levels of formal or experimental rigor to address similar sets of questions concerning institutions and governance.

Contributors address such factors as the physical nature of goods, the overall structure of public economies, tensions between agents and principals, majority voting and other procedures by which collective choices are made, the implications of alternative legal doctrines and property rights systems, the effects of government regulations on markets, and the origins of the normative expectations upon which individuals base their choices in field settings and in laboratory experiments. Each reading illustrates different ways in which the basic principles of game theory and modern political economy can be applied to deepen our understanding of particular empirical situations or to clarify the meaning of norms, rules, and institutions.

Polycentric Games

In a classic work of political economy, V. Ostrom, Tiebout, and Warren (1961) introduced the term *polycentric political system* to denote a political order in which multiple authorities serve overlapping jurisdictions. Their original referent was to patterns of metropolitan governance in the United States, but polycentricity has important implications for patterns of governance in all parts of the world (see McGinnis 1999a,b). For this volume I coined the term *polycentric games* to capture the distinctive nature of the Workshop approach to formal models of institutions.

This term generalizes and extends the logic behind the widely known and influential concepts of *two-level games* (Putnam 1988) and *nested games* (Tsebelis 1990). Putnam's model details how potential problems with the domestic ratification of international agreements can affect the process by which those agreements are reached in the first place. Tsebelis demonstrates that actions that seem irrational in one context may be perfectly understandable once analysts incorporate that actor's strategic interactions with other actors. In both cases, the important point is that analysts must expand the scope of their models to consider the cross-effects of concurrent games.

Polycentric games encompass a wider array of concurrent games linked into complex networks of interactions. A basic presumption is that actors in any single game can draw on a vast array of informational cues and strategic interactions to help them understand the behavior of other participants in that particular game. To do so they can draw on commonly understood norms or rules, and they can use institutional procedures to organize their interactions. Still, it is not enough to say that many games are occurring simultaneously; some guidance has to be given concerning how to handle the resulting complexity. As the readings included in this volume demonstrate, Workshop-affiliated scholars have developed conceptual frameworks and analytical tools that bring some coherence to the study of polycentric games.

Levels of Analysis and Arenas of Choice

Research on two-level, or nested, games directs attention to connections across domestic and international levels of analysis, defined as scales of aggregation. International relations scholars have identified several such levels of analysis, ranging from individuals to small groups to organizations to social orders to national governments to international systems and the global ecosystem (Russett and Starr 1996). International relations scholars use these levels to differentiate among supposedly distinct sets of explanatory factors, each of which brings a unique perspective to the study of international relations (see Waltz 1959). For example, personality or perceptual factors at the individual level of analysis can be understood without reference to the general tendencies of bureaucratic behavior. From this perspective, the diverse logics of democratic and autocratic governments would not be relevant for a systemic-level explanation of the global balance of power (Waltz 1979). Although some influential scholars (notably Waltz 1979) assert that attention to a single level of analysis is sufficient for the resolution of important puzzles in international relations theory, the recent popularity of two-level games (Putnam 1988) illustrates the growing realization that it is often essential to cross traditional levels of analysis.

Institutional analysis of polycentric games is based on a related distinction among different *arenas of choice:* operational choice, collective choice, and constitutional choice. This distinction is an essential component of the institutional analysis and development (IAD) framework, which is explained more fully in chapter 2 of this volume (and in E. Ostrom, Gardner, and Walker 1994).

In the operational choice arena, concrete actions are undertaken by those individuals most directly affected or by public officials. The outcomes of these actions directly impact the world in some demonstrable manner. The rules that define and constrain the operational activities of individual citizens and officials have been established by processes occurring in arenas of collective choice. Furthermore, agreement on the rules by which these rules themselves are subject to modification is the subject of deliberations of constitutional choice. In some circumstances, constitutional choice results in preparation of a written constitution, but more generally communities develop informal shared understandings about the ways in which a particular community organizes itself. These shared understandings are an essential component of the decisional context for collective choice and operational activities. Similarly, rules originating from the collective choice arena demarcate the range of permissible actions that is available to participants in the operational arena.

Authors of the essays reprinted in this volume originally used the term *level of analysis* to differentiate among these "arenas of choice." However, I find this usage confusing because these levels do not refer to observable scales of aggregation. Instead, the IAD framework points toward distinctions that signify differing modes of interaction, and each form of interaction can occur at all different scales of aggregation or among the same group of individuals at different points in time.

For example, constitutional choice typically involves a wider range of participants than routine operational choices, but a small community may commonly engage in all three forms of interaction. A community's efforts to decide how it will make future decisions can be analytically distinguished from times when it is simply debating the application of rules to the current situation. Periods of reflection on basic issues of constitutional order can be usefully distinguished from periods of routine application of rules. The crucial point is not the number of people involved but rather the nature of the activity occurring in different settings or at different times. For this reason, the term *arena of choice* is more appropriate than *levels of analysis*.

Important analytical similarities apply to all three arenas of choice. Actors in each arena confront an *action situation* with strategic options and role expectations as defined during periods of more encompassing interactions (analogous to "higher" levels of analysis). In this sense, individual and collective choice is constrained (by the institutional context) to some restricted range of strategic options. In the IAD framework, action situations are shaped by the configuration of physical conditions, attributes of the community, and the rules-in-use within that arena of choice. Applications also require a "model of the actor," that is, some representation (whether formal or informal) of the ways in which the relevant individuals or organizations decide. At the individual level of analysis this model would be based on some conceptualization of human cognition, whereas at the organizational level models of organizational inertia or bureaucratic politics might be most appropriate. Action situations at different levels of aggregation involve different kinds of actors, but these actors should exhibit different patterns of behavior depending on which arena of choice is being evaluated.

Outcomes of choice processes in one arena produce patterns of interactions and outcomes that shape the nature of interactions in the other arenas (especially in "lower" arenas). Influences move back and forth across arenas in complex but understandable patterns, as detailed in chapters 2 and 3 of this volume. In short, institutions matter because they link arenas of choice by defining the roles filled by individual or collective actors and shaping the perceived costs and benefits of alternative strategies. If one looks at political situations in this manner, then all three arenas (or centers) of choice are (at least potentially) involved in any particular application. The term *polycentric* (originally coined by V. Ostrom, Tiebout, and Warren 1961) nicely conveys an image of a network of overlapping and interlinked arenas of choice. The essence of a "game" in the technical sense defines interactions among participants with at least some difference of interest. Polycentric games encompass concurrent games of operational choice, collective choice, and constitutional choice that interweave among themselves at different levels of aggregation.

This broad conceptual framework provides an overall context for each of the more specialized analyses included in this volume. Foundational constitutional questions are rarely in doubt during the routine operational decisions that are most often the subject of game models. Yet, the outcome of processes of constitutional choice cannot be ignored, for such deliberations determine who has the capability or the responsibility to participate in operational decisions. Such situations are too complex to model in their entirety, but the essays collected here demonstrate that important insights can be garnered from partial models of specific situations, provided due consideration is given to concurrent or supplementary processes in other arenas of choice.

Organization of This Volume

The readings collected in this volume are organized into five parts. The essays in part I lay out inter-related frameworks for analysis: the IAD framework, a fuller representation of the action situations occurring in each arena of choice, and a conceptual grammar that differentiates among strategies, norms, and rules. This part begins with a more general essay, placing the Workshop approach to institutional analysis in the context of other influential research on political economy.

Each of the remaining parts addresses particular empirical settings or types of institutional arrangements. Part II focuses on voting institutions and other ways in which the interests of principals and agents can be (partially) reconciled. Part III investigates particular configurations of rules and regulations, especially those related to the management of natural resources. Workshop scholars have demonstrated the crucial importance of monitoring and sanctioning in maintaining effective systems of resource management and of governance, and the chapters included in part IV use different game models to investigate the expected performance of monitors with different incentive structures. The two essays in part V provide overviews of an extensive body of experimental research that indicates that the behavior of rational individuals is more nuanced and variated than is generally appreciated. The final essay in this part, Elinor Ostrom's recent presidential address to the American Political Science Association, returns to the general conceptual concerns laid out in chapter 1. In this essay she points toward future development of a "second generation" of rational choice models that can incorporate the behavioral determinants of individual action—and thus of the consequences of institutions.

As noted in the series foreword, this volume has been compiled in conjunction with two other edited volumes of previously published research by Workshop-affiliated scholars (McGinnis 1999a,b). This volume includes technical pieces of research selected from a much broader array of research methodologies used by Workshop scholars. It is primarily intended for use by scholars and students with some expertise in game theory and experimental research.

One unfortunate consequence of this decision to concentrate technical readings in a separate volume is that it may appear as if this formal or experimental work was tangential to the development of the extensive empirical research programs also associated with the Workshop. Nothing could be further from the truth, for game models and laboratory experiments have played important roles in the evolution of Workshop research programs on metropolitan governance, resource management, and development (McGinnis 1999a,b). Formal models highlight the fundamental structure of action situations, whereas the relatively controlled environment of the laboratory setting allows researchers to isolate the effects of particular factors on the ability of groups of individuals to cooperate for their common good. The benefits of cross-fertilization across formal models, experiments, and field research have been shown in E. Ostrom, Gardner, and Walker (1994). Formal and experimental methods supplement the strengths of field research; no one method can tell us everything we need to know.

Because of the diversity of topics covered in this volume, each part begins with an introductory essay that locates the works included in that part within the broader context of related research programs. These brief introductions focus on the essays included in this volume, along with a few closely associated works that could not be included here. The remainder of this general introduction provides a quick tour of the vast expanse of the field of modern political economy. This sketch cannot hope to be comprehensive; its only purpose is to highlight the unique contributions of institutional analysis to the study of institutions, public policy, and constitutional order.

Institutional Analysis and Modern Political Economy

Institutional analysis began as a variant of public choice or political economy. In fact, in early works *public choice* was the preferred term used by the founders of the Workshop (as in chap. 1 of this volume). Over time, however, public choice came to be understood as a particular type of research, focused on the use of microeconomic models to represent selected aspects of political interactions (Mueller 1989). A related line of research that focused more explicitly on voting as the central process of political decision making became associated with the term *social choice* (Riker 1982). Workshop scholars eventually settled on the term *institutional analysis* as a label for their unique mode of research.

These research traditions are associated with particular institutions of higher education in the United States. In a comparative review of the Virginia (public choice), Rochester (social choice), and Indiana (institutional analysis) schools, Mitchell (1988) concluded that the latter differed from the other two in its reliance on inductive empirical research rather than the development of formal models per se. Another major difference is that the work of Herbert Simon (1957, 1964, 1997) on public administration and on the cognitive limitations of human rationality had a major influence on institutional analysis. As a consequence, the game models used by Workshop-affiliated scholars typically include some aspects of "bounded rationality" (E. Ostrom, Gardner, and Walker 1994).

Laboratory research by Workshop-affiliated scholars has contributed toward development of a basic model of the individual as having access to a repertoire of alternative strategies and normative predispositions (chap. 16). In some situations the behavior of individuals or corporate actors will be subject to stringent forms of selection pressure, in which case the standard model of complete rationality would be appropriate. In other circumstances, behavior based on cognitive heuristics may suffice for an entity to survive until later iterations of the game. For if any actor is involved in multiple interactions simultaneously, then it becomes important to incorporate trade-offs between the allocation of that actor's time and effort to optimize behavior in each of these concurrent games. Different institutional contexts can elicit quite different forms of behavior from the same individual or organization.

Workshop scholars keep the broader institutional context in view by making use of inductive modes of research. Most of the models included in this volume were inspired by observation of the ingenious solutions devised by specific communities to address particular problems. Still, this research addresses the central problem of modern political economy, namely, the ever-present tension between the potential benefits of cooperation and the temptations of individual opportunism.

By focusing attention on strategic interactions among purposeful actors, game theory helps clarify how the independent actions of individual or other actors may result in collective outcomes that none would have preferred. In a still-influential early classic, Schelling (1960) defined game theory as a "theory of interdependent decision," but it has subsequently developed into a set of tools that can be applied to model a wide array of situations. The content of the game theoretic tool kit has changed over time, with alternative solution concepts developed for application to different types of games or based on different conceptualizations of the nature of human cognition and decision (Kreps 1990a,b). Workshop scholars have contributed to this ongoing dynamism, as shown both in this volume and in E. Ostrom, Gardner, and Walker (1994).

Markets, States, and Voting

Like other forms of modern political economy, institutional analysis builds on the basic principles of methodological individualism. The individual is taken to be the foundation of analysis, since individual choice is crucial to all social outcomes. Still, it is equally important to understand the institutional context within which individual choices occur, as well as the ways in which individual choices shape these institutional contexts. Institutions matter, in many different ways, but, ultimately, their impacts are filtered through choices made by individuals.

This emphasis on the joint importance of individual choice and institutions can be usefully contrasted with the widespread (but not universal) attitude among economists that markets are natural phenomena. The presumption is often made that individuals with differing endowments of goods or skills will automatically begin to engage in mutually beneficial exchanges. In the proper circumstances, voluntary exchanges of private goods certainly do result in collectively beneficial outcomes, as if guided by an "invisible hand" (Smith [1776] 1976). However, markets are themselves institutions that operate at optimal efficiency only if they are buttressed by secure political and legal institutions.

The standard point of departure for modern political economy is to treat governments as a response to market failures (see, e.g., Weimer and Vining 1989, chap. 3). Markets are an efficient way to organize the production and exchange of private goods, but it is widely recognized that competitive markets are less appropriate for the production of public goods or the management of common-pool resources. Workshop scholars have contributed to extensive empirical research programs on the provision of public goods and services in metropolitan areas and on the management of common-pool resources in mostly rural settings throughout the developing world (see McGinnis 1999a,b). In conjunction with these empirical investigations, Workshop scholars have also investigated the ways in which formal models can be used to expand the analytical repertoire of institutional analysts. The present volume focuses on these conceptual contributions.

The canonical alternative to privatization is to rely on "the state" (or "the government"), a Hobbesian sovereign that can readily impose policy decisions on recalcitrant individuals. Those who might seek to free ride can, in this way, be forced to pay taxes or similar extractions. However, hierarchical structures have their own dilemmas, specifically related to agent-principal relations. Can the public be sure that the ruler (as their agent) will serve the public interest? Can the ruler (as principal) insure that bureaucratic agents have the proper incentives to gather accurate information or to implement the sovereign's decisions properly? Tensions between the conflicting interests of agents and principals are impossible to resolve completely (Miller 1992).

From the point of view of political science, the limited ability of citizens to monitor or sanction rulers presents the fundamental problem of governance. Without this capacity, rulers are likely to shirk their responsibilities and extract more resources for their own use. Liberal democratic institutions attempt to make rulers accountable to the general population via the mechanism of elections. The basic idea sounds simple enough: those public officials who fall short of public expectations can be removed from office in the next election. Naturally, this idea does not turn out to be quite so simple in application.

Much of the modern political economy literature deals with various aspects of voting institutions. Social choice theorists have demonstrated that inconsistent policy can emerge from majority rule, or, for that matter, from any means of aggregating individual preferences into social outcomes in a way that satisfies basic criteria of fairness (Arrow 1963; Riker 1982). In some contexts, as when the relevant policy issues can be represented on a single dimension, competing candidates will offer nearly identical positions, in hopes of winning the approval of the "median voter" (Downs 1957). Conversely, under conditions of multidimensional issue spaces, practically any outcome could emerge from agenda manipulation or strategic voting (McKelvey 1976; Schofield 1978; Riker 1982). Myriad institutional arrangements have been devised to limit the volatility of majority rule. For example, committees may achieve coherence by limiting the range of alternatives considered (Shepsle 1979), and the nature of committees may help structure overall legislative outcomes (Krehbiel 1991).

From the Workshop perspective, voting is only one way in which collective decisions may be made. In many empirical contexts studied by Workshop scholars, specifically the management of common-pool resources, majority rule plays at best a minor role. Some Workshop scholars have focused on the implications of voting institutions (see part II), but, overall, a distinctive aspect of institutional analysis is the low profile given to studies of voting systems. Voting is treated as an institution for collective choice and not as an end in itself.

As mentioned earlier and detailed more fully in chapter 2, operational choice games are directly affected by the accepted procedures for collective choice processes, which were in turn determined by some process of constitutional choice. Buchanan and Tullock (1962) suggest one distinction between constitutional and operational choice, arguing that constitutional choice requires unanimity (at least in broad outlines) whereas choices in different issue areas might be determined according to simple or extraordinary majorities. Institutional analysts should not neglect other means of making collective decisions in any of these arenas.

However decisions are arrived at, one key issue is the process by which subsequent behavior is monitored and rule violations sanctioned. Principals have to be able to monitor the behavior of their agents before they can sanction them for poor performance. Such monitoring may be relatively transparent in a small community, but agents typically serve multiple principals with divergent interests. This monitoring function is perhaps the most important reason why diverse institutions are ubiquitous in complex societies.

New Institutional Economics, Rational Choice Institutionalism, and Institutional Analysis

Institutional analysis is closely related to other important traditions of research on institutions. "New institutional economics" (NIE) emphasizes the importance of transaction costs and property rights. Williamson (1975, 1985, 1996) has shown that nonmarket, nonhierarchical institutions are most appropriate for many empirical contexts. The basic problem is that potential gains from trade may not be realized if the actors do not have sufficient confidence that the other side will refrain from taking advantage of any temporary or limited dependency this exchange might imply. Different institutional forms minimize the costs of transacting in different settings. This is not a new problem. Milgrom, North, and Weingast (1990) model the medieval European institution of a "law merchant" to illustrate how the rulings of informal judges gave suspicious traders sufficient confidence to engage in long-distance trade, without establishing a central governing authority.

Influential research by Douglass North (1981, 1990; North and Thomas 1973) has demonstrated that a clear definition of property rights is essential before market processes can operate at anywhere near efficient levels. Economic growth requires investor confidence, for individuals or private corporations will make investments to improve the productive capacity of their assets only if they expect to enjoy the benefits of these investments. One theme common to Williamson's microlevel investigations of alternative organizational forms and North's macrolevel examination of property rights and economic growth is that efficiency is the most important criterion to be used in evaluating institutions (Eggertsson 1990).

These works have inspired many political scientists and policy analysts to study political institutions. The term *rational choice institutionalism* (RCI) denotes efforts to extend rational choice theory beyond individual choice or market exchange to encompass the entire range of political interactions. Three of the most influential scholars in this tradition are Bates, Miller, and Knight. Bates (1981, 1983, 1989) focuses on explaining the institutional foundations of the typically inefficient agricultural policies pursued by governments in postcolonial Africa. Miller (1992) stresses the principal-agent problems that bedevil all hierarchical organizations. Knight (1992) places unrelenting emphasis on the conflicts of interest that shape the design and operation of all institutions.

The basic thrust of RCI research is to insist that all politics takes place within institutionalized contexts. NIE scholars have similarly stressed the extent to which economic exchanges, even in competitive markets, take place within highly institutionalized contexts. One original contribution of institutional analysis (IA) is the extension of the analytical tools of methodological individualism to nonmarket and nonhierarchical institutions. Markets and states have received by far the bulk of attention among new institutional scholars in economics and political science, but many other kinds of institutions exist and play important roles in governance. E. Ostrom (1990), for example, shows that communities throughout the world use many alternatives to market exchange and hierarchical organization. In short, IA focuses on those institutions that NIE and RCI analysts overlook.

Common-Pool Resources and Institutions of Civil Society

Institutions involved in the management of *common-pool resources* have been one major focus of concentrated research by scholars associated with the Workshop. This type of resource combines characteristics of both private and public goods. It is like a private good in the sense that when one person appropriates part of a common-pool resource (CPR), that portion is no longer available for another person's use. At the same time, it resembles a public good in the high cost of excluding others from appropriation. Significant time and effort are required to establish rules, monitor compliance, and sanction rule violators.

If individual incentives to overexploit an open-access CPR are not restrained in some manner, then the expected outcome is exhaustion of the resource, an outcome identified by Hardin (1968) as the "tragedy of the commons." Each of the generic ways to avoid this outcome effectively changes the nature of the game (or "action situation") by imposing a structure of rights to that CPR. Different structures of rights give the relevant actors incentives to treat that CPR as if it were a different form of good. Privatization schemes parcel a resource into excludable units, which resource users then treat as if they were private goods. Their behavior can be expected to have all the benefits (and limitations) associated with markets in private (exclusive) property. If a resource is salient to a wider community beyond those directly affected by the conditions of this resource, then it might be treated as a public good fit for management by public officials. Of course, this solution entails all the agency and information problems typical of hierarchical organizations.

Scholars associated with the Workshop have shown the existence and common effectiveness of an alternative scheme, in which self-organized user groups manage the appropriate resource. For the most part, this research has been inductive in nature, using successful instances of self-governing groups as an "existence proof" of the reality of successful CPR management (E. Ostrom 1990). Claims are not made that user groups always achieve "efficient" outcomes. Instead, the remarkable thing is that many groups effectively manage common-pool resources over long periods.

Self-governing resource user groups are an important component of "civil society," which is taken to cover all activities not obviously part of either the private or public sector. Security of user-group rights can be justified in terms closely analogous to the protection of private property rights. Those groups of resource users who have successfully managed their common resources will have done so at the cost of establishing and enforcing rules that often call for self-sacrifice by individual members of that group. The members of this group are unlikely to continue to pay those costs if they expect governmental intervention to establish or enforce different rules. Similarly, owners of private property will be reluctant to invest in its future productivity if their property rights remain insecure.

For many of the same reasons, then, liberal governments should act to protect both private property rights and collective rights to manage common-pool resources. Just as individuals are presumed to be the best judge of their own tastes, user groups should be presumed to be capable managers of common property. A basic tenet of public policy should be that those groups who are able to manage CPRs effectively should be allowed (and encouraged) to do so. Government intervention should be undertaken only when user groups fail to manage their resources effectively or if user groups violate general standards of fairness, accountability, or other widely shared concerns. Instead of presuming that governmental officials or scientific experts know best how to manage CPRs, user groups should be given the initial benefit of the doubt.

Community management is not always successful, as shown by situations approximating Hardin's tragedy of the commons. In such cases of community failure, government intervention may be just as essential as it is for cases of market failure. The success of some user groups does not imply that governmental intervention is never appropriate. Since one of the most effective ways to facilitate cooperation among a group is to exclude people who have significantly different interests, there is no reason to presume that all groups will manage their resource equitably. Government intervention to prevent gross violations of general standards of fairness seems as reasonable as comparable restrictions in markets in certain private goods. For example, virtually all governments prohibit explicit trade in sexual services.

The analogy between market failure and user-group failure (or community failure) is quite close. In both cases the initial presumption argues against government intervention but allows such action to prohibit certain actions deemed unfair or unacceptable by society. Considerable room for debate remains open in the determination of which practices should be prohibited, and it is the responsibility of institutional analysts to provide a solid empirical foundation for these collective deliberations. The general point is that no one institutional solution will work in all empirical settings.

Markets, states, and user groups should be seen as complementary institutions. None can live up to their potential without support from the others (see Lichbach 1996). So far, however, markets and states have received the bulk of attention in the policy literature, and research by Workshop-affiliated scholars helps correct this imbalance.

Rent Seeking, Efficiency, and Self-Governance

Williamson (1996) defines governance as the efforts of public authorities to shape the transactions facing individuals in market exchange. Laws make some transactions easier to arrange, while making other, socially undesirable interactions more difficult for private actors to consummate. In the same way, public laws and policies shape the transaction costs facing elements of civil society. In this way, governments can play an essential role in laying the foundation for successful group management of common-pool resources and for the efficient operation of markets.

Unfortunately, in much of the world, governments actively undermine the ability of many groups to organize their own activities and introduce market distortions that further the interests of other, more influential groups. In totalitarian states the assertion of state control is taken to its logical extreme. In the petty autocracies so common in the developing world, governmental officials routinely assert the right to grant permits that individuals, corporations, or groups need to operate as a private business or as a user group. This right to grant permits is, of course, a lucrative source of bribes and corruption. In such circumstances, governmental office is treated as a private resource.

Whenever political authorities or bureaucratic officials are granted exclusive rights to the writing or enforcement of laws and regulations, the potential for rent creation is unavoidable. In this sense, rents are created whenever political authorities create artificial scarcities. This can be done by restricting production or entry into a market to a certain type of individuals, by requiring permits, or by putting restrictions on imports. Those who stand to benefit from such restrictions will engage in costly activities to obtain these benefits, and this "rent-seeking behavior" detracts from the overall welfare of society. The term *rent seeking* is used to cover a multitude of sins, and space does not allow a full discussion of the associated controversies (see Mueller 1989, 229–46; Tollison 1980, 1997). The key point is that rents prevent a market economy from attaining the efficiency of perfect competition. Recipients of the specific rents will benefit, but in the long term society as a whole will suffer.

The specific nature of the rents differs for different types of political order. In an autocratic regime, for example, the recipients of rents are likely to be members of the ruler's family or others with a close personal connection to the ruler. Those groups (such as the military or domestic security forces) upon which the ruler is most dependent may benefit greatly from the ruler's policies (see, e.g., Wintrobe 1998). In liberal democratic regimes, the recipients of protection are often interest groups that reward public officials with votes or campaign contributions (see, for e.g., Magee, Brock, and Young 1989).

Rents and other distortions to the operation of private markets are important topics for the field of public choice. In brief, public choice scholars apply economic methods of analysis to the behavior of public officials. In contrast, the still influential tradition of welfare economics presumes that some "benevolent social planner" is both willing and able to implement policies that are in the best interest of society as a whole (see Mueller 1989, chap. 19). In the public choice tradition, elected and appointed officials are presumed to be motivated by their own selfish interests, whether or not they are interested in maximizing social welfare. Niskanen (1971), for example, argues that agency heads act to maximize their agency's budgetary resources whenever possible. Such assumptions, while simplistic, can bring important insights to the interactions of public and private actors.

Public choice modelers often use the preferences of the "median voter" as an indicator of the likely outcome of democratic processes. In effect, the median voter's preferences act as surrogate for the social welfare function that was so deeply undermined by Arrow's theorem. In this way, modelers retain the ability to make assertions concerning the relative desirability of alternative policies or institutional structures for society as a whole.

Those of us who have spent time at the Workshop are skeptical whenever an administrator or a policy analyst claims to understand or implement policy for the greatest good of society. Vincent Ostrom (1989) identifies the pernicious effects of this widely shared attitude on the field of public administration, while laying out an alternative vision of democratic public administration within the context of polycentricity (to be discussed more fully subsequently).

Institutional analysis shares many characteristics of public choice, but its scope is wider. Self-governing groups need not delegate authority to particular individuals, not if they can handle the relevant tasks themselves. In other circumstances, delegation may be crucial, especially for the services of monitoring and sanctioning rule violators. Concerns about rent-seeking or agent-principal relations are essential in the latter context but may not be relevant in the former. Institutional analysis can be seen as a form of public choice where the focus is not restricted to the behavior of public officials and where citizens are not treated merely as voters or campaign contributors.

Public choice theory brings to political economy a relentless focus on the importance of efficiency in public policy. Clearly, the efficient allocation of resources to production or consumption is a goal worthy of careful pursuit. However, individuals or communities may decide, for whatever reason, to sacrifice efficiency for the pursuit of other goals, such as accountability, fairness, or sustainable development (however these terms are understood within the relevant communities).

One reaction is to expand the definition of efficiency to incorporate these additional goals into the underlying utility or production functions being modeled. This practice can be very useful for models of the long-term sustainability of resource management practices (see Dasgupta and Heal 1979). Workshop-affiliated scholars typically take a more indirect approach, to define economic efficiency more narrowly and to supplement analyses with explicit consideration of other collective goals. For example, Elinor Ostrom (1990) lays out eight "design principles" that were common to many instances of the successful long-term management of common-pool resources. Economic efficiency is not specifically included in this list, although it is implicit in the requirement for congruence between the rules-in-place and the physical nature of the goods. But the gist of this congruence principle refers instead to the importance of a meaningful correspondence between the rules-in-use and the normative beliefs of that community. That is, participants must feel that the rules are "fair" according to the standards of the relevant community, or else they will not put forth the effort required to overcome their dilemmas of collective action. This is not to say that economic efficiency is irrelevant, for too much sacrifice of efficiency may mean destruction of the resource. In some institutional arrangements, especially when competitive markets in private goods are available, the criterion of efficiency takes on a unique prominence. In other situations, multiple evaluative criteria must be given equal consideration.

Collective Action, Community, and Networks of Organizations

The ability of self-governing groups to resolve their own practical problems, as they themselves define the problems, is the central theme that unifies all of the many research programs undertaken by Workshop scholars. However, self-governance is not a simple matter of groups with homogeneous tastes forming a "club" (Sandler and Tschirhart 1980). Nonprofit organizations, for example, are an important expression of self-governance and an essential component of civil society. Hansmann (1980), Weisbrod (1977, 1988), and Salamon (1987, 1995) treat the nonprofit sector as a third sector of the economy that fills important roles left open by both the private and public sectors. From this point of view, nonprofit organizations are granted tax-exempt status to support their contribution to collectively desired goods and services that can be most efficiently provided in that manner. Still, nonprofit organizations cannot be fully understood when considered solely in efficiency terms. Participants in such groups often receive intangible benefits beyond the resolution of practical problems.

Whenever assets are held in common, some arrangements must be made for the monitoring of individual behavior and for the sanctioning of those individuals who do not fulfill their obligations under the agreed upon rules. At this point a contrast with the most influential approach to the study of collective action is in order. Olson (1965) argues that selective incentives are the key inducement for individual participation in collective action. This may be true for the types of groups Olson has in mind, that is, groups that lobby the government for the provision of some service. (Alternatively, groups may form to overthrow the existing government or to express their displeasure; incentives for group formation in this context are surveyed by Lichbach 1995.) In other words, these are voluntary groups that an individual may join if he or she chooses to do so, and they may or may not choose to contribute toward the production of the collective good in question. Many Workshop scholars focus on user groups (farmers, fishers, etc.) whose very livelihood is dependent on the success or failure of their efforts to manage a salient resource.

Such groups often see themselves as a community in which individuals with conflicting interests may still share common understandings and values. A sense of communal belonging can make an important contribution toward group success, but it must be consistent with the rational pursuit of individual interests (Hechter 1987). Lichbach (1996) concludes that community can play a contributing role in resolving dilemmas of collective action but that markets, contracts, and coercion also play essential roles. No one mode is sufficient unto itself.

March and Olsen (1984, 1989) emphasize one way in which a sense of shared understandings contributes toward the smooth operation of collective action. They stress the "logic of appropriateness" that delimits the range of behavior that is generally acceptable for each type of organization. Actions lying outside this range are deemed unacceptable or inappropriate for the agents of a particular organization or for interactions among members of that organization. This is one important way in which the "attributes of the community" (part of the IAD framework) directly impacts the nature of organizations and the rules by which they operate. In other words, the set of organizations in existence at any given time stands as a concrete embodiment of the cultural understandings shared by members of that community.

It is also important to realize that no organization stands completely on its own. Instead, organizations are linked together in complex networks of meaning. So are the individual members of a community. Thus, one important path toward an understanding of social interactions is to understand these networks, their inherent dynamics and the nature of their underlying structure (Nohria and Eccles 1992).

Market Competition and Polycentric Self-Governance

This concern for institutional context leads us back to the issue of market failures. If market processes are unable to produce the kinds and levels of public goods and services that a community seeks, but they are also reluctant to cede authority and sanctioning power to an unresponsive Leviathan, then this "collective consumption unit" will need to make some other arrangements. What types of arrangements will be made, and how will the community monitor the outcomes of these arrangements and apply sanctions to its representatives who do not deliver the goods? These are the basic questions confronting institutional analysts.

In Tiebout's (1956) "voting-with-the-feet" model, public entrepreneurs offer packages of public goods to their communities. Those individuals dissatisfied with the mix of public goods and services provided by their local officials have the option of moving to the jurisdiction of other public authorities whose policy package is more desirable than the first. If the costs of moving are low, and the range of alternative officials or jurisdictions is sufficiently wide, then the result would be a close approximation of the model of market provision of private goods. That is, groups of individuals with similar tastes for public goods and the taxes needed to finance them would gather in homogeneous communities serviced by public officials who provide for their needs.

There are, of course, significant problems with the application of this principle in practice. Individual tastes are diverse, and there exist far more public goods or services than jurisdictions, so the match between community preferences and public goods provision is sure to be inexact. Yet, this simple model does have some important implications. Capital, for example, is typically more mobile than labor, and so the owners of capital can exert considerable control over the policies of public officials. Lindblom (1977) argues that the ability of business leaders to credibly threaten to move their capital elsewhere places severe restraints on taxing and other economic policies. Those officials who enact especially distasteful policies will find their economic base undermined by capital flight, which will worsen economic conditions, which in turn is likely to result in removal of the offending officials in the next election. Since public officials are rational, they will be deterred by this prospect of losing office and will not enact policies too distasteful to the private sector. This situation can set up a "race to the bottom" as separate jurisdictions compete to provide favorable conditions for business. Lindblom is very concerned about this "market as prison" effect, because it greatly limits the ability of democratic officials to provide the policies desired by a majority of the public. Alternatively, this effect can be seen in a positive light, in the sense that it precludes the implementation of particularly ineffective economic policies.

In general, individuals should not have to move to other jurisdictions (or threaten to move their capital assets) to garner the benefits of competition among the providers of public goods or services. V. Ostrom, Tiebout, and Warren (1961) introduce the alternative conceptualization of a "polycentric political system" in which public officials representing a community (or the members of that community themselves) can select from alternative mechanisms for the production of public goods or services. These producers may be private firms, public agencies, or organizations that combine aspects of both. In this system the "providers" of public goods and services can arrange with those "producers" of the good or service that operate at the most efficient scale of production. This enables the system to achieve a degree of efficiency while still allowing communities a much wider range of choice. Vincent Ostrom and Elinor Ostrom (1977) use the term *public service industry* to denote the network of large and small organizations involved in the production and provision of any particular type of public good or service. In effect, then, a polycentric order consists of networks of interacting organizations (see McGinnis 1999a,b).

Another way in which polycentric order may lead to improved efficiency is by lowering rents. When multiple political authorities must coordinate the implementation of restrictions, then rent seekers will be hard pressed to protect their position against market and other pressures. Conversely, agents who have a monopoly on authority in a given issue area are likely to induce significant levels of rent-seeking behavior (see Mueller 1989).

Of course, efficiency is not the primary reason to advocate establishment and maintenance of a polycentric system of interacting authorities and citizens. Instead, polycentricity is a fundamental prerequisite for individual liberty and the ability of groups to govern their own affairs (V. Ostrom 1987, 1991, 1997). As stated in the series foreword, this connection between polycentricity and self-governance lies at the core of the theoretical and empirical research programs pursued by Workshop-affiliated scholars.

Self-governance at the local level is sustainable only in the context of a supportive political and cultural environment at the constitutional level. Market exchange can lead to efficient outcomes only when political and legal institutions guarantee secure property rights and provide general access to low-cost and effective means of conflict resolution. Even the coercion so often taken as the defining characteristic of the "state" can lose its effectiveness if the rulers lose contact with local traditions and expectations or if productive assets flow across national borders in search of more promising opportunities abroad. Markets, states, and user groups should be seen as complementary institutions. Each has its proper role to play in polycentric governance, its own unique configuration of strengths and weaknesses. In combination they constitute diverse institutional arrangements, such as those modeled and analyzed by the contributors to this volume.

REFERENCES

Arrow, Kenneth J. 1963. *Social Choice and Individual Values.* 2d ed. New Haven, CT: Yale University Press.

Bates, Robert H. 1981. *Markets and States in Tropical Africa: The Political Basis of Agricultural Policies.* Berkeley: University of California Press.

———. 1983. *Essays on the Political Economy of Rural Africa.* Cambridge: Cambridge University Press.

———. 1989. *Beyond the Miracle of the Market: The Political Economy of Agrarian Development in Kenya.* Cambridge: Cambridge University Press.

Buchanan, James M., and Gordon Tullock. 1962. *The Calculus of Consent: Logical Foundations of Constitutional Democracy.* Ann Arbor: University of Michigan Press.

Dasgupta, Partha S., and G. M. Heal. 1979. *Economic Theory and Exhaustible Resources.* Cambridge: Cambridge University Press.

Downs, Anthony. 1957. *An Economic Theory of Democracy.* New York: Harper and Row.

Eggertsson, Thráinn. 1990. *Economic Behavior and Institutions.* Cambridge: Cambridge University Press.

Hansmann, Henry B. 1980. "The Role of Nonprofit Enterprise." *Yale Law Review* 89 (5): 835–901.

Hardin, Garrett. 1968. "The Tragedy of the Commons." *Science* 162:1243–48.

Hechter, Michael. 1987. *Principles of Group Solidarity.* Berkeley: University of California Press.

Knight, Jack. 1992. *Institutions and Social Conflict.* Cambridge: Cambridge University Press.

Krehbiel, Keith. 1991. *Information and Legislative Organization.* Ann Arbor: University of Michigan Press.

Kreps, David M. 1990a. *A Course in Microeconomic Theory.* Princeton, NJ: Princeton University Press.

———. 1990b. *Game Theory and Economic Modelling.* New York: Oxford University Press.

Lichbach, Mark I. 1995. *The Rebel's Dilemma.* Ann Arbor: University of Michigan

Press.

———. 1996. *The Cooperator's Dilemma.* Ann Arbor: University of Michigan Press.

Lindblom, Charles E. 1977. *Politics and Markets: The World's Political Economic Systems.* New York: Basic Books.

Magee, Stephen P., William A. Brock, and Leslie Young. 1989. *Black Hole Tariffs and Endogenous Policy Theory: Political Economy in General Equilibrium.* Cambridge: Cambridge University Press.

March, James G., and Johan P. Olsen. 1984. "The New Institutionalism: Organizational Factors in Political Life." *American Political Science Review* 78: 734–49.

———. 1989. *Rediscovering Institutions: The Organizational Basis of Politics.* New York: Free Press.

McGinnis, Michael D. 1991. "Richardson, Rationality, and Restrictive Models of Arms Races." *Journal of Conflict Resolution* 35:443–73.

———, ed. 1999a. *Polycentric Governance and Development.* Ann Arbor: University of Michigan Press.

———, ed. 1999b. *Polycentricity and Local Public Economies.* Ann Arbor: University of Michigan Press.

McGinnis, Michael D., and John T. Williams. 2000. *Compound Dilemmas: Democracy, Collective Action, and Superpower Rivalry.* Draft book manuscript, Indiana University, Bloomington.

McKelvey, Richard D. 1976. "Intransitivities in Multi-dimensional Voting Models and Some Implications for Agenda Control." *Journal of Economic Theory* 12:472–82.

Milgrom, Paul, Douglass North, and Barry Weingast. 1990. "The Role of Institutions in the Revival of Trade: The Law Merchant, Private Judges, and the Champagne Fairs." *Economics and Politics* 2:1–23.

Miller, Gary J. 1992. *Managerial Dilemmas: The Political Economy of Hierarchy.* New York: Cambridge University Press.

Mitchell, William C. 1988. "Virginia, Rochester, and Bloomington: Twenty-Five Years of Public Choice and Political Science." *Public Choice* 56:101–19.

Mueller, Dennis C. 1989. *Public Choice II.* Cambridge: Cambridge University Press.

Niskanen. W. A., Jr. 1971. *Bureaucracy and Representative Government.* Chicago: AVC.

Nohria, Nitin, and Robert G. Eccles, eds. 1992. *Networks and Organizations: Structure, Form, and Action.* Boston: Harvard Business School Press.

North, Douglass C. 1981. *Structure and Change in Economic History.* New York: Norton.

———. 1990. *Institutions, Institutional Change, and Economic Performance.* Cambridge: Cambridge University Press.

North, Douglass C., and Robert Paul Thomas. 1973. *The Rise of the Western World: A New Economic History.* Cambridge: Cambridge University Press.

Olson, Mancur. 1965. *The Logic of Collective Action.* Cambridge, MA: Harvard University Press.

Ostrom, Elinor. 1990. *Governing the Commons: The Evolution of Institutions for*

Collective Action. New York: Cambridge University Press.

Ostrom, Elinor, Roy Gardner, and James Walker, with Arun Agrawal, William Blomquist, Edella Schlager, and Shui Yan Tang. 1994. *Rules, Games, and Common-Pool Resources.* Ann Arbor: University of Michigan Press.

Ostrom, Elinor, James Walker, and Roy Gardner. 1992. "Covenants with and without a Sword: Self-Governance Is Possible." *American Political Science Review* 86:404–17.

Ostrom, Vincent. 1987. *The Political Theory of a Compound Republic: Designing the American Experiment.* 2d rev. ed. San Francisco: ICS Press.

———. 1989. *The Intellectual Crisis in American Public Administration.* 2d ed. Tuscaloosa: University of Alabama Press.

———. 1991. *The Meaning of American Federalism: Constituting a Self-Governing Society.* San Francisco: ICS Press.

———. 1997. *The Meaning of Democracy and the Vulnerability of Democracies: A Response to Tocqueville's Challenge.* Ann Arbor: University of Michigan Press.

Ostrom, Vincent, and Elinor Ostrom. 1977. "Public Goods and Public Choices." In *Alternatives for Delivering Public Services. Toward Improved Performance,* ed. E. S. Savas, 7–49. Boulder, CO: Westview Press. Reprinted in McGinnis 1999b.

Ostrom, Vincent, Charles M. Tiebout, and Robert Warren. 1961. "The Organization of Government in Metropolitan Areas: A Theoretical Inquiry." *American Political Science Review* 55 (December): 831–42. Reprinted in McGinnis 1999b.

Putnam, Robert D. 1988. "Diplomacy and Domestic Politics: The Logic of Two-Level Games." *International Organization* 42:427–60.

Riker, William H. 1982. *Liberalism against Populism: A Confrontation between the Theory of Democracy and the Theory of Social Choice.* San Francisco: W. H. Freeman.

Russett, Bruce, and Harvey Starr. 1996. *World Politics: The Menu for Choice.* 5th ed. San Francisco: W. H. Freeman.

Salamon, Lester M. 1987. "Partners in Public Service: The Scope and Theory of Government-Nonprofit Relations." In *The Nonprofit Sector: A Research Handbook,* ed. Walter W. Powell, 99–117. New Haven, CT: Yale University Press.

———. 1995. *Partners in Public Service: Government-Nonprofit Relations in the Modern Welfare State.* Baltimore: Johns Hopkins University Press.

Sandler, Todd, and John T. Tshirhart. 1980. "The Economic Theory of Clubs: An Evaluation Survey." *Journal of Economic Literature* 18:1488–1521.

Schofield, Norman. 1978. "Instability of Simple Dynamic Games." *Review of Economic Studies* 45:575–94.

Schelling, Thomas C. 1960. *The Strategy of Conflict.* Cambridge, MA: Harvard University Press.

Shepsle, Kenneth. 1979. "Institutional Arrangements and Equilibrium in Multidimensional Voting Models." *American Journal of Political Science* 23:27–59.

Simon, Herbert A. 1957. *Models of Man.* New York: Wiley.

———. 1964. *Administrative Behavior: A Study of the Decision-Making Processes in Administrative Organizations.* New York: Macmillan.

———. 1997. *Models of Bounded Rationality: Empiricially Grounded Economic Reason.* Cambridge, MA: MIT Press.

Smith, Adam. [1776] 1976. *An Inquiry into the Nature and Causes of the Wealth of Nations, ed. Edwin Cannan.* Chicago: University of Chicago Press.

Tiebout, Charles M. 1956. "A Pure Theory of Public Expenditures." *Journal of Political Economy* 64 (October): 416–24.

Tocqueville, Alexis de. [1835] 1969. *Democracy in America.* Ed. J. P. Mayer. Trans. George Lawrence. Garden City, NY: Anchor Books.

Tollison, Robert D. 1980. "Rent Seeking: A Survey." *Kyklos* 35 (4): 575–602.

———. 1997. "Rent Seeking." In *Perspectives on Public Choice: A Handbook,* ed. Dennis C. Mueller, 506–25. Cambridge: Cambridge University Press.

Tsebelis, George. 1990. *Nested Games.* Berkeley: University of California Press.

Tullock, Gordon. 1965. *Politics of Bureaucracy.* Washington, DC: Public Affairs Press.

von Wright, Georg H. 1951. "Deontic Logic." *Mind* 60:58–74.

Waltz, Kenneth N. 1959. *Man, the State, and War.* New York: Columbia University Press.

———. 1979. *Theory of International Politics.* Reading, MA: Addison-Wesley.

Weimer, David L., and Aidan R. Vining. 1989. *Policy Analysis: Concepts and Practice.* 2d ed. Englewood Cliffs, NJ: Prentice-Hall.

Weisbrod, Burton A. 1977. *The Voluntary Nonprofit Sector: An Economic Analysis.* Lexington, MA: D. C. Heath.

———. 1988. *The Nonprofit Economy.* Cambridge, MA: Harvard University Press.

Williamson, Oliver E. 1975. *Markets and Hierarchies: Analysis and Antitrust Implications.* New York: Free Press.

———. 1985. *The Economic Institutions of Capitalism.* New York: Free Press.

———. 1996. *The Mechanisms of Governance.* New York: Oxford University Press.

Wintrobe, Ronald. 1998. *The Political Economy of Dictatorship.* Cambridge: Cambridge University Press.

Part I
Developing a Framework for the Analysis of Institutions

Introduction to Part I

In "Public Choice: A Different Approach to the Study of Public Administration" (chap. 1), Vincent Ostrom and Elinor Ostrom survey the analytical foundations of the Workshop approach to institutional analysis. In this essay from *Public Administration Review,* first published in 1971, they lay out a general method of analysis that has some important differences from what later became the mainstream approach to public choice. Mueller (1989, 1) defines public choice as "the economic study of nonmarket decision making, or simply the application of economics to political science." (Citations to many early works in this tradition are included in the Ostroms' long list of references.) Although this ever-growing body of public choice research has always been an important source of inspiration for scholars associated with the Workshop, many other sources push in somewhat different directions.

For example, Workshop scholars are skeptical of claims that a single model of rational behavior is valid for all individuals in all institutional contexts. Rational choice theory, grounded in methodological individualism, provides the general approach to research, but different models may be relevant for different situations. For example, the same individual might follow Simon's (1957, 1997) satisficing procedure in one decision context while engaging in more extensive information search and evaluation in other situations. Selective pressures may be strong enough to eliminate habit-driven behavior in some contexts but not in others. There is no reason to presume that any one individual acts exactly the same in all circumstances; still, there is merit in trying to locate the relevant range of decisional procedures within the context of a common explanation.

In chapter 1 the Ostroms review alternative approaches to the study of public administration. They begin by critiquing the standard view of public administration as an exercise in planning, a view most clearly articulated by Woodrow Wilson and other major figures in the U.S. reform tradition (see E. Ostrom 1972; V. Ostrom 1989). They then discuss Herbert Simon's (1957, 1964) influential critique of the assumption that individuals or organizations can reasonably be expected to have the capacity to plan at the necessary level of detail and comprehensiveness. They also discuss efforts by Tullock (1965) and other public choice theorists to interpret the behavior of bureaucratic officials as driven by an unremitting pursuit of their own selfish interests in

25

power, prestige, or resources. However, they argue that the same official, if placed in a different institutional context, might be less inclined (or less able) to take advantage of that position.

The authors argue that the "nature of the good" is an important analytical category that cannot be overlooked in political discourse or empirical evaluation. Different institutional arrangements will be most effective for the resolution of problems associated with the production, provision, or allocation of private goods, public goods, common-pool resources, and toll goods. (For an extended discussion of these distinctions and their implications for institutional analysis, see V. Ostrom and E. Ostrom 1977 and other essays reprinted in McGinnis 1999b.)

The authors conclude with a brief overview of the merits of polycentricity as an alternative approach to public administration. Rather than seeing governance as a question of the implementation of a single, uniform law, allowance should be made for local communities to form associations at whatever scale of aggregation they feel is most appropriate for the solution of their common problems. Public officials are then seen as *providers* of public goods and services to these communities. As providers, they do not need to produce the goods or services themselves but can instead choose to make arrangements with private producers or other public agencies or even encourage the communities to produce the relevant goods or services themselves. The potential benefits of such a system of polycentric order are described in more detail elsewhere (see especially V. Ostrom, Tiebout, and Warren 1961).

In chapter 1 the Ostroms briefly discuss their interpretation of the classic dilemmas of collective action as laid out by Olson (1965) and Hardin (1968). They emphasize the importance of including constitutional order within the purview of the analysis of particular policy problems, and they insist that voting is only one of many ways in which communities can arrive at collective decisions. All in all, they sketch *institutional analysis* as a distinctive approach to the study of public policy and political economy (see Mitchell 1988).

The Institutional Analysis and Development Framework

One of the distinctive emphases in institutional analysis is the recognition that rational choice theory is not the only way in which collective choices can be conceptualized. In addition, there may not be a single process of "rational choice" that applies equally to all institutional contexts. Because of these complexities, an institutional analyst seeking to model or analyze a particular empirical setting must first determine what explanatory factors are most likely to be useful in those circumstances. Through a long series of ongoing discussions among faculty, students, and practitioners, an overall "framework" of analysis has been developed. This framework lays out a general conceptual schema that encompasses all of the factors most relevant to alternative theories of choice and to particular models of specific empirical situations.

This sequence of framework-theory-model helps clarify the scope of disputes between advocates of different modes of analysis. In some cases analysts may share the same theory but disagree over what particular assumptions should be built into a model of a specific situation. Similarly, analysts who are less comfortable with rational choice theory as it is typically conceptualized may still share a common framework of analysis with rational choice modelers.

To craft this framework of analysis, emphasis has been placed on an understanding of the overall *action situation* confronting individuals and groups. In "The Three Worlds of Action: A Metatheoretical Synthesis of Institutional Approaches" (chap. 2), Larry L. Kiser and Elinor Ostrom summarize the *institutional analysis and development (IAD) framework,* an organizing schema that emerged out of extensive discussions at the Workshop. Two later formulations (Oakerson 1992; E. Ostrom, Gardner, and Walker 1994, chap. 2) include further refinements, but this essay remains the canonical presentation of the IAD framework precisely because it portrays the framework as a work in progress. Institutions themselves are in a constant state of flux. Institutional development occurs when changes in institutional arrangements interact with changes in physical conditions, individual behavior, and cultural understanding. Institutional analysis, then, requires consideration of a wide array of explanatory factors.

Distinctions among *action situations* in arenas of operational, collective, and constitutional choice recur throughout all Workshop research programs, as do discussions of interactions among physical or material conditions, attributes of the community, and the *rules-in-use* that shape the behavior of individuals or organizations. The IAD framework provides a shared language for a wide array of institutional analyses, thus facilitating comparisons among more specific theories and models of particular phenomena.

However, one aspect of this language can be potentially confusing. As discussed in the general introduction to this volume, I think the term *levels of analysis* should be reserved for processes operating at different scales of aggregation, as in international relations theory (Waltz 1959; Russett and Starr 1996). The term *arena of choice* seems a better reflection of the basic idea behind the "three worlds" in the title of this essay. Outcomes in one arena define the nature of the games being played concurrently in other arenas. For example, constitutional decisions define the processes by which organizations are expected to interact. Similarly, collective choices specify the operational rights and responsibilities of specific actors. All three arenas interact in any one situation.

A Subtle Difference of Emphasis

In her presidential address to the Public Choice Society, "An Agenda for the Study of Institutions" (chap. 3), Elinor Ostrom provides a fuller analysis of the "action situation" that lies at the core of the IAD framework. In doing so, she lays out a research program meant to generalize standard conceptualizations of game models. Rather than assuming that all actors are identical in interests or capabilities, Ostrom lays out a set of categories that jointly define the roles different types of players fill as well as the rules by which their interactions are structured. Although much of game theory has been focused on models of actors with symmetric interests or capabilities, Elinor Ostrom argues that it is crucially important to incorporate the ways in which actors differ. In particular, institutions define roles, thus imparting to certain individuals a set of interests, capabilities, and responsibilities that should be incorporated in models of action situations. She argues that all institutions share some foundational set of components; any one institution's consequences are determined by the configuration of these structural elements. In this way she hopes to move the field beyond a preoccupation with supposedly "institution-free" settings to fuller and more useful formal representations of institutions.

Ostrom conceptualizes an action situation as a generalization of standard game models. However, a typical reaction of game theorists is that they also have to define each of the components that Ostrom lists in her framework or else their game models would not be fully specified. Technically speaking this is correct, but there remains an important, but subtle, difference in emphasis. For Ostrom, and for other Workshop scholars, it is essential to keep in mind the extent to which actors' preferences as well as the choice options available to them are determined by the institutional arrangements that define their position or that shape their perceptions and options. It is essential to remember that concurrent games in other arenas of choice interact in complex ways with any ongoing process of interaction. The payoffs and menu of choices available to participants in operational games have been defined by collective choice processes. Games over collective deliberations are in turn shaped by the positions and interests defined or manifested in the constitutional choice arena. In a strict sense this may be no different from a complete specification of a game model, but in practice this concern with simultaneous consideration of multiple choice arenas inspires Workshop-affiliated scholars toward a more inductive mode of analysis.

Specifically, Ostrom calls for a shift away from the analytical practice of considering changes in the rules of the game in isolation. She argues that the entire set of rules must be subject to analysis, since rules act together in a configural manner. John R. Commons (1959) demonstrates that changes in one rule may directly imply changes in other rules, and Ostrom applies this idea to contemporary game theory. In this essay she develops several examples from the theoretical and experimental literature on public choice to illustrate the configural nature of rule changes.

This general point that rules need to be explicitly specified in game models is developed further in Roy Gardner and Elinor Ostrom (1991), in which several game models are used to represent the implications of a series of alternative rules for the assignment of fishing spots. Gardner and Ostrom begin by laying out Hobbes's conceptualization of a "state of nature" as a "default condition" that might apply to some situations, but they indicate that in nearly all social settings there are differences among actor interests and capabilities that can be attributed to the existence of institutional rules. This essay serves as a great follow-up piece to the more abstract presentation of the principles laid out in Elinor Ostrom's "Agenda for the Study of Institutions," but because the specific models discussed there were incorporated into the analysis presented in E. Ostrom, Gardner, and Walker (1994), it did not seem appropriate to reprint that material here.

Elinor Ostrom (1989) delves deeper into the nature of the actors who interact to form action situations. She assumes that individuals can learn but can do so only imperfectly. Individuals are subject to a multitude of influences, but individuals located within the supportive environment of polycentric order are more likely to successfully adapt to their situations. In effect, this essay can be seen as a more microlevel investigation of the core components of the IAD framework. Elinor Ostrom revisits these issues in a more comprehensive manner in her presidential address to the American Political Science Association, included as the last selection in the present volume.

These essays on interactions between institutions and individuals also reflect the influence of a community of scholars at the Center for Interdisciplinary Research at Bielefeld University in Germany, where the Ostroms served as visiting scholars during 1981. Elinor Ostrom later spent a semester there studying game theory with Reinhard Selten, a Nobel Prize laureate with a long-standing interest in innovative ways of interpreting the behavioral basis of game theory. His influence can be seen especially in the game models section of E. Ostrom, Gardner, and Walker (1994) and in essays Elinor Ostrom coauthored with Franz J. Weissing (including chap. 13 of this volume). A collection of essays edited by Kauffman, Majone, and V. Ostrom (1986) includes chapters by several policy analysts (including Paul Sabatier) and prominent game theorists Reinhard Selten and Martin Shubik. It includes two chapters by Elinor Ostrom that lay out the IAD framework as it was understood at that time and three chapters by Vincent Ostrom (including the essay coauthored with Roberta Herzberg reprinted as chap. 5 of this volume) on voting systems and the nature of constitutional order. This Bielefeld volume is a fascinating illustration of diverse but related efforts to comprehend interactions among systems, networks, hierarchies, institutions, and norms.

Incorporating Normative Considerations

Another distinction from the international relations literature is useful at this point. In the constructivist approach to international relations, a distinction is made between obligatory and constitutive norms (see Katzenstein 1996). Obligatory norms are moral prescriptions that specify what individuals or organizations *should* do in specific empirical contexts, whereas constitutive norms *define* the very nature of these organizations. In this sense both the organizations and the nature of their interactions are "socially constructed." In terms of the IAD framework, these constitutive norms are equivalent to the interactions between arenas of choice discussed earlier, whereas obligatory norms are manifested in the ways in which actors interact in any single choice arena. However, Workshop scholars would cringe at the term *obligatory,* since no norm is self-enforcing but must instead be supported by some form of monitoring and sanctioning.

Vincent Ostrom (1976) summarizes one of the most important inspirations behind this effort to bring normative expectations within the purview of institutional analysis. John R. Commons (1959) was concerned about directing attention to what he calls "going concerns" rather than remaining fixated on the details of organization charts or constitutional provisions. In a "going concern" individuals are engaged in a series of transactions, and these transactions are shaped by the overall structure of legal rules and moral expectations. Commons's basic framework has recently inspired an extensive body of research on the ways in which the costs associated with any form of transaction between two or more actors can be reduced in order to facilitate the likelihood that mutually beneficial transactions will be undertaken. This area of "transition cost economics" is most closely associated with Williamson (1975, 1985, 1996); related areas of "new institutional economics" are surveyed in Eggertsson (1990).

Vincent Ostrom asks us to step back to reexamine the "old institutional economics" of John R. Commons. Ostrom draws on distinctions Commons makes concerning relationships between the rights and responsibilities of individuals serving in different legal capacities. For example, any provision that a public official "must" take a certain action necessarily implies a responsibility for some other official, or for the citizenry as a whole, to undertake an effort to monitor the behavior of the first official, to make sure the original provision is being satisfied (see also V. Ostrom and E. Ostrom 1972). Without this additional structure, the first legal provision stands in isolation and cannot be expected to be realized in actual conditions. Mutually reinforced sets of rules, rights, and responsibilities constitute what Commons calls "working rules," and it is to this type of rules that Workshop scholars have directed their attention. Knowing the law is not enough—it is also essential to understand how those rules are implemented in specific action situations.

In "A Grammar of Institutions" (chap. 4), Sue E. S. Crawford and Elinor Ostrom push this line of argument further. They use a sequence of definitions to try to make sense out of the multiple meanings of such conceptually slippery terms as *institutions, rules,* and *norms.* This conceptualization builds on many sources, including the deontic logic of von Wright (1951) and the legal framework of John R. Commons (1959). One important component concerns the consequences of the extent to which individuals internalize the norms of their community. This internalization makes certain socially undesirable actions costly to the relevant individuals, but these internal costs are rarely sufficient to prevent completely the occurrence of proscribed behavior. Thus, finding some way to monitor and sanction rule-breaking behavior remains essential to the maintenance of social cooperation. In this way this conceptualization provides a logical foundation for the observed importance of monitoring and sanctioning in successful CPR regimes (E. Ostrom 1990; McGinnis 1999a).

The internal costs included in the Crawford-Ostrom representation of actor incentives act like the "obligatory norms" discussed previously, in the sense that they help define what actors consider to be acceptable and unacceptable behavior. Meanwhile, broader institutional arrangements (from other arenas of choice) define the options available to the actors themselves, by specifying which acts are prohibited, permitted, or required. In game models analysts select which actions are available to the various actors included in the model, but institutional analysts build on the realization that the set of options available to actors in any action situation is determined in games in other arenas of choice. Similarly, game theorists can posit whatever incentives they want for their modeled individuals to pursue, but in applications it is important to understand the source of those payoffs. Hence the need to consider the broader context of "polycentric games."

REFERENCES

Commons, John R. 1959. *The Legal Foundations of Capitalism.* Madison: University of Wisconsin Press.

Eggertsson, Thráinn. 1990. *Economic Behavior and Institutions.* Cambridge: Cambridge University Press.

Gardner, Roy, and Elinor Ostrom. 1991. "Rules and Games." *Public Choice* 70 (2) (May): 121–49.

Hardin, Garrett. 1968. "The Tragedy of the Commons." *Science* 162:1243–48.

Katzenstein, Peter J., ed. 1996. *The Culture of National Security.* New York: Columbia University Press.

Kaufmann, Franz-Xaver, Giandomenico Majone, and Vincent Ostrom, eds. 1986. *Guidance, Control, and Evaluation in the Public Sector.* Berlin and New York: Walter de Gruyter.

McGinnis, Michael D., ed. 1999a. *Polycentric Governance and Development.* Ann Arbor: University of Michigan Press.

———, ed. 1999b. *Polycentricity and Local Public Economies.* Ann Arbor: University of Michigan Press.

Mitchell, William C. 1988. "Virginia, Rochester, and Bloomington: Twenty-Five Years of Public Choice and Political Science." *Public Choice* 56:101–19.

Mueller, Dennis C. 1989. *Public Choice II.* Cambridge: Cambridge University Press.

Oakerson, Ronald J. 1992. "Analyzing the Commons: A Framework." In *Making the Commons Work: Theory, Practice, and Policy,* ed. Daniel W. Bromley et al., 41–59. San Francisco: ICS Press.

Olson, Mancur. 1965. *The Logic of Collective Action.* Cambridge, MA: Harvard University Press.

Ostrom, Elinor. 1972. "Metropolitan Reform: Propositions Derived from Two Traditions." *Social Science Quarterly* 53 (December): 474–93. Reprinted in McGinnis 1999b.

———. 1989. "Microconstitutional Change in Multiconstitutional Political Systems." *Rationality and Society* 1 (1) (July): 11–50.

———. 1990. *Governing the Commons: The Evolution of Institutions for Collective Action.* Cambridge: Cambridge University Press.

Ostrom, Elinor, Roy Gardner, and James Walker, with Arun Agrawal, William Blomquist, Edella Schlager, and Shui Yan Tang. 1994. *Rules, Games, and Common-Pool Resources.* Ann Arbor: University of Michigan Press.

Ostrom, Vincent. 1976. "John R. Commons's Foundation for Policy Analysis." *Journal of Economic Issues* 10 (4) (December): 839–57.

———. 1989. *The Intellectual Crisis in American Public Administration.* 2d ed. Tuscaloosa: University of Alabama Press.

Ostrom, Vincent, and Elinor Ostrom. 1972. "Legal and Political Conditions of Water Resource Development." *Land Economics* 48 (1) (February): 1–14. Reprinted in McGinnis 1999a.

———. 1977. "Public Goods and Public Choices." In *Alternatives for Delivering Public Services.* In *Toward Improved Performance,* ed. E. S. Savas, 7–49. Boulder, CO: Westview Press. Reprinted in McGinnis 1999b.

Ostrom, Vincent, Charles M. Tiebout, and Robert Warren. 1961. "The Organization of Government in Metropolitan Areas: A Theoretical Inquiry." *American Political Science Review* 55 (December): 831–42. Reprinted in McGinnis 1999b.

Russett, Bruce, and Harvey Starr. 1996. *World Politics: The Menu for Choice.* 5th ed. San Francisco: W. H. Freeman.

Simon, Herbert A. 1957. *Models of Man.* New York: Wiley.

———. 1964. *Administrative Behavior: A Study of the Decision-Making Processes in Administrative Organizations.* New York: Macmillan.

————. 1997. *Models of Bounded Rationality: Empiricially Grounded Economic Reason.* Cambridge, MA: MIT Press.

Tullock, Gordon. 1965. *Politics of Bureaucracy.* Washington, DC: Public Affairs Press.

von Wright, Georg H. 1951. "Deontic Logic." *Mind* 60:58–74.

Waltz, Kenneth N. 1959. *Man, the State, and War.* New York: Columbia University Press.

Williamson, Oliver E. 1975. *Markets and Hierarchies: Analysis and Antitrust Implications.* New York: Free Press.

————. 1985. *The Economic Institutions of Capitalism.* New York: Free Press.

————. 1996. *The Mechanisms of Governance.* New York: Oxford University Press.

CHAPTER 1

Public Choice: A Different Approach to the Study of Public Administration

Vincent Ostrom and Elinor Ostrom

In November 1963, a number of economists and a sprinkling of other social scientists were invited by James Buchanan and Gordon Tullock to explore a community of interest in the study of nonmarket decision making. That conference was reported in *Public Administration Review* (*PAR*) as "Developments in the 'No-Name' Fields of Public Administration" (*PAR*, March 1964, 62–63). A shared interest prevailed regarding the application of economic reasoning to "collective," "political," or "social" decision making, but no consensus developed on the choice of a name to characterize those interests. In December 1967 the decision was taken to form a Public Choice Society and to publish a journal, *Public Choice.*[1] The term *public choice* will be used here to refer to the work of this community of scholars. The bibliography following this essay refers to only a small fraction of the relevant literature but will serve to introduce the reader to public choice literature.

The public choice approach needs to be related to basic theoretical traditions in public administration. As background, we shall first examine the theoretical tradition as formulated by Wilson and those who followed. We shall next examine Herbert Simon's challenge to that tradition that has left the discipline in what Dwight Waldo has characterized as a "crisis of identity" (Waldo 1968, 118). The relevance of the public choice approach for dealing with the issues raised by Simon's challenge will then be considered.

The Traditional Theory of Public Administration

Woodrow Wilson's essay "The Study of Administration" called for a new science of administration based upon a radical distinction between politics

Originally published in *Public Administration Review* 31 (March/April 1971): 203–16. Reprinted by permission of the American Society for Public Administration and the authors.

Authors' note: The authors wish to acknowledge the support of the Office of Research and Advanced Studies at Indiana University.

34

and administration (Wilson 1887, 210). According to Wilson, governments may differ in the political principles reflected in their constitutions, but the principles of good administration are much the same in any system of government. There is "…but one rule of good administration for all governments alike" was Wilson's major thesis (218). "So far as administrative functions are concerned, all governments have a strong structural likeness; more than that, if they are to be uniformly useful and efficient, they must have a strong structural likeness" (218).

Good administration, according to Wilson, will be hierarchically ordered in a system of graded ranks subject to political direction by heads of departments at the center of government. The ranks of administration would be filled by a corps of technically trained civil servants "…prepared by a special schooling and drilled, after appointment, into a perfected organization, with an appropriate hierarchy and characteristic discipline…"(216). Perfection in administrative organization is attained in a hierarchically ordered and professionally trained public service. Efficiency is attained by perfection in the hierarchical ordering of a professionally trained public service. Wilson also conceptualizes efficiency in economic terms: "…the utmost possible efficiency and at the least possible cost of either money or of energy" (197).

For the next half century, the discipline of public administration developed within the framework set by Wilson. The ends of public administration were seen as the "management of men and material in the accomplishment of the purposes of the state" (White 1926, 6). Hierarchical structure was regarded as the ideal pattern of organization. According to L. D. White, "All large-scale organizations follow the same pattern, which in essence consists in the universal application of the superior-subordinate relationship through a number of levels of responsibility reaching from the *top to the bottom* of the structure" (33; emphasis added).

Simon's Challenge

Herbert Simon, drawing in part upon previous work by Luther Gulick (Gulick and Urwick 1937), sustained a devastating critique of the theory implicit in the traditional study of public administration. In his *Administrative Behavior,* Simon elucidated some of the accepted administrative principles and demonstrated the lack of logical coherence among them (Simon 1964). Simon characterized those principles as "proverbs." Like proverbs, incompatible principles allowed the administrative analyst to justify his position in relation to one or another principle. "No single one of these items is of sufficient importance," Simon concluded, "to suffice as a guiding principle for the administrative analyst" (36).

After his indictment of traditional administrative theory, Simon began an effort to reconstruct administrative theory. The first stage was "the construction of an adequate vocabulary and analytic scheme." Simon's reconstruction began with a distinction between facts and values. Individuals engage in a consideration of facts and of values in choosing among alternative possibilities bounded by the ordered rationality of an organization. Simon envisioned the subsequent stage of reconstruction to involve the establishment of a bridge between theory and empirical study "so that theory could provide a guide to the design of 'critical' experiments and studies, while experimental study could provide a sharp test and corrective of theory" (44).

One of Simon's central concerns was to establish the criterion of efficiency as a norm for evaluating alternative administrative actions. Simon argued that the "criterion of efficiency dictates that choice of alternatives which produce the largest result for the given application of resources" (179). In order to utilize the criterion of efficiency, the results of administrative actions must be defined and measured. Clear conceptual definitions of output are necessary before measures can be developed.

No necessary reason existed, Simon argued, for assuming that perfection in hierarchical ordering would always be the most efficient organizational arrangement (but see also Simon 1962, 1969). Alternative organizational forms needed to be empirically evaluated to determine their relative efficiency. Simon's own work with Ridley on the measurement of municipal activities represented a beginning effort to identify the output of government agencies and to develop indices to measure those outputs (Ridley and Simon 1938).

The Work of the Political Economists

During the period following Simon's challenge, another community of scholars has grappled with many of these same intellectual issues. This community of scholars has been composed predominantly of political economists who have been concerned with public investment and public expenditure decisions. One facet of this work has been manifest in benefit-cost analysis and the development of the Planning, Programming, and Budgeting (PPB) system (U.S. Congress 1969). PPB analysis rests upon much the same theoretical grounds as the traditional theory of public administration. The PPB analyst is essentially taking the methodological perspective of an "omniscient observer" or a "benevolent despot." Assuming that he knows the "will of the state," the PPB analyst selects a program for the efficient utilization of resources (i.e., men and material) in the accomplishment of those purposes. As Senator McClelland has correctly perceived, the assumption of omniscience may not hold; and, as a consequence, PPB analysis may involve radical errors and generate gross inefficiencies (see Wildavsky 1966).

Public choice represents another facet of work in political economy with more radical implications for the theory of public administration. Most political economists in the public choice tradition begin with the individual as the basic unit of analysis. Traditional "economic man" is replaced by "man: the decision maker."

The second concern in the public choice tradition is with the conceptualization of public goods as the type of event associated with the output of public agencies. These efforts are closely related to Simon's concern for the definition and measurement of the results of administrative action. In addition, public choice theory is concerned with the effect that different decision rules or decision-making arrangements will have upon the production of those events conceptualized as public goods and services. Thus, a model of man, the type of event characterized as public goods and services, and decision structures comprise the analytical variables in public choice theory. Our "man: the decision maker" will confront certain opportunities and possibilities in the world of events and will pursue his relative advantage within the strategic opportunities afforded by different types of decision rules or decision-making arrangements. The consequences are evaluated by whether or not the outcome is consistent with the efficiency criterion.

Work in public choice begins with methodological individualism where the perspective of a representative individual is used for analytical purposes (Brodbeck 1958). Since the individual is the basic unit of analysis, the assumptions made about individual behavior become critical in building a coherent theory (Coleman 1966a,b). Four basic assumptions about individual behavior are normally made.

First, individuals are assumed to be self-interested. The word *self-interest* is not equivalent to *selfish*. The assumption of self-interest implies primarily that individuals each have their own preferences that affect the decisions they make and that those preferences may differ from individual to individual (Buchanan and Tullock 1965).

Secondly, individuals are assumed to be rational. Rationality is defined as the ability to rank all known alternatives available to the individual in a transitive manner. Ranking implies that a rational individual either values alternative "A" more than alternative "B," or that he prefers alternative "B" to alternative "A," or that he is indifferent between them. Transitivity means that if he prefers alternative "A" to alternative "B," and "B" is preferred to "C," then "A" is necessarily preferred to "C" (see also Boulding 1966; Eulau 1964).

Third, individuals are assumed to adopt maximizing strategies. Maximization as a strategy implies the consistent choice of those alternatives that an individual thinks will provide the highest net benefit as weighed by his own preferences (Wade and Curry 1970). At times the assumption of maximization is related to that of satisficing, depending upon assumptions about the information available to an individual in a decision-making situation (Simon 1957).

Fourth, an explicit assumption needs to be stated concerning the level of information possessed by a representative individual. Three levels have been analytically defined as involving certainty, risk, and uncertainty (Knight 1965; Luce and Raiffa 1957). The condition of certainty is defined to exist when: (1) an individual knows all available strategies, (2) each strategy is known to lead invariably to only one specific outcome, and (3) the individual knows his or her own preferences for each outcome. Given this level of information, the decision of a maximizing individual is completely determined. He or she simply chooses that strategy that leads to the outcome for which he or she has the highest preference (Fishburn 1964).

Under conditions of risk, the individual is still assumed to know all available strategies. Any particular strategy may lead to a number of potential outcomes, and the individual is assumed to know the probability of each outcome (Arrow 1951). Thus, decision making becomes a weighting process whereby his or her preferences for different outcomes are combined with the probability of their occurrence prior to a selection of a strategy. Under risk, an individual may adopt mixed strategies in an effort to obtain the highest level of outcomes over a series of decisions in the long run.

Decision making under uncertainty is assumed to occur either where (1) an individual has a knowledge of all strategies and outcomes but lacks knowledge about the probabilities with which a strategy may lead to an outcome, or (2) an individual may not know all strategies or all outcomes that actually exist (Hart 1965; Radner 1970). Uncertainty is more characteristic of problematical situations than either certainty or risk. Under either certainty or risk, an analyst can project a relatively determinant solution to a particular problem. Under conditions of uncertainty, the determinateness of solutions is replaced by conclusions about the range of possible "solutions" (Shackle 1961).

Once uncertainty is postulated, a further assumption may be made that an individual learns about states of affairs as he or she develops and tests strategies (Simon 1959; Thompson 1966). Estimations are made about the consequences of strategies. If the predicted results follow, then a more reliable image of the world is established. If predicted results fail to occur, the individual is forced to change his or her image of the world and modify his strategies (E. Ostrom 1968). Individuals who learn may adopt a series of diverse strategies as they attempt to reduce the level of uncertainty in which they are operating (Cyert and March 1963).

The Nature of Public Goods and Services

Individuals who are self-interested and rational and who pursue maximizing strategies find themselves in a variety of situations. Such situations involve the production and consumption of a variety of goods. Political economists in the public choice tradition distinguish situations involving purely private goods as a logical category from purely public goods (Davis and Winston 1967; Samuelson 1954, 1955). Purely private goods are defined as those goods and services that are highly divisible and can be (1) packaged, contained, or measured in discrete units, and (2) provided under competitive market conditions where potential consumers or users can be excluded from enjoying the benefit unless they are willing to pay the price. Purely public goods, by contrast, are highly indivisible goods and services where potential consumers cannot be easily excluded from enjoying the benefits (Breton 1966). Once public goods are provided for some, they will be available for others to enjoy without reference to who pays the costs. National defense is a classic example of a public good. Once it is provided for some individuals living within a particular country, it is automatically provided for all individuals who are citizens of that country, whether they pay for it or not.

In addition to the two logical categories of purely private and purely public goods, most political economists would postulate the existence of an intermediate continuum. Within this continuum, the production or consumption of goods or services may involve spillover effects or externalities that are not isolated and contained within market transactions (Ayres and Kneese 1969; Buchanan and Stubblebine 1962; Davis and Winston 1962; Mishan 1967). Goods with appreciable spillovers are similar to private goods to the extent that some effects can be subject to the exclusion principle; but other effects are like public goods and spill over onto others not directly involved (Coase 1960). The air pollution that results from the production of private industry is an example of a negative externality. Efforts to reduce the cost of a negative externality are the equivalent of providing a public good. The benefits produced for a neighborhood by the location of a golf course or park are a positive externality.

The existence of public goods or significant externalities creates a number of critical problems for individuals affected by those circumstances (Buchanan 1968b; Head 1962; Phelps 1965). Each individual will maximize his or her net welfare if he or she takes advantage of a public good at minimum cost to his- or herself (Demsetz 1967). He or she will have little or no incentive to take individual action where the effect of individual action would be to conserve or maintain the quality of the good that each shares in common. Each individual is likely to adopt a "dog-in-the-manger" strategy by pursuing his or her own advantage and disregarding the consequences of his or her action upon others. Furthermore, individuals may not even be motivated to articulate their own honest preferences for a common good (Demsetz 1970; Musgrave 1959; Thompson 1968). If someone proposes an improvement in the quality of a public good, some individuals may have an incentive to withhold information about their preferences for such an improvement (Schelling 1963). If others were to make the improvement, the individual who had concealed his or her preference could then indicate that he or she was not a beneficiary and might avoid paying his or her share of the costs. If voluntary action is proposed, some individuals will have an incentive to "hold out," act as "free riders," and take advantage of the benefits provided by others (Cunningham 1967; Hirshleifer, DeHaven, and Milliman 1960). Garrett Hardin had indicated that these strategies typically give rise to "the tragedy of the commons" where increased individual effort leaves everyone worse off (Hardin 1968).

The Effect of Decision Structures upon Collective Action

The Problem of Collective Inaction

The problem arising from the indivisibility of a public good and the structure of individual incentives created by the failure of an exclusion principle is the basis for Mancur Olson's *The Logic of Collective Action* (1965). Olson concludes that individuals cannot be expected to form large voluntary associations to pursue matters of public interest unless special conditions exist (Musgrave 1939). Individuals will form voluntary associations in pursuit of public interests only when members will derive separable benefits of a sufficient magnitude to justify the cost of membership or where they can be coerced into bearing their share of the costs (Burgess and Robinson 1969; Garvey 1969; Leoni 1957). Thus, we cannot expect persons to organize themselves in a strictly voluntary association to realize their common interest in the provision of public goods and services. An individual's actions will be calculated by the probability that his or her efforts alone will make a difference. If that probability is nil, and if he or she is a rational person, we would expect his or her effort to be nil.

Constitutional Choice and Collective Action

The analysis of Mancur Olson would lead us to conclude that undertaking collective actions to provide public goods and services such as national defense, public parks, and education is not easily accomplished. If unanimity were the only decision rule that individuals utilized to undertake collective action, most public goods would not be provided (Bator 1958). Yet, individuals do surmount the problems of collective inaction to constitute enterprises that do not rely strictly upon the voluntary consent of all who are affected. Buchanan and Tullock begin to develop a logic that a representative individual might use in attempting to establish some method for gaining the benefits of collective action (Buchanan and Tullock 1965). While many students of public administration would not immediately see the relevance of a logic of constitutional decision making for the study of public administration, we feel that it provides an essential foundation for a different approach to the field. Using this logic, public agencies are not viewed simply as bureaucratic units that perform those services that someone at the top instructs them to perform. Rather, public agencies are viewed as means for allocating decision-making capabilities in order to provide public goods and services responsive to the preferences of individuals in different social contexts.

A constitutional choice is simply a choice of decision rules for making future collective decisions. Constitutional choice, as such, does not include the appropriation of funds or actions to alter events in the world except to provide the organizational structure for ordering the choices of future decision makers. A representative individual wanting to create an organization to provide a public good would, according to Buchanan and Tullock, need to take two types of costs into account: (1) external costs—those costs that an individual expects to bear as the result of decisions that deviate from his or her preferences and impose costs upon the individual—and (2) decision-making costs—the expenditure of resources, time, effort, and opportunities forgone in decision making (Buchanan 1969). Both types of costs are affected by the selection of decision rules that specify the proportion of individuals required to agree prior to future collective action (V. Ostrom 1968).

Expected external costs will be at their highest point where any one person can take action on behalf of the entire collectivity. Such costs would decline as the proportion of members participating in collective decision making increases. Expected external costs would reach zero where all were required to agree prior to collective action under a rule of unanimity. However, expected decision-making costs would have the opposite trend. Expenditures on decision making would be minimal if any one person could make future collective decisions for the whole group of affected individuals. Such costs would increase to their highest point with a rule of unanimity.

If our representative individual were a cost minimizer, and the two types of costs described above were an accurate representation of the costs he or she perceives, we would expect he or she to prefer the constitutional choice of a decision rule where the two cost curves intersect. When the two cost curves are roughly symmetrical, some form of simple majority vote would be a rational choice of a voting rule. If expected external costs were far greater than expected decision-making costs, an extraordinary majority would be a rational choice of a voting rule. On the other hand, if the opportunity costs inherent in decision making were expected to be very large in comparison to external costs, then reliance might be placed on a rule authorizing collective action by the decision of one person in the extreme case requiring rapid response. An optimal set of decision rules will vary with different situations, and we would not expect to find one good rule that would apply to the provision of all types of public goods and services.

Majority Vote and the Expression of Social Preferences

Scholars in the public choice tradition have also been concerned with the effect of decision rules upon the expression of individual preferences regarding the social welfare of a community of individuals (Arrow 1963; Bowen 1943; Bradford 1970; Downs 1960). Particular attention has been paid to majority vote as a means of expressing such preferences (Downs 1957; Farquharson 1969; Hinich and Ordeshook 1970; Plott 1967; Riker 1958, 1962; Rothenberg 1970). Duncan Black in *The Theory of Committees and Elections* has demonstrated that if a community is assumed to have a single-peaked preference ordering, then a choice reflecting the median preference position will dominate all others under majority vote, providing the numbers are odd (Black 1958; Black and Newing 1951). Edwin Haefele and others have pointed out that this solution has interesting implications for the strategy of those who must win the approval of an electorate (Davis and Hinich 1966, 1967; Haefele 1972). If representatives are aware of their constituents' preferences, the task of developing a winning coalition depends upon the formulation of a program that will occupy the median position of voter preferences, providing that voters are making a choice between two alternatives. Under these circumstances, persons in political or administrative leadership would have an incentive to formulate a program oriented to the median preference position of their constituents. Voters would then choose the alternative, if presented with a choice, that most closely approximates the median position (Downs 1961; Hotelling 1929). Single peakedness implies a substantial homogeneity in social preference with the bulk of preferences clustering around a single central tendency. Such conditions might reasonably apply to a public good for which there is a relatively uniform demand in relation to any particular community of interest. Substantial variations in demand for different mixes of public goods, as might be reflected in the differences between wealthy neighborhoods and ghetto neighborhoods in a big city, would most likely not meet the condition of single peakedness when applied to the provision of educational services, police services, or welfare services. Majority voting under such conditions would fail to reflect the social preference of such diverse neighborhoods if they were subsumed in the same constituency.

Bureaucratic Organization

If the essential characteristic of a bureaucracy is an ordered structure of authority where command is unified in one position and all other positions are ranked in a series of one-many relationships, then we would assume from the Buchanan and Tullock cost calculus that a constitutional system based exclusively upon a bureaucratic ordering would be an extremely costly affair. Presumably, an ordered system of one-man rule might sustain considerable speed and dispatch in some decision making. However, the level of potential deprivations or external social costs would be very high. If external costs can be reduced to a low order of magnitude, then reliance upon a bureaucratic ordering would have considerable advantage.

The possibility of reducing expected external costs to a low order of magnitude so that advantage might be taken of the low decision costs potentially inherent in a bureaucratic ordering can be realized only if (1) appropriate decision-making arrangements are available to assure the integrity of substantial unanimity at the level of constitutional choice, and (2) methods of collective choice are continuously available to reflect the social preferences of members of the community for different public goods and services. The rationale for bureaucratic organization in a democratic society can be sustained only if both of these conditions are met.

In the political economy tradition, two different approaches have been taken in the analysis of bureaucratic organizations. R. H. Coase in an article entitled "The Nature of the Firm" has developed an explanation for bureaucratic organization in business firms (Coase 1937). According to Coase, rational individuals might be expected to organize a firm where management responsibilities would be assumed by an entrepreneur, and others would be willing to become employees if the firm could conduct business under direction of the entrepreneur at a lesser cost than if each and every transaction were to be organized as market transactions. The firm would be organized on the basis of long-term employment contracts, rather than short-term market transactions. Each employee would agree, for certain remuneration, to work in accordance with the directions of an entrepreneur within certain limits. The employment contract is analogous to a constitution in defining decision-making arrangements between employer and employee.

Coase anticipates limits to the size of firms where the costs of using a factor of production purchased in the market would be less than adding a new component to the firm to produce that added factor of production. As more employees are added, management costs would be expected to increase. A point would be reached where the saving on the marginal employee would not exceed the added costs of managing that employee. No net savings would accrue to the entrepreneur. If a firm became too large, an entrepreneur might also fail to see some of his opportunities and not take best advantage of potential opportunities in the reallocation of his workforce. Another entrepreneur with a smaller, more efficient firm would thus have a competitive advantage over the larger firm that had exceeded the limit of scale economy in firm size.

Gordon Tullock in *The Politics of Bureaucracy* develops another analysis using a model of "economic man" to discern the consequences that can be expected to follow from rational behavior in large public bureaucracies (Tullock 1965). Tullock's "economic man" is an ambitious public employee who seeks to advance his career opportunities for promotions within the bureaucracy. Since career advancement depends upon favorable recommendations by his or her superiors, a career-oriented public servant will act so as to please his or her superiors. Favorable information will be forwarded; unfavorable information will be repressed. Distortion of information will diminish control and generate expectations that diverge from events sustained by actions (Niskanen 1971). Large-scale bureaucracies will, thus, become error prone and cumbersome in adapting to rapidly changing conditions. Efforts to correct the malfunctioning of bureaucracies by tightening control will simply magnify errors.

Coase's analysis would indicate that elements of bureaucratic organization can enhance efficiency if the rule-making authority of an entrepreneur is constrained by mutually agreeable limits and he is free to take best advantage of opportunities in reallocating work assignments within those constraints. Both Coase and Tullock recognize limits to economies of scale in bureaucratic organization (Williamson 1967). No such limits were recognized in the traditional theory of public administration. Bureaucratic organization is as subject to institutional weaknesses and institutional failures as any other form of organizational arrangement.

Producer Performance and Consumer Interests in the Provision of Public Goods and Services

The problem of collective inaction can be overcome under somewhat optimal conditions provided that (1) substantial unanimity can be sustained at the level of constitutional choice, (2) political and administrative leadership is led to search out median solutions within a community of people that has a single-peaked order of preferences, and (3) a public service can be produced by an enterprise subject to those constraints. Some difficult problems will always remain to plague those concerned with the provision of public goods and services.

Once a public good is provided, the absence of an exclusion principle means that each individual will have little or no choice but to take advantage of whatever is provided unless he is able to move or is wealthy enough to provide for himself (Tiebout 1956). Under these conditions, the producer of a public good may be relatively free to induce savings in production costs by increasing the burden or cost to the user or consumer of public goods and services. Shifts of producer costs to consumers may result in an aggregate loss of efficiency, where the savings on the production side are exceeded by added costs on the consumption side. Public agencies rarely, if ever, calculate the value of users' time and inconvenience when they engage in studies of how to make better use of their employees' time. What is the value of the time of citizens who stand in line waiting for service as against the value of a clerk's time who is servicing them? (Kafoglis 1968; Warren 1970; Weschler 1968). If the citizen has no place else to go and if he or she is one in a million other citizens, the probability of his interest being taken into account is negligible. The most impoverished members of a community are the most exposed to deprivations under these circumstances. A preoccupation with producer efficiency in public administration may have contributed to the impoverishment of ghettos.

This problem is further complicated by conditions of changing preferences among any community of people and the problem of changing levels of demand in relation to the available supply of a public good or service. No one can know the preferences or values of other persons apart from giving those persons an opportunity to express their preferences or values. If constituencies and collectivities are organized in a way that does not reflect the diversity of interests among different groups of people, then producers of public goods and services will be taking action without information as to the changing preferences of the persons they serve. Expenditures may be made with little reference to consumer utility. Producer efficiency in the absence of consumer utility is without meaning. Large per capita expenditures for educational services that are not conceived by the recipients to enhance their life prospects may be grossly unproductive. Education can be a sound investment in human development only when individuals perceive the effort as enhancing their life prospects.

Similar difficulties may be engendered when conditions of demand for a public good or service increase in relation to the available supply. When demands begin to exceed supply, the dynamics inherent in "the tragedy of the commons" may arise all over again (Mohring 1970; Nelson 1962; Rothenberg 1970; Seneca 1970; Williamson 1966). A congested highway carries less and less traffic as the demand grows. What was once a public good for local residents may now become a public "bad" as congested and noisy traffic precludes a growing number of opportunities for the use of streets by local residents (Buchanan 1970). In short, the value of public goods may be subject to serious erosion under conditions of changing demand.

Finally, producer performance and consumer interests are closely tied together when we recognize that the capacities to levy taxes, to make appropriate expenditure decisions, and to provide the necessary public facilities are insufficient for optimality in the use of public facilities. One pattern of use may impair the value of a public facility for other patterns of use. The construction of a public street or highway, for example, would be insufficient to enhance the welfare potential for members of a community without attention to an extensive body of regulations controlling the use of such facilities by pedestrian and vehicular traffic. As demand for automobile traffic turns streets into a flood of vehicles, who is to articulate the interests of pedestrians and other potential users and allocate the good among all potential users?

The interests of the users of public goods and services will be taken into account only to the extent that producers of public goods and services stand exposed to the potential demands of those users. If producers fail to adapt to changing demands or fail to modify conditions of supply to meet changing demands, then the availability of alternative administrative, political, judicial, and constitutional remedies may be necessary for the maintenance of an efficient and responsive system of public administration. Efficiency in public administrations will depend upon the sense of constitutional decision making that public administrators bring to the task of constituting the conditions of public life in a community.

Most political economists in the public choice tradition would anticipate that no single form of organization is good for all social circumstances (see also Ashby 1960, 1962). Different forms of organization will give rise to some capabilities and will be subject to other limitations. Market organization will be subject to limitations that will give rise to institutional weaknesses or institutional failure. Bureaucratic organization will provide opportunities to develop some capabilities and will be subject to other limitations. Those limitations will in turn generate institutional weakness or institutional failure if they are exceeded. A knowledge of the capabilities and limitations of diverse forms of organizational arrangements will be necessary for both the future study and practice of public administration.

Toward New Perspectives in the Study of Public Administration

Our prior analysis has been largely, though not exclusively, oriented toward a circumstance involving calculations relative to the provision of a single public good. If we proceed with an assumption that we live in a world involving a large variety of potential public goods that come in different shapes and forms, we may want to consider what our representative individual as a self-interested calculator pursuing maximizing strategies would search out as an appropriate way for organizing an administrative system to provide an optimal mix of different public goods and services. Would a representative individual expect to get the best results by having all public goods and services provided by a single integrated bureaucratic structure subject to the control and direction of a single chief executive? Or would he expect to get better results by having access to a number of different collectivities capable of providing public services in response to a diversity of communities of interest? (Bish 1968; Duggal 1966; Haefele 1970; Hirschman 1967; McKean 1965; Pauly 1970).

If the answer to the first question is "no," then the presumptions inherent in Wilson's theory of administration and in the traditional principles of public administration will not stand as a satisfactory basis for a theory of administration in a democratic society. If the answer to the second question is "yes," then we are confronted with the task of developing an alternative theory of public administration that is appropriate for citizens living in a democratic society (Lindblom 1965).

If a domain that is relevant to the provision of a public good or service can be specified so that those who are potentially affected can be contained within the boundaries of an appropriate jurisdiction and externalities do not spill over onto others, then a public enterprise can be operated with substantial autonomy, provided that an appropriate structure of legal and political remedies is available to assure that some are not able to use the coercive powers of a collectivity to deprive others of lawful rights or claims (Hirsch 1968, 1963). Even where such conditions could not be met, solutions can be devised by reference to overlapping jurisdictions so that the larger jurisdictions are able to control for externalities while allowing substantial autonomy for the same people organized as small collectivities to make provision for their own public welfare.

In the traditional theory of public administration, the existence of overlapping jurisdictions has often been taken as prima facie evidence of duplication of effort, inefficiency, and waste. If we contemplate the possibility that different scales of organization may be appropriate to different levels of operation in providing a particular type of public service, substantial advantage may derive from the provision of services by overlapping jurisdictions (Dawson 1970; Hirsch 1964; Williams 1966). For example, local police may not be very proficient in dealing with organized crime operating on a state, interstate, national, or international basis. Large-scale national police agencies may be a necessary but not sufficient condition for dealing with such problems. Control over the movement of traffic in and out of urban centers may pose problems of an intermediate scale in policing operations. Crimes in the street, however, may reflect the absence of police services responsive to local neighborhood interests.

Once we contemplate the possibility that public administration can be organized in relation to diverse collectivities organized as concurrent political regimes, we might further contemplate the possibility that there will not be one rule of good administration for all governments alike. Instead of a single integrated hierarchy of authority coordinating all public services, we might anticipate the existence of multiorganizational arrangements in the public sector that tend to take on the characteristics of public service industries composed of many public agencies operating with substantial independence of one another (V. Ostrom and E. Ostrom 1965). Should we not begin to look at the police industry (E. Ostrom 1971; Shoup 1964), the education industry (Barlow 1970; Holtmann 1966; Machlup 1962), the water industry (Bain, Caves, and Margolis 1966; Schmid 1967; Weschler 1968), and other public service industries on the assumption that these industries have a structure that allows for coordination without primary reliance upon hierarchical structures (Warren 1966)? Once we begin to look for order among multiorganizational arrangements in the public sector, important new vistas will become relevant to the study of public administration (Bish 1971). So-called grants-in-aid may take on the attributes of a transfer of funds related to the purchase of a mix of public services to take appropriate account of externalities that spill over from one jurisdiction to another (Breton 1965; Weldon 1966).

A combination of user taxes, service charges, intergovernmental transfers of funds, and voucher systems may evoke some of the characteristics of market arrangements among public service agencies (Breton 1967; Buchanan 1968a; Dales 1968; V. Ostrom 1969; V. Ostrom, Tiebout, and Warren 1961). Instead of a bureaucratic hierarchy serving as the primary means for sustaining legal rationality in a political order as Max Weber has suggested, we should not be surprised to find that legal rationality can be sustained by recourse to judicial determination of issues arising from conflicts over jurisdiction among administrative agencies. Given the high potential cost of political stalemate for the continuity and survival of any administrative enterprise, we should not be surprised to find rational, self-interested public administrators consciously bargaining among themselves and mobilizing political support from their clientele in order to avoid political stalemate and sustain the political feasibility of their agencies (Lindblom 1955). Perhaps a system of public administration composed of a variety of multiorganizational arrangements and highly dependent upon mobilizing clientele support will come reasonably close to sustaining a high level of performance in advancing the public welfare.

NOTE

1. Originally issued as *Papers on Non-Market Decision Making, Public Choice* first appeared in spring 1968.

REFERENCES

Abbreviations for journal references most frequently cited:

AER	*American Economic Review*
AJS	*American Journal of Sociology*
APSR	*American Political Science Review*
CJE&PS	*Canadian Journal of Economics and Political Science*
JLE	*Journal of Law and Economics*
JPE	*Journal of Political Economy*
MJ of PS	*Midwest Journal of Political Science*
Papers	*Papers on Non-Market Decision Making*
PAR	*Public Administration Review*
PC	*Public Choice*
PF	*Public Finance*
QJE	*Quarterly Journal of Economics*

Arrow, Kenneth F., "Alternative Approaches to the Theory of Choice in Risk-Taking Situations," *Econometrica*, 19 (October 1951), 404–37.

———, *Social Choice and Individual Values*, second edition (New York: John Wiley & Sons, 1963).

Ashby, W. Ross, "Principles of the Self-Organizing System," in *Principles of Self-Organization*, H. Von Foerster and G. W. Zopf (eds.). (New York: The Macmillan Co., 1962), 255–78.

———, *Design for a Brain*, second edition (New York: John Wiley & Sons, 1960).

Ayres, Robert U. and Allen V. Kneese, "Production, Consumption and Externalities," *AER*, 59 (June 1969), 282–97.

Bain, Joe S., Richard E. Caves, and Julius Margolis, *Northern California's Water Industry: The Comparative Efficiency of Public Enterprise in Developing a Scarce Natural Resource* (Baltimore: The Johns Hopkins Press, 1966).

Barlow, Robin, "Efficiency Aspects of Local School Finance," *JPE*, 78 (October 1970), 1028–40.

Bator, Francis, "The Anatomy of Market Failure," *QJE*, LXXII (August 1958), 351–79.

Bish, Robert L., "A Comment on V. P. Duggal's 'Is There an Unseen Hand in Government?' ," *Annals of Public and Cooperative Economy*, XXXIX (January/March 1968), 89–94.

———, *The Public Economy of Metropolitan Areas* (Chicago: Markham Publishing Company, 1971).

Black, Duncan, *The Theory of Committees and Elections* (Cambridge, England: Cambridge University Press, 1958).

——— and R. A. Newing, *Committee Decisions with Complementary Valuation* (London: William Hodge, 1951).

Boulding, Kenneth E., "The Ethics of Rational Decision," *Management Science*, 12 (February 1966), 161–69.

———, "Towards a Pure Theory of Threat Systems," *AER,* 53 (May 1963), 424–34.

Bowen, Howard R., "The Interpretation of Voting in the Allocation of Economic Resources," *QJE,* LVIII (November 1943), 27–48.

Bradford, D. V., "Constraints on Public Action and Rules for Social Decision," *AER,* 60 (September 1970), 642–54.

Breton, Albert, "A Theory of the Demand for Public Goods," *CJE&PS,* XXXII (November 1966), 455–67.

———, "A Theory of Government Grants," *CJE&PS,* XXXI (May 1965), 175–87.

———, *Discriminatory Government Policies in Federal Countries* (Montreal: The Canadian Trade Committee, Private Planning Association of Canada, 1967).

Brodbeck, May, "Methodological Individualism: Definition and Reduction," *Philosophy of Science,* 25 (January 1958), 1–22.

Buchanan, James M., "A Public Choice Approach to Public Utility Pricing," *PC,* 5 (Fall) (1968a), 1–17.

———, *The Demand and Supply of Public Goods* (Chicago: Rand McNally, 1968b).

———, *Cost and Choice: An Inquiry in Economic Theory* (Chicago: Markham Publishing Company, 1969).

———, "Public Goods and Public Bads," in *Financing the Metropolis,* John P. Crecine (ed.) (Beverly Hills: Sage Publications, 1970).

——— and W. Craig Stubblebine, "Externality," *Economica,* XXIX (November 1962), 371–84.

——— and Gordon Tullock, *The Calculus of Consent: Logical Foundations of Constitutional Democracy* (Ann Arbor, Mich.: University of Michigan Press, 1965).

Burgess, Philip M. and James A. Robinson, "Alliances and the Theory of Collective Action: A Simulation of Coalition Processes," *MJ of PS,* XII (May 1969), 194–219.

Campbell, Colin D. and Gordon Tullock, "A Measure of the Importance of Cyclical Majorities," *Economic Journal,* 75 (1965), 853–57.

——— and Gordon Tullock, "The Paradox of Voting: A Possible Method of Calculation," *APSR,* LX (September 1966), 684–85.

Coase, R. H., "The Nature of the Firm," *Economica,* 4 (1937), 386–85.

———, "The Problem of Social Cost," *JLE,* III (October 1960) 1–44.

Coleman, James S., "Foundations for a Theory of Collective Decisions," *AJS,* 71 (May 1966), 615–27.

———, "Individual Interests and Collective Action," *Papers,* 1 (1966), 49–63.

Cunningham, R. L., "Ethics and Game Theory: The Prisoners Dilemma," *Papers,* 2 (1967), 11–26.

Cyert, Richard M. and James G. March, *A Behavioral Theory of the Firm* (Englewood Cliffs, N.J.: Prentice Hall, 1963).

Dales, J. H., *Pollution, Property and Prices: An Essay in Policy-Making and Economics* (Toronto: University of Toronto Press, 1968).

Davis, Otto A. and Melvin Hinich, "A Mathematical Model of Policy Formation in a

Democratic Society," *Mathematical Applications in Political Science,* 11 (1966), 175–208.

———— and Melvin J. Hinich, "Some Results Related to a Mathematical Model of Policy Formation in a Democratic Society," *Mathematical Applications and Political Science,* III (1967), 14–38.

———— and Andrew Whinston, "Externalities, Welfare and the Theory of Games," *JPE,* 60 (June 1962), 241–62.

———— and Andrew Whinston, "On the Distinction Between Public and Private Goods," *AER,* 57 (May 1967), 360–73.

Dawson, D. A., *Economies of Scale in the Public Secondary School Education Sector in Ontario,* (Hamilton, Ontario: McMaster University, Department of Economics, Working Paper No. 70–04, 1970).

Demsetz, Harold, "Private Property, Information and Efficiency," *AER,* LVII (May 1967), 347–60.

————"The Private Production of Public Goods" *JLE,* XII (October 1970), 293–306.

Downs, Anthony, *An Economic Theory of Democracy* (New York: Harper & Row, 1957).

————, "In Defense of Majority Voting," *JPE,* LXIX (April 1961), 192–99.

————, "Why the Government Budget is Too Small in a Democracy," *World Politics,* XII (July 1960), 541–64.

Duggal, V. P., "Is There an Unseen Hand in Government?" *Annals of Public and Comparative Economy,* 37 (April/June 1966), 145–50.

Eulau, Heinz, "Logic of Rationality in Unanimous Decision Making," in *Nomos VII: Rational Decision,* Carl I. Friedrich (ed.) (New York: Atherton Press, 1964), 26–54.

Farquharson, Robin, *Theory of Voting* (New Haven: Yale University Press, 1969).

Fishburn, Peter C., *Decision and Value Theory* (New York: John Wiley & Sons, Inc., 1964).

Garvey, Gerald, "The Political Economy of Patronal Groups," *PC,* VII (Fall 1969), 33–45.

————, "The Theory of Party Equilibrium," *APSR,* LX (March 1966), 29–38.

Gulick, Luther and L. Urwick (eds.), *Papers on the Science of Administration* (New York: Columbia University, Institute of Public Administration, 1937).

Haefele, Edwin T., "Coalitions, Minority Representation, and Vote-Trading Probabilities," *PC,* VII (Spring 1970), 75–90.

————, "Environmental Quality as a Problem of Social Choice," in *Environmental Quality Analysis: Theory and Method in the Social Sciences,* in Allen V. Kneese and Blair T. Bower (eds.), (Baltimore, MD: Resources for the Future, 1972), 281–331.

Hardin, Garrett, "The Tragedy of the Commons," *Science,* 162 (December 13, 1968), 1243–48.

Hart, Albert G., *Anticipations, Uncertainty and Dynamic Planning* (New York: Augustus M. Kelley, 1965).

Head, J. G., "Public Goods and Public Policy," *Public Finance,* 17 (1962), 197–219.

Hinich, Melvin J. and Peter Ordeshook, "Plurality Maximization vs. Vote Maximization: A Spatial Analysis with Variable Participation," *APSR*, 64 (September 1970), 772–91.

Hirsch, Werner Z., "Local versus Areawide Urban Government Services," *National Tax Journal*, 17 (December 1964), 331–39.

———, "The Supply of Urban Public Services," in *Issues in Urban Economics*, Harvey S. Perloff and Lowden Wingo (eds.) (Baltimore: Johns Hopkins Press, 1968), 477–526.

———, "Urban Government Services and Their Financing," in *Urban Life and Form*, Werner Z. Hirsch (ed.) (New York: Holt, Rinehart and Winston, 1963), 129–66.

Hirschman, A. O., "The Principles of the Hiding Hand," *The Public Interest*, 6 (Winter 1967).

Hirshleifer, Jack, James C. DeHaven, and Jerome W. Milliman, *Water Supply Economics, Technology, and Policy* (Chicago: The University of Chicago Press, 1960).

Holtmann, A. G., "A Note on Public Education and Spillovers through Migration," *JPE*, 74 (October 1966), 524–25.

Hotelling, Harold, "Stability in Competition," *Economic Journal*, 39 (1929), 41–57.

Kafoglis, Madelyn L., "Participatory Democracy in the Community Action Program," *PC*, V (Fall 1968), 73–85.

Knight, Frank H., *Risk, Uncertainty and Profit* (New York: Harper and Row, reissued 1965).

Leoni, Bruno, "The Meaning of 'Political' in Political Decisions," *Political Studies*, V (October 1957), 225–39.

Lindblom, Charles E., *Bargaining: The Hidden Hand in Government* (Santa Monica: The RAND Corporation, 1955).

———, *The Intelligence of Democracy: Decision Making through Mutual Adjustment* (New York: The Free Press, 1965).

Luce, R. Duncan and Howard Raiffa, *Games and Decisions: Introduction and Critical Survey* (New York: John Wiley & Sons, 1957).

McKean, Roland L., "The Unseen Hand in Government," *AER*, 55 (June 1965), 496–506.

Machlup, Fritz, *The Production and Distribution of Knowledge in the U.S.* (Princeton, N.J.: Princeton University Press, 1962).

Mishan, Ezra J., *The Costs of Economic Growth* (New York: Frederick A. Praeger, 1967).

Mohring, H., "The Peak Load Problem with Increasing Returns and Pricing Constraints," *AER*, 60 (September 1970), 693–705.

Musgrave, Richard, "The Voluntary Exchange Theory of Public Economy," *QJE*, LIII (February 1939), 213–37.

———, *The Theory of Public Finance* (New York: McGraw-Hill, 1959).

Nelson, James C., "The Pricing of Highway, Waterway, and Airway Facilities," *AER*, 52 (May 1962), 426–35.

Niskanen, William A., *Bureaucracy and Representative Government* (Chicago: Aldine, Atherton, 1971).

Olson, Mancur, *The Logic of Collective Action* (Cambridge, Mass.: Harvard University Press, 1965).

Ostrom, Elinor, "Institutional Arrangements and the Measurement of Policy Consequences in Urban Affairs," *Urban Affairs Quarterly* (June 1971), 447–75.

———, "Some Postulated Effects of Learning on Constitutional Behavior," *PC*, V (Fall 1968), 87–104.

Ostrom, Vincent, "Water Resource Development: Some Problems in Economic and Political Analysis of Public Policy," in *Political Science and Public Policy*, Austin Ranney (ed.) (Chicago: Markham Publishing Company, 1968).

———, "Operational Federalism: Organization for the Provision of Public Services in the American Federal System," *PC*, VI (Spring 1969), 1–17.

——— and Elinor Ostrom. "A Behavioral Approach to the Study of Intergovernmental Relations," *Annals of the American Academy of Political and Social Science*, 359 (May 1965), 137–46. (Reprinted in Michael D. McGinnis, ed., *Polycentricity and Local Public Economies* [Ann Arbor: University of Michigan Press, 1999].)

———, Charles M. Tiebout, and Robert Warren, "The Organization of Government in Metropolitan Areas: A Theoretical Inquiry," *APSR*, 55 (December 1961), 831–42. (Reprinted in Michael D. McGinnis, ed., *Polycentricity and Local Public Economies* [Ann Arbor: University of Michigan Press, 1999].)

Pauly, Mark V., "Optimality, 'Public' Goods and Local Governments: A General Theoretical Analysis," *JPE*, 78 (May/June 1970), 572–85.

Phelps, Edmund, *Private Wants and Public Needs* (New York: W. W. Norton, 1965).

Plott, Charles R., "A Notion of Equilibrium and Its Possibility under Majority Rule," *AER*, 57 (September 1967), 787–806.

Radner, Roy, "Problems in the Theory of Markets under Uncertainty," *AER*, LX (May 1970), 454–60.

Ridley, Clarence E. and Herbert A. Simon, *Measuring Municipal Activities* (Chicago: The International City Managers' Association, 1938).

Riker, William H., "Arrow's Theorem and Some Examples of the Paradox of Voting," *Mathematical Applications in Political Science*, 1 (1965).

———, "The Paradox of Voting and Congressional Rules for Voting on Amendments," *APSR*, 52 (1958), 349–66.

———, *The Theory of Political Coalitions* (New Haven: Yale University Press, 1962).

Rothenberg, Jerome, "The Economics of Congestion and Pollutions: An Integrated View," *AER*, LX (May 1970), 114–21.

Samuelson, Paul A., "Diagrammatic Exposition of a Theory of Public Expenditure," *Review of Economics and Statistics*, XXXVII (November 1955), 350–56.

———, "The Pure Theory of Public Expenditure," *Review of Economics and Statistics*, XXXVI (November 1954), 387–89.

Schelling, Thomas C., *The Strategy of Conflict* (Cambridge, Mass.: Harvard University Press, 1963).

Schmid, A. Allan, "Nonmarket Values and Efficiency of Public Investments in Water

Resources," *AER,* LVII (May 1967), 158–68.

Seneca, Joseph J., "The Welfare Effects of Zero Pricing of Public Goods," *PC,* VII (Spring 1970), 101–10.

Shackle, G. L. S., *Decision, Order and Time in Human Affairs* (Cambridge, England: Cambridge University Press, 1961).

Shoup, Carl S., "Standard for Distributing of Free Government Service: Crime Prevention," *PF,* 19 (1964), 383–92.

Simon, Herbert A., *Administrative Behavior: A Study of Decision-Making Processes in Administrative Organizations* (New York: Macmillan, 1964).

———, *Models of Men, Social and Rational* (New York: John Wiley & Sons, 1957).

———, "The Architecture of Complexity," *Proceedings of the American Philosophical Society,* 106 (December 1962), 467–82.

———, *The Sciences of the Artificial* (Cambridge, Mass.: The M.I.T. Press, 1969).

———, "Theories of Decision-Making in Economics and Behavioral Science," *AER,* XLIX (June 1959), 253–83.

Stigler, George J., "The Economics of Information," *JPE,* LXIX (June 1961), 213–25.

Thompson, E. A., "A Pareto-Optimal Group Decision Process," *Papers,* 1 (1966), 133–40.

Thompson, Wilbur, "The City as a Distorted Price System," *Psychology Today* (August 1968), 28–33.

Tiebout, Charles M., "A Pure Theory of Local Expenditures," *JPE,* 64 (October 1956), 416–24.

Tullock, Gordon, "A Simple Algebraic Logrolling Model," *AER,* LX (June 1970), 419–26.

———, *Politics of Bureaucracy* (Washington, D.C.: The Public Affairs Press, 1965).

———, "The General Irrelevance of the General Impossibility Theorem," *QJE,* 81 (May 1967), 256–70.

U.S. Congress, Joint Economic Committee, Subcommittee on Economy in Government, *A Compendium of Papers on the Analysis and Evaluation of Public Expenditures: The PPB System,* three volumes (Washington, D.C.: U.S. Government Printing Office, 1969).

Wade, L. L. and R. L. Curry, Jr., *A Logic of Public Policy: Aspects of Political Economy* (Belmont, Calif.: Wadsworth Publishing Company, 1970).

Waldo, Dwight, "Scope of the Theory of Public Administration" in *Theory and Practice of Public Administration: Scope, Objectives, and Methods,* James C. Charlesworth (ed.) (Philadelphia: The American Academy of Political and Social Science, 1968).

Warren, Robert, "Federal-Local Development Planning: Scale Effects in Representation and Policy Making," *PAR,* XXX (November/December 1970), 584–95.

———, *Government in Metropolitan Regions: A Reappraisal of Fractionated Political Organization,* (Davis, Calif.: University of California, Davis, Institute of Governmental Affairs, 1966).

Weldon, J. C., "Public Goods and Federalism," *CJE&PS,* 32 (1966), 230–38.

Weschler, Louis F., *Water Resources Management: The Orange County Experience* (Davis, Calif.: University of California, Davis, Institute of Governmental Affairs, 1968).

———, Paul D. Marr, and Bruce M. Hackett, *California Service Center Program* (Davis, Calif.: University of California, Davis, Institute of Governmental Affairs, 1968).

Wheeler, H. J., "Alternative Voting Rules and Local Expenditures: The Town Meeting vs. City," *Papers,* 2 (1967), 61–70.

White, Leonard D., *Introduction to the Study of Public Administration* (New York: The Macmillan Company, 1926).

Wildavsky, Aaron, "The Political Economy of Efficiency," *PAR,* XXVI (December 1966), 292–310.

Williams, Alan, "The Optimal Provision of Public Goods in a System of Local Government," *JPE,* 74 (February 1966), 18–33.

Williamson, Oliver E., "Hierarchical Control and Optimum Firm Size," *JPE,* 75 (April 1967), 123–38.

———, "Peak Load Pricing and Optimal Capacity under Indivisibility Constraints," *AER,* LVI (September 1966), 810–27.

Wilson, Woodrow, "The Study of Administration," *Political Science Quarterly,* II (June 1887), 197–222.

CHAPTER 2

The Three Worlds of Action: A Metatheoretical Synthesis of Institutional Approaches

Larry L. Kiser and Elinor Ostrom

Studying the effects of institutional arrangements on patterns of human behavior and the resulting patterns of outcomes is a strategy of inquiry used by
scholars straddling the academic disciplines of political science and economics. Institutional arrangements are the rules used by individuals for determining who and what are included in decision situations, how information is
structured, what actions can be taken and in what sequence, and how individual actions will be aggregated into collective decisions. Institutional arrangements are thus complex composites of rules, all of which exist in a language
shared by some community of individuals rather than as the physical parts of
some external environment.

As language-based phenomena, institutional rules do not impinge directly on the world (V. Ostrom 1980). A change in a decision rule (for example, adopting a two-thirds instead of a majority voting rule) cannot have an
immediate and direct effect on some physical distribution of things. Institutional change impinges on the world by affecting the shared understandings
of individuals making choices within decision situations affected by the rules.
Effects in the world will result, if they do, in three steps. First, the individuals
affected by a change in rules must be cognizant of and abide by the change.
Second, institutional change has to affect the strategies they adopt. Third, the

Originally published in Elinor Ostrom, ed., *Strategies of Political Inquiry* (Beverly Hills,
CA, 1982), 179–222. Copyright © 1982 by Sage Publications, Inc. Reprinted by permission of
Sage Publications, Inc. and the authors.

Authors' note: Major portions of this manuscript were written while Kiser was a postdoctoral
trainee at the Workshop in Political Theory and Policy Analysis at Indiana University, under an
NIMH traineeship on "Research Training in Institutional Analysis and Design" (Grant No. 5 T32
MH15222). The essay draws on a long tradition of work and discussion at the Workshop and could
not have been written without the extended discussions and comments on earlier, related manu-
scripts among participants including Deby Dean, Vernon Greene, John Hamilton, Roger B. Parks,
Stephen L. Percy, Ron Oakerson, Vincent Ostrom, Rick Wilson, and Susan Wynne. We appreciate
the helpful comments we received on the first draft of this manuscript from David Austin, Judy
Gillespie, David Kessler, Roger B. Parks, Cyrus Reed, and Rick Wilson. We also appreciate
Teresa Therrien's careful work in editing and typing this manuscript.

aggregation of changed individual strategies must lead to different results. Not all changes in behavior lead to changes in outcomes. A change in the institutional arrangement for deciding energy policy, for example, does not change patterns of energy use until individuals in their everyday lives lower thermostat settings, join a car pool, buy a solar heater, and so on.

This complex set of transformations from rules through common understanding of decision situations to individual behavior is difficult to analyze and understand. The purpose of this essay is to provide a metatheoretical framework for understanding the complex set of transformations that links institutional arrangements to individual behavior and aggregate results occurring in the "real" world. The framework is *meta*theoretical, because it describes the array of elements that is used in specific theories about institutions rather than presenting a particular theory.

This metatheoretical framework focuses on the scholarly work in political science and economics that uses a microinstitutional approach to the analysis of political phenomena. The microinstitutional approach is "micro" because it starts from the individual as a basic unit of analysis to explain and predict individual behavior and resulting aggregated outcomes. It is an "institutional" approach because major explanatory variables include the set of institutional arrangements that individuals use to affect the incentive systems of a social order and the impact of incentive systems on human behavior.

Patterns of human action and the results that occur in interdependent choice-making situations are the phenomena to be explained using this approach. Interdependence occurs whenever results are dependent on the actions of more than one individual. Interdependence can at times produce aggregated results that vary dramatically from individual expectations or preferences. Interdependence describes hundreds of everyday experiences as individuals interact with one another, trying to achieve a better life in the marketplace, within families, in clubs, in neighborhoods, in legislatures, in bureaucracies, and in other collective arrangements (see Schelling 1978 for a discussion of such situations).

Interdependent decision situations can be incredibly complex. Institutional analysts abstract from this complexity by modeling typical decision situations and the behavior of individuals involved in such situations. Analysts attempt to identify the smallest set of working parts to yield a coherent and testable theory for observed patterns of actions and results. Given the complexity of the theoretical models used in institutional analysis, social scientists wishing to understand this literature are frequently confused about how the various elements of the theories are linked together and about how theorists use key terms. We try to provide a synthesis and overview of the political-economy literature that uses the individual as a unit of analysis while asking how institutional arrangements affect the level, type, and distribution of outcomes. This essay is like a schematic road map to a complex, difficult, and changeable academic terrain. We hope it will help the neophyte interested in gaining an overview before approaching the specifics. We also hope it will help those who have traveled the hills and valleys of institutional analysis and who may wish to step back with us and look at the slippery terrain they have traveled from a more general perspective.

While many authors have contributed to this approach, we have been particularly influenced in offering this synthesis by the work of Kenneth Arrow, James Buchanan, John R. Commons, Anthony Downs, Frank Knight, Mancur Olson, Vincent Ostrom, William Riker, Kenneth Shepsle, Herbert Simon, Thomas Schelling, Gordon Tullock, and Oliver Williamson. Instead of peppering this essay with extensive and disruptive citations, we have provided the reader with a general bibliography of many of the key articles and books related to the synthesis presented here.

The Theoretical Working Parts of Institutional Analysis

Implicit or explicit in the theories explaining individual behavior within institutional structures are five working parts, including (1) the decision maker, (2) the community affected by interdependent decision making, (3) events (or goods and services) that interacting individuals seek to produce and consume, (4) institutional arrangements guiding individual decisions, and (5) the decision situation in which individuals make choices. Political economists select appropriate assumptions about the attributes of each working part. The set of assumptions about the working parts explains actions by individual decision makers and aggregated outcomes. Figure 2.1 shows the relationship among the five working parts, actions, and outcomes.

The framework displayed in figure 2.1 rests on a methodological individualist perspective. Attributes of the individual decision maker constitute the core of the analysis. Assumptions about the individual animate all particular models based on this microinstitutional frame (Popper 1967; Simon 1978). However, several alternative assumptions may be made within the model of the individual used by a particular theorist. This approach is distinct from macroinstitutional political economy, which animates theoretical models with social forces beyond the influence of individuals. Individuals in macroinstitutional political economy have little choice but to obey these overriding social forces.

While methodological individualism places the attributes of individual decision makers at the core of the analytical framework, the other working parts are equally important to the explanations derived from the framework. The other working parts establish the environment in which individuals

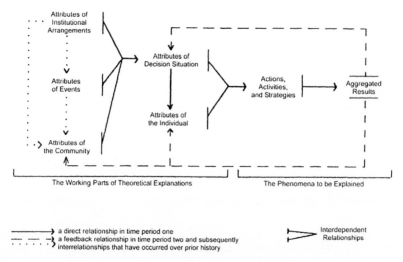

Fig. 2.1. The working parts of institutional analysis

make choices. Combining assumptions about all five working parts enables political economists to predict two types of results. One prediction addresses individual decision makers' strategies or actions, and the second prediction addresses the aggregation of individual actions into outcomes for the community.

Political economists rarely develop all five working parts within a given analysis. Writers tend to focus, instead, on changes within one or two working parts, assuming that all other working parts remain unchanged. Unfortunately, such assumptions are often implicit and lead to error. Holding other factors constant is not equivalent to ignoring them. The configuration of working parts constitutes the environment in which individuals make choices; thus, analysis must recognize the condition of the unchanging factors (see Boynton 1982). Ignoring any of the working parts can misrepresent the decision maker's environment and produce misleading predictions.

To use an analogy, one can think of a specific model developed to explain the effect of institutional arrangements as a large Tinkertoy composed of at least five components, each of which also contains a number of elements or subcomponents. The large Tinkertoy model is built to resemble a more complex and indeterminate machine that operates in the real world. The Tinkertoy is a successful model for some purpose if it helps predict what the complex machine will do under particular circumstances. No single Tinkertoy model is used to predict the actions for all complex machines that might be modeled by institutional theorists. As different machines are modeled, major changes may be made in some subcomponents, while other working parts may remain relatively stable. However, changes in some of the working parts may require changes in others, since considerable interdependence exists among the particular assumptions made about different working parts.

Political economists apply the working parts of microinstitutional analysis at three related but distinct levels of analysis. One is the *operational level,* which explains the world of action. The second is the *collective choice level,* which explains the world of authoritative decision making. The third is the *constitutional level,* which explains the design of collective-choice mechanisms. The operational, collective-choice, and constitutional levels of analysis are the "three worlds" referred to in our title. Figure 2.2 displays the three levels of analysis and the components of each level. It shows that the same working parts make up all three worlds. This essay, therefore, concentrates first on each of the working parts, applying them to the world of action. We then apply the working parts to the other two levels of analysis.

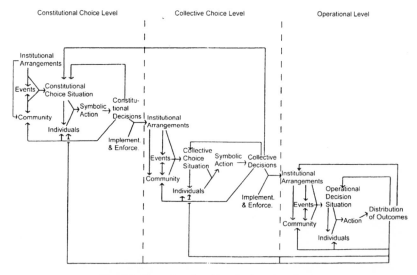

Fig. 2.2. Three levels of institutional analysis

Attributes of the Individual

Microinstitutional political economy emphasizes that only individuals, not groups, act. To predict that action, theorists must at least make assumptions about

1. the individual's level of information about the decision situation,
2. the individual's valuation of potential outcomes and of alternative

actions possible within the decision situation, and

3. the individual's calculation process for selecting among alternative actions or strategies.

Individuals act to achieve valued outcomes; therefore, individuals must have some notion of how actions link up with outcomes and how values are realized by selecting among alternative actions.

Neoclassical economists handle the three minimum assumptions by modeling an individual producer in a competitive market as having perfect information about relevant goods and resource prices, as valuing only profits, and as calculating to maximize expected profits. Herbert Simon and others influenced by him (see Williamson 1975) develop alternative models by assuming that the individual producer possesses incomplete information about the decision situation. These political economists also assume that individuals value multiple outcomes rather than just profits and that individuals sometimes have difficulty discriminating among the rankings of these outcomes. The producer is much less precise in decision making than the producer in the neoclassical model, searching for information and learning the characteristics of the decision situation. The Simon-type producer is less apt than the neoclassical producer to maximize values and is more likely to seek satisficing levels.

The individual in both the neoclassical and the Simon approaches is rational, although scholars within the neoclassical tradition frequently criticize Simon's work as "antirational." Simon (1957, 1972, 1978) defends his model of the individual as intendedly rational. The difference between the two specific models is the amount of information that individuals possess about themselves and their environment. The difference causes the individual to use a different calculating process for deciding how to act within a given decision situation.

Specific assumptions about information, valuation, and calculation in the model of the individual help to shape the overall model of interdependent choice. Neoclassical assumptions of complete information and profit maximization yield, when combined with the other working parts, a highly determinate, machinelike general theory. Assumptions of incomplete information, multiple values, and satisficing yield a less determinate, more flexible theory of choice (see V. Ostrom and Hennessey 1975; V. Ostrom 1976). The neoclassical assumptions are useful when modeling uncomplicated market exchange situations, but Simon-type assumptions are apt to be more useful when modeling more complex situations where options continually shift.

Attributes of the Decision Situation

Figure 2.1 shows that the attributes of the individuals combine with the attributes of the decision situation to yield individuals' actions or strategies.

The decision maker, in other words, chooses a strategy that is suitable to the decision situation. An individual with given attributes will make different choices as decision situations change. Also, as noted earlier, changes in the attributes of decision situations can require changes in the attributes of the individual decision maker. Particular assumptions about the individual's valuation of outcomes and calculation process must be compatible with the decision situation.

The literature on game theory investigates the attributes of decision situations, developing attributes such as

1. the number of decision makers involved;
2. the types of choices available to the decision maker;
3. the linkages between actions and results;
4. complexity;
5. repetitiveness;
6. types of outcomes (zero-sum, positive sum, and so on);
7. durability, stability, reversibility, and vulnerability to threat of outcomes; and
8. possibilities for communication among interacting decision makers.

Decision theorists who are not interested in the study of institutional arrangements stop with a model of the decision situation. Most game theorists, for example, do not inquire about the variables that affect a decision situation. The decision situation is itself taken as a given rather than as the result of other variables. Considerable theoretical advances in decision theory have been based on an in-depth examination of the behavior of differently modeled individuals in differently structured decision situations. Mancur Olson (1965), for example, develops the concept of the free rider within the context of alternative decision situations, showing that an individual is more apt to choose a free-riding strategy in large-group rather than small-group situations. Other writers note that the tragedy of the commons is more likely to occur in large-group rather than in small-group situations.

The linkages between actions and results are especially important in describing decision situations, as this is where the concept of the degree of certainty is described. To describe a situation as certain is to assert that each alternative action available to a decision maker links to one and only one result. Risk and uncertainty, on the other hand, recognize the possibility of multiple results linked to at least some of the actions. Risk describes a situation where probability distributions across alternative results can be known, while uncertainty describes a situation where probability distributions across results cannot be known.

This essay does not permit a full review of the attributes for decision situations. Hamburger (1979) presents an elementary introduction to this literature, outlining alternative descriptions of gaming situations and developing the relationship between the decision situation, the individual decision maker, and the choice among strategies. In general, the decision situation describes the array of choices confronting the decision maker and the pattern of consequences likely to flow from alternative choices. The attributes of the individual describe the capabilities and motives of the individual in the decision situation. The mix of the decision situation's attributes and the individual's attributes causes the individual to select particular actions.

Actions and Strategies

Microinstitutional political economy predicts individuals' actions and strategies and aggregated results flowing from models of decision makers within defined decision situations. An action in tennis, for example, can be either charging the net or defending the baseline. The player's choice depends on the player and the situation.

Political economists distinguish between actions and strategies. An action is like a "move" in a game, while a strategy is a sequence of moves. Thus, situations permitting only a single act at a particular decision point prohibit strategy, while repetitive situations encourage it. The decision maker in repetitive situations may make the same move every time the decision situation repeats, or the decision maker may randomly switch from one move to another. The consistent approach is called a "pure" strategy and the switching approach a "mixed" strategy.

Strategy implies commitment to a method for selecting actions over time. A strategy is a plan of action, often contingent upon other decision makers' actions. The strategy of "tit for tat" in extended Prisoners' Dilemma games is a contingent strategy. One player cooperates on the second play of the game if the other player cooperates on the first play of game. Complex decision situations such as those in chess require contingent strategy, because a player can select the best move only after the opposing player makes a move.

Strategy also involves activities such as information gathering, giving, and withholding. Also, strategies include threats, guile, bargaining, and force. Some strategies are "sophisticated" plans where an individual makes real or apparent sacrifices on initial rounds of a decision situation to increase the possibility for gains in later rounds. Farquharson (1969) discusses the strategy of voting for a less favored option early in a contest to increase the probability of being able to vote for a more favored option later. Another sophisticated plan is the act of binding oneself irrevocably to particular future actions. Individuals often precommit when they fear they will succumb to short-term temptations that counter long-term interests (Elster 1979; Thaler and Sheffrin 1981).

Aggregated Results

The mathematical transformation of individual acts into group results often surprises the people whose acts produced the results. Individuals sometimes do not connect aggregate results with their own individual actions. Thomas Schelling (1978) gives an example of aggregation in the children's game of musical chairs, where every child rushes to sit in a chair when the music stops. The result is that one child is always left standing. Regardless of how determined each child is to sit and how aggressively each child plays the game, the mathematical transformation of the actions to results is always the same. There is always one more child in the game than there are chairs; therefore, someone must stand.

The result in musical chairs is intended, but the results in many group situations are unintended. Schelling gives another example: individual drivers slowing down to look at a rush-hour automobile accident in the opposite lane of an expressway. Drivers would be happy with a quick, ten-second look; but the resulting traffic jam gives them an extra ten-minute drive home. Each driver acts to participate, at least vicariously, in the excitement of the accident; but the result is boredom. When complaining about the traffic jam later, many drivers probably fail to admit that their own actions helped to produce their misery. Thomas Hobbes, writing long before the age of the automobile, recognized the problem when he described men in the state of nature striving for their own good and finding only misery.

Why do individuals take actions that bring on misery? How can they avoid the misery and realize happiness instead? These kinds of questions lead us to focus on the left side of figure 2.1, to examine attributes of institutional arrangements, attributes of events, and attributes of community as they affect decision situations.

Institutional Arrangements

Economists and political scientists rejected for some time any self-conscious study of institutional arrangements. Both disciplines placed these "skeletons" in the academic closet for many years.[1] Only recently has work on institutional arrangements reappeared in respectable academic journals and publishing houses (Blumstein 1981; Riker 1980; Shepsle 1979). The relatively recent attention to institutions means that the conception of institutional arrangements is necessarily incomplete. To streamline work in this area and to encourage additional analysis, we propose a new conceptualization of institutional arrangements. Our purpose is to develop a typology that can be applied to a wide range of institutional conditions.

Institutional arrangements are the sets of rules governing the number of decision makers, allowable actions and strategies, authorized results, transformations internal to decision situations, and linkages among decision situations. Examining institutional arrangements involves a stepping back in the process to look at one of the three major factors affecting the structure of decision situations. The givens are no longer givens but rather the results of the interaction of rules, events, and community.

Hurwicz (1973) uses the term *decision mechanism* to describe institutional arrangements, conveying the image of a device constraining and guiding the choices that individuals make. Institutional arrangements simultaneously restrict and expand the freedom for individual action. As John R. Commons (1950, 21) wrote, "an institution is collective action in control, liberation, and expansion of individual action" (see also Knight 1965, 304). Persons interacting in a market situation are restricted, because they can only conduct particular types of voluntary transactions. Persons in legislative situations are also restricted, because they can vote only on amendments and legislation according to highly formalized procedures.

But institutional rules also expand freedom by increasing predictability in decision situations. Decision makers knowing that rules restrict actions by other decision makers can predict responses to their own actions better than they could without those rules. An auto salesperson, for example, is freer to grant credit to a buyer, knowing that rules require future payment and that courts exist to enforce those rules. Otherwise, buyers and sellers would limit themselves to transactions with immediate payment in full.

Rules, of course, require enforcement to be effective; but much theoretical work overlooks the enforcement aspect by assuming that decision makers take only lawful actions. The assumption simplifies analysis, enabling analysts to concentrate on issues unrelated to enforcement. But if the assumption is unconscious, the analysis can yield implications incongruent with a real world of illegal behavior. Sometimes analysis needs to include enforcement variables to explain real-world actions. Interaction with complicated games like football is more predictable when on-site referees aggressively mete out penalties to keep play within the rules.

This does not imply that enforcement is the only reason that individuals follow rules.[2] People also follow rules because those in a community share a belief that the rules are fair. Individuals within such a community, expecting others to follow the rules, see opportunities to pursue their own interests by following the rules too. Without this shared belief, enforcement would become too expensive to maintain regularity and predictability in ongoing human relationships.

Institutional arrangements transform an individual's actions into outcomes affecting both the actor and others. Predictability of action-outcome linkages, therefore, depends upon both transformations in the physical world and transformations within decision mechanisms in the institutional world. A person drives safely to work because the physical operation of cars on the road is predictable and because other motorists' behavior according to traffic rules is predictable. The driver's control over outcomes increases because each motorist possesses rights of way on the road and commensurate duties to observe equivalent rights of others.

Scholars frequently refer to institutional arrangements by organization names such as markets, legislatures, courts, private firms, public bureaus, contracts, and international treaties. But we propose a more abstract reference that facilitates comparison among organizational structures. The proposal looks to the rules that characterize all institutional arrangements—rules delineating the participants and allowable actions, rules distributing authority among participants, rules aggregating participants' choices into collective decisions, rules outlining procedure and information flows, and rules distributing payoffs among participants. These rules may be formal, that is, detailed and written, or informal, that is, simply understood by participants in the arrangement. Informal rules are what Commons (1959, 138) calls the "working rules of a going concern," rules that evolve perhaps unconsciously with the functioning of the organization.

We distinguish between institutional arrangements and organizations. Political economists frequently interchange the terms, but in our attempt to abstract comparisons among organizations we define institutional arrangements as distinct from organizations. We define organizations as composites of participants following rules governing activities and transactions to realize particular outputs. These activities occur within specific facilities. The rules, which are components of all organizations, are the institutional arrangements.

The distinction between institutional arrangements and organizations avoids problems of classifying organizations by proper name. National legislatures, for example, make up a single class; but legislatures can function very differently from one another. Identifying the rules or the institutional arrangements of the legislatures helps one to understand those important differences, whereas classifying by the term *legislature* obscures vital information about the way those organizations operate (see Shepsle 1979).

Our typology of institutional rules classifies rules according to their effect on decision situations and on individual choice behavior within decision situations.[3] The typology concentrates on (1) the entry and exit conditions for participating in organizations, (2) allowable actions and allowable outcomes from interaction within organizations, (3) the distribution of authority among positions within organizations to take particular actions, (4) the aggregation of joint decisions within organizations, (5) procedural rules in complex situations linking decision situations together, and (6) information constraints within organizations.

The types of rules that address each of these six issues are (1) boundary rules, (2) scope rules, (3) position and authority rules, (4) aggregation rules, (5) procedural rules, and (6) information rules. Each set of rules helps to shape incentives facing individuals within decision situations. Other aspects of institutional arrangements may also affect decision makers' incentives, but we think these six include aspects theorists identify as most important to explain behavior within and outcomes from interdependent decision situations.

The rules in intercollegiate tennis provide an example of institutional arrangements. The rules delineate

1. who can enter tennis as a competitive sport within the member universities and the conditions under which players lose eligibility (boundary rules);
2. the size of the court, the height of the net, the physical actions that a player is allowed to take, the number of times the ball can bounce on each side of the net, and the allowed set of outcomes (scope rules);
3. the rights and duties assigned to players and to referees (position and authority rules);
4. how specific acts and physical results are scored and aggregated into wins and losses (aggregation rules);
5. how players will proceed through tournament competition (procedural rules); and
6. how information about opponents' strategies, tournament rules, specific calls, and other matters is conveyed to participants (information rules).

Any particular model of an institutional arrangement will need to make particular assumptions about boundary rules, scope rules, position and authority rules, aggregation rules, procedure rules, and information rules. Given the large number of combinations of specific rules that can be designed, we are unlikely ever to have a complete theory of institutions.[4] However, substantial theoretical and empirical work has recently focused on the effect of one or more of these types of rules or the behavior of individuals in particular types of decision situations.

The Attributes of Events

People have long been aware that the nature of goods affects calculations bearing upon human welfare. Aristotle (1962, 58), for example, observed, "that which is common to the greatest number has the least care bestowed upon it." This aspect of decision making has attracted even more attention with the growth of public-goods analysis in recent times.

The variety in goods and events is vast. But most of the variety is inconsequential to the effects of institutional arrangements on the results of human interaction. A given institutional arrangement can effectively guide resource allocation toward notably dissimilar goods. Arrangements within competitive markets, for example, effectively make both wheat and television sets available to consumers. But competitive markets are ineffective in making police protection or environmental cleanup available.

This section focuses on attributes of goods that distinguish wheat and television sets from police protection and environmental cleanup and that affect the ability of markets to make those goods available. The discussion identifies attributes of goods and events that help to shape individuals' decision-making incentives. The attributes include jointness of use, exclusion, degree of choice, and measurability.[5]

Jointness of Use or Consumption

Howard Bowen (1943–44) distinguished between goods consumed simultaneously by more than one individual (joint consumption) and goods consumed by a single individual (separable consumption). The distinction is the effect that use by one individual has on the goods' availability to others. An individual's use of separable consumption goods reduces the goods' availability to others, while an individual's use of joint consumption goods permits simultaneous use by others. Paul Samuelson (1954, 1955) repeated this distinction a decade later with the classic series of articles on pure public goods. Samuelson called joint consumption goods public goods, defining "a public good as a good which is subject to joint or collective consumption where any one individual's consumption is *nonsubstractible* from any other individual's consumption of a good" (1954, 387; our emphasis). Samuelson (1954, 877) also noted that "a public consumption good differs from a private consumption good in that each man's consumption of it is related to the total by a condition of equality rather than that of summation."

Few, if any, joint consumption goods are perfectly nonsubtractible. The gravitational pull of the earth is an example: Regardless of the number of people using gravitational pull, the force is equally available to additional users. Most joint consumption goods are only partially nonsubtractible. When the number of users reaches some critical point, one person's use partially reduces the good's availability for simultaneous use by others. Highway congestion is an example. The highway is available for simultaneous use by many motorists, but as traffic builds during rush hours, one motorist's use of the highway causes delays and inconvenience for others. Buchanan (1970) identified the effects of congestion in public goods as erosion, noting the role of supply in contending against the degradation of public goods.

Exclusion

Scholars have extended the analysis of public goods, adding other characteristics to distinguish public from private goods. Musgrave (1959), Head (1962), and Olson (1965) identified the attribute of exclusion, noting that a public good makes excluding additional consumers infeasible once the good is available to any consumer. A private good, on the other hand, makes exclusion feasible. An individual can use the good without sharing it with other consumers. Anyone can benefit from a nonexcludable public good or event so long as nature or producers supply it. Nature supplies the air we breathe, and it is freely available to all. Producers supply views of buildings, and the views (whether obnoxious or enjoyable) are also freely available to all.

Both exclusion and jointness of consumption vary continuously. Goods are rarely totally excludable or totally nonexcludable, and they are rarely totally subtractible or totally nonsubtractible. But analysis frequently presumes pure categories to sharpen the distinction between public and private goods.

Exclusion and jointness of consumption are independent attributes, permitting a classification of goods according to table 2.1. The table shows jointness of consumption with two columns, one for highly subtractible goods and one for less subtractible goods. The table shows exclusion with two rows, one for low-cost exclusion goods and one for high-cost exclusion goods. The result is four cells distinguishing private goods, which are highly subtractible and where exclusion is inexpensive, from public goods, which are less subtractible and where exclusion is expensive. Toll goods and common-pool resources range between these two extremes with both types possessing some attributes of publicness. Toll goods are less subtractible, but exclusion is cheap. Common-pool resources are highly subtractible, but exclusion is expensive.

Measurement

Since public goods are difficult to package or unitize, they are usually also difficult to measure. Measures such as hectares, meters, and kilograms do not apply. While other kinds of indexes are possible, such as decibels, degrees,

TABLE 2.1. Jointness of Consumption

Exclusion	Highly Subtractible	Less Subtractible
Low cost	Private goods: bread, milk, automobiles, haircuts	Toll goods: theaters, night clubs, telephone service, cable TV, electric power, library
High cost	Common-pool resources: water pumped from a ground water basin, fish taken from an ocean, crude oil extracted from	Public goods: peace and security, of a community, national defense, mosquito abatement, air pollution

Note: Table 2.1 is adapted from V. Ostrom and E. Ostrom (1977a, 12).

and parts per million, aggregation is difficult with these measures. Sometimes indexes do not even exist. Thus, producers often know only imprecisely what they are producing, and consumers know only imprecisely what they are consuming. Cost calculations, valuation, and pricing are, therefore, often crude.

Private goods, on the other hand, are easier to package or unitize and, thus, to measure. This helps both producers and consumers make more informed decisions regarding such goods. Producers apply more precise cost-accounting procedures and management controls when producing private goods, and consumers make more precise budgeting decisions when consuming private goods.

Degree of Choice

Nonsubtractible and nonexcludable goods frequently give individuals little choice about consumption. The mere existence of a commodity may force individuals to consume it or at least pay much to avoid consuming the commodity. Some forced consumers may not even value the commodity. Congested streets, for example, inconvenience local residents and shoppers, who are required to cope with the traffic whether they like it or not. Private goods, however, do not pose this problem. The individual who does not value a private good simply need not purchase it.

Institutional Arrangements and Public Goods

The capacity of individuals to achieve a relatively optimal level of any type of goods is dependent upon the type of institutional rules developed to relate individuals to one another and to events in the world. However, the design of effective institutional arrangements to produce public goods is more difficult than the design of effective institutional arrangements to produce private goods. Mancur Olson (1965) demonstrates that nonsubtractible and nonexcludable public goods present serious problems in human organization. Once a public good is supplied, all individuals within the relevant domain are free to use the good without paying toward the cost of producing it. Cost-minimizing individuals have an incentive to become "free riders," holding out on others who pay. The free riders' success encourages others to stop payment, too, causing a drop in the amount of the good supplied. Unless institutional changes create new incentives, each person responding to self-interest ignores the interests of other, and ultimately all suffer.

Economists have developed the implications of the public-private goods distinction for the market system, concluding that competitive markets fail to allocate resources toward public goods efficiently. In some cases the failure may be complete, with no production and consumption of the public good resulting. Welfare is maximized where the sum of all consumers' marginal rates of substitution (that is, valuation of marginal consumption units) equals the social marginal cost of producing the good. Voluntary market transactions fail to achieve this condition. The market distribution of prices for a public good among consumers is unlikely to correspond to the distribution of consumer marginal valuations, because markets generally require the same price from all consumers.

This problem does not develop in a market for private goods. Each consumer independently purchases more or less of the good at the market price, depending on consumer preferences. Consumers with marginal valuations exceeding the price of the good purchase additional units until marginal values fall to equal the price. Consumers with marginal valuations below the price reduce purchases until marginal values rise to equal the price. The result is that while consumers with different preferences pay the same price, the market permits efficient adjustment among consumers. Furthermore, competition among producers assures that producers use economically efficient means for producing the good.

Public-good situations usually require some form of collective action with sanctions compelling each individual to share in the production costs. Individuals in small groups can monitor each other's costs. Individuals in small groups can monitor each other's actions and personally coerce each other to share costs. Thus, families can effectively provide joint consumption goods for members. But large groups are not so effective with casual arrangements. Each individual is more anonymous, and each individual's share of the good seems insignificant. Thus, each is tempted to free ride on payments by others, unless the group coerces payment from users. The lack of coercion leads to Aristotle's contention that the good or property shared by "the greatest number has the least care bestowed upon it."

Institutional arrangements with coercive sanctions are not sufficient to guarantee optimal amounts of public goods. Instruments of coercion may harm as well as help. As governmental institutions permit officials to deprive some citizens and reward others, institutional arrangements designed to advance the common good might become an instrument of tyranny instead. Moreover, difficulties in measuring the output of public goods and services imply that government officials have difficulties monitoring the performance of public employees. Service delivery, therefore, departs from desired levels.

Where citizens have little choice about the quality of public services supplied to them, they also have little incentive to do anything about those services. Citizens' costs of trying to change the quality can exceed increased benefits that individual citizens can expect. The structure of institutional arrangements may have some effect on the degree of choice that individuals have. Different electoral systems, such as the plurality-vote or single-member constituencies used in the United States, give more voice to individual constituents than other systems, such as the proportional-representation, party-list arrangements characteristic of Western Europe. Councilpersons representing local wards would be more sensitive to protests by local residents about how streets are used in those wards than are councilpersons elected at large. Voucher systems where individuals use a pro rata share of tax funds to obtain services from alternative vendors of educational services, for example, may allow for a much greater degree of choice on the part of individual users.

Our tennis example illustrates the influence of the attributes of goods on decision situations. Tennis players and spectators jointly consume the benefits of a game, but exclusion is relatively easy. Fencing tennis courts is legally allowable and is inexpensive. The fence helps to exclude individuals who do not qualify, by membership in a group or by admission price, for entry to the game. Thus, tennis is a toll good, fitting the northeast cell in table 2.1. Private voluntary arrangements and collective arrangements both effectively make this good available to consumers.

Fencing also affects players' strategies during a game. A fence retains the ball, causing players to play more aggressively than they would without the fence. Aggressive play does not interfere with games in neighboring courts or require time out to chase an errant ball.

The Community

The community is the third cluster of variables constituting the decision situation. The community includes all individuals directly or indirectly affected by the decision situation. The attributes of the community relevant to institutional analysis include the level of common understanding, similarity in individuals' preferences, and the distribution of resources among those affected by a decision situation.

The relevant community in our tennis example includes the players and others affected by the game results, particularly by a continued iteration of the tennis games. In intercollegiate tennis, the community includes players, referees, judges, spectators, students, parents, and alumni. College administrators who govern student activities and who participate in intercollegiate athletic associations are also part of the community. In intramural tennis the community is much smaller, often including only the opposing players. The differences in the community between intercollegiate and intramural tennis alter the competition and change decision situations for players. A player will probably use a strategy in an intercollegiate game that differs from his or her strategy in a friendly intramural game.

Level of Common Understanding
Common understanding among community members is necessary for interde-

pendent decision situations. Without common understanding a community of individuals dissolves into a disorganized aggregate with no social cohesion. Common understanding among community members may grow from a history of shared experience, enhanced by common terms and resulting in common expectations.

Individuals cannot play a game without coming to a common understanding of the rules. Players must share a similar view of the range of allowable actions, or the distribution of rights and duties among players, of likely consequences, and of preferences among players for alternative outcomes. Common understanding, however, does not imply equal distribution of information among community members. Some common knowledge of the institutional constraints is necessary for interdependent decision making, but participants may vary in their level of knowledge. Incentives, therefore, may differ among individuals (even individuals with similar preferences) choosing within similar decision situations.

Rule-ordered transformations do not, however, provide the level of predictability that physical mechanisms can provide. Part of the ambiguity arises from problems with human language. Montias (1976, 18) defines a rule as "a message stipulating and constraining the actions of a set of participants for an indefinite period and under specified states of these individuals' environments." The messages, relying on language, are sometimes garbled or ambiguous. Because words are "symbols that name, and thus, stand for classes of things and relationships" (V. Ostrom 1980, 10), words are always simpler than the phenomena to which they refer.

Stability in rule-ordered transformations depends upon shared definitions of the words formulating the rules. Even if all community members conscientiously try to implement a confusing law, it will yield irregular behavior when individuals interpret the law in a variety of circumstances. As "rules are not self-formulating, self-determining, or self-enforcing" (V. Ostrom 1980, 11), individuals must formulate the rules, apply them to particular circumstances, and enforce performance consistent with them. Ambiguity can arise at any of these points.

Changing circumstances aggravate the instability in rule-ordered transformations. Vincent Ostrom (1980, 1) writes,

the exigencies to which rules apply are themselves subject to change. Applying language to changing configurations of development increases the ambiguities and threaten the shared criteria of choice with erosion of their appropriate meaning.

Predicting behavior, therefore, on the basis of rules is necessarily imprecise, with the degree of imprecision depending on the existence of mechanisms for resolving conflicting interpretations of rules. Such mechanisms enhance the shared meaning of rules and reduce variation in behavior.

When individuals frequently interact directly with one another in a particular decision situation, as in tennis, the level of common understanding is higher than when individuals participate infrequently and in widely disparate locations. However, a geographically dispersed community can interact in a market when its members share a common understanding of the market's structure through frequent interaction and through enforcement of property law.

Level of Common Agreement about Values

A second community characteristic affecting the decision situation is the level of agreement among members when evaluating actions and results in interdependent decision making. This is especially important in situations that force consumption upon individuals in a community. Lack of agreement usually distorts the distribution of cost among community members, causing ferment among those assuming disproportionate burdens. This helps to explain the general unrest within the United States during the war in Vietnam, which forced many citizens to participate in what they considered to be an immoral war. Common agreement about values is less important in situations involving private goods.

Community agreement about institutional arrangements, however, is always important. Agreement about the moral correctness and the fairness of rules constraining decision making reduces the need for enforcement. Otherwise, individuals seek to evade and change the rules. Individuals' actions become less predictable and the order of the system breaks down, causing authorities to invest heavily to monitor community members' actions and to impose sanctions on members taking unauthorized actions.

Distribution of Resources

A third attribute of the community relevant to institutional analysis is the resource distribution among community members. A competitive market is defined to be a situation of nearly equal distribution among producers. No single producer possesses enough resources to manipulate market prices or to influence market transactions.

Individuals controlling disproportionate resource shares for participating in particular decision structures fundamentally affect the nature of the decision situation. The decision situation differs from the equal distribution case, even when institutional rules and attributes of events remain the same. Thus, if a few firms gain market control, the situation changes from one involving many decision makers (a competitive situation) to one involving a few (an oligopolistic situation). Economic theory predicts that producer decisions in a competitive situation will differ from decisions in an oligopoly. Decisions in legislatures and other political organizations also change when power distributions, say from leadership qualities, grow less equal.

Feedback among the Working Parts

The discussion so far addresses only the rightward flow of relationships among the working parts. Figure 2.1 shows that institutional arrangements, attributes of goods and events, and attributes of community combine to determine attributes of the decision situation. The attributes of the decision situation combine with the attributes of the individual decision maker to yield actions and strategies. Actions and strategies aggregate finally into results or outcomes.

Figure 2.1, however, also shows feedback relationships moving from right to left. Aggregated results subsequently affect the community attributes, attributes of the decision situation, and attributes of the individual. Reasoning along the feedback paths presumes that the working parts of the framework have cycled at least once, with individuals choosing in a given decision situation, acting on those choices, and realizing outcomes. The feedbacks show the dynamic quality of the framework.

Consider the feedback from results to the attribute of the community. Perhaps the most direct effect of outcomes is to increase the community's shared experience from interaction. This alters the community and decisions by members of the community. Our tennis example illustrates the change. If a player decides during a particular volley to charge the net, both players experience the result. The next time the opportunity to charge the net arises, even though the physical situation is exactly like the first occasion, the decision situation is not the same. The opposing player has some insight into the first player's strategy and ability and has some insight into his or her own ability to counter the charge to the net. The links between alternative actions and their consequences, therefore, change, perhaps causing the offensive player this time to play toward the baseline.

Feedback from results to the attributes of the individual shows the effects of experience on the individual decision maker rather than on the community. Whereas feedback to the community alters subsequent decision situations, feedback to the individual alters the decision maker's understanding of the decision situation. The decision maker, therefore, may change strategies in the situation and may even change objectives.

The tennis player, having charged the net during an earlier volley and scored, may decide that scoring is not as enjoyable as expected if the tactic demoralizes the opposing player. The game ceases to be fun, so the next time the player may change strategy by challenging the opposing player with a returnable volley.

The Three Levels of Analysis

The analytic components of the world of action discussed in the preceding apply to all three levels of analysis. Figure 2.2 shows the constitutional choice, collective choice, and operational levels of analysis and shows the same working parts in each level. Each section of the diagram includes attributes of institutional arrangements, events, community, the decision situation, and the individual decision maker. Each section also includes either symbolic or real action and aggregated results.

The World of Action

The first level of analysis is the world of action or the operational level, shown at the extreme right of the figure. Individuals functioning at this level either take direct action or adopt a strategy for future actions, depending on expected contingencies. Decisions in a market or in a tennis game occur at this level.

In a highly organized and free society, individuals are authorized to take a wide variety of actions at this level without prior agreement with other individuals. Authority derives from institutional arrangements, including property law, business and corporate law, and constitutional guarantees of individual freedom.

The World of Collective Choice

The second level of analysis, shown in the center of figure 2.2, is the world of collective decisions. Collective decisions are made by officials (including citizens acting as officials) to determine, enforce, continue, or alter actions authorized within institutional arrangements. Like individual strategies in the world of action, collective decisions are plans for future action. Unlike individual strategies, collective decisions are enforceable against nonconforming individuals. An individual failing to abide by a personal diet strategy, for example, may feel guilty, but no official has the power to enforce the diet plan. Officials have the power, however, to enforce a collective plan such as a city ordinance. City officials can impose sanctions against individuals who violate the ordinance. The authority to impose sanctions is a key attribute of the collective-choice level of decision making.

The World of Constitutional Choice

The third level, shown in the left section of figure 2.1, is the world of constitutional decision making. Constitutional decisions are collective choices about rules governing future collective decisions to authorize actions. Constitutional choices, in other words, are decisions about decision rules. Organizing an enterprise is a constitutional decision about rules to constrain future collective choices within the enterprise. But constitutional choice also continues beyond the initial organizing period, for as individuals react to consequences of earlier rules for collective decision making, participants may change the rules to improve the result.

While components of the framework are similar across the three levels of analysis, attributes may vary in importance from one level to another. The framework, however, shows the potential for institutional models to explain a broad range of situations, with institutional arrangements linking each level of decision making to the next level. Constitutional decisions establish institutional arrangements and their enforcement for collective choice. Collective decisions, in turn, establish institutional arrangements and their enforcement for individual action.

Differences among the Levels of Decision Making

The decision at the operational level differs fundamentally from the decisions at the other two levels. But the case with which scholars have adapted similar analytical models to all three levels of decision making have masked this difference. *The operational level is the only level of analysis where an action in the physical world flows directly from a decision.* Individuals do not act directly upon decisions in the worlds of collective or constitutional choice. Decision makers, instead, select symbols conveying information about preferred future actions.

In the world of collective choice, decision makers select among proposals to authorize future actions or select among candidates for official positions in future collective choice situations. But the only immediate results from the decisions are symbolic expressions, like votes. Aggregated votes, then, produce a collective decision about a plan for future action or about the selection of public officials. The individuals making the collective decisions sometimes also undertake the actions authorized by the collective decision, but frequently, the collective decision makers are different individuals from those taking the authorized action. Legislators, for example, make collective decisions governing actions by administrative employees.

Another difference between the operational level and the other two levels of decision making is the role of implementation and enforcement. Figure 2.2 shows this difference by including a working part for implementation and enforcement in the constitutional and collective choice level but not in the operational level. Constitutional and collective decisions must be implemented and enforced to ensure controls over the world of action, although analysts frequently assume that this aspect of political economy occurs auto-matically. Studies such as those by Pressman and Wildavsky (1973) illustrate the folly of such assumptions.

Our framework also calls attention to the distinction between constitu-tional and collective decision making. This is another point of confusion in political-economic literature. Writers frequently call both kinds of decisions "politics," missing the point that constitutional choices precede and constrain collective choices. Calling both kinds of decisions by the same name brings us, as Buchanan (1977, 71) asserts, "deep in the cardinal sin of using the same word, politics, to mean two quite different things, and indeed this dou-ble use has been one of the major sources of modern confusion."

Our tennis example illustrates the distinctions and the relationships among the three levels of analysis. The Intercollegiate Tennis Association determines the rules for college tennis. The association's members make up the community at the collective choice level of analysis, and individual representatives of the association members are the decision makers. The collective decision situation describes the alternatives and the conse-quences regarding the selection of rules for intercollegiate tennis.

Members of the association interact and representatives individually de-cide upon tournament rules. Representatives confront decision situations at the collective choice level, analogous to the tennis player deciding whether to charge the net or to play the baseline at the operational level. The repre-sentative makes the decision and voices that decision according to insti-tutional arrangements at the collective choice level. These symbolic actions by the several representatives are then aggregated, again according to the institutional arrangement, into a collective decision. This collective decision becomes part of the rule configuration that constrains interaction among col-lege tennis players in the future.

Imagine that the Intercollegiate Tennis Association must decide whether to raise the net by six inches for all intercollegiate tennis games after some future date. When deciding, association members will consider the makeup of the Intercollegiate Tennis Association (that is, the community), the con-straint on the range of choices members can make (institutional arrangements), and the nature of the good over which the Intercollegiate Tennis Association members are interacting (the game or the event). The decision situation re-sulting from these three elements necessarily differs from decision situations regarding intercollegiate tennis at the operational level, because the elements shaping the decision situation change from one level to the next. The attributes of the good, for example, differ between the collective choice and operational levels of analysis. A tennis game, at the operational level, may be a private means for two rivals to challenge each other. At the collective choice level, ten-nis extends beyond a private matter between two players to matters involving all members of the association. Tennis, at the collective choice level, becomes a joint consumption good.

Feedback Mechanism in the Three Worlds

Feedback paths at the operational level also exist at the collective choice and the constitutional choice levels. Collective decisions resulting from the aggregated individual vote increase the shared experience of the community. The shared experience affects collective decision situations on subsequent rounds of the process. The results also affect the awareness of individual decision makers, causing them to vote differently in subsequent rounds.

Feedback paths also cross the levels of analysis. The representation of the linkages can become quite complex, depending on the channel of the feedback. The feedback may be as simple as having the distribution of outcomes at the operational level change the attributes of the individual member who participates in the collective choice. On the other hand, the feedback may be complex, working through an individual who participates in the operational decision situation and later pressures other individuals at the collective choice level to alter their decisions. This individual, then, operating at a collective choice level by voting for members of the Intercollegiate Tennis Association or otherwise pressuring the members of the association, helps produce a collective decision that affects the attributes of the individual member of the association. Perhaps the combined pressure from a number of disgruntled players makes the association member more aware of the outcomes of a particular tennis rule and causes the member to vote differently in future collective choice situations regarding the troublesome rule.

The rules that govern the collective choice situation are determined at the third level of political-economy framework in the constitutional choice situation. The constitutional level determines the formal composition of the collective choice body and how the members of the body are selected. The constitutional level determines the issues that the collective choice body can address and how the body can address those issues. The constitutional level determines the relationship among the members of the collective choice body.

Individual administrators and members of athletic departments in a wide variety of colleges and universities with potential for playing intercollegiate tennis can form a body to establish an Intercollegiate Tennis Association. The members of that body have to construct the association, determining its size, how the members of the association are to be selected, the responsibilities of the association, and how the association is to function. Decisions regarding each one of these aspects about the association must be made by individuals in the constitutional body functioning in a constitutional choice situation.

That constitutional choice situation, as at other levels of analysis, is determined by the composition of the community from the colleges and universities interested in forming an Intercollegiate Tennis Association, the rules governing the interaction that will establish the Intercollegiate Tennis Association, and the good that intercollegiate tennis represents. The schools may agree that all interested schools have one vote in the constitution of the association or that the larger universities have more votes in constituting the association than the smaller colleges. The members may bar some schools from participating in the constitutional level of choice.

The good may be seen as competition among various colleges and universities, of which tennis is only one part of that competition. The schools may be interested in establishing contact and rivalry with each other in a variety of ways and see the constituting of a tennis association as a means to begin that rivalry. They foresee this rivalry and contact as eventually spilling over into other sports and into academic areas.

This view of the good, combined with the limits on the schools that can participate in the constitutional choices and the rules that govern the constitutional choice process, determines the choice situation that each individual participating in constitutional choices confronts. Change the representation by the schools from heavy influence by administrators to heavy influence by athletic departments, and the alternatives open to individual choice makers will change. Change the good from an attempt to foster contact and rivalry among the schools to an attempt to increase the fund-raising potential of the various schools, and the choice situations confronting the individual choice makers will change. Changes in these working parts of the system will reverberate through the redesign of the Intercollegiate Tennis Association, through changes in the rules governing the play of intercollegiate tennis, through changes in the decision situation that encourages charging the net to a situation that encourages playing the baseline, to less aggressive games in intercollegiate tennis.

Feedback occurs within the constitutional level of analysis, as it does within the other levels. Individuals participating in the design of institutional arrangements make a choice regarding a rule within the institutional configuration on the basis of certain levels of knowledge and awareness about the consequences of alternative votes. The aggregation of individual votes on the rule alternatives is a constitutional decision, which may be a consequence that the individual voter did not expect, given the constitutional choice situation. But having gone through the process, the individual is now more aware of the relationships among the alternatives up for vote and the likely consequences. In similar rounds in the future at the constitutional level, that individual could vote differently because of this new level of awareness.

A participant from one of the small colleges in the process of constructing the Intercollegiate Tennis Association may learn from a vote about the representation on the association that the large universities are agreeing on a united position before the vote. In the meantime, the smaller colleges entered the vote with little foreknowledge of each other's preferences and no agreement among themselves on a position to take. The result is a rule that gives the universities disproportionately heavy representation in the association. The participant from the small college may have expected that the universities would be more heavily represented but not by as much as the actual constitutional decision allowed. On the next vote college participants may circulate proposals prior to a meeting with each other and work out a united position before the vote. On a similar issue, the individual's vote this time is different, even though the constitutional choice situation may be similar to the first situation.

Feedback also crosses from one level to the next. Figure 2.2 shows that outcomes at the operational level can feed back to the individual participating at the collective choice level and to the individual participating at the constitutional choice level. These individuals may or may not be the same individual as the one who decides at the operational level, and even when the three individuals are the same individual, the attributes that are brought to bear on the choice situation at the three levels may be very different. In effect, the individual enters the choice situations at the three levels as though the individual were three different individuals.

Figure 2.2 and this discussion purposely keep the linkages among the three levels of analysis and the feedback routes simple. The levels and their interrelationships are often very complicated. The decision situation will often be affected simultaneously by the rules of many institutional arrangements, rather than one arrangement as represented in figures 2.1 and 2.2. The rule, for example, that a tennis player cannot shoot and kill an opposing player because he charges the net or otherwise plays aggressively comes from a state government rather than the Intercollegiate Tennis Association. Thus, more than one set of collective choices bears upon the decision situation at the operational level. Moreover, each set of collective choices is made within the rules of different constitutions.

Another complication is the problem of deciding where to stop adding levels to figure 2.2. Even at the constitutional choice level, there are rules that affect the choice situation. Thus, some means of determining those rules must exist. That is, there must be a preconstitutional level and a preconstitutional level. The problem is the one of infinite regress that Buchanan and Tullock (1962) speak of in *The Calculus of Consent,* when they begin their analysis at the constitutional level with the rule of unanimity.

Such complications, however, add little to the explanatory and predictive powers of the framework, only serving to make it more cumbersome. We are trying here to provide a framework to enable scholars to understand the bare bones of this approach. Once this minimal structure is understood, the analyst can proceed to add still more complexity to the framework.

Conclusion

In this essay we have taken an ecumenical rather than a parochial view of a family of theories that all use the individual as a basic unit of analysis. These theories examine the effect of institutional arrangements on the decision situations in which individuals act, producing results for themselves and others. The number of variables and complex linkages involved when scholars attempt to develop and test theories of institutions has confused many pursuing the attempt as well as their readers. The schematic framework presented above develops what we consider to be the essential elements and key linkages among three levels of analysis—constitutional, collective, and operational—and within any one of these levels. It is our hope that this metatheoretical approach helps scholars to understand the working parts of theories that have the surface appearance of being totally unrelated.

Using this framework, neoclassical economics can be viewed as a particular theory focusing primarily on the world of action. The definition of the individual used is relatively narrow. The decision situation constituted by institutional arrangements, goods, and community is limited in scope and presents individuals with little real choice. Tight linkages between individual actions and group results exist and are easy to understand for participants as well as observers. Analytical methods have been successful in predicting results when real-world situations closely resemble the highly limited situations posited by the theory.

The assumptions developed in neoclassical economics about the individual are the gears that drive the analytical machine. Because of this, many have viewed the assumptions as having an independent existence of their own, unrelated to the assumptions made about the other working parts of a particular theory. Thus, considerable work in political science has adopted the narrowly defined model of the individual used in neoclassical economics as "the" definition of a rational individual, no matter what type of situation is being modeled. However, we would argue that it is the nature of the situation being modeled that produces the opportunity to model the individual in this narrow fashion. The competitive market as modeled produces highly specific information about prices sufficient for an individual to have complete information. But complete information in this case is focused on one piece of information—current market price.

We see no reason to assume that the specific model of the individual that has been so successful as part of a particular theory should be as successful an analytic device when picked up and thrust into theories attempting to explain dramatically different types of decision situations. In a legislature or an election, the number of relevant variables, the degree of direct effect on participants, the level of information, and the tightness of the linkage structure all differ substantially from a market.

As Boynton (1982) points out, all theories specify relationships between some variables under well-defined conditions. The conditions of the competitive market are well specified and describe only a small segment of the decision situation one could analyze using the framework presented above. Under these conditions, analytical solutions can be derived that have nice equilibrium properties. Because of the type of institutional arrangements, goods, and community, the constituted decision situation in a competitive market can be modeled as relatively mechanistic.

Theories that apply this framework to other types of decision situations may attempt to gain analytical solutions but will frequently lose explanatory power as a result of the unreality of the assumptions needed to drive analytical solutions. To assume that individuals perform calculation processes requiring high levels of information when they are acting in uncertain environments may lead to the formulation of beautiful analytical models that are empirically irrelevant. A major question to be pursued is how institutional arrangements help to structure decision situations in complex arenas so that individuals are able to achieve productive outcomes even when we cannot derive analytic solutions. If we limit the way we model individuals and situations to those models that have been successful in explaining market behavior, we may continually fail to show how *different* institutional arrangements help fallible and less than fully informed persons to achieve relatively satisfactory outcomes.

NOTES

1. The extent of closeting institutional variables is illustrated in examining the table of contents and index of Dahl and Lindblom's major study, *Politics, Economics and Welfare* (1953). The term *institution* does not appear in any form in the contents or index. The term *law* appears in the index paired with the term *command,* and two pages are cited—both of which focus on a sociological definition of the power of a leader to command the performance of a subordinate. The term *rule* is paired in the index with the term *regulation.* The three pages cited all discuss bureaucratic rules and regulations as conceptualized as *red tape.* It is difficult to contemplate how difficult it must have been to do a comparative study of markets, voting, and hierarchical decision mechanisms without examining self-consciously the rule systems affecting these different processes.

2. Commons (1959, 54) recognized that the predictability of behavior depended both on the willingness of participants to follow rules as ethical systems and on the probability that officials will enforce their claims to rights if remedial action was requested. He articulated both in the following way.

> Now the distinction between ethical rights and duties and legal rights and duties is the distinction between two causes of probability respecting human conduct. Legal rights and duties are none other than the probability that officials will act in a certain way respecting the claims that citizens make against each other... But there is also an ethical ideal not relating directly to the state, and an ethical probability. In most of the transactions of modern society respecting the rights of property, liberty, domestic relations, and so on, scarcely one transaction in a billion gets before the courts or in the hands of public officials. These ethical transactions are guided, nevertheless, to an indefinite extent, by the probabilities of official behavior, but the bulk of transactions are on an ethical level guided by ethical ideals considerably above the minimum legal probabilities of what officials will do.

3. For a classification of rules for the purpose of identifying whether the rules relate solely to (1) interactions among individuals, (2) relations of individuals to objects, or (3) relations among individuals with respect to objects, see Montias (1976, 24–25). Douglas Rae (1971) partially classified institutional rules for the purpose of asking how they might contribute to the degree of political democracy within a system.

4. Recent work in experimental economics is extremely important for the purpose of isolating the effect of specific institutional rules on the type of behavior adopted and results produced in controlled decision situations. See Smith (1978); Coppinger, Smith, and Titus (1980); and Cox, Robertson, and Smith (1981).

5. This section draws heavily on V. Ostrom and E. Ostrom (1977a). See also the discussion in Benjamin (1982).

REFERENCES

Aristotle (1962) *The Politics.* Baltimore: Penguin.

Alchian, A. A. and H. Demsetz (1972) "Production, information costs, and economic organizations." *American Economic Review* 62, 5: 777–95.

Arrow, K. (1966) *Social Choice and Individual Values* (2d ed.). New York: John Wiley.

———. (1969) "The organization of economic activity: issues pertinent to the choice of market versus nonmarket allocation," in *The Analysis and Evaluation of Public Expenditures: The PPB System.* Washington, DC: U.S. Congress, Joint Economic Committee, Subcommittee on Economy in Government.

Auster, R. D. and M. Silver (1979) *The State as a Firm: Economic Forces in Political Development.* Boston: Martinus Nijhoff.

Becker, G. S. (1976) *The Economic Approach to Human Behavior.* Chicago: University of Chicago Press.

Benjamin, R. (1982) "The Historical Nature of Social-Scientific Knowledge: The Case of Comparative Political Inquiry," 69–98 in E. Ostrom (ed.) *Strategies of Political Inquiry.* Beverly Hills, CA: Sage Publications.

Black, D. (1958) *The Theory of Committees and Elections.* Cambridge, MA: Cambridge University Press.

Blumstein, J. (1981) "The resurgence of institutionalism." *Journal of Policy Analysis and Management* 1, 1: 129–32.

Borcherding, T. E. (ed.) (1977) *Budgets and Bureaucrats.* Durham, NC: Duke University Press.

Boulding, K. E. (1963) "Towards a pure theory of threat systems." *American Economic Review* 53, 2: 424–34.

Bowen, H. R. (1943–44) "The interpretation of voting in the allocation of economic resources." *Quarterly Journal of Economics* 58 (November): 27–48.

Boynton, G. R. (1982) "On Getting from Here to There: Reflections on Two Paragraphs and Other Things," 29–68 in E. Ostrom (ed.) *Strategies of Political Inquiry.* Beverly Hills, CA: Sage Publications.

Brunner, K. and W. H. Meckling (1977) "The perception of man and the conception of government." *Journal of Money, Credit and Banking* 9, 1 (Pt. 1): 70–85.

Buchanan, J. M. (1970) "Public goods and public bads," 51–71 in J. P. Crecine (ed.) *Financing the Metropolis.* Beverly Hills, CA: Sage Publications.

——— (1972) "Towards analysis of closed behavioral systems," in J. M. Buchanan and R. D. Tollison (eds.) *Theory of Public Choice. Political Applications of Economics.* Ann Arbor: University of Michigan Press.

——— (1975) "Individual choice in voting and the market." *Journal of Political Economy* 62, 3: 334–43.

——— (1977) *Freedom in Constitutional Contract: Perspectives of a Political Economist.* College Station: Texas A&M University Press.

——— and G. Tullock (1962) *The Calculus of Consent: Logical Foundations of Constitutional Democracy.* Ann Arbor: University of Michigan Press.

Caves, R. (1977) *American Industry: Structure, Conduct, and Performance* (4th ed.). Englewood Cliffs, NJ: Prentice-Hall.

Coase, R. H. (1937) "The nature of the firm." *Economica* (new series) 4, 16: 386–405.

Coleman, J. (1973) *The Mathematics of Collective Action.* Chicago: Aldine.

Commons, J. R. (1950) *The Economics of Collective Action.* New York: Macmillan.

——— (1959) *Legal Foundations of Capitalism.* Madison: University of Wisconsin Press.

Coppinger, V., V. L. Smith, and J. Titus (1980) "Incentives and behavior in English, Dutch, and sealed-bid auctions." *Economic Inquiry* 18 (January): 1–22.

Cox, J. C., B. Robertson, and V. L. Smith (1981) "Theory and behavior of single object auctions." Tucson: University of Arizona.

Dahl, R. A. and C. E. Lindblom (1953) *Politics, Economics, and Welfare, Planning, and Politics—Economic Systems Resolved into Basic Social Processes.* New

York: Harper & Row.

Downs, A. (1957) *An Economic Theory of Democracy.* New York: Harper & Row.

—— (1967) *Inside Bureaucracy.* Boston: Little, Brown.

Elster, I. (1979) *Ulysses and the Sirens.* New York: Cambridge University Press.

Farquharson, R. (1969) *Theory of Voting.* New Haven, CT: Yale University Press.

Hamburger, H. (1979) *Games as Models of Social Phenomena.* San Francisco: Freeman.

Head, J. G. (1962) "Public goods and public policy." *Public Finance* 17, 3: 197–219.

Hirschman, A. O. (1970) *Exit, Voice, and Loyalty. Responses to Declines in Firms,
Organizations, and States.* Cambridge, MA: Harvard University Press.

Hurwicz, L. (1973) "The design of mechanisms for resource allocation." *American
Economic Review* 63, 2: 1–30.

Jensen, M. C. and W. H. Meckling (1976) "Theory of the firm: managerial behavior,
agency costs and ownership structure." *Journal of Financial Economics* 3, 4:
305–60.

Katz, D. and R. L. Kahn (1966) *The Social Psychology of Organizations.* New York:
John Wiley.

Kirzner, I. M. (1973) *Competition and Entrepreneurship.* Chicago: University of
Chicago Press.

Knight, F. H. (1921) *Risk, Uncertainty, and Profit.* New York: Houghton Mifflin.

—— (1965) *Freedom and Reform.* New York: Harper & Row.

Kramer, G. H. and J. Hertzberg (1975) "Formal theory," 351–404 in F. L. Greenstein
and N. W. Polsby (eds.) *Handbook of Political Science.* Reading, MA:
Addison-Wesley.

Leibenstein, H. (1976) *Beyond Economic Man.* Cambridge, MA: Harvard University
Press.

Luce, R. and H. Raiffa (1957) *Games and Decisions: An Introduction and Critical
Survey.* New York: John Wiley.

March, J. G. (1978) "Bounded rationality, ambiguity, and the engineering of choice."
Bell Journal of Economics 9, 2: 587–608.

—— and H. A. Simon (1958) *Organizations.* New York: John Wiley.

Marschak, J. (1968) "Economics of inquiring, communicating, deciding." *American
Economic Review* 58 (May): 1–18.

Montias, J. M. (1976) *The Structure of Economic Systems.* New Haven, CT: Yale
University Press.

Mueller, D. C. (1979) *Public Choice.* Cambridge, MA: Cambridge University Press.

Musgrave, R. A. (1959) *The Theory of Public Finance: A Study in Public Economics.* New York: McGraw-Hill.

Niskanen, W. A. (1971) *Bureaucracy and Representative Government.* Chicago:
Aldine.

Olson, M. (1965) *The Logic of Collective Action. Public Goods and the Theory of
Groups.* Cambridge, MA: Harvard University Press.

Ostrom, V. (1976) "Some paradoxes for planners: human knowledge and its limitations," 243–54 in A. L. Chickering (ed.) *The Politics of Planning: A Review and
Critique of Centralized Economic Planning.* San Francisco: Institute for Con-

temporary Studies.

———— (1980) "Artisanship and artifact." *Public Administration Review* 40, 4: 309–17. (Reprinted in Michael D. McGinnis, ed., *Polycentric Governance and Development* [Ann Arbor: University of Michigan Press, 1999].)

———— (1987) *The Political Theory of a Compound Republic.* San Francisco, CA: ICS Press.

———— (1989) *The Intellectual Crisis in American Public Administration.* 2d ed. Tuscaloosa: University of Alabama Press.

———— and T. Hennessey (1975) *Conjectures of Institutional Analysis and Design: An Inquiry into Principles of Human Governance.* Bloomington: Indiana University, Workshop in Political Theory and Policy Analysis.

Ostrom, V. and E. Ostrom (1977a) "Public goods and public choices," 7–49 in E. S. Savas (ed.) *Alternatives for Delivering Public Services. Toward Improved Performance.* Boulder, CO: Westview. (Reprinted in Michael D. McGinnis, ed., *Polycentricity and Local Public Economies* [Ann Arbor: University of Michigan Press, 1999].)

———— (1977b) "A theory for institutional analysis of common pool problems," 157–72 in G. Hardin and J. Baden (eds.) *Managing the Commons.* San Francisco: Freeman.

Parks, R. B. et al. (1981) "Consumers as coproducers of public services: some economic and institutional considerations." *Policy Studies Journal* 9, 7: 1001–11. (Reprinted in Michael D. McGinnis, ed., *Polycentricity and Local Public Economies* [Ann Arbor: University of Michigan Press].)

Popper, K. R. (1967) "La rationalité et le statut du principe de rationalité," 145–50 in E. M. Classen (ed.) *Les Foundements Philosophiques des Systémes Economiques: Textes de Jacques Rueff et Essais Rediges en son Monneur 23 aout 1966.* Paris: Payot.

Pressman, J. L. and A. B. Wildavsky (1973) *Implementation: How Great Expectations in Washington Are Dashed in Oakland.* Berkeley: University of California Press.

Rae, D. W. (1971) "Political democracy as a property of political institutions." *American Political Science Review* 65, 1: 111–19.

Riker, W. H. (1980) "Implications from the disequilibrium of majority rule for the study of institutions." *American Political Science Review* 74, 2: 432–45.

———— and P. C. Ordeshook (1973) *An Introduction to Positive Political Theory.* Englewood Cliffs, NJ: Prentice-Hall.

Ross, S. A. (1973) "The economic theory of agencies: the principal's problem." *American Economic Review* 63, 2: 134–39.

Samuelson, P. A. (1954) "The pure theory of public expenditure." *Review of Economics and Statistics* 36 (November): 387–89.

———— (1955) "Diagammatic exposition of a theory of public expenditure." *Review of Economics and Statistics* 37, 4: 350–56.

Savas, F. S. (1978) "The institutional structure of local government services: a conceptual model." *Public Administration Review* 38, 5: 412–19.

Schelling, T. C. (1978) *Micromotives and Macrobehavior.* New York: Norton.

Sen, A. K. (1970) *Collective Choice and Social Welfare.* San Francisco: Holden-Day.

Shepsle, K. A. (1974) "Theories of collective choice," 1–87 in C. P. Cotter (ed.) *Political Science Annual* (Vol. 5). Indianapolis: Bobbs-Merrill.

——— (1979) "Institutional arrangements and equilibrium in multidimensional voting models." *American Journal of Political Science* 23: 27–59.

——— and B. R. Weingast (1981) "Political preferences for the pork barrel: a generalization." *American Journal of Political Science* 25, 1: 96–111.

Shubik, M. (1959) *Strategy and Market Structure: Competition, Oligopoly, and the Theory of Games.* New York: John Wiley.

——— (1975) "Oligopoly theory, communication, and information." *American Economic Review* 65 (May): 280–83.

Simon, H. A. (1957) *Models of Man: Social and Rational.* New York: John Wiley.

——— (1972) "Theories of bounded rationality," in C. McGuire and R. Radner (eds.) *Decision and Organization.* Amsterdam: Elsevier North Holland.

——— (1978) "Rationality as process and as product of thought." *American Economic Review* 68, 2: 1–16.

Simon, H. A. (1981) *The Sciences of the Artificial* (2d ed.). Cambridge, MA: MIT Press.

Smith, V. L. (1976) "Experimental economics: induced value theory." *American Economic Review* 66 (May): 274–79.

——— (1978) "Experimental mechanisms for public choice," 323–55 in P. C. Ordeshook (ed.) *Game Theory and Political Science.* New York: New York University Press.

Taylor, M. (1976) *Anarchy and Cooperation.* New York: John Wiley.

Thaler, R. H. and H. M. Sheffrin (1981) "A theory of self-control." *Journal of Political Economy* 89, 2: 392–406.

Tullock, G. (1965) *The Politics of Bureaucracy.* Washington, DC: Public Affairs Press.

Williamson, O. E. (1975) *Markets and Hierarchies: Analysis and Anti-Trust Implications. A Study in the Economics of Internal Organizations.* New York: Macmillan.

CHAPTER 3

An Agenda for the Study of Institutions

Elinor Ostrom

1. The Multiple Meanings of Institutions

Recently, public choice theorists have evidenced considerable interest in the study of institutions. William Riker (1982, 20) recently observed, for example, that "we cannot study simply tastes and values, but must study institutions as well." Little agreement exists, however, on what the term *institution* means, whether the study of institutions is an appropriate endeavor, and how to undertake a cumulative study of institutions.

Riker defines institutions as "rules about behavior, especially about making decisions" (1982, 4). Charles Plott also defines institutions to mean "the rules for individual expression, information transmittal, and social choice..." (1979, 156). Plott uses the term *institutions* in his effort to state the fundamental equation of public choice theory. Using ⊕ as an unspecified abstract operator, Plott's fundamental equation is

preferences ⊕ institutions ⊕ physical possibilities = outcomes   (1)

Plott himself points out, however, that the term *institution* refers to different concepts. He ponders:

> Could it be, for example, that preferences and opportunities *alone* determine the structure of institutions (including the constitution)? These questions might be addressed without changing "the fundamental equation" but before that can be done, *a lot of work must be done on deter-*

Originally published in *Public Choice* 48 (1986): 3–25. Copyright © 1986 by Kluwer Academic Publishers. Reprinted with kind permission of Kluwer Academic Publishers and the author.

Author's note: This essay was delivered as the presidential address at the Public Choice Society meetings, Hilton Hotel, Phoenix, Arizona, March 30, 1984. I appreciate the support of the National Science Foundation in the form of Grant No. SES 83–09829 and of William Erickson-Blomquist, Roy Gardner, Judith Gillespie, Gerd-Michael Hellstern, Roberta Herzberg, Larry Kiser, Vincent Ostrom, Roger Parks, Paul Sabatier, Reinhard Selten, Kenneth Shepsle, and York Willbern, who commented on earlier drafts.

mining exactly what goes under the title of an "institution." Are customs and ethics to be regarded as institutions? What about organizations such as coalitions? These are embarrassing questions which suggest the "fundamental equation" is perhaps not as fundamental as we would like. (Plott 1979, 160; my emphasis)

Plott's questions are indeed embarrassing. No scientific field can advance far if the participants do not share a common understanding of key terms in their field. In a volume entitled *The Economic Theory of Social Institutions,* Andrew Schotter specifically views social institutions as standards of behavior rather than the rules of the game. What Schotter calls "social institutions"

> are not rules of the game but rather the alternative equilibrium standards of behavior or conventions of behavior that evolve from a given game described by its rules. In other words, for us, institutions are properties of the equilibrium of games and not properties of the game's description. We care about what the agents do with the rules of the game, not what the rules are. (Schotter 1981, 155)

Schotter sees his enterprise as a positive analysis of the regularities in *behavior* that will emerge from a set of rules and contrasts this with a normative approach that attempts to examine which rules lead to which types of behavioral regularities. Schotter draws on a rich intellectual tradition that stresses the evolution of learned strategies among individuals who interact with one another repeatedly over a long period of time (Menger 1963; Hayek 1976, 1978; see also Ullman-Margalit 1978; Taylor 1976; Nozick 1975). Rawls characterizes this view of how individuals come to follow similar strategies over time as "the summary view of rules" (Rawls 1968, 321).

Still another way of viewing "institutions" is equivalent to the term *political structure.* This view differs from that of Schotter in that it does not equate institutions with behavioral regularities. It differs from that of Riker and Plott in that it does not focus on underlying rules. Institutions, defined as political structure, refer to attributes of the current system such as size (Dahl and Tufte 1973), degree of competition (Dye 1966; Dawson and Robinson 1963), extent of overlap (ACIR 1974), and other attributes of a current system.

The multiplicity of uses for a key term like *institution* signals a problem in the general conception held by scholars of how preferences, rules, individual strategies, customs and norms, and the current structural aspects of ongoing political systems are related to one another. Over time we have reached general agreement about how we will use such key theoretical terms as *preferences, actions, outcomes, coalitions,* and *games.* Further, we have a general agreement about how these concepts are used in our theories to generate predicted outcomes.

The multiple referents for the term *institutions* indicate that multiple concepts need to be separately identified and treated as separate terms. We cannot communicate effectively if signs used by one scholar in a field have different referents than the same sign used by another scholar in the same field. As scholars, we are in our own game situation—a language generating game. The "solution" is the result of our choice of strategies about the use of a set of terms to refer to the objects and relations of interest in our field.

No one can legislate a language for a scientific community. Scholars begin to use a language consistently when terms are carefully defined in a manner perceived by other scholars as useful in helping to explain important phenomena. In this essay, I do not try to resolve the debate over *which* of the definitions of institution is the "right definition." Instead, one concept—that of rules—is used as a referent for the term *institution* and defined. I distinguish rules from physical or behavioral laws and discuss the prescriptive nature of rules. Then I show how theorists use rules in public choice analysis. Two methodological issues are raised. One relates to the configurational character of rules. A second relates to the multiple levels of analysis needed for the systematic study of rules. In the last section, I propose an alternative strategy that takes into account the configurational character of rules and the need for a self-conscious study of multiple levels of analysis.

2. What Is Meant by Rules

Focusing specifically on the term *rule* does not immediately help us. Even this narrower term is used variously. Shimanoff (1980, 57) identified over 100 synonyms for the term *rule* (see also Ganz 1971). Even among political economists the term is used to refer to personal routines or strategies (e.g., Heiner 1983) as well as to a set of rules used by more than one person to order decision making in interdependent situations. In game theory, "the rules of the game include not only the move and information structure and the physical consequences of all decisions, but also the preference systems of all the players" (Shubik 1982, 8).

Rules, as I wish to use the term, are potentially linguistic entities (Ganz 1971; V. Ostrom 1980; Commons 1957) that refer to prescriptions commonly known and used by a set of participants to order repetitive, interdependent relationships. Prescriptions refer to which actions (or states of the world) are *required, prohibited,* or *permitted.* Rules are the result of implicit or explicit efforts by a set of individuals to achieve order and predictability within defined situations by (1) creating positions (e.g., member, convener, agent, etc.); (2) stating how participants enter or leave positions; (3) stating which actions participants in these positions are required, permitted, or forbidden to take; and (4) stating which outcome participants are required, permitted, or forbidden to affect.

Rules are thus artifacts that are subject to human intervention and change (V. Ostrom 1980). Rules, as I wish to use the term, are distinct from physical and behavioral laws. I use the term differently than a game theorist who considers linguistic prescriptions as well as physical and behavioral laws to be "the rules of the game." If a theorist wants only to analyze a given game or situation, no advantage is gained by distinguishing between rules, on the one hand, and physical or behavioral laws, on the other hand. To change the outcomes of a situation, however, it is essential to distinguish rules from behavioral or physical laws. Rules are the means by which we intervene to change the structure of incentives in situations. It is, of course, frequently difficult in practice to change the rules participants use to order their relationships. Theoretically, rules can be changed while physical and behavioral laws cannot. Rules are interesting variables precisely because they are potentially subject to change. That rules can be changed by humans is one of their key characteristics.

That rules have prescriptive force is another characteristic. Prescriptive force means that knowledge and acceptance of a rule leads individuals to recognize that, if they break the rule, other individuals may hold them accountable (see Harré 1974). One may be held accountable directly by fellow participants, who call rule infraction to one's attention, or by specialists—referees or public officials—who monitor performance. The term *rules* should *not* be equated with formal laws. Formal laws may become rules when participants understand a law, at least tacitly, and are held accountable for breaking a law. Enforcement is necessary for a law to become a rule. Participants may design or evolve their own rules or follow rules designed by others.

An unstated assumption of almost all formal models is that individuals are, in general, rule followers. Even when theorists like Becker (1976) have overtly modeled illegal behavior, some probability is presumed to exist that illegal actions will be observed, and if observed by an enforcer, that penalties will be extracted. Most public choice analysis is of the rules in use—or working rules as John R. Commons (1957) called them. Many interesting questions need exploration concerning the origin of rules, the relationship of formal laws to rules, and processes for changing rules. But, these topics cannot be addressed here.

Considerable dispute exists over the prescriptive force of "permission." Ganz (1971) and Shimanoff (1980) argue that prescriptive force is restricted to "obligation" and "prohibition" and does not include "permission," while Commons (1957), von Wright (1968), V. Ostrom (1980), and Toulmin (1974) all overtly include "permission" in their conception of rules. Part of this difficulty stems from efforts to predict behavior directly from specifying rules rather than viewing rules as a set of variables defining a structured situation. In this rule-structured situation, individuals select actions from a *set of allowable actions* in light of the full set of incentives existing in the situation.

Instead of viewing rules as directly affecting behavior, I view rules as directly affecting the structure of a situation in which actions are selected. Rules rarely prescribe one and only one action or outcome. Rules specify sets of actions or sets of outcomes in three ways.

1. A rule states that some particular actions or outcomes are forbidden. The remaining physically possible or attainable actions and outcomes are then permitted. The rule states what is forbidden. A residual class of actions or outcomes is permitted. (Most traffic laws regarding speed are of this type. The upper and lower bounds of the permitted speed are delimited by forbidding transit above and below specific speeds.)
2. A rule enumerates specific actions or outcomes or states the upper and lower bounds of permitted actions or outcomes and forbids those that are not specifically included. (Most public agencies are authorized to engage in only those activities specifically enumerated in the organic or special legislation that establishes them.)
3. A rule requires a particular action or outcome. (Recent efforts to constrain judicial discretion are rules of this type. A judge must impose a particular sentence if a jury concludes that a defendant is guilty of a particular crime.)

Only the third type of rule requires that an individual take one and only one action rather than choose from a set of actions. The third type of rule is used much less frequently to structure situations than the first two.

In the everyday world, rules are stated in words and must be understood (at least implicitly) for participants to use them in complex chains of actions. For analysis, however, rules can be viewed as relations operating on the structure of a situation. Rules can be formally represented as relations, whose domain are the set of physically possible variables and their values, and whose range are the values of the variables, in the situation under analysis. (See below for further elaboration.) Viewing rules as directly affecting the structure of a situation, rather than as directly producing behavior, is a subtle but extremely important distinction.

3. How Rules Are Used in Public Choice Theory

Most public choice theorists "know" that multiple levels of analysis are involved in understanding how rules affect behavior. But this tacit knowledge of the multiple levels of analysis and how they intertwine is not self-consciously built into the way we pursue our work. Plott, for example, has been engaged in a sophisticated research program related to the theoretical and experimental study of rules. Yet, as discussed above, he poses the central question of our discipline as a *single* equation rather than as a set of equations. We have not yet developed a self-conscious awareness of the methodological consequences of the multiple levels of analysis needed to study the effects of rules on behavior and outcomes.

Most public choice theorists also "know" that *configurations* of rules, rather than single rules, jointly affect the structure of the situations we analyze. Again, this tacit knowledge is not reflected in the way we proceed. Most of our theoretical work has proved theorems about the expected results of the use of one rule in isolation of other rules as if rules operated separably rather than configurationally.

To illustrate the multiple levels of analysis and the configurational character of rules, I will use several examples from public choice literature. The first example combines the work of several scholars who have studied how citizens' preferences for public goods are translated through two arenas—an electoral arena and a bargaining arena—into an agreement that a bureau will produce a particular quantity of goods for a particular budget. The second example is from an experimental study of Grether, Isaac, and Plott (1979) of the combination of default condition rules used in conjunction with aggregation rules. The third example is from McKelvey and Ordeshook (1984), who conducted an experimental study of the conjunction of three rules.

3.1. Rules as They Affect Outcomes
in Electoral and Bargaining Arenas

In a classic model of the election arena, Anthony Downs (1957) concludes that electoral procedures based on plurality vote will constrain a governing party to select (and therefore produce) the output-cost combination most preferred by a median voter within a community. The Downsian model predicts an optimal equilibrium in terms of allocative efficiency. Downs's prediction of optimal performance results from his analysis of the behavior of elected officials under the threat of being voted out of office by a competing party. It is the presence of a competitor ready to snatch any advantage that pushes the government party toward constant attention to what citizens prefer.

When William Niskanen (1971) examines how bureaucracy affects the linkage between citizen preferences and government performance, he focuses on the process of bargaining between the team of elected officials (called the *sponsor* by Niskanen) and *bureau chiefs* assigned the responsibility to direct agencies producing the desired goods and services. Niskanen assumes that a bureau chief attempts to obtain as large a budget as possible in order to secure the most private gain and to produce the most goods and services for a community. Niskanen's elected officials, like Downs's, know the preferences of the citizens that elect them. So do the bureau chiefs. However, elected officials do not know the production costs of the bureau. The equilibrium predicted by Niskanen is not responsive to citizen preferences since more than optimal levels of output are produced. The predicted result is technically efficient but unresponsive to the preferences of those served.

Niskanen's model is based on an assumption that bureau chiefs could threaten elected officials with *no* output if the officials did not agree to the initial demand. Romer and Rosenthal (1978) argue that a more realistic assumption would be that the budget reverts to the status quo budget (the one used for the previous year) if the officials (or, the general public in a referendum) did not agree to the initial budgetary request. Changing this assumption in the model, Romer and Rosenthal continue to predict that the equilibrium budget-output combination represents a nonoptimal oversupply. Their predicted outcome is, however, less than that predicted by Niskanen.

A dramatic change in assumptions is made by McGuire, Coiner, and Spancake (1979), who introduce a second bureau to compete with the monopoly bureau chief in the bargaining arena.[1] Whatever offer is made by one bureau can then be challenged by the second bureau. Over time the offers will approach the same optimal level as predicted by Downs. If one bureau proposes too high a budget, the other will be motivated to make a counteroffer of a more optimal budget-output combination. As the number of bureaus increases beyond two, the pressure on all bureaus to offer an optimal budget-output combination also increases.

The above models focus primarily on the structure of an operational situation and only indirectly on the rules yielding that structure. Without *explicit* analysis of the rules and other factors affecting the structure of a situation—such as the attributes of goods and the community—*implicit* assumptions underlying the overt analysis may be the most important assumptions generating predicted results.[2] In the analysis of electoral and bargaining arenas, all theorists used similar assumptions about the nature of goods and community norms. Goods are modeled as divisible in production and subject to a known technology. In regard to norms, all presume a high level of cutthroat competition is acceptable. These assumptions are not responsible for the differences among predicted outcomes.

The models have, however, different implicit or explicit assumptions about some of the rules affecting the situation. The models developed by Niskanen and by Romer and Rosenthal both give the bureau chief the capacity to make a "take it or leave it" offer. Both of these models assume an authority rule giving the bureau chief full control over the agenda. Both models also assume that the aggregation rule between the bureau chief and the sponsors is unanimity. The models differ, however, in regard to the default specified in the aggregation rule. Niskanen presumed this rule would allow the budget to revert to zero. No agreement—no funds! An aggregation rule with such a default condition can be formally stated as

$$B_{t+1} = \{B_{bc} \text{ iff } B_{bc} = B_s; 0 \text{ otherwise}\}, \tag{2}$$

where

$B_{t+1} =$ the budget-output combination for the next period,
$B_{bc} =$ the budget-output proposal of the bureau chief,
$B_s =$ the budget-output proposal accepted by the sponsor.

In other words, the aggregation rule affecting the structure of this situation requires unanimity among the participants and sets the budget for the next time period to zero if such agreement is not reached. The first part of this rule states the outcome when there is unanimous agreement. The second part of this rule states the outcome when there is no agreement, or the default condition.

Romer and Rosenthal presumed the rule would be to continue the budget in effect for the previous year. No agreement—continuance of the status quo! Their rule can be formally stated as:

$$B_{t+1} = \{B_{bc} \text{ iff } B_{bc} = B_s; B_t \text{ otherwise}\}, \tag{3}$$

where B_t is the level of the current budget-output combination.

Niskanen and McGuire, Coiner, and Spancake agree on unanimity and the default condition of the aggregation rule but differ on the boundary rules allowing entry of potential producers into the bargaining arena. Once a position rule has defined a position, S_i, such as a bureau chief, a formal boundary rule consistent with the Niskanen model could be stated as

$$\text{Let } S_i = \{1\}. \tag{4}$$

A boundary rule consistent with the McGuire, Coiner, and Spancake model would be the following:
$$\text{Let } S_i = \{1, ..., n\}. \tag{5}$$

Assuming that the other rules are similar, we can array the configuration of rules that differ in the various analyses as shown in table 3.1. The Downsian model is placed in the upper left cell since he made a similar assumption about the default condition of the aggregation rule as Romer and Rosenthal (see Downs 1957, 69), but had to assume implicitly that elected officials controlled the agenda in their bargaining relationships with bureau chiefs (see Mackay and Weaver 1978). Consequently, the difference in the results predicted by Downs, by Niskanen, and by Romer and Rosenthal can be related to changes in authority rules and aggregation rules holding other rules constant.

McGuire, Coiner, and Spancake accepted the Niskanen presumption of a zero reversion level while changing the boundary rules allowing producers to enter the bargaining process. This change in boundary rules generates a different situation leading to a prediction of relatively optimal performance as

TABLE. 3.1. Predicted Equilibrium Budget/Output Combinations under Different Rule Configurations

Authority Rules	Boundary Rules	
Aggregation rules	Entry to bargaining process restricted to one bureau	Allow multiple bureaus to enter bargaining process
Open agenda Reversion level is status quo	Downs (1957) Equilibrium is the most preferred budget/output combination of the median voter. Thus, preferences of median voter dominate decision.	Parks and E. Ostrom (1981) Even if no direct competition between two producers serving same jurisdiction, presence of comparison agencies in same urban area will reduce costs of monitoring and increase pressure toward an equilibrium producing the highest net value for the community
Reversion level is zero budget	No model yet developed for this combination of rules	No model yet developed for this combination of rules
Restricted agenda controlled by bureau chief Reversion level is status quo	Romer and Rosenthal (1978) Equilibrium is the highest budget/output combination that provides the median voter with at least as much value as the status quo.	No model yet developed for this combination of rules, but given McGuire, Coiner, and Spancake (1979) status quo reversion level can only enhance tendency of equilibrium to move toward highest net value for the community
Reversion level is zero budget	Niskanen (1971) Equilibrium is the largest budget/output combination capable of winning majority approval in an all-or-nothing vote. Preference of median voter is only a constraint.	McGuire, Coiner, and Spancake (1979) Equilibrium tends over time toward budget/output combination producing the highest net value for the community

contrasted to Niskanen's prediction of nonoptimality. The change in boundary rules opens up a new column of potential operational situations under varying conditions of authority and aggregation rules. An effort that Parks and E. Ostrom (1981) made to examine the effect of multiple producers in metropolitan areas upon the efficiency of public agencies is closely related to the rule conditions specified in the upper right-hand cell. The implications of the situations created by the other combinations of rules represented in the second column have not yet been explored.

In this discussion I wanted to illustrate what I meant by a "rule configuration." Table 3.1 presents a visual display of the configuration of rules that are consistent with the models of Downs, Niskanen, Romer and Rosenthal, and McGuire, Coiner, and Spancake. The results predicted in a situation, using one rule, are dependent upon the other rules simultaneously in force. Both Niskanen, and Romer and Rosenthal assume that only one bureau can be present in the bargaining. The boundary rule is the same. Their different results stem from the variation in the default condition of the aggregation rule. Both Niskanen and McGuire, Coiner, and Spancake agree on the default condition, but differ in regard to the boundary rule. Different results are predicted dependent on the configuration of rules, rather than any single rule, underlying the operational situation.

Second, I wanted to illustrate the multiple levels of analysis involved. The overt models presented by these theorists are all at one level. By examining the rules affecting the structure of these models, I have focused on a second level of analysis.

3.2. Committee Decisions under Unanimity and Varying Default Conditions

A second example of the study of rules by public choice theorists is a recent set of experiments conducted by Grether, Isaac, and Plott (1979), who examine the effect of various rules for assigning airport slots. Under one experimental condition, the Grether, Isaac, and Plott situation involves a committee of 9 or 14 individuals that had to divide a discrete set of objects ("cards" or "flags") using a unanimity rule. Three default conditions are used if unanimity is not reached.

    a. If the committee defaulted, each committee member received his/her "initial allocation" of slots that was unambiguously specified and known before the meeting began.

    b. If the committee defaulted, slots were allocated randomly.

    c. If the committee defaulted, slots were taken at random only from those with large initial allocations and given to those with small or no initial allocation (Grether, Isaac, and Plott 1979, V–2).

All three of these rules can be stated in a form similar to that of equations (2) and (3) above.

While Romer and Rosenthal make a theoretical argument that the particular default condition used as part of an unanimity rule affects the predicted outcomes, Grether, Isaac, and Plott provide evidence that default conditions markedly affect behavior. The decisions about slot allocations reached by committees tended to shift directly to the value specified in each of the default conditions.

> In summary, the committee decisions are substantially influenced if not completely determined by the consequences of default. Under the grandfather arrangement, "hardnosed" committee members will simply default rather than take less than the default value. Social pressures do exist for those with "large" initial endowments to give to those with "small" endowments, but even if there is no default because of concessions to social pressure the final outcome is not "far" from the "grandfather" alternative. On the other hand, when the consequence of default is an equal chance lottery, the slots will be divided equally, independent of the initial allocation....Default values literally determine the outcomes in processes such as these. (Grether, Isaac, and Plott 1979, V-7)

It has frequently been presumed that aggregation rules varied unidimensionally across one continuum from an "any one" rule to a unanimity rule (Buchanan and Tullock 1962). What should now be recognized is that most prior analysis of aggregation rules has implicitly or explicitly assumed only one of the possible default conditions that work in combination with the voting rule to yield incentives in the operational situation. There is nothing inherently conservative about a unanimity rule unless the default condition is the status quo.

Cumulative knowledge from the analysis of these diverse situations requires that we understand that Romer and Rosenthal and Grether, Isaac, and Plott are examining the effect of variations of the same rule given the preferences of participants. If some participants strongly prefer other outcomes to that stated in a default rule, a strong bargainer can threaten them with the default unless the final outcome is moved closer to his own preferred outcome. But when some participants prefer the outcome stated in the default rule, they can afford to block any proposals that do not approach this condition (see Wilson and Herzberg 1984).

To enhance cumulation, we need to develop formal representations for rules themselves as well as for the action situations on which rules operate. Most formal analyses loosely state the rules affecting the structure of the action situation: (1) in the written paragraphs leading up to the formal representation of the situation; (2) in footnotes justifying why the presentation of the situation is modeled in a particular manner; or (3) even worse, leave them unstated, as implicit assumptions underlying the formal analysis of the situation itself.[3]

3.3. PMR, Germaneness, and Open versus Closed Information Rules

An experiment conducted by McKelvey and Ordeshook strongly demonstrates the configurational relationships when pure majority rule (PMR) is combined with one "germaneness" rule and two information rules. PMR and a loose operationalization of a germaneness rule—a change in outcome can be made in only one dimension on any one move—is used throughout the experiment.[4] McKelvey and Ordeshook use a closed or an open information rule. Under their "closed" rule, members of a five-person committee can speak only if recognized by the chair, can address only the chair, and can make comments solely related to the particular motion immediately being considered. Under their "open" rule, participants can speak without being recognized, can talk to anyone, and can discuss future as well as present motions.

McKelvey and Ordeshook find that the distribution of outcomes reached under the closed information rule, when used in combination with PMR and their germaneness rule, to be significantly different than the distributions of outcomes reached under the open information rule. The experiment is a good example of how rules operate configurationally.

Rules affecting communication flow and content affect the type of outcomes that will be produced from PMR combined with a particular germaneness rule. McKelvey and Ordeshook, however, interpret their own results rather strangely. Their overt hypothesis is "that the ability to communicate facilitates circumventing formal procedural rules" (8). A close examination of their series of experiments finds no evidence of participants breaking the rules laid down by the experimenters. What they *do* test is whether a rule giving capabilities or assigning limitations on communication patterns changes the way in which PMR and their germaneness rule operate. They test the configurational operation of rule systems. And, they find that the operation of one rule depends upon the operation of other rules in a rule configuration.

4. Consequences of the Configurational Character of Rules on the Appropriate Strategies of Inquiry

These three examples provide strong evidence for the configurational, or nonseparable, attribute of rules. This leads me to argue against an implicitly held belief of some scholars that what we learn about the operation of one rule in "isolation" from other rules will hold across all situations in which that rule is used. I will characterize this view as a belief in the separable character of rules. I presume that rules combine in a configurational or inter-active manner. If rules combine configurationally rather than separably, this dramatically affects the scientific strategy we should take in the study of rules and their effects.

A key example of the problems resulting from the view of the separable character of rules is the way theorists have approached the study of PMR as an aggregation rule. Many scholars, who have studied PMR, have self-consciously formulated their models in as general a manner as possible. By proving a theorem in a general case, it is presumed that the theorem will hold in all specific cases that contain PMR.

The penchant for generality has been interpreted to mean a formulation devoid of the specification of any rule, other than PMR. A set of N in-dividuals somehow forms a committee or legislature. Position rules are rarely mentioned. The implicit assumption of most of these models is one and only one position exists—that of member. No information is presented concerning boundary rules. We do not know how the participants were se-lected, how they will be retained, whether they can leave, and how they are replaced. The participants compare points in n-dimensional space against one point in the same space called the status quo. We have no idea how that policy space came into being and what limits there may be on the poli-cies that could be adopted. (One might presume from the way such general models are formulated that no constitutional rules protect against the tak-ing of property without due process or prohibiting infringements on free-dom of speech.) Authority rules are left unstated. We must guess at what actions individual participants are authorized to take. From the way that the models are described, it appears that any participant can make any pro-posal concerning movement to any place in policy space. We do not know anything about the information rules. Everyone appears to be able to talk to everyone and provides information about their personal preferences to everyone. PMR is the only rule specified.

In this general case, in which only a single rule is formulated, theorists typically make specific assumptions about preference orderings. This sug-gests that the concepts of "generality" and "specificity" are used arbitrarily. Specific assumptions about preference orderings are accepted as appropriate in general models, while efforts to increase the specificity of the rules in these same models are criticized because they are too specific.

The search for equilibria has occurred predominantly within the context of such "general" models. And, in such "general" models, equilibria are virtually nonexistent and are fragile to slight movements of preferences or the willingness of participants to dissemble (Riker 1982). McKelvey and Ordeshook (1984, 1) are willing to state that "the principal lesson of social choice theory is that preference configurations which yield majority undominated outcomes are rare and almost always are fragile and thus are unlikely to be found in reality."

If rules combine in a configurational manner, however, theorems proved about a "zero" institutional arrangement will not necessarily be true when other rules are fully specified. Shepsle and his colleagues have repeatedly shown that when several other rules are overtly combined with PMR, equilibria outcomes are more likely. Shepsle and Weingast (1981) have summarized the effects of

1. Scope rules that operate to limit the set of outcomes that can be affected at a node in a process, for example, amendment control rules (Shepsle 1979a,b), "small change" rules (Tullock 1981), rules requiring the status quo outcome to be considered at the last decision node, and rules requiring a committee proposal to be considered at the penultimate decision node.
2. Authority rules that operate to create and/or limit the action sets available to participants in positions, for instance, rules that assign a convener special powers to order the agenda (McKelvey 1979; Plott and Levine 1978; Isaac and Plott 1978); rules that assign a full committee, such as the Rules Committee in the House of Representatives, authority to set the procedures for debate and even to exclude a bill from consideration; and rules that constrain the action sets of members in regard to striking part of a motion, adding a part of a motion, and/or substituting a part of a motion (Fiorina 1980).

Structure-induced equilibria are present in many situations where scope rules, which limit the outcomes that can be reached, or authority rules, which constrain the actions of the participants in particular positions, are combined with PMR. This leads to an optimistic conclusion that equilibria are more likely, than previously argued, in committees and assemblies using majority rule to aggregate individual votes. This substantive optimism is tempered somewhat when one recognizes the methodological consequences of rejecting the belief that rules can be studied as separable phenomena.

The methodological problem rests in the logic of combinatorics. If we were fortunate enough to be studying separable phenomena, then we could simply proceed to study individual rules out of context as we have done with PMR. We could then proceed to study other rules, out of context, and derive separable conclusions for each type of rule. Eventually, we could add our results together to build more complex models. This is an appropriate scientific method for the study of separable phenomena.

However, if the way one rule operates is affected by other rules, then we cannot continue to study each rule in isolation from others. A simple, scientific program is more difficult to envision once the configurational nature of rules is accepted. A configurational approach affects the way we do comparative statics. Instead of studying the effect of change of one rule on outcomes, regardless of the other rules in effect, we need to carefully state which other rules are in effect that condition the relationships produced by a change in any particular rule. We cannot just assume that other variables are controlled and unchanging. We need to know the value of the other variables affecting the relationship examined in a comparative statics framework.

Thus, we have much to do! It is more comforting to think about proving theorems about the effects of using one particular rule out of context of the other rules simultaneously in effect. If, however, combinations of rules work differently than isolated rules, we had better recognize the type of phenomena with which we are working and readjust our scientific agenda. We do, however, need a coherent strategy for analyzing and testing the effects of combinations of rules. How can we isolate a key set of generally formulated rules that provides the core of the rules to be studied? How can we build on the results of previous analytical work in our field?

5. Multiple Levels of Analysis and an Alternative Strategy of Inquiry

I have no final answers to these questions, but I do have an initial strategy to propose. This strategy relates to my earlier stress on the multiple levels of analysis involved in the study of rules. We have a relatively well-developed body of theory related to the study of situations such as markets, committees, elections, and games in general. Thus, we already know what variables we must identify to represent one level of analysis. We can build on this knowledge as we develop the second level of analysis.

5.1. The Structure of an Action Situation

The particular form of representation differs for neoclassical market theory, committee structures, and games in extensive form. However, in order to analyze any of these situations, an analyst specifies and relates together seven variables that form the structure of a situation.

1. The set of positions to be held by participants.
2. The set of participants (including a random actor where relevant) in each position.
3. The set of actions that participants in positions can take at different

nodes in a decision tree.

4. The set of outcomes that participants jointly affect through their actions.

5. A set of functions that map participant and random actions at decision nodes into intermediate or final outcomes.

6. The amount of information available at a decision node.

7. The benefits and costs to be assigned to actions and outcomes.

These seven variables plus a model of the decision maker must be explicitly stated (or are implicitly assumed) in order to construct any formal model of an interdependent situation. We can consider these seven to be a universal set of necessary variables for the construction of formal decision models where outcomes are dependent on the acts of more than a single individual. This is a minimal set in that it is not possible to generate a prediction about behavior in an interdependent situation without having explicitly or implicitly specified something about each of these seven variables and related them together into a coherent structure. I call the analytical entity created when a theorist specifies these seven variables an action situation.

The most complete and general mathematical structure for representing an action situation is a game in extensive form (Selten 1975; Shubik 1982). The set of instructions given to participants in a well constructed laboratory experiment is also a means of representing an action situation. Using these variables, the simplest possible working model of any particular type of situation—whether a committee, a market, or a hierarchy—can be constructed.[5] A change in any of these variables produces a different action situation and may lead to very different outcomes. More complex models of committees, markets, or other interdependent situations are constructed by adding to the complexity of the variables used to construct the simplest possible situations.[6]

5.2. An Action Arena: Models of the Situation and the Individual

In addition to the seven universal variables of an action situation, an analyst must also utilize a model of the individual, which specifies how individuals process information, how they assign values to actions and outcomes, how they select an action, and what resources they have available. The model of the individual is the animating force that allows the analyst to generate predictions about likely outcomes given the structure of the situation (Popper 1967). When a specific model of the individual is added to the action situation, I call the resulting analytical entity an "action arena." An action arena thus consists of a model of the situation and a model of the individual in the situation (see E. Ostrom 1991).

When a theorist analyzes an action arena, the model of the situation and the model of the individual are assumed as givens. At this level of analysis, the task of the analyst is viewed as one of predicting the type of behavior and results, given this structure. Questions concerning the presence or absence of retentive, attractive, and/or stable equilibria and evaluations of the efficiency and equity of these results are pursued at this level. The key question at this level is: Given the analytical structure assumed, how does this situation work to produce outcomes?

5.3. Rules as Relations

Let me return now to the point I made above that all rules can be represented as relations. I can now be more specific. From sets of physically possible actions, outcomes, decision functions, information, positions, payoffs, and participants, rules select the feasible sets of the values of these variables. The action situation is the intersection of these feasible sets. In regard to driving a car for example, it is physically possible for a 13 year old to drive a car at 120 miles per hour on a freeway. If one were to model the action situation of a freeway in a state with well enforced traffic laws, one would posit the position of licensed drivers traveling an average of 60 to 65 miles per hour (depending on the enforcement patterns of the state). The values of the variables in the action situation are constrained by physical and behavior laws and then further contained by the rules in use. Most formal analyses, to date, are of action situations; this is the surface structure that our representations model. The rules are part of the underlying structure that shapes the representations we use.

But how do we overtly examine this part of the underlying structure? What rules should be examined when we conduct analysis at a deeper level? The approach I recommend is that we focus on those rules that can directly affect the structure of an action situation. This strategy helps us identify seven broad types of rules that operate configurationally to affect the structure of an action situation. These rules include

1. *Position rules* that specify a set of positions and how many participants hold each position.
2. *Boundary rules* that specify how participants are chosen to hold these positions and how participants leave these positions.
3. *Scope rules* that specify the set of outcomes that may be affected and the external inducements and/or costs assigned to each of these outcomes.
4. *Authority rules* that specify the set of actions assigned to a position at a particular node.

5. *Aggregation rules* that specify the decision function to be used at a particular node to map actions into intermediate or final outcomes.
6. *Information rules* that authorize channels of communication among participants in positions and specify the language and form in which communication will take place.
7. *Payoff rules* prescribe how benefits and costs are to be distributed to participants in positions.

Given the wide diversity of rules that are found in everyday life, social rules could be classified in many ways. The method I am recommending has several advantages. First, rules are tied directly to the variables of an analytical entity familiar to all public choice theorists, economists, and game theorists. From this comes a strategy, or a heuristic, for identifying the rules affecting the structure of that situation. For each variable identified in the action situation, the theorist interested in rules needs to ask what rules produced the variable as specified in the situation. For example, in regard to the number of participants, the rule analyst would be led to ask: Why are there N participants? How did they enter? Under what conditions can they leave? Are there costs, incentives, or penalties associated with entering or exiting? Are some participants forced into entry because of their residence or occupation?

In regard to the actions that can be taken, the rule analyst would ask: Why these actions rather than others? Are all participants in positions assigned the same action set? Or, is some convener, or other position, assigned an action set containing options not available to the remaining participants? Are sets of actions time or path dependent?

In regard to the outcomes that can be affected, the rule analyst would ask: Why these outcomes rather than others? Are the participants all principals who can affect any state variable they are defined to own? Or, are the participants fiduciaries who are authorized to affect particular state variables within specified ranges but not beyond? Similar questions can be asked about each variable overtly placed in a model of an action situation.

Answers to these sets of questions can then be formalized as a set of relations that, combined with physical and behavioral laws, produce the particular values of the variables of the situation. I am not arguing that there is a unique set of relations that produces any particular model of a situation. Given the pervasiveness of situations with the structure of a Prisoners' Dilemma, one can expect that multiple sets of rules may produce action situations with the same structure. This is not problematic when one focuses exclusively on predicting behavior within the situation. It poses a serious problem when the question of how to change that structure arises. To change a situation, one must know which set of rules produces the situation.

Other factors also affect this structure. We know, for example, that rules that generate a competitive market produce relatively optimal equilibria when used to allocate homogeneous, divisible goods from which potential consumers can be excluded. The same rules generate less optimal situations when goods are jointly consumed and it is difficult to exclude consumers. But the theorist interested in how changes in rules affect behavior within situations must hold other factors constant while an analysis is conducted of changes in the rules.

Besides providing a general heuristic for identifying the relevant rules that affect the structure of a situation, a second advantage of this approach is that it leads to a relatively natural classification system for sets of rules. Classifying rules by what they affect enables us to identify sets of rules that all directly affect the same working part of the situation. This should enhance our capabilities for developing a formal language for representing rules themselves. Specific rules used in everyday life are named in a nontheoretical manner—frequently referring to the number of the rule in some written rule book or piece of legislation. Theorists studying rules tend to name the rule they are examining for some feature related to the particular type of situation in which the rule occurs.

For systematic cumulation to occur, we need to identify when rules, called by different names, are really the same rule. It is important that scholars understand, for example, that Romer and Rosenthal and Grether, Isaac, and Plott all examined consequences of default conditions of aggregation rules. Proceeding to formalize the rules used by Grether, Isaac, and Plott in their series of experiments would help other scholars identify which rules, called by other everyday terms, are similar to the "grandfather" default condition, to the random default condition, or to the "taking from the large and giving to the small" default condition.

By paying as much care to the formalization of the rules affecting an action situation as we do to formalizing the action situation itself, we will eventually establish rigorous theoretical propositions concerning the completeness and consistency of rules themselves. From Romer and Rosenthal and from Grether, Isaac, and Plott, we now know that any specification of a unanimity rule without an explicit default condition is incomplete. I am willing to speculate that any aggregation rule without a default condition is incomplete.

6. Some Concluding Thoughts

Given the multiple referents for the term *institutions,* our first need is for a consistent language if public choice scholars are going to return to a major study of institutions. To begin this task, I have focused on one term—that of *rules*—used by some theorists as a referent for the term *institutions.* My effort is intended to clarify what we mean by rules, how rules differ from physical or behavioral laws, how we can classify rules in a theoretically interesting manner, and how we can begin to formalize rule configurations. I have not answered the question, "What are institutions?" This involves an argument over which referent is "the" right or preferred referent. Rather, I try to clarify one referent and leave the clarification of other referents to other scholars.

Secondly, I provided several examples of how public choice analysts have studied rules. These examples illustrate two points. First, rules operate configurationally rather than separably. Second, the study of rules involves multiple levels of analysis rather than a single level of analysis. The configurational character of rules significantly affects the strategies we use to analyze rules. One approach has been to posit a single rule and examine the type of equilibria, or absence of equilibria, likely to result from the operation of this single rule. Scholars have concluded that stable equilibria do not exist in situations in which individuals use majority rule aggregation procedures. This is not consistent with empirical observation. Further, when scholars introduce rules constraining actions and outcomes into majority rule models, it is then possible to predict stable equilibria. The methodological consequence of the configurational character of rules is that theorists need to specify a set of rules, rather than a single rule, when attempting to ask what consequences are produced by changes in a particular rule.

Once this conclusion is accepted, a method to identify sets of rules is essential if we hope to develop any cumulative knowledge about the effects of rules. If more than one rule need be specified, the key question is how many different rules must be specified to know that we have identified a rule configuration. My preliminary answer is that we need to identify seven types of rules that directly affect the seven types of variables we use to construct most of the action situations we analyze. When we analyze changes in one of these rules, we should identify the specific setting of the other variables that condition how the changes in the first rule affects outcomes.

The analysis of rules needs at least two levels. We can represent these levels by reformulating Plott's fundamental equation into two equations:

$$\begin{array}{ccccc} \text{Structure of an} & \oplus & \text{Model of a} & = \text{Outcomes,} & (6) \\ \text{Action Situation} & & \text{Decision Maker} & & \end{array}$$

$$\begin{array}{ccccc} \text{Rules} & \oplus & \text{Physical} & \oplus & \text{Behavioral} & = \text{Structure of an} & (7) \\ & & \text{Laws} & & \text{Laws} & \text{Action Situation.} \end{array}$$

Equation (6) is the one most public choice theorists use in their work. As we delve somewhat deeper into the analysis of rules themselves, previous work that has focused on action situations themselves can be integrated into a broader framework. Equation (7) involves the specification of the rules, as well as the physical and behavioral laws, that affect the values of the variables in an action situation. The seventh equation is the one we must use when we want to analyze how rules change the structure of a situation leading, in turn, to a change in outcomes (see V. Ostrom 1982, 1991). The seventh equation makes apparent the need to study the effects of rules where physical and behavioral laws are invariant.

In light of these characteristics, much future work needs to be done. We need a formal language for the representation of rules as functions affecting the variables in an action situation. We also need to address questions concerning the origin and change of rule configurations in use. How do individuals evolve a particular rule configuration? What factors affect the likelihood of their following a set of rules? What affects the enforcement of rules? How is the level of enforcement related to rule conformance? What factors affect the reproducibility and reliability of a rule system? When is it possible to develop new rules through self-conscious choice? And, when are new rules bound to fail?

NOTES

1. Niskanen had himself suggested that an important structural change that could be made to improve bureau performance was to increase the competition between bureaus.
2. See Kiser and E. Ostrom (1982) for a discussion of how rules, goods, and attributes of a community all contribute to the structure of a situation.
3. It is surprising how often one reads in a public choice article that prior models had implicit assumptions that drove the analysis. A recent example is in Mackay and Weaver (1978, 143) where they argue that

> Standard demand side models of the collective choice process, in which fiscal outcomes are considered representative of broad-based citizen demands, implicitly assume not only that a "democratic" voting rule is employed to aggregate citizen-voters' demands but also that the agenda formation process is characterized by both free access and unrestricted scope.

4. They do not specifically mention that they intend to operationalize the concept of germaneness, but it would appear from the "dicta" that they think they have done so. However, as Shepsle (1979a) conceptualized this rule, decisions about one dimension of a policy space would be made sequentially. Once a decision about a particular dimension had been reached, no further action on that dimension would be possible. Allowing members to zigzag all over the policy space, one dimension at a time, is hardly a reasonable operationalization of the germaneness rule as specified by Shepsle.
5. The simplest possible representation of a committee, for example, can be con-

structed using the following assumptions

1. One position exists—that of member.
2. Three participants are members.
3. The set of outcomes that can be affected by the member contains two elements, one of which is designated as the status quo.
4. A member is assigned an action set containing two elements: (a) vote for the status quo and (b) vote for the alternative outcome.
5. If two members vote for the alternative outcome, it is obtained; otherwise, the status quo outcome is obtained.
6. Payoffs are assigned to each participant depending on individual actions and joint outcomes.
7. Complete information is available about elements (1) through (6).

For this simplest possible representation of a committee, and using a well-defined model of the rational actor, we know that an equilibrium outcome exists. Unless two of the members prefer the alternative outcome to the status quo and both vote, the status quo is the equilibrium outcome. If two members do prefer and vote for the alternative outcome, it is the equilibrium outcome. The prediction of outcome is more problematic as soon as a third outcome is added. Only when the valuation patterns of participants meet restricted conditions can an equilibrium outcome be predicted for such a simple committee situation with three members and three potential outcomes using majority rule (Arrow 1963, Plott 1967).

6. A more complex committee situation is created, for example, if a second position, that of a convener, is added to the situation, and the action set of the convener includes actions not available to the other members (e.g., Isaac and Plott 1978; Eavey and Miller 1982). See also Gardner (1983) for an analysis of purges of recruitment to committees. Gardner's approach is very similar to the general strategy I am recommending.

REFERENCES

Advisory Commission on Intergovernmental Relations (ACIR) (1974). *Governmental functions and processes: local and areawide. Volume IV of substate regionalism and the federal system.* Washington, D.C.: U.S. Government Printing Office.

Arrow, K. (1963). *Social choice and individual values,* 2d ed. New York: Wiley.

Becker, G. S. (1976). *The economic approach to human behavior.* Chicago: The University of Chicago Press.

Buchanan, J. M., and Tullock, G. (1962). *The calculus of consent.* Ann Arbor: University of Michigan Press.

Commons, J. R. (1957). *Legal foundations of capitalism.* Madison: University of Wisconsin Press.

Dahl, R. A., and Tufte, E. R. (1973). *Size and democracy.* Stanford, Calif.: Stanford University Press.

Dawson, R. E., and Robinson, J. A. (1963). Interparty competition, economic variables and welfare policies in the American states. *Journal of Politics* 25: 265–89.

Downs, A. (1957). *An economic theory of democracy.* New York: Harper and Row.

Dye, T. R. (1966). *Politics, economics, and the public.* Chicago: Rand McNally.

Eavey, C. L., and Miller, G. J. (1982). Committee leadership and the chairman's power. Paper delivered at the Annual Meetings of the American Political Science Association, Denver, Colo., September 2–5.

Fiorina, M. P. (1980). Legislative facilitation of government growth: Universalism and reciprocity practices in majority rule institutions. *Research in Public Policy Analysis and Management* 1: 197–221.

Ganz, J. S. (1971). *Rules. A systematic study.* The Hague: Mouton.

Gardner, R. (1983). Variation of the electorate: Veto and purge. *Public Choice* 40(3): 237–47.

Grether, D. M., Isaac, R. M., and Plott, C. R. (1979). Alternative methods of allocating airport slots: Performance and evaluation. A report prepared for the Civil Aeronautics Board.

Harré, R. (1974). Some remarks on "rule" as a scientific concept. In T. Mischel (Ed.), *Understanding other persons,* 143–83. Oxford, England: Basil-Blackwell.

Hayek, F. A. (1976). *The mirage of social justice.* Chicago: University of Chicago Press.

Hayek, F. A. (1978). *New studies in philosophy, politics, economics, and the history of ideas.* Chicago: University of Chicago Press.

Heiner, R. A. (1983). The origin of predictable behavior. *American Economic Review* 83(4): 560–97.

Isaac, R. M., and Plott, C. R. (1978). Comparative game models of the influence of the closed rule in three person, majority rule committees: Theory and experiment. In P. C. Ordeshook (Ed.), *Game theory and political science,* 283–322. New York: New York University Press.

Kiser, L., and Ostrom, E. (1982). The three worlds of action: A meta-theoretical synthesis of institutional approaches. In E. Ostrom (Ed.), *Strategies of political inquiry,* 179–222. Beverly Hills: Sage Publications. (Chapter 2 in this volume.)

Mackay, R. J., and Weaver, C. (1978). Monopoly bureaus and fiscal outcomes: Deductive models and implications for reform. In G. Tullock and R. E. Wagner (Eds.), *Policy analysis and deductive reasoning,* 141–65. Lexington, Mass.: Lexington Books.

McGuire, T., Coiner, M., and Spancake, L. (1979). Budget maximizing agencies and efficiency in government. *Public Choice* 34(3/4): 333–59.

McKelvey, R. D. (1979). General conditions for global intransitivities. *Econometrica* 47: 1,085–111.

McKelvey, R. D., and Ordeshook, P. C. (1984). An experimental study of the effects of procedural rules on committee behavior. *Journal of Politics* 46(1) 185–205.

Menger, K. (1963). *Problems in economics and sociology.* (Originally published in 1883 and translated by Francis J. Nock.) Urbana: University of Illinois Press.

Niskanen, W. A. (1971). *Bureaucracy and representative government.* Chicago: Aldine Atherton.

Nozick, R. (1975). *Anarchy, state, and utopia.* New York: Basic Books.

Ostrom, E. (1991) A method of institutional analysis and an application to multiorganizational arrangements. In Franz-Xaver Kaufmann (Ed.), *The public sector—challenge for coordination and learning,* 501–23. Berlin and New York: Walter de Gruyter.

Ostrom, V. (1980). Artisanship and artifact. *Public Administration Review* 40(4): 309–17. (Reprinted in Michael D. McGinnis, ed., *Polycentric Governance and Development* [Ann Arbor: University of Michigan Press, 1999].)

Ostrom, V. (1982). A forgotten tradition: The constitutional level of analysis. In J. A. Gillespie and D. A. Zinnes (Eds.), *Missing elements in political inquiry: Logic and levels of analysis,* 237–52. Beverly Hills: Sage Publications. (Reprinted in Michael D. McGinnis, ed., *Polycentric Governance and Development.* [Ann Arbor: University of Michigan Press, 1999].)

Ostrom, Vincent (1991) Constitutional considerations with particular reference to federal systems. In Franz-Xaver Kaufmann, ed. *The Public Sector—Challenge for Coordination and Learning,* 141–50. Berlin and New York: Walter de Gruyter.

Parks, R. B., and Ostrom, E. (1981). Complex models of urban service systems. In T. N. Clark (Ed.), *Urban policy analysis. Directions for future research.* Urban Affairs Annual Reviews 21: 171–99. Beverly Hills: Sage Publications. (Reprinted in Michael D. McGinnis, ed., *Polycentricity and Local Public Economies.* [Ann Arbor: University of Michigan Press, 1999].)

Plott, C. R. (1967). A notion of equilibrium and its possibility under majority rule. *American Economic Review* 57(4): 787–807.

Plott, C. R. (1979). The application of laboratory experimental methods to public choice. In C. S. Russell (Ed.), *Collective decision making: Applications from public choice theory,* 137–60. Baltimore, Md.: Johns Hopkins University Press.

Plott, C. R., and Levine, M. E. (1978). A model for agenda influence on committee decisions. *American Economic Review* 68: 146–60.

Popper, K. R. (1967). La rationalité et le statut du principle de rationalité. In E. M. Classen (Ed.), *Les foundements philosophiques des systemes economiques: Textes de Jacques Rueff et essais redigés en son honneur 23 acut* 1966, 145–50. Paris, France: Payot.

Rawls, J. (1968). Two concepts of rules. In N. S. Care and C. Landesman (Eds.). *Readings in the theory of action,* 306–40. Bloomington, Ind.: Indiana University Press. Originally printed in the *Philosophical Review* 4(1955).

Riker, W. H. (1982). Implications from the disequilibrium of majority rule for the study of institutions. In P. C. Ordeshook and K. A. Shepsle (Eds.), *Political equilibrium,* 3–24. Boston: Kluwer-Nijhoff. Originally published in the *American Political Science Review* 74(June 1980): 432–47.

Romer, T., and Rosenthal, H. (1978). Political resource allocation, controlled agendas, and the status quo. *Public Choice* 33(4): 27–43.

Schotter, A. (1981). *The economic theory of social institutions.* Cambridge, England: Cambridge University Press.

Seiten, R. (1975). Reexamination of the perfectness concept for equilibrium points in extensive games. *International Journal of Game Theory* 4: 25–55.

Shepsle, K. A. (1979a). Institutional arrangements and equilibrium in multidimensional voting models. *American Journal of Political Science,* 23(1): 27–59.

Shepsle, K. A. (1979b). The role of institutional structure in the creation of policy equilibrium. In D. W. Rae and T. J. Eismeier (Eds.), *Public policy and public choice,* 249–83. *Sage Yearbooks in Politics and Public Policy* 6. Beverly Hills: Sage Publications.

Shepsle, K. A., and Weingast, B. R. (1981). Structure-induced equilibrium and legislative choice. *Public Choice* 37(3): 503–20.

Shimanoff, S. B. (1980). *Communication rules: Theory and research.* Beverly Hills: Sage Publications.

Shubik, M. (1982). *Game theory in the social sciences. Concepts and solutions.* Cambridge, Mass.: MIT Press.

Taylor, M. (1976). *Anarchy and cooperation.* New York: Wiley.

Toulmin, S. (1974). Rules and their relevance for understanding human behavior. In T. Mischel (Ed.), *Understanding other persons,* 185–215. Oxford, England: Basil-Blackwell.

Tullock, G. (1981). Why so much stability? *Public Choice* 37(2): 189–205.

Ullman-Margalit, E. (1978). *The emergence of norms.* New York: Oxford University Press.

von Wright, G. H. (1968). The logic of practical discourse. In Raymond Klibansky (Ed.), *Contemporary philosophy,* 141–67. Italy: La Nuava Italia Editrice.

Wilson, R., and Herzberg, R. (1984). Voting is only a block away: Theory and experiments on blocking coalitions. Paper presented at the Public Choice Society meetings, Phoenix, Ariz., March 29–31.

CHAPTER 4

A Grammar of Institutions

Sue E. S. Crawford and Elinor Ostrom

The core of traditional political science was the study of institutions and of political philosophy. The behavioral revolution swept both aside and focused political scientists on the study of political behavior. Then came the onslaught of rational choice theory. To the surprise of political scientists, these efforts concluded that many fond beliefs about the effects of institutions lacked firm theoretical foundations based on individual choice. The past decade has witnessed a flurry of new scholarship in the rational choice tradition flying the banner of "The New Institutionalism." The flood of work produced in this tradition facilitates a return to the core of the discipline, namely institutions and the nature of political orders.

Renewed interest in institutions has, however, generated a simmering theoretical debate about what institutions are. The debate is healthy and likely to lead to a clarification of core concepts used in political science. The institutional grammar introduced here is based on a view that institutions are enduring regularities of human action in situations structured by rules, norms, and shared strategies, as well as by the physical world. The rules, norms, and shared strategies are constituted and reconstituted by human interaction in

Originally published in *American Political Science Review* 89, 3 (September 1995): 582–600. Reprinted by permission of the American Political Science Association and the authors.

Authors' note: An earlier draft of this article was presented at the Conference on Individual and System held at the Merriam Lab, Department of Political Science, University of Illinois, Urbana, in February 1992. A longer version was presented to the ECPR Planning Session on "Contextual Solutions to Problems of Political Action," Limerick, Ireland, March 30–April 3, 1992. We have presented versions at Workshop in Political Theory and Policy Analysis colloquia, at the Fall 1993 Workshop MiniConference, and the June 1994 Workshop on the Workshop. We thank Mike McGinnis for the detailed comments he has made on several drafts. We also appreciate the comments made by Piotr Chmielewski, Roy Gardner, Louis Helling, Robert Keohane, Vincent Ostrom, James Walker, Tjip Walker, and John Williams on earlier drafts. We have benefited from reading the work of many colleagues at the Workshop, particularly Ron Oakerson. Thanks also to Patty Dalecki for her incredibly accurate and insightful editorial and graphics assistance. Financial support from the National Science Foundation (Grant No. SBR–9319835) is gratefully acknowledged.

frequently occurring or repetitive situations.[1] Where one draws the boundaries of an institution depends on the theoretical question of interest, the timescale posited, and the pragmatics of a research project. One example of an institution encompasses the regularities of action and outcomes in the U.S. Congress. Another limits the focus to only one house or one committee or subcommittee. Likewise, analysis of institutional statements can target broad prescriptions (e.g., Congress may pass legislation regulating interstate commerce) or focus on a narrower prescription (e.g., a cloture rule). We shall introduce a grammar that provides a theoretical structure for analysis of the humanly constituted elements of institutions (i.e., rules, norms, and shared strategies).

Institutions

Several approaches have been taken to answer the question, What is an institution? One major approach to this question is the *institutions-as-equilibria* approach drawing on the work of Menger (1963) and von Hayek (1945, 1967) for its intellectual foundations and Schotter (1981), Riker (1980), and Calvert (1992) for contemporary views. Works in this approach focus on the stability that can arise from mutually understood actor preferences and optimizing behavior. Scholars in this tradition treat these stable patterns of behavior as institutions. Lewis (1969), Ullmann-Margalit (1977), and Coleman (1987) articulate an *institutions-as-norms* approach. A third approach, the *institutions-as-rules* approach, draws on Hohfeld (1913) and Commons (1968) for its roots and Shepsle (1975, 1979, 1989), Shepsle and Weingast (1984, 1987), Plott (1986), Oakerson and Parks (1988), North (1986, 1990), E. Ostrom (1986, 1990), V. Ostrom (1980, 1987, 1991, 1997), Williamson (1985), and Knight (1992) for contemporary developments. The institutions-as-norms and institutions-as-rules approaches both focus on linguistic constraints (spoken, written, or tacitly understood prescriptions or advice) that influence mutually understood actor preferences and optimizing behavior.

All three approaches offer institutional explanations for observed regularities in the patterns of human behavior. The differences among the approaches relate primarily to the grounds on which explanations for observed regularities rest. Viewing institutions as equilibrium behavior rests on an assumption that rational individuals, interacting with other rational individuals, continue to change their planned responses to the actions of others until no improvement can be obtained in their expected outcomes from independent action. To understand why regularized patterns of interaction exist, one needs to ask why all actors would be motivated to produce a particular equilibrium.

There are important philosophical reasons for looking at institutions as equilibria. This view places the responsibility for a social order on the individuals who are part of that order, rather than on some external "state" or "third-party enforcer." It integrates analysis of behavior *within* institutions with analysis of how institutions come into being. By focusing on mutually understood actor expectations, preferences, and optimizing behavior, the analyst can examine institutions that do not require outside enforcement or irrational commitments to following rules. An institution can be viewed as no more than a regular behavior pattern sustained by mutual expectations about the actions that others will take: "The institution is just an equilibrium" (Calvert 1992, 17).[2] Starting from this position, one avoids the trap of reifying institutions by treating them as things that exist apart from the shared understandings and resulting behavior of participants.

While the institutions-as-equilibria approach offers advantages, disadvantages exist as well. Stable outcomes resulting from shared understandings about the appropriate actions for a particular situation are all treated as if they had similar foundations. Lumping together all shared understandings fails to clarify the difference between shared advice based on prudence, shared obligations based on normative judgments, and shared commitments based on rules created and enforced by a community.

The grounds for looking at institutions as norms rest on an assumption that many observed patterns of interaction are based on the shared perceptions among a group of individuals of proper and improper behavior in particular situations. To understand why some regularized patterns of interaction exist, one needs to go beyond immediate means-ends relationships to analyze the shared beliefs of a group about normative obligations. The grounds for looking at institutions as rules rest on an assumption that many observed patterns of interaction are based on a common understanding that actions inconsistent with those that are proscribed or required are likely to be sanctioned or rendered ineffective if actors with the authority to impose punishment are informed about them. To understand regularized patterns of interaction affected by rules, one needs to examine the actions and outcomes that rules allow, require, or forbid and the mechanisms that exist to enforce those rules.

The three approaches are not mutually exclusive. All start from the individual and build social orders on individualistic and situational foundations. The first approach focuses more on the regularities of outcomes in structured situations. The other two focus on linguistic constraints that influence these regularities. If one slightly refocuses the institutions-as-equilibria approach to one of institutions-as-equilibrium-strategies or, more generally, institutions-as-shared-strategies, the three approaches can be directly compared. Focusing on shared strategies shifts the emphasis away from the outcomes obtained and back to the shared understandings and expectations that influence behavior leading to outcomes. In all three cases, then, constraints and opportunities can be articulated as institutional statements. Each approach, in our view, focuses on a different type of constraint or opportunity.

In light of the merits of each approach, we propose not to argue about whether institutions are rules, norms, or strategies. We use the broad term *institutional statement* to encompass all three concepts. An *institutional statement* refers to a shared linguistic constraint or opportunity that prescribes, permits, or advises actions or outcomes for actors (both individual and corporate).[3] Institutional statements are spoken, written, or tacitly understood in a form intelligible to actors in an empirical setting. In theoretical analyses, institutional statements will often be interpretations or abstractions of empirical constraints and opportunities. We develop a grammar of institutions as a theory that generates structural descriptions of institutional statements.[4] The syntax of this grammar operationalizes the structural descriptions; it identifies common components of institutional statements and establishes the set of components that comprises each type of institutional statement. We shall focus primarily on the syntax of the institutional grammar and only briefly discuss semantics and pragmatics.

Three disclaimers apply. First, we do not argue that the institutional statements explaining behavior are always articulated easily and fully by participants. Knowledge of institutional statements is often habituated and part of the tacit knowledge of a community. Moreover, we do not assume that all individuals recognize the existence of an institutional grammar and explicitly use it to formulate institutional statements. Second, a statement that fits the grammatical structure is not necessarily a meaningful or significant institutional statement.[5] Third, to explain behavior in institutions, one needs to combine the grammar of institutions with a theory of action, just as to understand a language one requires a grammar and a theory of language use (Chomsky 1965, 9). The theory of action animates the structural model of a situation generated by the grammar and by relevant attributes of a physical world (Popper 1967).

The grammar facilitates analysis of the content of institutional statements, of distinctions among types of institutional statements, and of the evolution of institutional statements. As with any grammar, its application to existing statements sometimes yields tough judgment calls and counterexamples. The principles that guide the identification of components and the classification of statements cannot eliminate all ambiguity.

The Syntax of a Grammar of Institutions

The general syntax of the grammar of institutions contains five components: ATTRIBUTES, DEONTIC, AIM, CONDITIONS, and OR ELSE. The five letters (ADICO) provide a shorthand way of referring to the components. Regardless of how institutional statements are expressed in natural language, they can be rewritten in the ADICO format, where

A ATTRIBUTES is a holder for any value of a participant-level variable that distinguishes to whom the institutional statement applies (e.g., 18 years of age, female, college-educated, 1 year experience, or a specific position, such as employee or supervisor).

D DEONTIC is a holder for the three modal verbs using deontic logic: *may* (permitted), *must* (obliged), and *must not* (forbidden).

I AIM is a holder that describes particular actions or outcomes to which the deontic is assigned.

C CONDITIONS is a holder for those variables that define when, where, how, and to what extent an AIM is permitted, obligatory, or forbidden.

O OR ELSE is a holder for those variables that define the sanctions to be imposed for not following a rule.

All shared strategies can be written as [ATTRIBUTES] [AIM] [CONDITIONS] (AIC); all norms can be written as [ATTRIBUTES] [DEONTIC] [AIM] [CONDITIONS] (ADIC); and all rules can be written as [ATTRIBUTES] [DEONTIC] [AIM] [CONDITIONS] [OR ELSE] (ADICO). The syntax is cumulative: norms contain all of the components of a shared strategy plus a DEONTIC; rules contain all the components of a norm plus an OR ELSE.[6]

In linguistic terms, the components operate as phrasemarkers, and the AIC, ADIC, and ADICO are the basis of shared strategies, norms, and rules respectively (see Chomsky 1965, 17). Linguists focus on the transformations that create spoken sentences from base strings. Institutional analysts will focus more on transforming rules and norms as practiced in ongoing situations to base strings in order to capture the core advice or prescription of the institutional statement in use. The ability to translate existing institutional statements and formal representations of institutional statements into an established format provides many advantages for comparison, analysis, and synthesis that we discuss throughout.

To ease the discussion of the syntax components, we refer throughout to the following examples.

1. All male U.S. citizens, 18 years of age and older, must register with the Selective Service by filling out a form at the U.S. Post Office or else face arrest for evading registration.

2. All senators may move to amend a bill after a bill has been introduced, or else the senator attempting to forbid another senator from taking this action by calling him or her out of order will be called out of order or ignored.

3. All villagers must not let their animals trample the irrigation channels, or else the villager who owns the livestock will be levied a fine.

4. All neighborhood residents must clean their yard when the neighborhood organization organizes a major neighborhood cleanup day.
5. The person who places a phone call calls back when the call gets disconnected.

ATTRIBUTES

All institutional statements for a group apply to a subset of participants from that group. The subset can range from one participant to all participants of the group. A set of ATTRIBUTES establishes the subset of the group affected by a particular statement. If individuals make up the group, the ATTRIBUTES will be individual-level values. Individual-level ATTRIBUTES include values assigned to variables such as age, residence, gender, citizenship, and position. When members of a group governed by a set of institutions are corporate actors, rather than individuals, the ATTRIBUTES are organizational variables (e.g., size of membership, geographic location, or ownership of the residuals). In the first example, the relevant ATTRIBUTES are male, citizen of the United States, and over 18 years old. In the last example, the ATTRIBUTES is the caller who placed the call. The other examples list no specific attribute. When no specific attribute is listed, the default value for the ATTRIBUTES component is all members of the group.[7] This means that the ATTRIBUTES component always has something in it, even when a specific attribute is not contained in the statement. Thus the second example applies to all senators in a legislature, and the third example applies to all villagers in a particular village.

DEONTIC

The DEONTIC component draws on the modal operations used in deontic logic to distinguish prescriptive from nonprescriptive statements (see Hilpinen 1971, 1981; von Wright 1951). The complete set of DEONTIC operators (*D*) consists of permitted (*P*), obliged (*O*), and forbidden (*F*). Institutional statements use the operative phrases *may, must,* and *must not* to assign these operators to actions and outcomes. Thus the statement that all members *may* vote assigns the DEONTIC permitted, *P*, to the action of voting. We represent the assignment of a DEONTIC operator to an action $[a_i]$ as $[D][a_i]$, where *D* stands for *P*, *O*, or *F*. Similarly, $[D][o_i]$ represents the assignment of a deontic to an outcome.[8]

The three DEONTIC operators are interdefinable (von Wright 1968, 143). If one of the deontics is taken as a primitive, the other two can be defined in terms of that primitive. For example, let us use $[P]$ as a primitive. $[P][a_i]$ would be read, "One is permitted to do $[a_i]$" or "One *may* do $[a_i]$." The statement that an act is forbidden $[F][a_i]$ can be restated using $[P]$ as the primitive as $[\sim P][a_i]$. In other words, when $[a_i]$ is forbidden, one is not permitted to do $[a_i]$. On the other hand, if the negation of an action $[\sim a_i]$ is forbidden, one is obliged to take the action. The statement that an act must be done, $[O][a_i]$, can be defined as $[\sim P][\sim a_i]$. If an action is obligatory, one is not permitted not to do $[a_i]$. Alternatively, we could use $[F]$ as the primitive. Then $P[a_i]$ can be defined as $[\sim F][a_i]$, and $O[a_i]$ can be defined as $[F][\sim a_i]$. With $[O]$ as the primitive, $[P][a_i]$ can be defined as $[\sim O][a_i$ or $\sim a_i]$, while $[F][a_i]$ can be defined as $[O][\sim a_i]$. Interdefinability also exists for prescriptions that refer to outcomes instead of actions. Any prescription with a DEONTIC assigned to some outcome, $[o_i]$, can be restated using either of the other two DEONTIC operators.

The meanings of the deontics obliged (*must*) and forbidden (*must not*) fit well into most conceptions of normative statements. The meaning of permitted (*may*) is more perplexing (Schauer 1991). Susan Shimanoff, for example, concludes that "it is incongruous to talk of rules prescribing behavior which is merely permitted" (Shimanoff 1980, 44). However, statements that assign permission (P) to an action are meaningful in at least two ways. In many instances, assigning a *may* to an action is the equivalent of "constituting" that action (Searle 1969). For example, a law stating that an individual may vote in an election is meaningful. It creates an action—voting—that did not exist before. Permission establishes qualifications for positions (i.e., who may vote and who may hold offices). Furthermore, when one speaks of a "right" to take an action or affect an outcome, the person with the right is permitted (rather than required or forbidden) to take that action or affect an outcome. Others, who have a duty to recognize that right, are the ones who are forbidden or required to take actions or affect outcomes.[9]

The interdefinability of deontics plays a key role in unraveling another meaning of permission. Any rule or norm that assigns permission, P, can be restated using the logical equivalent $\sim F$. If an action is permitted, then that same action is not forbidden. Those who share a common understanding of that institutional statement recognize that the action is not forbidden. Measures that would be taken to prevent that action—or to react to that action as if it were forbidden—are no longer acceptable. In other words, any institutional statement that assigns *may* or *not forbidden* to an action implies that at least some action that could otherwise be taken by others to prevent or punish the permitted action is forbidden. In the second example, senators must not treat the action of amending a bill as a forbidden action. The OR ELSE indicates the institutional consequence of attempting to treat the permitted action as forbidden (the senator is called out of order or ignored).

The assignment of $\sim F$ operates in a similar manner in institutional statements without an OR ELSE. Consider a legislative body that shares an institutional statement like the following: [All junior members] [P] [contest senior members] [in committee hearings]. This is the equivalent to [All junior members] [~*F*] [contest senior members] [in committee hearings]. This prescription implies a prescription on senior members not to reprimand or castigate junior members who challenge them in committee hearings. The existence of such a norm does not ensure that all senior members will follow it. However, there will be a shared notion that a rebuke based on seniority alone is inappropriate or unacceptable. If a senior member reprimands a junior member for challenging him or her, then we would expect the junior member to invoke the grant of permission to defend against the senior member's actions. It means something for the junior member to say, "Everyone here knows that I am permitted to challenge senior members in committee hearings."

AIM

The AIM is the specific action or outcome to which an institutional statement refers. Anything thought of as an AIM must be physically possible (von Wright 1963). An individual cannot logically be required to undertake a physically impossible action or affect a physically impossible outcome. Further, in order for any action to be conceived of as an AIM, its negation [$\sim a_i$ or o_i] must also be physically possible. In other words, including an action or outcome in the AIM component implies that it is avoidable. Thus the capability of voting implies the capability of not voting. Voting for candidate A implies the option of not voting for candidate A.

In the first example given earlier, the AIM is the action of registering for the Selective Service and the DEONTIC operator required, *O*, is assigned to the action for all individuals with the ATTRIBUTES listed in the rule. In the second example, the AIM is the action of offering a motion to amend a bill; the DEONTIC operator is *P*, or permitted for all senators. The AIM in the fourth example also includes an action. The norm requires the action of cleaning up the yard. The third example assigns the DEONTIC *F* to the outcome of livestock damage. The AIM of the rule does not specify actions that an irrigator must, may, or may not take. The AIM only specifies the forbidden outcome. Villagers may select any actions that are not forbidden by another rule to keep their livestock from damaging the irrigation channel. In the fifth example, the AIM is the action of calling back.

CONDITIONS

CONDITIONS indicate the set of variables that define when, where, and how an institutional statement applies. The CONDITIONS for a statement might indicate when a statement applies, such as during certain weather conditions or at a particular stage in the legislative process. Likewise, the CONDITIONS might indicate where a statement applies (e.g., within a particular jurisdictional area) or how a statement is to be followed (e.g., through a defined process). If an institutional statement does not specify a particular condition, the default value for the condition is "at all times and in all places." Thus, like the ATTRIBUTES, the CONDITIONS component always has some value, regardless of whether an institutional statement overtly specifies conditions.

The CONDITIONS component in the first example states how the statement is to be followed, namely, by filling out a form at the U.S. Post Office. It does not restrict when or where the statement applies. The CONDITIONS component in the second, fourth, and fifth examples indicates when the prescription applies. After a bill has been introduced, the prescription of the second rule applies. Any senator may move to amend a bill after it has been introduced. The fourth example applies on the days designated as cleanup days by a neighborhood organization, and the fifth example applies when a phone call is disconnected. The third example does not specify any specific CONDITIONS. We assume that the rule applies for members of the village at all times and in all places with no additional restrictions on how the rule is to be followed.

OR ELSE

The final component of our institutional syntax, the OR ELSE, is the sanction assigned to detected noncompliance with an institutional statement. Only rules include this component. The content of the OR ELSE affects the very nature of a rule. Rowe (1989) discusses the difference between a speed-limit law with minor sanctions and a speed-limit law with a severe penalty as the sanction. The prescription is the same. For both laws, the severity of the sanction in the OR ELSE is the only difference, but expected behavior is different.[10]

In order for threats associated with prescriptions to qualify as an OR ELSE, they must meet three qualifications. First, the threat must be backed by another rule or norm that changes the DEONTIC assigned to some AIM, for at least one actor, under the CONDITIONS that an individual fails to follow the rule to which the threat applies. The rule or norm backing an OR ELSE both establishes a range of punishments available and assigns the authority and procedures for imposing an OR ELSE. Often the actions threatened in the OR ELSE are forbidden under most CONDITIONS (e.g., imposing a fine, incarcerating a citizen, or putting someone's livestock in a village pen). The prescription backing the OR ELSE makes these actions permitted in the CONDITIONS that someone breaks a rule. However, the shift in the DEONTIC is not always from F to P or O. The OR ELSE might involve forbidding some action that is usually permitted or obligatory, a shift from P or O to F. For example, a license branch official could be forbidden to issue a driver's license to a citizen after that citizen has been convicted of repeated drunken driving offenses. Although the OR ELSE often refers to physical punishments, the OR

ELSE may also refer to institutional actions such as taking away a position or refusing to accept an amendment as legal. For example, one of the rules governing the amendment process could state that legislators with [ATTRIBUTES] [must] [obtain positive votes from at least a majority] [when voting for an amendment] or [the amendment fails].

Second, institutional statements must exist that affect the constraints and opportunities facing actors who monitor conformance to the prescription. Monitoring institutions may assign these tasks to specialized individuals or authorize all participants to engage in these activities. Although the actors who monitor also frequently sanction nonconforming actors, they may only report nonconformance to someone else who is responsible for sanctioning.

Third, the OR ELSE must be crafted in an arena used for discussing, prescribing, and arranging for the enforcement of rules. This arena may be a legislature or a court, but we do not consider government sponsorship or government backing to be a necessary condition for an institutional statement to be a rule. Many self-organized, communal, or private organizations develop their own rules that include (1) sanctions backed by another rule or norm that change the DEONTIC assigned to some AIM for at least one actor under the condition that individuals fail to follow the rule and (2) institutional statements that affect the constraints and opportunities of actors to take the responsibility to monitor the conformance of others to the prescription.[11] For some research questions, it may be useful to divide rules into those made within government arenas and those made outside of government arenas. The syntax in no way prevents this distinction.

The first four examples have the syntax of a rule, if we assume that all three qualifications for the OR ELSES in these statements are met. For example, the potential punishment for villagers who let their livestock trample the irrigation channels qualifies as an OR ELSE only when a second rule or norm accepted in that village prescribes others to employ the sanctions defined in the OR ELSE, an institutional statement prescribes or advises monitoring, and these arrangements were made in a setting considered by the villagers to be a rule-making arena.

Operationalizing the Syntax

All linguistic statements that explain shared strategies contain three of the syntax components (AIC), those that explain norms contain four (ADIC), and those that explain rules contain all five (ADICO). Parsing the five examples using F as the primitive DEONTIC yields the following.

1. U.S. citizens [age $\geq$ 18, gender = male] [F] [~(register with Selective Service)] [filling out form at U.S. Post Office] [face arrest].

2. Senators [all] [~*F*] [move an amendment] [after bill introduced] [called out of order or ignored].
3. Villagers [all] [*F*] [irrigation channel trampled by their animals] [at all times] [fine].
4. Neighborhood residents [all] [*F*] [~(clean yard)] [organized cleanup day] [ ].
5. Person [placed call] [ ] [calls back] [call disconnected] [ ].

Now, how can the syntax be operationalized in empirical or theoretic research? In empirical studies, the researcher's task is to discover the linguistic statements that form the institutional basis for shared expectations that influence observed regularities in behavior. Essentially, this entails discovering which ADICO components exist in these statements and the contents of those components. Determining the existence of a component requires knowledge about whether individuals in a group share some level of common understanding of its content. In the case of an OR ELSE, complementary institutional statements are needed, as well as evidence that individuals crafted the OR ELSE in a rule-making arena. Uncovering components requires qualitative research methods, including in-depth interviews or archival retrieval. Using the syntax to structure game-theoretic analysis requires the researcher to identify how the components of the institutional statements assumed in the game are reflected in the structure of a game and in the payoffs assigned to actions and outcomes.

Operationalizing the ATTRIBUTES, AIM, and CONDITIONS

The AIM usually supplies the focus for theoretical and empirical studies. Scholars decide to study the impact of institutions on behavior for some subset of actions or outcomes. Studies of agenda setting and voting institutions, for example, focus on institutional statements with AIMS that relate to setting agendas and voting. Once researchers select the focal set of actions or outcomes for their studies, the next step is developing the analysis of institutional statements with AIMS that relate to those actions or outcomes.

The existence of the ATTRIBUTES and CONDITIONS components is fairly straightforward; they always exist if an institutional statement exists. In some cases they exist in their most general form—all individuals or all organizations and under all conditions.[12] Discovering the contents of the ATTRIBUTES and CONDITIONS components requires the empirical researcher to inquire about the actors to whom particular institutional statements apply and the conditions under which they apply. The syntax reminds formal theorists that there are assumptions about ATTRIBUTES and CONDITIONS built into the games that they model. Game-theoretical models that are constructed to provide an institutional explanation for observed regularities of behavior make explicit assumptions about the ATTRIBUTES and the CONDITIONS included in the institutional statements of interest. Recognizing the implicit existence of these components helps scholars to discover and explicitly state those assumptions.

Operationalizing the DEONTIC

The first step in operationalizing the DEONTIC is deciding whether the institutional statement contains one. In empirical studies, the researcher listens for normative discourse. Is there an articulated sense of moral or social obligation expressed? If the individuals in a study share only AIC statements, then their discussion of why they would follow such advice focuses only on prudence or wise judgement: "The best thing to do when faced with a choice between A and B under condition Y is to choose A, because one is usually better off with this choice." When individuals shift to a language of obligation, they use terms such as *must, must not, should,* or *should not* to describe appropriate behavior (Orbell, van de Kragt, and Dawes 1991): "The obligatory action when faced with a choice between A and B under condition Y is to choose A, because this is the proper action." In other words, one looks for institutional statements that are used to evaluate behavior (Collett 1977). When social and moral obligations are discussed, an empirical researcher initially assumes that it is appropriate to include a DEONTIC in institutional statements used to explain behavior.

When a DEONTIC exists, one expects negative repercussions to follow behaviors that do not conform to the prescription and positive rewards to follow compliance. We use a delta parameter concept to capture the costs or benefits of these inducements and deterrents for any particular DEONTIC. Usually, we expect the severity of the delta parameters to provide some indication of the importance or valence of the DEONTIC to the community.[13]

Delta parameters may reflect internal and/or external effects of a DEONTIC. The distinction between external and internal sources of delta parameters is similar to the one that Coleman (1987) makes between "internalized norms" and "externally sanctioned norms" and that Kerr et al. (1997) make between a "personal" and a "social" norm. Delta parameters originating from external sources are a way to represent the benefits and costs of establishing a reputation (see Kreps 1990). The delta parameters originating from internal sources can be thought of as the guilt or shame felt when breaking a prescription and the pride or "warm glow" felt when following a prescription, particularly if it is costly to follow in a particular situation (Andreoni 1989; Frank 1988; Ledyard 1995).

If an action is forbidden by a norm and an individual engages in that action, then we expect that person to experience some type of cost represented by at least one delta parameter. If norms indicate that an AIM is *permitted,* then we expect others, who treat that action as if it were forbidden, to experience some cost. In other words, norms or rules that forbid or require some AIM will be reflected in a cost parameter to the individuals to whom the prescription applies, while prescriptions that establish permission for an action place the cost parameter on others. Others may experience a cost if they try to obstruct an individual when a shared norm indicates that the individual is permitted to take that action.[14]

Using delta parameters brings normative considerations overtly into the analysis of action and consequences, rather than employing ad hoc interpretations. Adding overt consideration of nonmaterial costs and benefits is not the same as assuming that individuals incorporate the welfare or utility of other actors into their own calculus; nor is it the same as substituting the collective benefits for individual utility. We prefer the delta parameter interpretation of "other" or "normative" motivation because it ties that motivation to institutions by representing individual commitments to the norms or rules of a situation. Our preference for the delta parameter interpretation is bolstered by recent evidence that group discussion affects cooperation in social dilemma experiments in a manner that supports the logic of a delta effect and disputes a group identification effect (Kerr 1996).

In field research, the deltas are frequently not observable, especially in situations where behavior prescribed by a norm is habituated. First, one may not observe any instances of nonconformance so that no opportunity exists to witness reactions to deviations. Second, few external rewards for conformance occur in situations where everyone expects conformance. In an empirical setting where one cannot observe repercussions or see tangible rewards, one has to rely primarily on the accounts that participants give as to the normative content of their actions and the repercussions of nonconformance.

In formal analyses, one can model the DEONTIC by adding delta parameters to the players' payoffs to represent the perceived costs and rewards of obeying (δ^o) or breaking (δ^b) a prescription. The delta parameters can be defined as

$$\Delta = \delta^o + \delta^b,$$

where

Δ = the sum of all delta parameters
δ^o = the change in expected payoffs from obeying a prescription
δ^b = the change in expected payoffs from breaking a prescription.

One can further divide these rewards and costs into those that arise from external versus strictly internal sources of valuation. Thus:

$$\delta^o = \delta^{oe} + \delta^{oi} \text{ and } \delta^b = \delta^{be} + \delta^{bi},$$

where

e = the change in expected payoffs originating from external sources
i = the change in expected payoffs originating from internal sources.

The analyst may not wish to focus on all four parameters in any particular analysis. Three of the four delta parameters could be assigned a zero value in a game-theoretic analysis involving a norm or a rule. In order to analyze the impact of a DEONTIC on expected outcomes of a game, however, at least one of the delta parameters must have a nonzero value. The payoff structures for individuals who share prescriptions must differ from those of players in a similar situation in which players merely accept a shared understanding of prudent action.

In situations where it is reasonable to assume that all players who break the prescription feel the same cost, the delta parameters can be modeled as if they were the same for all players and as if their magnitude is public information. Alternatively, the theorist can model players as being different types who react differentially to breaking prescriptions (see Harsanyi 1967–68). One type of player can perceive the costs of breaking a prescription (δ^{bi} or δ^{be}) to be high while another type perceives costs (δ^{bi} or δ^{be}) to be low. Coleman's (1988) zealot, for example, is a player with high external deltas for obeying norms (a high positive δ^{oe}).

The existence of a DEONTIC implies the presence of additional information that individuals use in developing their expectations about others' behavior and thus their own best response. If players all adopt a norm, the payoff structure differs from the payoff structure for a similar situation in which the players do not adopt a norm. The payoffs may even change enough so that the predicted outcome of the game changes. Uncertainty about the presence of actors who have accepted certain norms may be sufficient grounds for changing the behavior of other players. Kreps et al. (1982) have analyzed repeated Prisoner's Dilemma games where information asymmetries exist among players concerning the probability that other players will play Tit-for-Tat.[15] In such games, players who are "perfectly rational" (i.e., the players' payoff functions have a zero value for all four delta components associated with playing Tit-for-Tat) will adopt behavior consistent with the norm for most of the game, due to their changed expectations about the behavior of others.

Operationalizing the Concept of OR ELSE

The criteria for indicating that an OR ELSE exists are (1) a known range of sanctions, (2) a norm or rule that prescribes sanctions, and (3) some provision for monitoring—all emanating from a rule-making arena. We wish to emphasize that the sanction in an OR ELSE differs from a logical consequence or an interpersonal threat or retaliation. Neither of these repercussions fits the OR ELSE criteria. In empirical analyses, researchers determine whether observed regularities in behavior are based on a shared understanding about the sanctions associated with a rule and whether there is a shared norm or rule that changes the DEONTIC associated with the sanctioning act for some player that can be traced to a rule-making arena. If so, explanations of behavior would consider the attitudes or feelings about following rules (the value of internal delta parameters), the reputation effects of compliance or noncompliance (the external delta parameters), the chance of being observed, the probability that the OR ELSE sanction would be imposed, and the size of the OR ELSE sanction.

In the ADICO syntax, a prescription used to explain the behavior that fails to meet the OR ELSE test is a norm. In other words, a written law could be a norm rather than a rule if the law does not meet the OR ELSE requirements. Analysis of the decisions that individuals make to follow or break such a norm (written as a law) would be based primarily on their own feelings about conformance (δ^{oi} and δ^{bi}) and their appraisal of potential reactions from others (δ^{oe} and δ^{be}), their estimates of the values that other players place on conformance (the other players' δ^{oi} and δ^{bi}), and their estimates of the values that other players place on feedback (the other players' δ^{oe} and δ^{be}).

To incorporate the syntax into formal analysis of rules and behavior, the payoffs for actions governed by rules need to include delta parameters that reflect the DEONTIC and a parameter representing the sanction defined in the OR ELSE. If the enforcing players are brought into the analysis, the enforcing players would have delta parameters in their payoffs since there is a rule or norm that prescribes sanctioning. If the OR ELSE is backed by rule, then we expect the payoffs for sanctioning or not sanctioning to include delta parameters and a variable representing the punishment defined in the OR ELSE of the sanctioning rule. If it is costly to monitor the actions of others and/or to impose sanctions on them, those assigned these tasks may not be motivated to undertake these assignments unless (1) the monitor and sanctioner face a probability of an OR ELSE, (2) social pressure to monitor and sanction is large and is salient to the monitor and sanctioner (large δ^{oe} or δ^{be}), (3) the monitor and sanctioner hold some strong moral commitment to their responsibilities (large δ^{oi} or δ^{bi}), and/or (4) the payment schemes for the monitor and sanctioner create prudent rewards high enough to offset the costs. When a norm backs an OR ELSE, enforcement rests solely on the value of the delta parameters and on the payment scheme for the monitor and sanctioner.

Using the Syntax for Synthesis

In any science, understanding what others have already discovered is an important part of research. Synthesizing findings from the different subfields that relate to each type of institutional statement is an important task for those interested in institutions. The similarities between rules, norms, and shared strategies in the ADICO syntax illustrate why the literature so frequently mixes the concepts or uses them interchangeably. They do share several of the same features. Table 4.1 sorts the concepts that other authors use into the types of institutional statements created by our syntax. In other words, all of the terms used by other authors, shown in the top section of the table, appear to describe institutional statements that are shared strategies according to the syntax; they contain AIC components. The need for a consensus in the use of terms is vividly illustrated by examining the number of different terms in each section of the table. Moreover, several terms appear in all three sections. Notice that some of the terms used by an author appear in more than one of the categories. This means that the term could fit either of the categories in the syntax given the diverse ways that the author uses the term in the work cited. That these terms have been used in so many different ways is not a criticism of past work. Rather, it portrays the difficulty of untangling these concepts so that a less ambiguous set of definitions can be used as the foundation for further theoretical work.

A growing body of work considers the mix of normative and material motivations that individuals consider when faced with choices (Coleman 1988; Ellickson 1991; Elster 1989a,b; Etzioni 1988; Hirschman 1985; Knack 1992; Mansbridge 1990, 1994; Margolis 1991; Offe and Wiesenthal 1980; E. Ostrom 1990; V. Ostrom 1986; Udéhn 1993). These works treat the normative aspects of decisions up front as a significant part of the analysis. Margolis argues for the necessity of such an approach: "If we analyze everything in terms of strict self-interest and then include some social motivation only if we get stuck or if there is something left over, it is not likely to lead to nearly as powerful a social theory as if the two things are built in at the base of the analysis" (1991, 130). Delta parameters provide a conceptual language with which to build normative considerations into analysis from the beginning and to discuss differences in studies that incorporate normative incentives.

Table 4.2 lists studies that have addressed three types of questions about normative motivations. The top section of the table lists different assumptions regarding the meaning and sign of delta parameters. Knack's (1992) analysis of voter turnout, for example, illustrates the insight possible from a careful study of internal and external normative influences. He offers empirical evidence of the substantive content of internal and external pressures associated with a turnout norm and of the influence of these pres-

TABLE 4.1. Shared Strategies, Norms, and Rules as Used in ADICO Syntax and in Recent Literature

Terms	Citations
AIC	
Shared Strategies	*The Present Article*
Norms	Axelrod 1986
Rules, strategies	Axelrod 1981
Doxic elements of action	Bourdieu 1977
Equilibrium strategies	Calvert 1992
Norms	Levi 1990
Rules	March and Olsen 1989
Taken-for-granted actions	Meyer and Rowan 1991
Rules	Myerson 1991
Rules of action	Rowe 1989
Scripts	Schank and Abelson 1977
Focal points	Schelling 1978
Institutions	Schotter 1981
Conventions	Ullman-Margalit 1977
ADIC	
Norms	*The Present Article*
Conventions	Braybrooke 1987
Rules	Braybrooke 1994
Norms	Coleman 1987
Institutions	DiMaggio and Powell 1991
Conventions	Lewis 1969
Norms	Levi 1990
Rules	March and Olsen 1989
Taken-for-granted actions	Meyer and Rowan 1991
Ethical codes	North 1981
Obligations	Rowe 1989
Institutions	Schotter 1981
Conventions	Sugden 1986
Social norms	Ullmann-Margalit 1977
Conventions	Weber 1947
ADICO	
Rules	*The Present Article*
Norms backed by metanorms	Axelrod 1986
Working rules	Commons 1968
Norms	Coleman 1987
Legalistic institutions	Levi 1990
Rules	Knight 1992
Rules	North 1990
Rules	Shepsle 1979, 1989
PD norms, decrees	Ullmann-Margalit 1977
Laws	Weber 1947
Rules of the game-form	Hurwicz 1994

sures on the probability that an individual will vote. Interestingly, his interpretation of the declines in voter turnout echoes the importance of monitoring. He finds that social sanctions (external deltas) are a key influence on voting turnout. Social pressure only operates, however, when voters

TABLE 4.2. Delta Parameters and Normative Concepts Used in Recent Literature

Delta Parameters	Concepts Used by Other Authors	Citations
Size, sign, and interpretation of delta parameters		
$+\delta^{oi}$	warm glow	Andreoni 1989; Ledyard 1995
$+\delta^{oe}$	encouragement	Coleman 1988
	status improvement/ reputation enhancement	Coleman 1988
	honor	Ullmann-Margalit 1977
$-\delta^{bi}$	duty	Knack 1992
$-\delta^{be}$	cost of being punished (P and P')	Axelrod 1986[a]
	social sanctions	Knack 1992
	third-party sanctions	Bendor and Mookherjee 1990[b]
$+\delta^{oi}$ and $-\delta^{bi}$	internalized norms	Coleman 1987
	public spiritedness	Mansbridge 1994
	moral duty	Etzioni 1988
	duty	Commons 1968
$+\delta^{oe}$ and $-\delta^{be}$	externally sanctioned norms	Coleman 1987
	reputation	Kreps 1990
	responsibility	Commons 1968
	moral judgment	Sugden 1986
Types of players		
$+\delta^{oe}$ large	zealot	Coleman 1988
$\Delta = 0$	selfish rational individual	Elster 1989a
$+\delta^{oi}$ and/or $-\delta^{bi}$ large	everyday Kantian	Elster 1989a
$+\delta^{o}$ large when number of cooperators low	elite participationists	Elster 1989a
$+\delta^{o}$ large when number of cooperators high	mass participationists	Elster 1989a
Δ larger when number of cooperators > threshold	people motivated by fairness	Elster 1989a
Creation and maintenance of delta parameters		
Δ affected by labor union activities		Offe and Wiesenthal 1980
Δ are scarce resources that erode with use.		Olson 1991
Δ are resources that increase with use.		Hirschman 1985; Mansbridge 1994
Δ affected by external fines		Frey 1994
Δ lower when rules come from outside authority		Frey 1994

[a]In some cases, these sanctions may meet the criteria of an OR ELSE.
[b]As with the P and P' of Axelrod's, the third-party sanctions may at times meet the criteria of an OR ELSE.

expect to be in situations where someone will ask them if they voted. As the percentage of individuals in organizations, in relationships with neighborhood residents, in extended family situations, and in marriages decreases, this monitoring decreases and the power of the social sanction (an external delta) diminishes.

The middle section of table 4.2 cites work that addresses the implications of various mixes of different types of individuals or players. For example, a selfish rational individual would be a type that assigns a zero value to praise or blame for obeying or breaking prescriptions. One interesting variant of types ties the size of the delta to the number of others who conform to the prescription; its size is conditional. Elster (1989a) examines the consequences of communities with various mixes of types with conditional and constant deltas.

Scholars cited in the bottom section of table 4.2 discuss variables that influence the direction and size of delta parameters. Offe and Wiesenthal (1980) examine the costs that labor unions face to build and maintain a shared commitment to participation norms as this in turn affects their ability to compete with other interest groups. Several other authors ask whether normative incentives increase or decrease with use. Olson (1991) views the delta parameters as scarce resources that can be dissipated with too much use, while Hirschman (1985) and Mansbridge (1994) come to the opposite conclusion. They argue that the normative constraints increase in size as they are used repeatedly by individuals in a group. Frey (1994) contends that external interventions, such as fines, adversely affect the size of delta parameters, particularly the internal deltas. He also speculates that the deltas associated with rules will be higher when individuals participate in making their own rules than when rules are made by higher authorities.

Using the Grammar for Analysis

Now our applications of the grammar move from synthesis to analysis of collective-action problems. The issue of how individuals solve collective-action problems is at the forefront of attention across the social sciences. The disjunction between theoretical predictions of complete free riding in Prisoner's Dilemma situations and the rates of cooperation that case studies and laboratory experiments reveal provokes much discussion (Udéhn 1993; Weiss 1991). The scholarly discourse about these issues involves terms such as *common understandings, shared beliefs, scripts, norms, rules, procedures, institutions, informal rules, informal institutions, conventions, internal solutions, external solutions,* as well as a wide diversity of highly technical terms related to particular solution theories.

To illustrate how the ADICO syntax clarifies such analysis, we model the effect on behavior of three structural adjustments (represented by additions of an ADICO component) to a social dilemma situation. This application of the grammar employs game theory and stresses formal analysis. We use existing game theory as our illustrative theory of action. However, the logic discussed in the application applies to nonmathematical analyses as well.

Collective-action problems can be represented by many different game structures (see Taylor 1987). However, because most social scientists know the Prisoner's Dilemma game well, we use a Prisoner's Dilemma game so that we can more easily jump into existing debates and rely on extensive prior work. We start with a simple two-person Prisoner's Dilemma game and use the ADICO format to illustrate the research issues, the game structures, and the predicted outcomes that arise from Prisoner's Dilemma situations when we add (1) shared expectations of other players' behavior only (AIC statements), (2) normative views of the appropriate actions to be taken (ADIC statements), and (3) rules (ADICO statements). Table 4.3 summarizes the institutional and payoff characteristics of four games based on a two-person Prisoner's Dilemma situation.[16] The first game is the base two-person Prisoner's Dilemma game. The *shared strategies* game adds a set of shared strategies that equate to the *grim trigger* strategy. The *norms* game adds a cooperating norm to the base situation. The *rules* game adds a cooperating rule, a monitoring norm, and a sanctioning norm to the base Prisoner's Dilemma game. These four examples represent only one way to add the ADICO statements to a two-person Prisoner's Dilemma game.

The Base Game

In the two-person Prisoner's Dilemma base game, we assume that $1 > c > d > 0$ (see fig. 4.A1 in the appendix). The best payoff (1) comes from choosing D while the other player chooses C; the second best payoff (c) occurs for both players choosing C; third is the payoff (d) resulting from both players choosing D; and the worst payoff (0) comes from choosing C while the other player chooses D. Given these payoffs, both players are better off choosing D no matter what the other player chooses. The solution to this game, if played only once, is for both players to choose D and receive d, instead of the more desirable c that they could have received if they had both chosen C. Even if repeated a finite number of times, the solution is for both players to always choose D.

The Shared Strategies Game

Predictions that individuals will select C rather than D in a Prisoner's Dilemma game based on shared strategies rely upon changes in players' expectations about each other's future behavior. In order to incorporate those expectations

TABLE 4.3. Game Summaries

Institutional Statements	Payoffs
Base game	
None (physical world)	*Player 1 or 2*
	$C = c$ if other C
	$\quad = 0$ if other D
	$D = 1$ if other C
	$\quad = d$ if other D
Shared strategies game	
AIC Statements	*Player 1 or 2*
[All players] [ ] [C] [first round] [ ]	$C = c + t(c)$ if other C
[All players] [ ] [C] [if all C in	$\quad = 0 + t(d)$ if other D
previous round] [ ]	$D = 1 + t(d)$ if other C
[All players] [ ] [D] [all rounds	$\quad = d + t(d)$ if other D
after a D] [	
Norms Game	
ADIC statement	*Player 1 or 2*
[P1[a] and P2] [must] [C] [always] [ ]	$C =$ base game payoffs $+ \delta^{oi} + \delta^{oe}$ if P3 $\Rightarrow$ M^b
	$\quad =$ base game payoffs $+ \delta^{oi}$ if P3 $\Rightarrow$ $\sim M$
	$D =$ base game payoffs $+ \delta^{bi} + \delta^{be}$ if P3 $\Rightarrow$ M
	$\quad =$ base game payoffs $+ \delta^{bi}$ if P3 $\Rightarrow$ $\sim M$
	Player 3
	$M = E$ if (P1 and P2) $\Rightarrow C$
	$\quad = R - E$ if (P1 or P2) $\Rightarrow D$
	$\sim M = 0$
Rules game	
ADICO statement	*Players 1 and 2*
[P1 and P2] [must] [C] [always] [f]	$C =$ norm game payoffs
	$D =$ norm game payoffs $+ f$ if (P3 $\Rightarrow M$) and
ADIC statements	(P4 $\Rightarrow S$)
[P3] [must] [monitor] [always] [ ]	$\quad =$ norm game payoffs if (P3 $\Rightarrow \sim M$) or
[P4] [must] [impose f on defector]	(P4 $\Rightarrow \sim S$)
[when P3 reports a D] [ ]	
	Player 3
	$M =$ norm game payoffs $+ \delta''_m$
	$\sim M =$ norm game payoffs $- \delta^b_m$
	Player 4
	Only plays if P3 $\Rightarrow M$
	$S = \delta''_s - E_s$
	$\sim S = -\delta^b_s$

[a]P1 refers to player 1, and so on.
[b](P3 $\rightarrow M$) indicates that player 3 chooses M.

into formal analysis, we use an indefinitely repeated version of the base game in which future expected payoffs are part of a player's calculation at any one round (see fig. 4.A2 in the appendix). In the shared strategies game in table

4.3, the players all use a grim trigger: all players cooperate in each round of the game until someone defects, after which players defect for the rest of the game.[17] Assuming that players do not discount future payoffs, both players cooperating in every round is predicted if and only if the expected payoffs from cooperating are greater than the expected payoffs for defecting: $c + t(c) > 1 + t(d)$. This simplifies to

$$c > \frac{1+d}{1+t},$$

where t is the expected number of future rounds.[18]

All players must share common knowledge that all have adopted a trigger in order for it to be an institutional statement. If all players do not consider it prudent to defect for all rounds after someone initially defects, the trigger strategy is not shared and will not work. Little empirical evidence exists that individuals share a belief in the prudence of a grim trigger (E. Ostrom, Gardner, and Walker 1994). Herein lies the frailty of the grim trigger as either a resolution of the Prisoner's Dilemma game or a possible institution.

The Norms Game

Predictions that individuals will select C rather than D in a Prisoner's Dilemma base game based on norms rely upon changes in players' payoffs because of the addition of at least one delta parameter. To avoid the case of predicting cooperation as a dominant strategy through only internal commitments to norms, we add both external and internal delta parameters. We allow nonzero internal and external delta parameters for both obeying the cooperation norm (δ^{oi} and δ^{oe}) and for breaking the cooperation norm (δ^{bi} and δ^{be}).

In order to make the discussion more applicable to situations with more than two players, we add a third player, a *monitor,* who chooses to monitor (M) or not to monitor ($\sim M$) (see fig. 4.A3). We assume that external reinforcements for obeying (δ^{oe}) or for breaking (δ^{be}) a norm only occur when the monitor detects the choices of a player. In the simple two-person game, this assumption is not necessary. Players 1 and 2 know whether the other player cooperated by simply looking at their own payoffs. However, as soon as the number of players in a Prisoner's Dilemma is larger than 2, identifying who cooperates and who defects is no longer trivial.[19] The monitor in this game is motivated solely by prudential rewards associated with discovering defection.[20] There is no monitoring norm or rule.

In the norms game, predictions about players' strategies depend on the relationships among the original payoffs in the base game, the added delta parameters, and the benefits that the monitor receives for reporting nonconformance. This game has many equilibria. Assuming that all of the delta parameters are symmetric (player 1 and player 2 have the same values for each delta parameter) and that the sum of the external parameters is greater than the sum of the internal parameters (i.e., the social pressure to follow the prescription is greater than the internal pressure), four equilibrium regions exist as shown in figure 4.1.

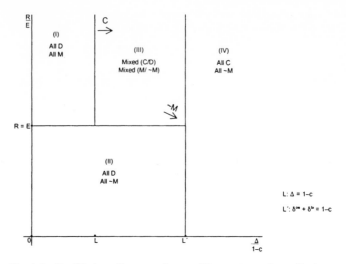

Fig. 4.1. Equilibrium diagram: Game with a norm and monitoring

The vertical axis is the ratio of the monitor's reward for detecting defection to the expense of monitoring (*R/E*). The vertical axis is divided into regions above and below the point at which the reward to the monitor equals the expense (*R = E*). The reward is higher than the cost above this point and lower than the cost below this point.

The horizontal axis is the ratio of the sum of all delta parameters to the advantage of choosing *D* [Δ/(1 − *c*)]. Since we assume that players 1 and 2 have the same values for each of the delta parameters and that both internal and external delta parameters increase monotonically from left to right, a single axis represents the relationship between the sum of all the delta parameters and the advantage of defecting. As one moves to the right, the size of the delta parameters relative to the advantage of *D* increases. At point *L*, the sum of the delta parameters equals the advantage of defection when the other player cooperates (Δ = 1 − *c*). Point *L'* on the horizontal axis represents that point where the portion of the sum of delta parameters derived from the internal deltas equals the advantage of defection ($\delta^{oi} + \delta^{bi} = 1 - c$). Thus the area to the right of *L'* represents the region where the internal deltas offset the advantage of defecting and the value of the external deltas becomes irrelevant. In this area (equilibrium region IV on fig. 4.1), both players choose *C*. This is the trivial case in which *C* is the dominant strategy because the internal costs and rewards for players 1 and 2 are sufficiently high to offset the fact that 1 is greater than *c*.

In region III, both players 1 and 2 select a mixed strategy between *C* and *D*.[21] As one moves from left to right in region III, the sum of the delta parameters increases and thus the probability that players 1 and 2 assign to selecting *C* increases. In this region, the monitor also selects a mixed strategy. The *relative amount* of the monitor's reward decreases as one moves from north to south. Since the monitor receives a reward only if defection is detected, the *probability* of obtaining a reward decreases as one moves from west to east because the probability of defection decreases. The combined effect is that the monitor has the least incentive to monitor when relative rewards are low and the probability of defection is low (in the southeast corner) and the greatest incentive to monitor when the relative rewards are high and the probability of defection is high (in the northwest corner of this region). Thus, as one moves from northwest to southeast in this region, the probability assigned to $\sim M$ increases.

In equilibrium regions I and II of figure 4.1, *D* remains the dominant strategy for players 1 and 2 as it is in the base game, but for different reasons. In region II, the expenses of monitoring are higher than the expected rewards to the monitor. Thus the monitor will choose $\sim M$ in region II. Since $\sim M$ is the dominant strategy, players 1 and 2 need not consider the external cost parameters (δ^{oe} and δ^{be}). Given that the internal deltas are relatively low in relation to the advantage of defecting $[(\delta^{oi} + \delta^{bi}) < (1 - c)]$, *D* is the dominant strategy for players 1 and 2. Region I, on the other hand, represents a socially perverse outcome whereby players 1 and 2 always defect because the advantage of defecting $(1 - c)$ is greater than the sum of all delta parameters. At the same time, the monitor has a dominant strategy of *M* because the rewards received from detecting defection exceed the monitoring costs and are guaranteed to occur (assuming perfect detection), because players 1 and 2 face the dominant strategy of *D*.

This analysis demonstrates that (1) introducing norms and monitoring is not sufficient to change predicted results in a Prisoner's Dilemma base game and (2) a change in predicted results may not be socially beneficial. No set of equilibria exists where the actions of the monitor totally prevent defection. The only equilibrium region in which players 1 and 2 select a pure strategy of cooperating occurs where the value of the internal normative parameters exceeds the advantage of defecting (IV). The presence of a monitor who is motivated to select a mixed strategy boosts the level of cooperation in one region (III) but does not ensure cooperation. Moreover, the lower the probability of defection, the higher the monitoring rewards (*R*) need to be to offset the reduced probability of receiving the reward.

The results in figure 4.1 hold because the reward to the monitor comes only if there is defection to be reported. If the monitor is rewarded specifically for monitoring, regardless of whether defection is discovered, there is an additional equilibrium region in which the monitor ensures cooperation. In this region, the reward for monitoring exceeds the cost of monitoring regardless of whether defection occurs. Thus the monitor chooses a pure strategy of M, which makes C a dominant strategy for players 1 and 2 whenever the sum of their internal and external delta parameters exceeds the advantage of D.

Empirical studies and formal models suggest several other motivational schemes for monitors. Some motivate monitors by embedding them in a series of nested institutions that rewards monitors who actively and reliably monitor with positive returns from the increased productivity that the rules generate (see Greif, Milgrom, and Weingast 1990; Milgrom, North, and Weingast 1990). Monitors may also be direct participants in ongoing relationships where efforts are made to reward one another for monitoring and to ensure that monitors participate in the greater returns that all achieve when temptations to defect are reduced. In such situations, monitors may achieve sufficient benefits from monitoring to induce a high level of conformance (but never 100 percent) in an isolated system without recourse to central authorities (Weissing and Ostrom 1991a, 1993). If one wanted to analyze the incentive structure found in many field settings where monitors are hired as external, disinterested guards, one could change the game so that the monitor receives a salary regardless of whether he or she catches a defection or shirks. In such a setting, the monitor has little incentive to monitor, so that the rate of cooperation depends heavily on the size of the internal delta parameters for players 1 and 2.

The Rules Game

Predictions that individuals will select C rather than D in the rules game rely upon (1) the base game payoffs, (2) changes in players' payoffs because of the addition of at least one delta parameter, (3) changes in the players' payoffs due to the addition of an objective and institutionally created consequence for breaking a rule, (4) the probability of detection, and (5) the probability of sanctioning (see fig. 4.A4 in the appendix).

The rules game shifts the norms game to a game with a rule backed by two norms. The rule that structures this game is: [Players 1 and 2] [must] [C] [always] [f]. The rule adds a fine (f) to the payoffs for players 1 and 2 for D if their defection is monitored and sanctioned. A sanctioning norm creates this fine and assigns the responsibility of imposing it on player 4: [Player 4] [must] [impose f on a defector] [when player 3 reports a D] [ ]. The rule is also backed by a monitoring norm: [Player 3] [must] [monitor] [always] [ ].

The monitoring norm adds delta parameters to player 3's payoffs. Whereas the monitor was rewarded only by prudential rewards in the norms game, now the monitor considers both prudential rewards and normative rewards as well as the costs of monitoring. The sanctioning norm adds another player, player 4 (the *sanctioner*), whose payoffs include delta parameters reflecting the sanctioning norm DEONTIC.[22] The cost of sanctioning is the only other parameter in player 4's payoffs. In other words, player 4 is a volunteer sanctioner "rewarded" solely by normative interests.

A wide variety of mixed strategy equilibria is possible, with these equilibria depending on (1) the relative expected value of the fine and the relative size of the external delta parameters for players 1 and 2, (2) the size of the reward and deltas associated with monitoring as compared to the costs of monitoring, and (3) the sign and size of the deltas associated with sanctioning minus the costs of sanctioning as compared to the value of the delta parameters for not sanctioning.[23]

We set aside the tasks of analyzing the many possible equilibrium regions of the rules game and focus here on the simpler task of establishing conditions for equilibria in which players 1 and 2 always cooperate. The rule adds a fine and a new player, yet the monitor still plays a crucial role. The parts of the game that come from the OR ELSE (the fine and the sanctioning norm) do not even enter the payoffs for players 1 and 2 unless the monitor chooses M.[24] As in the norms game, we assume that player 1 and player 2 do not see each other's choices and that the external delta components occur only when player 3 monitors.

Cooperation is a pure strategy for players 1 and 2 if and only if either of the following conditions is met.

$$\delta^{oi} + \delta^{bi} > 1 - c$$

or

$$[(\delta^{oi} + \delta^{bi}) + (p(M)*(\delta^{oe} + \delta^{be}) + (p(S)*f)) > 1 - c],$$

where

f = fine for breaking the rule
$p(M)$ = probability of being monitored
$p(S)$ = probability of being sanctioned.

The first condition is the same as in the norms game. The second condition figures in the effect of the monitor and sanctioner. The probability of monitoring [$p(M)$] and sanctioning [$p(S)$] depends upon the payoffs to player 3 (the monitor) and player 4 (the sanctioner). In order for the monitor and the sanctioner to be motivated, the values of following the monitoring and sanctioning norms have to be greater than the relative cost of doing their jobs. In the case of the sanctioner, the value of the delta parameters needs to be greater than the expense for imposing the sanction (E_s): $(\delta^o_s + \delta^b_s) > E_s$. The monitor balances both delta parameters and material rewards against the expense of observing players 1 and 2 (E_m). Player 3 monitors when $p(D)R_m + \delta^o_m + \delta^b_m > E_m$, where $p(D)$ is the probability that either player 1 or player 2 will choose D.

In settings where players develop a strong internalization of norms (high internal delta parameters), the presence of even a low-to-moderate sanction (f) may be sufficient when combined with reliable monitors and sanctioners [high $p(M)$ and $p(S)$], to encourage a high rate of cooperation. If players 1 and 2 expect either the monitor or the sanctioner to break their respective norms, then the expected probability of the sanctioner choosing to sanction $[p(S)]$ tends toward zero, and f drops out of the decision calculus for players 1 and 2. Even in a rule-governed game, if monitors are not motivated to monitor and sanctioners are not motivated to sanction, cooperation rests substantially on internalized norms of the players. Clearly, recognizing rules in formal analysis of dilemma situations does not automatically "solve" the dilemma and end analysis. Instead, adding rules suggests a whole new set of research questions. For example, how do changes in the level of internalization of rules (δ^{bi} and δ^{oi}) affect the levels of monitoring and sanctioning required to bolster cooperation at given levels of social pressure (δ^{oe} and δ^{be})? What size do external delta parameters need to be in order to ensure cooperation at various rates of monitoring and sanctioning with a given value of f? How do the incentives to monitor and sanction differ if we assume that players 3 and 4 are the same person? What are the empirical equivalents of delta parameters and OR ELSES in field situations?

Moving beyond Syntax

Although our focus is mainly on the syntax of a grammar of institutions, the grammar of institutions, as is the case with other grammars, also includes semantics (i.e., the meaning of statements) and pragmatics (i.e., how to apply the syntax and semantics in practice). The delta parameters provide one example of the semantics of the grammar. Discussing how to operationalize the syntax components begins to address the pragmatics of the grammar. Analyses of many issues of interest to scholars of institutions will delve more deeply into pragmatic issues such as (1) the effects of configurations of institutional statements; (2) the consistency and completeness of the institutional statements of an institution; and (3) legitimacy and compliance.

Institutional Configurations

The descriptions of the components of rule and norm statements (except the OR ELSE) have focused on single statements as if the contents of the institutional statements were independent. The focus on single statements is for expository reasons only. When we examine the interactions of individuals in an empirical situation, we often find that a configuration of rules, norms, and shared strategies influences the choices of individuals at any one point in time. In fact, we often find institutional configurations, such as shared strategies and professional norms, nested within enforced government regulations.[25]

In some cases, the CONDITIONS component of an institutional statement explicitly states the linkages between statements in a configuration. For example, a rule permitting some action may state, as CONDITIONS of the rule, that the individual must follow a procedure outlined in another rule. In other cases, the linkage between statements is implicit. For example, the CONDITIONS component of a voting rule for legislation may not overtly make reference to the quorum rule, but the specific quorum rule in place strongly influences the effects of the voting rule. A rule stating that a majority must approve before a bill becomes a law differentially affects behavior depending upon (1) the quorum rule that states how many members must be present and voting for a vote to be legal and (2) the rule that states what happens if no positive action is taken (e.g., the OR ELSE rule that states whether the outcome is a return to the status quo or some other alternative).

Analyzing complex institutional configurations will require new capabilities. We are currently in the process of developing a classification system that identifies types of rules or norms according to their function in a configuration. Our current efforts integrate the institutional grammar with extensive work that outlines levels of action and the elements of an action situation that have been elucidated elsewhere (Kiser and Ostrom 1982; E. Ostrom 1986; E. Ostrom, Gardner, and Walker 1994).[26] The classification system seeks to enhance efforts to place individual institutional statements in the context of the mix of institutional statements that structures a situation.

Institutional Consistency and Completeness

When an institutional statement assigns a DEONTIC operator to an action or outcome, then that action or outcome becomes obliged, permitted, or forbidden for those with the listed ATTRIBUTES under the specified CONDITIONS. Thus, institutional statements with DEONTICS can be thought of as transformations that partition sets of possible actions or outcomes into subsets of obligatory, permitted, and forbidden actions and outcomes. Recognizing this set logic based on DEONTICS provides a tie to the works in philosophy of law and logic (Braybrooke 1994; Ellickson 1991; Gibbard 1990; V. Ostrom 1997; Tyler 1990; von Wright 1951, 1963). Moreover, the rigor of the logic-based system disciplines discourse by making inconsistencies more apparent.

Set logic, combined with the organizational capacity of the syntax, identifies inconsistencies and incompleteness in rule systems. We presume that most rule systems are incomplete. The existing rules do not cover all possible combinations of ATTRIBUTES, CONDITIONS, and AIMS. The incompleteness of most existing rule systems becomes even more apparent when we recognize that the attributes and conditions of situations may be interpreted in different ways.[27] In complex situations, we presume that inconsistent rules develop; an action may be required under one rule and forbidden under another. Inconsistent systems may snare the unwary and trap others in Catch-22 situations. In political arenas, conflict plays an important role in generating information about inconsistencies. With effective conflict-resolution arenas, some inconsistencies can be tackled and resolved. Conflict also plays a role in addressing incompleteness. Much of political persuasion involves efforts by actors to convince others that new circumstances match the CONDITIONS of a rule that assigns a DEONTIC to the action that is in their best interest (McGinnis 1993).

Legitimacy and Compliance

If rulers enforce rules primarily by physical force and fiat, individuals subject to these rules are unlikely to develop internal deltas associated with breaking the rules; nor are external deltas likely to enhance the rate of rule conformance (Tyler 1990). If those who are supposed to follow a rule view it as illegitimate, they may even reward one another for actions that break the rules (a positive δ^{be}) instead of adopting the type of metanorm envisioned by Axelrod (1986) (a negative δ^{be}).

The complementarity of deltas and the OR ELSES emerges in the analysis of compliance. When delta parameters are close to zero, the costs of maintaining compliance through the OR ELSE drastically increase (Ayers and Braithwaite 1992; Levi 1988; Margolis 1991). Without a relatively high level of voluntary, contingent compliance to rules, Levi (1988) explains that rulers can rarely afford the continuing costs involved in hiring enough monitors and sanctioners, motivating them, and imposing sanctions in a manner sufficient to ensure that citizens conform to the rules rather than risk the chance of detection and punishment (see also Frey 1994). If violators can expect to reap the benefits of violating prescriptions without facing the probability of some established punishment (if there is no OR ELSE), then the experience of feeling like a "sucker" may erode the value of the delta parameters (Levi 1988, 1990; Mansbridge 1994; E. Ostrom 1990; E. Ostrom, Gardner, and Walker 1994).

The ADICO syntax illustrates the cumulative manner in which institutional statements affect individuals' expectations about the actions of others and the consequences of their own actions. We started this enterprise as an effort to define clearly the concept of rules and to create a grammar of rules.[28] We found that in order to do this we needed to clarify how rules were related to norms and shared strategies. Fitting this all into a syntax helped us to catch inconsistencies, which further tested and refined our understanding of each type of institutional statement and each component.

Delving deeper into each of the syntax components brought to light connections among these concepts that, at least for us, had not previously been linked. Further applications and developments of the grammar can strengthen our understanding of the links between institutions and theories of action. Attending to a grammar of institutions equips us to return to the core issues of institutions and political order with a respect for the complexities involved.

APPENDIX

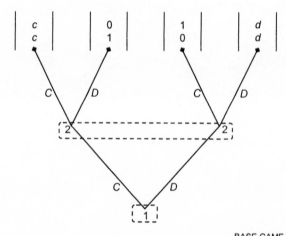

BASE GAME

	C	D
C	c,c	0,1
D	1,0	d,d

Fig. 4.A1. Base game

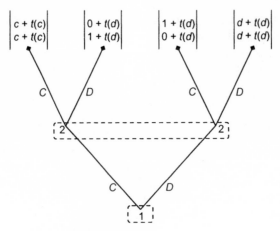

t = number of expected future rounds
$t()$ = expectation of payoffs from future rounds of the game

---

Institutional Statements

Shared Strategies

[All Players] [C] [First Round]
[All Players] [C] [All Rounds in which All Players Play C in the Previous Round]
[All Players] [D] [All Rounds after a D]

Fig. 4.A2. Repeated game with shared strategies

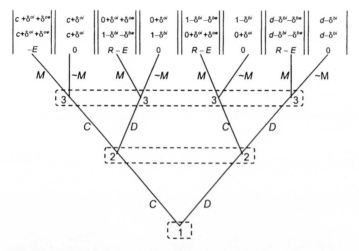

$$\left|\begin{array}{cc} c+\delta^{oi}+\delta^{oe} & c+\delta^{oi} \\ c+\delta^{oi}+\delta^{oe} & c+\delta^{oi} \\ -E & 0 \end{array}\right| \left|\begin{array}{cc} 0+\delta^{oi}+\delta^{oe} & 0+\delta^{oi} \\ 1-\delta^{bi}-\delta^{be} & 1-\delta^{bi} \\ R-E & 0 \end{array}\right| \left|\begin{array}{cc} 1-\delta^{bi}-\delta^{be} & 1-\delta^{bi} \\ 0+\delta^{oi}+\delta^{oe} & 0+\delta^{oi} \\ R-E & 0 \end{array}\right| \left|\begin{array}{cc} d-\delta^{bi}-\delta^{be} & d-\delta^{bi} \\ d-\delta^{bi}-\delta^{be} & d-\delta^{bi} \\ R-E & 0 \end{array}\right|$$

Payoffs to Monitor (Player 3)

R = reward for detecting defector

E = expense of monitoring

Deltas

δ^{oe} = external changes in payoffs from obeying prescription

δ^{oi} = internal changes in payoffs from obeying prescription

δ^{be} = external changes in payoffs from breaking prescription

δ^{bi} = internal changes in payoffs from breaking prescription

Institutional Statements

Norm

[Players 1 & 2] [Must] [C] [Always]

Fig. 4.A3. Game with a norm and monitoring

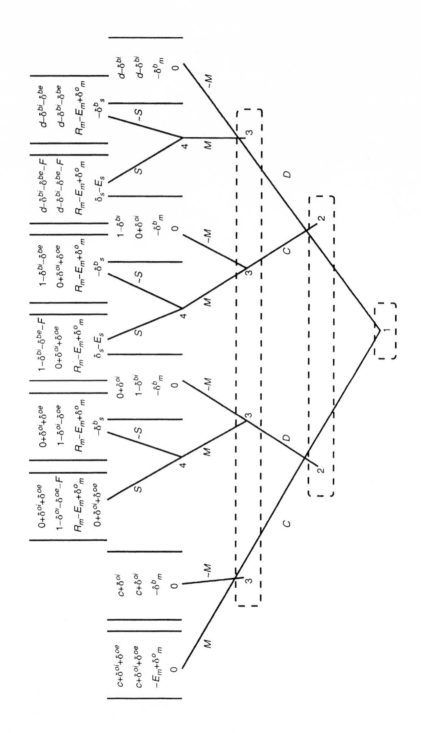

Payoffs to Monitor (Player 3)

R_m = reward for detecting defection

E_m = transaction cost of monitoring

δ^o_m = change in payoff from obeying monitoring norm

δ^b_m = change in payoff from breaking monitoring norm

Payoffs to Sanctioner (Player 4)

E_s = expense of sanctioning

δ^o_s = change in payoff from obeying sanctioning norm

δ^b_s = change in payoff from breaking sanctioning norm

Deltas for Player 1 or 2

δ^{oe} = external changes in payoffs from obeying prescription

δ^{oi} = internal changes in payoffs from obeying prescription

δ^{be} = external changes in payoffs from breaking prescription

δ^{bi} = internal changes in payoffs from breaking prescription

OR ELSE for Player 1 or 2

f = fine for Defection

Institutional Statements

Rule: [Players 1 & 2] [Must] [C] [All Rounds] [OR ELSE f]

Monitoring Norm: [Player 3] [Must] [Monitor] [Always]

Sanctioning Norm: [Player 4] [Must] [Impose f on defecting player] [When Monitor Reports D]

Fig. 4.A4. Game with a rule

NOTES

1. Our concept of institutions parallels Giddens's (1979, 1984) concept of system. Another similarity to Giddens's work is our stress on the constitutive character of rules (see also V. Ostrom 1980). Our concept of situations draws on Popper (1967).

2. Calvert indicates: "There is, strictly speaking, no separate animal that we can identify as an institution. . . . 'Institution' is just a name we give to certain parts of certain kinds of equilibria" (1992, 18).

3. These are "forms of *talk about* experience" in the sense that Burke views the study of human relations in terms of linguistic instruments (1969, 317).

4. Chomsky defines the grammar of a language as the theory of that language (1957, 49). Burke explains his use of the term *grammar* as a theory to explain "what people are doing and why they are doing it" and uses a syntax composed of five terms that he posits to be the generating principles of his investigation (1969, xv).

5. This also applies to grammars in linguistic studies as Chomsky indicates (1957, 15; 1965, 11).

6. A formal syntax can be compared with that of other scholars. We have just learned about the Dalhousie logic of rules, which uses a syntax without an OR ELSE and requires that prescriptive statements be recast to use the forbidden DEONTIC operator (see Braybrooke 1994). Our concept of ATTRIBUTES is the same as the *wenn* component of the Dalhousie syntax.

7. The largest group to which a prescriptive statement could apply is the *folk* component in the Dalhousie system (Braybrooke 1994).

8. The logical relationships among the deontic operators include the following: (1) $D = P \cup O \cup F$; (2) $F \cap P = \emptyset$, $F \cap O = \emptyset$, and $O \cap P = O$; and (3) O implies P.

9. John R. Commons (1968) stresses the correlative nature of rights. To state that someone has a right, someone must have a duty to observe that right. The person with a right is permitted to do something, while those with a duty are forbidden or required to do something.

10. Tsebelis (1989, 1991) argues that in a game with only mixed strategy equilibria, increasing the size of the OR ELSE does not reduce the level of rule infraction but rather reduces the level of monitoring. Weissing and E. Ostrom (1991b) have shown that Tsebelis's results hold in many but not all cases.

11. The complexity of modern economies is so great that centralized law creation cannot effectively cope with the need to achieve normative regulation among communities of individuals who repeatedly face collective-action problems (Cooter 1993; Ellickson 1991).

12. Universal statements are relatively rare explanations for human behavior. The shared strategy that all rational players should adopt a dominant strategy and the norm that all actors are always obliged to tell the truth exemplify unqualified institutional statements.

13. The logic of the delta parameter is similar to Etzioni's (1988) discussion of deontology.

14. Societies undergoing substantial "liberalization" could be thought of as developing shared understandings that individuals who had been forbidden to take certain

actions in earlier times are now permitted to do so. When the new norm is shared, individuals who still attempt to obstruct the previously restricted actions now face a cost for breaking the new norm. Thus, changes in norms over time will be reflected both in the particular DEONTIC assigned to an action and in to whom the cost of breaking the norm is assigned.

15. Kreps et al. do not assume that the basis for one actor playing Tit-for-Tat is necessarily the acceptance of a norm. They simply assume that some players only have available to them a Tit-for-Tat strategy or that there is some probability that one player's payoffs are such that Tit-for-Tat is strongly dominant (1982, 247). The latter condition would be the case if some players in a population have some combination of delta parameters associated with playing Tit-for-Tat whose values are high.

16. See the appendix for the extensive form of each game.

17. The punishment in the grim trigger might cause one to wonder whether the retaliation is an OR ELSE. The shared strategy is not a rule, using the ADICO syntax, unless the prescription to punish someone who deviates is backed by another rule or norm. An institutional statement that could back a trigger rule might be a sanctioning rule such as, "All other players must defect for the rest of the game (or, in not-so-grim triggers, for n rounds) when one player defects in any of the rounds, OR ELSE the other players face the probability of a further sanction," and a monitoring norm that "all players must monitor all other players." Notice that the sanctioning rule changes the DEONTIC assigned to C from *permitted* or *obligatory* to *forbidden* in the CONDITION of a defection in the prior round. If the advice to cooperate to avoid a trigger response is not backed by monitoring and sanctioning institutions, the massive defection that is threatened by the trigger is either a prudent response to defection or a norm of retaliation.

18. It would, of course, be possible to include discount rates in the analysis, but we assume they are zero here to keep the focus on other questions and not those related to the size of the discount rate. For a discussion of the importance of discount rates in the analysis of cooperation, see Axelrod (1981, 1986).

19. It is possible to illustrate the addition of norms without a player who is assigned the specialized role of a monitor by simply assuming that players monitor each other (Weissing and Ostrom 1991a, 1993). To do this, however, one needs to model a sequential structure that introduces more complexity than we desire in this initial application.

20. Freelance reporters are an example of this type of monitor. They receive payment for detecting and reporting nonconformance with accepted norms. Rewards include fees for stories and increased probabilities of receiving prizes for good reporting.

21. A mixed strategy is a probability distribution over the pure strategies for a player. In a static game, one may view the mixed strategy as the probability of choosing one or the other pure strategies. One can also interpret mixed strategies as behavioral tendencies in a repeated context where the probability of choosing a pure strategy, say C, is viewed as a cooperation rate.

22. The delta parameters for players 3 or 4 could be disaggregated into their internal and external components for an analysis that wished to focus on questions that distinguished between internal and external pressures on the monitor or sanctioner.

23. In this game, player 3 always correctly detects whether defection has occurred, and player 4 only has the option of sanctioning players who have defected. If players 1 and 2 cooperate, player 4 does not have a choice of whether or not to sanction. This eliminates issues of false detection and corrupt sanctioners from the current analysis. Future efforts will address questions related to monitoring and sanctioning errors.

24. One could argue that when prescriptions are rules, individuals will place higher values on the deltas than when the prescriptions are norms (see Braybrooke 1987). If one assumes that the presence of rules influences the internal deltas, that δ^{oi} are higher in the rules game than in the norms game, then the rule would influence the structure of the game even when the monitor fails.

25. See E. Ostrom (1986) for a discussion of the configurational aspect of rules.

26. The element of the action situation that best fits the contents of the AIM of the prescription is the first part of the classification; the match between the elements and the CONDITIONS provides the second part of the classification. Further refinements in the classification focus either on the other components in the institutional statements or on creating subdivisions within the basic categories provided by the elements of an action situation.

27. For a philosophical discussion of uncertainty and incompleteness in the applications of rules, see Heritage (1984).

28. For earlier efforts to address the concept of rules, see Black (1962), Collett (1977), Ganz (1971), and Harré (1974).

REFERENCES

Andreoni, James. 1989. "Giving with Impure Altruism: Applications to Charity and Ricardian Equivalence." *Journal of Political Economy* 97:447–58.

Axelrod, Robert. 1981. "The Emergence of Cooperation among Egoists." *American Political Science Review* 75:306–18.

Axelrod, Robert. 1986. "An Evolutionary Approach to Norms." *American Political Science Review* 80:1095–111.

Ayers, Ian, and John Braithwaite. 1992. *Responsive Regulation: Transcending the Regulation Debate.* Oxford: Oxford University Press.

Bendor, Jonathan, and Dilip Mookherjee. 1990. "Norms, Third-Party Sanctions, and Cooperation." *Journal of Law, Economics, and Organization* 6:33–63.

Black, Max. 1962. *Models and Metaphors.* Ithaca: Cornell University Press.

Bourdieu, Pierre. 1977. *Outline of a Theory of Practice.* New York, Cambridge University Press.

Braybrooke, David. 1987. *Philosophy of Social Science.* Englewood Cliffs, NJ: Prentice-Hall.

Braybrooke, David. 1994. "The Logic of Political Discussion: An Introduction to Issue-Processing." Presented at the annual meeting of the Midwest Political Science Association, Chicago.

Burke, Kenneth. 1969. A *Grammar of Motives.* Berkeley: University of California Press.

Calvert, Randall. 1992. "Rational Actors, Equilibrium, and Social Institutions." University of Rochester. Typescript.

Chomsky, Noam. 1957. *Syntactic Structures.* The Hague: Mouton.

Chomsky, Noam. 1965. *Aspects of the Theory of Syntax.* Cambridge: MIT Press.

Coleman, James S. 1987. "Norms as Social Capital." In *Economic Imperialism,* ed. Gerard Radnitzky and Peter Bernholz. New York: Paragon.

Coleman, James S. 1988. "Free Riders and Zealots: The Role of Social Networks." *Sociological Theory* 6:52–57.

Collett, Peter. 1977. "The Rules of Conduct." In *Social Rules and Social Behaviour,* ed. Peter Collett. Totowa, NJ: Rowman & Littlefield.

Commons, John R. 1968. *Legal Foundations of Capitalism.* Madison: University of Wisconsin Press.

Cooter, Robert D. 1993. "Structural Adjudication and the New Law Merchant: A Model of Decentralized Law." University of California, Berkeley. Typescript.

DiMaggio, Paul J., and Walter W. Powell. 1991. Introduction to *The New Institutionalism in Organizational Analysis,* ed. Walter W. Powell and Paul J. DiMaggio. Chicago: University of Chicago Press.

Ellickson, Robert C. 1991. *Order without Law: How Neighbors Settle Disputes.* Cambridge: Harvard University Press.

Elster, Jon. 1989a. *The Cement of Society: A Study of Social Order.* Cambridge: Cambridge University Press.

Elster, Jon. 1989b. *Solomonic Judgements: Studies in the Limitations of Rationality.* Cambridge: Cambridge University Press.

Etzioni, Amitai. 1988. *The Moral Dimension: Towards a New Economics.* New York. Free Press.

Frank, Robert H. 1988. *Passions within Reason.* New York. Norton.

Frey, Bruno S. 1994. "How Intrinsic Motivation Is Crowded out and In." *Rationality and Society* 6:334–52.

Ganz, Joan Safran. 1971. *Rules: A Systematic Study.* The Hague: Mouton.

Gibbard, Allan. 1990. *Wise Choices, Apt Feelings: A Theory of Normative Judgment.* Cambridge: Harvard University Press.

Giddens, Anthony. 1979. *Central Problems in Social Theory.* Berkeley: University of California Press.

Giddens, Anthony. 1984. *The Constitution of Society: Outline of the Theory of Structuration.* Berkeley: University of California Press.

Greif, Avner, Paul Milgrom, and Barry Weingast. 1990. "The Merchant Guild as a Nexus of Contracts." Working Papers in Economics E–90–23. Stanford, CA: Stanford University, Hoover Institution.

Harré, R. 1974. "Some Remarks on 'Rule' as a Scientific Concept." In *Understanding Other Persons,* ed. Theodore Mischel. Oxford: Basil Blackwell.

Harsanyi, John C. 1967–68. "Games with Incomplete Information Played by 'Bayesian' Players." *Management Science* 14:159–82, 320–34, 486–502.

Hayek, Friedrich A. von. 1945. "The Use of Knowledge in Society." *American Economic Review* 35:519–30.

Hayek, Friedrich A. von. 1967. "Notes on the Evolution of Rules of Conduct." In

Studies in Philosophy, Politics, and Economics, ed. Friedrich A. von Hayek. Chicago: University of Chicago Press.

Heritage, John. 1984. *Garfinkel and Ethnomethodology.* Cambridge: Polity.

Hilpinen, Risto, ed. 1971. *Deontic Logic: Introductory and Systematic Readings.* Dordrecht: Reidel.

Hilpinen, Risto. 1981. *New Studies in Deontic Logic: Norms, Actions, and the Foundations of Ethics.* Dordrecht: Reidel.

Hirschman, Albert O. 1985. "Against Parsimony: Three Easy Ways of Complicating Some Categories of Economic Discourse." *Economics and Philosophy* 1:16–19.

Hohfeld, Wesley N. 1913. "Some Fundamental Legal Concepts as Applied in the Study of Primitive Law." *Yale Law Journal* 23:16–59.

Hurwicz, Leonid. 1994. "Institutional Change and the Theory of Mechanism Design." *Academia Economic Papers* 22:1–26.

Kerr, Norbert L. 1996. "Does My Contribution Really Matter? Efficacy in Social Dilemmas." In *European Review of Social Psychology,* vol. 7, ed. W. Stroebe and M. Hewstone. Chichester: Wiley.

Kerr, Norbert L., Jennifer Garst, Donna A. Gerard, and Susan E. Harris. 1997. "That Still, Small Voice: Commitment to Cooperate as an Internalized versus a Social Norm." *Personality and Social Psychology Bulletin* 23:1300–11.

Kiser, Larry L., and Elinor Ostrom. 1982. "The Three Worlds of Action: A Metatheoretical Synthesis of Institutional Approaches." In *Strategies of Political Inquiry,* ed. Elinor Ostrom. Beverly Hills: Sage. (Chapter 2 in this volume.)

Knack, Stephen. 1992. "Civic Norms, Social Sanctions, and Voter Turnout." *Rationality and Society* 4:133–56.

Knight, Jack. 1992. *Institutions and Social Conflict.* New York: Cambridge University Press.

Kreps, David M. 1990. "Corporate Culture and Economic Theory." In *Perspectives on Positive Political Economy,* ed. James E. Alt and Kenneth A. Shepsle. New York: Cambridge University Press.

Kreps, David M., Paul Milgrom, John Roberts, and Robert Wilson. 1982. "Rational Cooperation in the Finitely Repeated Prisoners' Dilemma." *Journal of Economic Theory* 27:245–52.

Ledyard, John. 1995. "Is There a Problem with Public Goods Provision?" In *The Handbook of Experimental Economics,* ed. John Kagel and Alvin Roth. Princeton: Princeton University Press.

Levi, Margaret. 1988. *Of Rule and Revenue.* Berkeley: University of California Press.

Levi, Margaret. 1990. "A Logic of Institutional Change." In *The Limits of Rationality,* ed. Karen Schweers Cook and Margaret Levi. Chicago: University of Chicago Press.

Lewis, David K. 1969. *Convention: A Philosophical Study.* Cambridge: Harvard University Press.

McGinnis, Michael D. 1993. "Heterogeneity of Interpretation: Rule Repertoires, Strategies, and Regime Maintenance." Presented at the conference on Heterogeneity and Collective Action, Indiana University, Bloomington.

Mansbridge, Jane. 1990. "Expanding the Range of Formal Modeling." In *Beyond Self-Interest,* ed. Jane Mansbridge. Chicago: University of Chicago Press.

Mansbridge, Jane. 1994. "Public Spirit in Political Systems." In *Values and Public Policy,* ed. Henry J. Aaron, Thomas E. Mann, and Timothy Taylor. Washington: Brookings Institution.

March, James G., and Johan P. Olsen. 1989. *Rediscovering Institutions: The Organizational Basis of Politics.* New York: Free Press.

Margolis, Howard. 1991. "Self-Interest and Social Motivation." In *Individuality and Cooperative Action,* ed. Joseph E. Earley. Washington: Georgetown University Press.

Menger, Karl. 1963. *Problems in Economics and Sociology.* Trans. Francis J. Nock. Urbana: University of Illinois Press.

Meyer, John W., and Brian Rowan. 1991. "Institutionalized Organizations: Formal Structure as Myth and Ceremony." In *The New Institutionalism in Organizational Analysis,* ed. Walter W. Powell and Paul J. Dimaggio. Chicago: University of Chicago Press.

Milgrom, Paul R., Douglass C. North, and Barry R. Weingast. 1990. "The Role of Institutions in the Revival of Trade: The Law Merchant, Private Judges, and the Champagne Fairs." *Economics and Politics* 2:1–23.

Myerson, Roger B. 1991. *Game Theory: Analysis of Conflict.* Cambridge: Harvard University Press.

North, Douglass C. 1981. *Structure and Change in Economic History.* New York: Norton.

North, Douglass. 1986. "The New Institutional Economics." *Journal of Institutional and Theoretical Economics* 142:230–37.

North, Douglass. 1990. *Institutions, Institutional Change, and Economic Performance.* New York: Cambridge University Press.

Oakerson, Ronald J., and Roger B. Parks. 1988. "Citizen Voice and Public Entrepreneurship: The Organizational Dynamic of a Complex Metropolitan County." *Publius* 18:91–112.

Offe, Claus, and Helmut Wiesenthal. 1980. "Two Logics of Collective Action: Theoretical Notes on Social Class and Organizational Form." *Political Power and Social Theory* 1:67–115.

Olson, Mancur. 1991. "The Role of Morals and Incentives in Society." In *Individuality and Cooperative Action,* ed. Joseph E. Earley. Washington: Georgetown University Press.

Orbell, John M., Alphons J. van de Kragt, and Robyn M. Dawes. 1991. "Covenants without the Sword: The Role of Promises in Social Dilemma Circumstances." In *Social Norms and Economic Institutions,* ed. Kenneth J. Koford and Jeffrey B. Miller. Ann Arbor: University of Michigan Press.

Ostrom, Elinor. 1986. "An Agenda for the Study of Institutions." *Public Choice* 48:3–25. (Chapter 3 in this volume.)

Ostrom, Elinor. 1990. *Governing the Commons: The Evolution of Institutions for Collective Action.* New York: Cambridge University Press.

Ostrom, Elinor, Roy Gardner, and James Walker. 1994. *Rules, Games, and Common-*

Pool Resources. Ann Arbor: University of Michigan Press.

Ostrom, Vincent. 1980. "Artisanship and Artifact." *Public Administration Review* 40:309–17. (Reprinted in Michael D. McGinnis, ed., *Polycentric Governance and Development* [Ann Arbor: University of Michigan Press, 1999].)

Ostrom, Vincent. 1986. "A Fallabilist's Approach to Norms and Criteria of Choice." In *Guidance, Control, and Evaluation in the Public Sector,* ed. Franz X. Kaufmann, Giandomenico Majone, and Vincent Ostrom. Berlin: de Gruyter.

Ostrom, Vincent. 1987. *The Political Theory of a Compound Republic: Designing the American Experiment.* 2d rev. ed. San Francisco: Institute for Contemporary Studies Press.

Ostrom, Vincent. 1991. *The Meaning of American Federalism: Constituting a Self-Governing Society.* San Francisco: Institute for Contemporary Studies Press.

Ostrom, Vincent. 1997. *The Meaning of Democracy and the Vulnerability of Democracies.* Ann Arbor: University of Michigan Press.

Plott, Charles R. 1986. "Rational Choice in Experimental Markets." *Journal of Business* 59:301–27.

Popper, Karl R. 1967. "Rationality and the Status of the Rationality Principle." In *Les fondements philosophiques des systems economiques,* ed E. M. Classen. Paris: Payot.

Riker, William. 1980. "Implications from the Disequilibrium of Majority Rule for the Study of Institutions." *American Political Science Review* 74:432–46.

Rowe, Nicholas. 1989. *Rules and Institutions.* Ann Arbor: University of Michigan Press.

Schank, Roger C., and Robert P. Abelson. 1977. *Scripts, Plans, Goals, and Understanding: An Inquiry into Human Knowledge Structures.* Yale, NJ: Erlbaum.

Schauer, Frederick. 1991. *Playing by the Rules.* Oxford: Clarendon.

Schelling, Thomas C. 1978. *Micromotives and Macrobehavior.* New York: Norton.

Schotter, Andrew. 1981. *The Economic Theory of Social Institutions.* Cambridge: Cambridge University Press.

Searle, John R. 1969. *Speech Acts: An Essay in the Philosophy of Language.* London: Cambridge University Press.

Shepsle, Kenneth A. 1975. "Congressional Committee Assignments: An Optimization Model with Institutional Constraints." *Public Choice* 22:55–78.

Shepsle, Kenneth A. 1979. "Institutional Arrangements and Equilibrium in Multidimensional Voting Models." *American Journal of Political Science* 23:27–59.

Shepsle, Kenneth A. 1989. "Studying Institutions: Some Lessons from the Rational Choice Approach." *Journal of Theoretical Politics* 1:131–47.

Shepsle, Kenneth A., and Barry R. Weingast. 1984. "When Do Rules of Procedure Matter?" *Journal of Politics* 46:206–21.

Shepsle, Kenneth A., and Barry R. Weingast. 1987. "The Institutional Foundations of Committee Power." *American Political Science Review* 81:85–104.

Shimanoff, Susan B. 1980. *Communication Rules: Theory and Research.* Beverly Hills: Sage.

Sugden, Robert. 1986. *The Economics of Rights, Co-operation, and Welfare.* Oxford: Basil Blackwell.

Taylor, Michael. 1987. *The Possibility of Cooperation.* New York: Cambridge University Press.

Tsebelis, George. 1989. "The Abuse of Probability in Political Analysis: The Robinson Crusoe Fallacy." *American Political Science Review* 83:77–91.

Tsebelis, George. 1991. "The Effect of Fines on Regulated Industries: Game Theory Versus Decision Theory." *Journal of Theoretical Politics* 3:81–101.

Tyler, Tom R. 1990. *Why People Obey the Law.* New Haven: Yale University Press.

Udéhn, Lars. 1993. "Twenty-Five Years with *The Logic of Collective Action.*" *Acta Sociologica* 36:239–61.

Ullmann-Margalit, Edna. 1977. *The Emergence of Norms.* Oxford: Clarendon.

Weber, Max. 1947. *The Theory of Social and Economic Organization.* Trans. A. M. Henderson and Talcott Parsons. New York: Free Press.

Weiss, Richard M. 1991. "Alternative Social Science Perspectives on 'Social Dilemma' Laboratory Research." In *Social Norms and Economic Institutions,* ed. Kenneth J. Koford and Jeffrey B. Miller. Ann Arbor: University of Michigan Press.

Weissing, Franz J., and Elinor Ostrom. 1991a. "Irrigation Institutions and the Games Irrigators Play: Rule Enforcement without Guards." In *Game Equilibrium Models II: Methods, Morals, and Markets,* ed. Reinhard Selten. Berlin: Springer-Verlag.

Weissing, Franz J., and Elinor Ostrom. 1991b. "Crime and Punishment." *Journal of Theoretical Politics* 3:343–49.

Weissing, Franz J., and Elinor Ostrom. 1993. "Irrigation Institutions and the Games Irrigators Play: Rule Enforcement on Government- and Farmer-managed Systems." In *Games in Hierarchies and Networks,* ed. Fritz W. Scharpf. Boulder: Westview. (Chapter 13 in this volume.)

Williamson, Oliver E. 1985. *The Economic Institutions of Capitalism: Firms, Markets, Relational Contracting.* New York: Free Press.

von Wright, Georg H. 1951. "Deontic Logic." *Mind* 60:58–74.

von Wright, Georg H. 1963. *Norm and Action: A Logical Enquiry.* London: Routledge & Kegan Paul.

von Wright, Georg H. 1968. "The Logic of Practical Discourse." In *Contemporary Philosophy,* ed. Raymond Klibansky. Florence, Italy: La Nuova Italia Editrice.

Part II
Voting, Conflict, and Leadership

Introduction to Part II

The essays in part II show how the general frameworks presented in part I can be used to address specific questions of voting, conflict, and leadership. Despite their many differences in substance and method, all of these essays deal with important aspects of the problems faced by any community seeking to agree upon some collective decision.

Voting and Constitutional Order

When North Americans of the revolutionary era engaged in an explicit exercise in constitutional choice, they designed a multifaceted system of checks and balances. Vincent Ostrom (1987) details the basic conceptual framework upon which the founders built this constitution as an act of creative artisanship (V. Ostrom 1980). In "Votes and Vetoes" (chap. 5), Roberta Herzberg and Vincent Ostrom demonstrate the natural connection between voting as a positive act and the many veto points included in the constitutional order of checks and balances. They argue that the ability of diverse minority interests to block the implementation of policies that they strongly oppose has played an important role in sustaining the American experiment.

Voting is typically taken to be the defining characteristic of modern democracy. Several voting institutions are included in the U.S. Constitution, but the essential foundation for the subsequent success of the American experiment in democracy lies in the polycentric nature of the U.S. constitutional order (V. Ostrom 1991, 1997). Voting institutions, per se, receive considerably less attention in this book than would be expected by most political scientists. Voting is one important way in which individual preferences can be said to be aggregated into public policy, but communities need not rely on formal mechanisms of voting. Normative theorists have long emphasized that an attitude of respect toward the perceptions and interests of other citizens is a fundamental requisite of a sound democracy, and discussions in informal settings are often sufficient for small groups to arrive at a consensus. In short, democratic governance requires more than voting procedures and representative institutions.

One of the most important lessons of the Workshop for political science in general has been to highlight the importance of forms of political interaction that do not involve voting. The point is not that voting is inconsequential but rather that it may have received proportionally too much attention by political scientists (see McGinnis 1996). Reliance on majority rule has a tendency to direct political agents to conceptualize politics in terms of a contest in which one either wins or loses, rather than approaching politics from a problem-solving orientation. In so doing, political scientists run the risk of losing sight of the ways in which real communities of people manage to work out their own problems. It is this latter form of politics that lies at the core of the Workshop approach to policy analysis.

An overemphasis on majority voting can contribute to the reinforcement of a Hobbesian conceptualization of sovereignty that is fundamentally at odds with sovereignty as conceptualized by the founders of the American republic. As Vincent Ostrom (1987, 1991, 1997) emphasizes, the capacity of self-governing communities of individuals to resolve their own problems is the key ingredient in the perpetuation of a democratic system of governance. Still, voting remains one way in which groups can make collective decisions, and as such voting institutions are fully deserving of careful analysis. And, indeed, voting has played a central role in the research programs of several scholars associated with the Workshop. Full consideration of this topic would take us too far from the central theme of this volume, but to ignore it entirely would result in an incomplete portrayal of Workshop research programs.

Social Choice and Laboratory Experiments

The central result of social choice theory identifies the problems of logical coherence that plague any effort to use majority voting to resolve questions of public policy. Arrow's (1963) theorem tells us that voting cycles can occur in any system of aggregating preferences that satisfies the full set of fairness conditions laid out in that work. Another central result of social choice theory concerns the instability of majority voting processes (see Riker 1982). This instability can result from the general ability of participants to find some policy proposal that will defeat the status quo.

Since very little in the real world of legislative behavior corresponds to this impression of endless chaos, many researchers have sought to uncover the ways in which chaotic outcomes are avoided. Perhaps the most well-known effort is the "structure-induced equilibrium" of Shepsle (1979), in which specific legislative rules are shown to limit the ability of agenda setters to raise certain issues. Those scholars who study voting institutions from the Workshop perspective have focused on more subtle forms of institutional arrangements that might have many of the same effects as the details of legislative procedures.

In "Negative Decision Powers and Institutional Equilibrium: Experiments on Blocking Coalitions" (chap. 6), Rick K. Wilson and Roberta Herzberg represent negative voting power in terms of a spatial voting model. They report on laboratory experiments that demonstrate that giving one individual blocking power tends to help that individual get a more desirable outcome from the voting game. They did not detect any evidence that blocking power was able to uniquely define any particular point in the policy space as the "natural" outcome of the game, except to reinforce a tendency to remain at the status quo point. For if the individual with blocking power prefers the status quo to any offered alternative, then the status quo is going to be preserved.

In another work, Herzberg (1992) uses a spatial voting model to explore the logical consistency of an argument made by John Calhoun concerning the effects of concurrent majorities. That essay stands as a good example of the ways in which formal game representations can help clarify the strengths and weaknesses of arguments typically made in a more informal manner. Jillson and Wilson (1994) apply modern spatial voting models to the decisions made by the first U.S. Congress, during the Articles of Confederation era, thus demonstrating that these technical advances can also help improve our understanding of our past historical eras.

In a series of related works, Herzberg and Wilson used laboratory experiments to explore several other aspects of voting institutions. Herzberg and Wilson (1991) demonstrate that the potential for agenda manipulation can be limited if legislators incur costs by adding issues to the agenda. By treating the gathering of certain kinds of information as costly, a degree of stability can be imposed on collective outcomes. It is often argued that the capacity of individuals to misrepresent their preferences via strategic or sophisticated voting can serve to counteract the manipulations of agenda setters (see Riker 1982). The experimental results of Herzberg and Wilson (1988) show how difficult it is for experimental subjects to achieve their desired results via sophisticated voting. Of course, this result does not necessarily imply the nonexistence of sophisticated voting in other contexts, when participants have at their disposal a wider range of ways to gather information or to coordinate their behavior on matters of great concern to them at the time. Still, these results do point out the difficulty of sophisticated behavior in a relatively sparse institutional context. In a related vein, Haney, Herzberg, and Wilson (1992) explore some of the difficulties associated with the efforts of leaders to garner useful advice from a small group of advisers. Again, however, the specific nature of the laboratory setting may not have captured the rich institutional context of foreign policymaking in the real world. Nonetheless, laboratory experiments have proven to be a very useful means of investigating the implications of alternative institutional arrangements.

Conflict and Governance

The dilemmas of collective action in the international arena take center stage in the next selection. In "Policy Uncertainty in Two-Level Games: Examples of Correlated Equilibria" (chap. 7), John T. Williams and I examine situations in which a diverse array of information is readily available for policymakers to consider. We argue for a broader interpretation of relationships between national and international politics by justifying a formulation of "two-level games" that differs in substantial ways from Putnam's (1988) more influential formulation. Whereas Putnam specifies a particular sequence of events in which the results of international negotiations are presented to domestic legislatures for approval or ratification, we argue that policy advocates in all states involved in international interactions are going to have access to extensive amounts of information that should help them predict the likely behavior of other governments. As a consequence, we argue that Aumann's (1987) correlated equilibrium may be a more appropriate tool for modeling the outcomes of these two-level interactions. This essay differs from much of the game theoretic tradition by not specifying the steps by which individuals obtain information, but it does present an interesting alternative to more standard representations based on games of incomplete information (or signaling games).

Aumann's notion is disarmingly simple. Even though game theorists may be uncertain which of the many possible equilibria will be selected by the players, the participants have access to a wide array of informational cues that they can use to help them coordinate on a particular solution. This is not to say that coordination problems are trivial; far from it. As we show in this chapter, a failure of coordination may occur with some positive probability even for the case of correlated equilibrium.

We do not specify the sources of informational cues in our general analysis of the relevance of correlated equilibrium to two-level games, precisely because the nature of the relevant cues will differ in different contexts. However, we could have placed more emphasis on the importance of institutions in providing cues to the players as they seek to coordinate their behavior, either explicitly or implicitly. For example, CIA estimates of Soviet military spending played a central (and controversial) role in helping sustain the behavior patterns underlying rivalry between the two superpowers during the Cold War (see McGinnis and Williams 2000).

The concept of correlated equilibrium is equally relevant at both of the levels of analysis included in a domestic-international game. Competing policy advocates within any one country have a common interest in arriving at a collectively sound policy but conflicting interests over the content of that policy. A similar mixture of common and conflicting interests recurs at the level of international interactions, for the leaders of rival powers may share a common interest in avoiding war or excessively high arms expenditures. Elsewhere (McGinnis and Williams 2000), we demonstrate that under conditions of international rivalry we can expect the efficient use of information in policy debates to resolve domestic coordination problems while simultaneously exacerbating problems of international cooperation. From the perspective of polycentric games, it is important to understand multilevel dilemmas of collective action and other cross-level linkages.

Under our interpretation of Aumann's correlated equilibrium solution concept, participants are conditioning their behavior on the common signals generated from the vast amount of information that becomes available in debates over contentious public issues. No one individual has access to all relevant information, and reasonable people are likely to interpret this same information as support of contradictory policy initiatives. Still, at the aggregate level, there may be circumstances under which information use may be so efficient as to approximate the behavior expected of a unitary rational actor. Game models in the international relations literature often make this assumption, and our research was one effort to delineate the circumstances under which that assumption is most warranted.

On the other hand, if very little information is generated, then we are inclined to be much more skeptical of the appropriateness of treating collective entities as fully rational actors. There are also reasons to wonder whether individual humans are fully capable of the prodigious feats of rational calculation routinely attributed to them by many game theorists. Under conditions of competitive markets it may prove reasonable to assume that only the most efficient actors survive, but in many other decisional contexts there remains plenty of room for individuals to indulge other predilections. (See the essays in part V of this volume.)

The potential instability of processes of majority voting (Arrow 1963; Riker 1982) is a particularly pernicious source of coordination dilemmas in democratic polities. For this reason, we focused on voting examples in our discussion of the implications of correlated equilibria in two-level games.

Voting Over Resource Management Issues

A natural next step at this point would be to draw out explicit linkages between voting institutions and the management of common-pool resources (CPRs). In chapter 1, Vincent and Elinor Ostrom discuss Buchanan and

Tullock's (1962) classic analysis of constitutional choice, and in V. Ostrom and E. Ostrom (1977) they apply this model to the management of common-pool resources. Their basic question is this: If a community is confronted with the problem of managing a CPR, how might the community go about setting up a means of voting on these collective problems? Buchanan and Tullock point out the costs associated with finding out the preferences of all individuals (as is necessary under a rule of unanimity) and the dangers of having a collective decision imposed against one's will (as happens to minorities under a system of majority rule). It turned out that voting per se played a more minor role in the CPR management than was originally expected. Instead, a wide array of cases studied by field researchers demonstrated that communities used other, less formal mechanisms to arrive at collective decisions.

After investigating many cases from the field and from laboratory experiments, it became apparent that majority voting may have even more negative connotations than was originally suggested in this early work. In Walker, Gardner, Herr, and E. Ostrom (2000), the authors report on an experimental analysis of the ways in which majority voting schemes can be used to resolve the distributional conflicts associated with the management of common-pool resources. The results of these experiments are troubling, for those subjects who were in the minority felt they were dealt with unfairly. Basically, members of a winning coalition divided all of the positive gains among themselves, leaving nothing for the other players. Subject comments make it clear that the ways in which majority voting was used in this experimental setting violated deeply felt notions of fairness and justice on the part of many of the participants. These experiments demonstrate that simple majority rule voting is not the best way to deal with some problems.

Leadership and Principal-Agent Relations

Disputes concerning which rules are appropriate for application to specific decisions are an important component of interactions in all arenas of collective choice. In the specific context of groups dealing with the management of common-pool resources, discussions often center on which of the relevant allocation schemes is most appropriate or fair in any given situation. In more fully articulated constitutional orders, alternative allocation schemes are enshrined into law (see chap. 10). Concern for interactions between law and alternative institutions for resource management harkens back to research completed by Workshop scholars long before the Workshop was even established (e.g., V. Ostrom 1953, 1976; V. Ostrom and E. Ostrom 1972).

In the developing world, legal arrangements are less well established, but many of the same concerns are relevant. In "Shepherds and Their Leaders among the *Raikas* of India: A Principal-Agent Perspective" (chap. 8), Arun Agrawal shows how a self-governed group of pastoralists structures the process of choosing and monitoring an individual to act on the group's behalf. This agent is authorized to select times and locations for temporary settlement and migration patterns and other matters for which some means of coordination is required. Agrawal uses a normal-form game matrix to examine the circumstances under which the group as a whole will be able to monitor the activities of this agent. He concludes that situations that facilitate direct monitoring by the group as a whole are most conducive to long-term success of the group, both in economic terms and in terms of fostering the sense of community that such groups need.

It may be useful to conclude the introduction to this part by acknowledging the diversity of the essays included here. The substantive topics of these essays range from majority voting (primarily in the United States) to leadership of pastoral communities (in India) to the multilevel nature of international relations (throughout the world). The methods include formal models illustrated with abstract examples or tested against laboratory experiments and field research, as well as philosophical investigations. As a set, these essays represent a microcosm of the Workshop approach to institutional analysis. Despite this diversity in methods and substantive topics, certain common concerns permeate all of these works. How do groups make collective decisions? What information is relevant to making these decisions? How are the interests of potentially aggrieved groups taken into account when evaluating policy changes? And, especially, how can communities restrain the opportunistic behavior of their leaders and agents? None of these questions is answered fully in any of these readings, but each provides a piece of the overall picture.

REFERENCES

Arrow, Kenneth J. 1963. *Social Choice and Individual Values.* 2d ed. New Haven: Yale University Press.
Aumann, Robert J. 1987. "Correlated Equilibrium as an Expression of Bayesian Rationality." *Econometrica* 55:1–18.
Buchanan, James M., and Gordon Tullock. 1962. *The Calculus of Consent: Logical Foundations of Constitutional Democracy.* Ann Arbor: University of Michigan Press.
Haney, Patrick J., Roberta Herzberg, and Rick K. Wilson. 1992. "Advice and Consent: Unitary Actors, Advisory Models, and Experimental Tests." *Journal of Conflict Resolution* 36 (4): 603–33.

———. 1992. "An Analytic Choice Approach to Concurrent Majorities: The Relevance of John C. Calhoun's Theory for Institutional Design." *Journal of Politics* 54 (1): 54–81.

———, and Rick K. Wilson. 1988. "Results of Sophisticated Voting in an Experimental Setting." *Journal of Politics* 50 (2): 471–86.

———, and Rick K. Wilson. 1991. "Costly Agendas and Spatial Voting Games: Theory and Experiments on Agenda Access Costs." In *Laboratory Research in Political Economy*, ed. Thomas R. Palfrey, 169–99. Ann Arbor: University of Michigan Press.

Jillson, Calvin, and Rick K. Wilson. 1994. *Congressional Dynamics: Structure, Coordination, and Choice in the First American Congress, 1774–1789.* Stanford: Stanford University Press.

McGinnis, Michael D. 1992. "Bridging or Broadening the Gap? A Comment on Wagner's 'Rationality and Misperception in Deterrence Theory'." *Journal of Theoretical Politics* 4 (4): 443–57.

———. 1996. "Elinor Ostrom: A Career in Institutional Analysis." *PS: Political Science and Politics* 29 (4): 737–41.

———, ed. 1999a. *Polycentric Governance and Development.* Ann Arbor: University of Michigan Press.

———, ed. 1999b. *Polycentricity and Local Public Economies.* Ann Arbor: University of Michigan Press.

McGinnis, Michael D., and John T. Williams. 2000. *Compound Dilemmas: Democracy, Collective Action, and Superpower Rivalry.* Draft book manuscript. Indiana University, Bloomington.

Ostrom, Vincent. 1953. "State Administration of Natural Resources in the West." *American Political Science Review* 47 (2) (June): 478–93.

———. 1976. "John R. Commons's Foundation for Policy Analysis." *Journal of Economic Issues* 10 (4) (December): 839–57.

———. 1980. "Artisanship and Artifact." *Public Administration Review* 40(4) (July–Aug.): 309–17. Reprinted in McGinnis 1999a.

———. 1987. *The Political Theory of a Compound Republic: Designing the American Experiment.* 2d rev. ed. San Francisco: ICS Press.

———. 1991. *The Meaning of American Federalism: Constituting a Self-Governing Society.* San Francisco: ICS Press.

———. 1997. *The Meaning of Democracy and the Vulnerability of Democracies: A Response to Tocqueville's Challenge.* Ann Arbor: University of Michigan Press.

Ostrom, Vincent, and Elinor Ostrom. 1972. "Legal and Political Conditions of Water Resource Development." *Land Economics* 48 (1) (February): 1–14. Reprinted in McGinnis 1999a.

———. 1977. "A Theory for Institutional Analysis of Common Pool Problems." In Garett Hardin and John Baden, eds. *Managing the Commons*, 157–72. San Francisco: W. H. Freeman.

Putnam, Robert D. 1988. "Diplomacy and Domestic Politics: The Logic of Two-Level Games." *International Organization*, 42, 427–60.

Riker, William H. 1982. *Liberalism against Populism: A Confrontation between the*

Theory of Democracy and the Theory of Social Choice. San Francisco: W. H. Freeman.

Shepsle, Kenneth. 1979. "Institutional Arrangements and Equilibrium in Multi-dimensional Voting Models." *American Journal of Political Science* 23: 27–59.

Walker, James, Roy Gardner, Andy Herr, and Elinor Ostrom. 2000. "Collective Choice in the Commons: Experimental Results on Proposed Allocation Rules and Votes." *Economic Journal* 110 (460): 212–34.

CHAPTER 5

Votes and Vetoes

Roberta Herzberg and Vincent Ostrom

I. Introduction

A critical factor in the constitution of a democracy is the use of voting to take collective decisions and the use of vetoes to interpose limits in the exercise of authority. The laws that apply to the exercise of suffrage are of fundamental constitutional importance in any democratic society whether or not such laws are formulated in a specific document identified as the "constitution." A democratic republic is that form of government where the people of a republic exercise basic prerogatives of government. Voting as a way of signaling choice is necessary for people to participate in the taking of collective decisions; and laws of suffrage are one of the most fundamental elements in the constitution of democratic republics (Rae 1967, 3–7).

People cannot, however, discharge all of the prerogatives of government while acting in a collective capacity. Different institutional arrangements are necessary for assigning different decision-making responsibilities to officials who exercise a fiduciary responsibility in the discharge of political responsibilities in relation to the larger community of people in a democratic republic. Linkages that establish patterns of accountability are expressed in structures that reflect dependencies, autonomy, limits, and interdependencies. Patterns of dependency require those who exercise authority to be accountable to others. The exercise of discretion, however, always implies some degree of autonomy. Those who have the responsibility to exercise discretion on the behalf of the larger community of people must have an autonomous authority to do so. Where complex divisions of labor occur in the exercise of governmental

Originally published in F. X. Kaufmann, G. Majone, and V. Ostrom, eds., *Guidance, Control, and Evaluation in the Public Sector* (Berlin and New York: Walter de Gruyter, 1986, 431–43). Reprinted by permission of Walter de Gruyter and the authors.

Authors' note: Theo Toonen was especially helpful in his critique of an earlier draft of this essay. What we say needs to be appropriately qualified in addressing the characteristics of particular electoral systems, and the way they are linked to structures of government and to the longer-term cultural heritage and patterns of settlement that have established the basic cultural and spatial patterns in different societies.

authority, limits can either be formally specified or informally exercised. Limits are manifest as vetoes and express the way that authority is shared among diverse decision structures. Patterns of interdependencies are reflected by the way that dependent, autonomous, and limited relationships get linked together.

Some political systems make explicit use of vetoes as an exercise of constitutional prerogative. The U.S. system of checks and balances is a complex structure of authorizations and vetoes to both distribute the power of government and establish conditions for the separable and joint use of those powers. The withholding of essential capabilities in the operation of society may also function as a veto mechanism. A strike by labor unions or a general suspension of regular activities as an expression of solidarity by a population as a whole can be used as vetoes to place limits upon the authority of governments.

Processes of collective decision making depend upon elucidating information both about preferences that are held by people in the larger society and the means-ends calculations that apply to collective endeavors that are being taken to affect the joint welfare that members of a society share with one another. The communication of preferences through voting mechanisms is an essential way for people to inform one another about what is valued and how to establish priorities among different possibilities that are the objects of collective action. Seeking public office and voting for candidates to public office provide the occasion for a dialogue in democratic societies about basic priorities in public life.

Beyond establishing priorities in public life there is a problem of fashioning mutual understandings about what is to be done and how the burdens and benefits of collective action are to be shared among members of a society. Whatever is done involves costs. To pursue some possibilities forecloses other possibilities. To gain best advantage of the opportunities that are created often requires intelligent use of public services and public facilities on the part of citizens and ordinary members of a society as well as public servants. Only when there is a reciprocal understanding among producers and consumers of public goods and services can best advantage be taken of the opportunities that are made available. Good performance depends upon proportioning the supply of public services to effective demand in a way that yields a good fit between what is produced and how that which is produced is used or enjoyed by people in the larger society.

Many essential public services are intermediate products that enter effectively into the economic life of a society only if properly used by members of the larger society. Those members of that larger society become essential coproducers of many public services. Teachers cannot produce education alone. Education requires the coproductive efforts of students and their parents as well as the efforts of teachers. The dialogue between candidates for public office and voters can be an occasion then for fashioning mutual understandings that pertain to the way that public endeavors are organized and operated and how the reciprocal responsibilities of officials, voters, taxpayers, those who render public services, and those who use public services relate to one another in mutually interdependent chains of actions. Votes and vetoes, thus, shape the conditions of entrepreneurship in the public sector and provide a way that voters can hold officials accountable for the discharge of their offices as a public trust.

Electoral arrangements are fashioned in different ways in different societies. The method of the vote, the nature of the election, who is subject to election, and under what circumstances, in turn, have profound effects upon the nature of the dialogue that occurs between voters and candidates for public office in different societies. The way in which a democracy performs is, thus, significantly influenced by patterns of suffrage, structure of electoral arrangements, and how elections are linked to other decision structures in the organization of a system of government.

2. Comparing Votes and Vetoes in Different Electoral Systems

Given the limitations and constraints upon human communication in collective decision making, institutions are designed to relate citizen preferences to political action. Voting and veto mechanisms are ways of transmitting information about citizen interests for the purpose of political order. To understand the degree to which official channels of political action provide citizen input, we concern ourselves with three aspects of the process. First, we examine the way that different voting arrangements structure communication and affect the way that preferences are articulated and aggregated. Second, we consider how deliberations inform decisions, and finally, we outline how the consequences of decisions impinge upon those who are subject to collective decisions. Comparisons will be limited to two different types of competitive electoral arrangements: the single-member constituency where one representative is elected from each of a series of geographically defined constituencies and the at-large constituency where all representatives are elected from the same constituency in proportion to the votes received in an election. Most democratic systems include some form or variation of these two basic types of electoral arrangements. In concentrating on these two basic types, we hope to isolate some general conclusions regarding the effect of electoral rules on the linkages between popular preferences and public decisions. Constraining the analysis to these two forms makes it difficult to capture the diversity of forms included in each of the basic types. Only in this general framework, however, can we examine the impact of these electoral rules apart from the other factors of particular importance in determining linkages in each nation.

Elections usually involve the selection of someone to function in some collective decision structure. Elections are, thus, always linked in critical ways to other decision structures. We assume voters base their electoral decisions on expected policy outcomes at later stages in the decision-making process. Thus, electoral support can play an important part in the difficulty or ease with which future governmental decisions are made. These linkages always imply that those who are subject to selections are players in a series of simultaneous and sequential games where a move in any one game potentially affects the opportunities that are available in the other linked games. The electoral connection between voting and decision structures can be expected to affect the structure of decisions made by representatives and elected officials in particular ways. The strong regional perspective of members of the U.S. Congress can be attributed, in part, to the electoral structure that designates representation on the basis of specific geographical constituencies and the interests of those constituents. Voters reward representatives for making policies that take their regional interests into account (Mayhew 1974; Arnold 1979). To the extent that voters select representatives within smaller geographical boundaries, they can protect local interests that might otherwise be lost in the broader, national arena. The cost of regional protection, however, may be conflict and stalemate with regard to truly national policy objectives if there is some fundamental antithesis between local and national interests.

Elections serve to regularize patterns of citizen input in the decision-making process so as to limit other forms for participation, such as political protest or mass demonstrations (Ginsberg 1982). The frequency and multiplicity of electoral opportunities provide electors with mechanisms to both channel and constrain collective choices in relation to the different communities of interest that are related to different units of government. If interests are not being served by currently elected representatives, voters may indicate their displeasure and throw them out. Thus, elections act as a crude veto mechanism over collective decisions.

The effectiveness of this veto relates to the rules and structure of the electoral system. Major electoral losses are strong ways of informing officials of citizen discontent with existing policies. The effectiveness of this veto mechanism depends, in part, on how direct the relationship is between the electoral decision and the formation of a government. In a single-member district system where structure drives the process toward two parties, a single party has the potential for exercising leadership in particular decision structures or in a government as a whole. A major electoral loss for the majority party is a clear message of discontent with leadership policies. In a proportional representation system, voters may be less clear about how to use their vote to indicate concern about the policies of a coalition government. In this latter case, voters may depend on protest movements and mass demonstrations in the interim between elections. More generally, veto opportunities arise whenever decisions n one context function as constraints on decisions in other contexts. Vetoes exist as a negative expression of electoral opportunities but may also occur when one set of public decision makers constrains the actions of another set. Voters exercise an indirect veto whenever they are able to elect government officials to separate positions in the decision process so that those officials act as a check on the actions of others. Citizens, thus, participate in the exercise of both direct and indirect veto actions on collective decisions.

3. Single-Member Constituencies

Electoral arrangements that rely upon single-member constituencies require a process of apportionment where constituencies are assigned by some specifiable criteria. The most common criterion is one of equal proportions so that each constituency contains an approximately equal population. Other criteria may apply that take account of the boundaries of other political jurisdictions. In many federal systems of government the assignment of constituencies in the national legislature, for example, takes account of cantonal, provincial, or state boundaries. The mechanism for determining constituency boundaries affects the relative equality of political representation (Dixon 1968).

Most single-member-constituency electoral systems function under a plurality rule that designates the candidate who receives the largest number of votes as winner. Alternatively, simple majority rule would require a run-off election between the two leading candidates when no single candidate has received a majority of the total votes. Strict majority rule variations are relatively rare: the candidate, receiving the largest vote wins.

Designating the winner as the one receiving the largest number of votes yields certain anomalies for political systems relying upon single-member-constituency electoral systems. If four candidates stand for election, the minimum winning vote would be one fourth of the total votes plus one. A minority of the population can, under that circumstance, successfully elect a representative for that constituency. If similar circumstances existed in each other constituency, a minority of the aggregate population might be successful in electing a majority of members to a legislature.

Although the above example is possible, it is highly unlikely for a number of reasons that significantly ameliorate such tendencies. First, since only the largest number of votes counts, there is an incentive for voters to shy away from any candidate whose chance of winning is not strong. Unless your candidate receives the greatest margin of support, you stand among the losers. Second, since only majority support can guarantee a victory, incentives exist for candidates to take positions that appeal to the greatest number of voters. These incentives reduce the probability of a four-way split.

When two candidates run for an election in a single-member constituency, each has a relatively equal chance of winning. When a third candidate enters the race, more complex calculations arise. The appeal of a third candidate is likely to detract more heavily from one of the two principal contending candidates than the other. The votes gained by the third candidate are likely to be at the cost of that candidate who is most closely associated with the appeal being made by the third candidate. Two scenarios are possible. First, inclusion of a third-party candidate may split the vote of a potential majority coalition and contribute to victory of the least preferred candidate. Alternatively, the minority coalition may be split resulting in a landslide for the winning candidate. The latter may be less serious with respect to outcome, but it can have the effect of justifying serious underrepresentation of fairly sizable minorities. Voters may, thus, fear the consequences of voting for a third candidate. Ony if a candidate has a serious chance of being one of the two major contenders are voters likely to give their support.

When national elections are organized into single-member districts, parties with support spread evenly across all districts may be seriously underrepresented in the seats-to-votes ratio while parties with concentrated regional strength may be greatly overrepresented in the national arena (Tufte 1973). The potential for underrepresentation in single-member constituencies was most recently indicated by the electoral fate of the Liberal/Social Democratic Parties' Alliance in Great Britain. Gaining 26.1 percent of the popular vote nationally, the Alliance captured only 23 seats in Parliament compared to Labour's 209 seats with 28.3 percent of the vote. The Alliance suffered from its consistent and strong second place finish in districts won by the Conservative majority. Unlike Labour with its regional strongholds, the Alliance could not depend on "safe" districts. For a party to be advantaged in a single-member constituency arrangement, electoral support must be sufficiently concentrated to result in winning votes in particular districts.

The tendency toward underrepresentation of mainstream third parties accounts for the strong bias toward a two-party system in electoral systems that rely upon single-member constituencies. To maintain national support and attention requires being kept in the public eye. With only 23 seats, it is unlikely that the Alliance will be able to turn broad electoral support into a highly visible place in parliamentary proceedings. Regional concentration of support is a necessary condition for maintenance of third parties in a single-member system.

There is, however, another tendency at work in single-member constituencies that requires attention. If the constituency shares preferences so that the largest portion of the population is clustered around a median between two extremes, the contending candidates will have incentives to make their principal appeal to the median voter (Black 1958; Downs 1957). This yields a tendency for the contenders to make appeals that are not easily distinguished from one another. Voters are presented with a choice between two candidates who advocate much the same program.

When such circumstances arise and there is little risk of voter abstention, incentives exist for third candidates to enter electoral contests. This, in turn, creates incentives for the two leading candidates to differentiate their appeal to the electorate. Concern about throwing away one's vote once the two leading candidates have differentiated their appeal contributes to a lack of viability for a third candidate. This, in part, explains why third parties often are important in raising new issues but are virtually powerless in gaining electoral victories. They may create sufficient public concern for new issues to be placed on the agenda of public deliberations, but the rules of the electoral system seriously constrain their chances for electoral success (Mazmanian 1974). They affect the agenda of collective action without being able to win office and sponsor legislation. Single-member-constituency systems are thus marked by tendencies toward two leading contenders but with recurrent entry of third contenders who maintain only short-term viability. Such a system biases appeals to voter preferences that reflect central tendencies in a population but leads to a challenge by other contenders whenever the appeal of leading contenders fails to provide distinguishable options.

In summary, single-member-constituency systems place major emphasis on maximizing relative vote totals and in so doing limit the potential for minority representation. The policy focus expected from such a system is toward more broad, and sometimes ambiguous, positions. To the extent that support must be broad, each individual voter exercises less power in potentially vetoing public programs. Only if the margin of support changes significantly are changes in representatives and policy likely to occur.

The working out of these tendencies is significantly influenced by the linkages to political parties, interest groups, and the relevant decision structures of government. Incentives exist to gain the advantage of teamwork by forming a joint slate of candidates who cooperate in making a joint appeal to the electorate and in getting out the vote on election day. Information dissemination, where the party acts as a signaling device, is an important factor in single-member-constituency elections. The slating of candidates affects the options that are available to voters. The availability of ways and means to appeal to voters and get out the vote has an effect upon who wins and who loses. Who wins and loses elections in turn affects how governmental prerogatives are exercised in taking of collective decisions. The relationship between candidates and later decisions is affected by the degree of party cohesion with respect to issue positions. Party may be a meaningful force in electoral considerations but far less effective in structuring the remainder of the decision process. The way these structures are linked establishes who and what are taken into account in making collective decisions (Ostrogorski 1964). If party discipline is strong, as in Great Britain, electors vote for candidates who reflect a party position, and the linkage between an electoral decision and policy decisions is directly associated. Control over the government goes to the majority party in Parliament. Alternatively, where parties serve as electoral tags, but individual candidates control their own issue positions and future policy decisions, the linkage associated with party is less clear (Crotty and Jacobson 1980).

When a party, for whatever reason, fails to provide strong links between elections and decisions, interest groups may assume this role. The difficulty with interest groups providing the link between citizens and official representatives is their tendency to concentrate on specialized policy decisions without concern for the broader interests and support necessary to win elections in single-member constituencies. Rather than giving coherence to the linkage process, interest groups may generate conflicting signals that confound the voting decision and its relation to future governmental decisions. When such groups are powerful forces in politics, interests may be well protected at the cost of collective action. Such groups may gain access to official agendas and control key points where they exercise veto power over decisions that threaten their specific interests.

Success in future elections requires representatives in single-member constituencies to place emphasis upon the demands and interests of constituents. Each representative faces the multiple realities of constituency, party organization, and a voice in the formal structure of government. Wherever there are independent opportunities to affect outcomes, a potential veto point exists. The structure of linkages establishes the potential combination of votes and vetoes and the incentives that people have to either support or constrain choice in linked structures of votes and vetoes.

In the British parliamentary system, control both over legislative policy and over the executive apparatus of government accrues to the leadership that maintains a voting majority in the House of Commons. This creates strong incentives to maintain a disciplined coalition that will vote together on legislative matters. Its future success depends in turn upon maintaining a strong organization that is capable of slating candidates, conducting a campaign, and getting out the votes with a few weeks' notice when the timing of an election is usually determined by the leadership of the majority party or coalition.

The structure is organized to facilitate dominance by majority leadership that can preempt consideration of policy issues during its tenure in office.

Opportunities to influence policy must await a new election when a similar preemptive advantage accrues to the new majority leadership. The absence of vetoes within the government itself means that the creation of veto potentials depends upon a realignment of parties, a modification in the constitution of particular parties to constrain the dominance of parliamentary leadership, ways to withhold essential economic services, or a combination of all three strategies. The implicit negotiation of what might be referred to as "social contracts" then serves as constraints upon the "unlimited" power of Parliament.

The U.S. system of government is marked by many different structures that are linked by explicit veto mechanisms internal to the structure of government. The legislative process in the national government is organized by reference to three different decision structures that rely upon quite different constituency relationships. The House of Representatives is elected from 435 single-member constituencies. The members of the Senate are elected from states that constitute single-member constituencies for any particular senatorial election. The president is elected through a mechanism in which each state votes as a constituency to elect presidential electors equal to the number of representatives and senators for that state, and these electors, in turn, elect the president.

Enactment of legislation depends upon the concurrence of the House of Representatives, the Senate, and the president. A veto by the president can be overridden by a two-thirds majority in the House of Representatives and the Senate. The electoral arrangements mean that the national constituency is aggregated in three different ways and that legislation requires the concurrence of those who represent three different types of constituency relationships to the same population. The institutional rules of each of these bodies of government reflect the different constituency basis of each branch. The smaller the base of constituency concerns, the less centralized official decision mechanisms are. Although each deals with national interests, the differences in electoral structure create different conditions for the consideration of these national issues. As constituency size is altered, some change in the agenda is likely to occur. Redundancy in deliberation occurs by taking account of different communities of interests shared by the same population.

The organization of political parties in the United States is largely controlled by state laws. Open primary elections have increasingly replaced the old party caucus as a closed selection process. Under primary election procedures, any party member is free to challenge any other party member in a publicly conducted election for a party's nomination to a particular office. Slating under these rules is primarily controlled by the party members in each constituency. Party leadership exercises only weak influence in establishing party slates and in how individual representatives vote in making collective decisions.

The competence of the judiciary to consider the constitutional standing of legislation and the discharge of governmental prerogative in accordance with constitutional due process means that courts can also interpose potential vetoes. Since courts act only on the initiative of those who seek redress of a grievance, the exercise of judicial vetoes is potentially available to any aggrieved party. Factors such as education and wealth of the individual or the calendar of the courts may limit the degree to which a veto through the judiciary is available to any one individual. However, controlling for resource constraints, any person (and that person need not be a citizen) has potential access to veto capabilities in relation to collective decisions.

4. At-Large Constituencies with Proportional Representation

Parliamentary systems of government in continental Europe have traditionally relied upon some form of at-large constituencies with proportional representation. In its most simple form, a nation-state forms a single constituency. The voter's choice in that simple form is a choice among the parties that submit lists of candidates standing for election. The proportion of votes cast for each party is used to determine the proportion of representatives that are elected to the national legislative assembly. The organization of government follows the parliamentary tradition of assigning political leadership to those who can gain and maintain majority support.

In the traditional party-list system, the slating of candidates is done by a national party organization. Although formally the national party controls final lists, regional and local party officials may in fact play a major part in recruitment and in determining the general content of the list. The extent to which regional officials influence the selection process will influence how important regional considerations may be in the strategy of electoral politics. The order in which the names of candidates is listed is of critical importance. The higher on the list, the greater the probability of election.

Since success depends, in large part, upon an appeal to a national constituency in relation to identifiable communities of opinion, each political party has an incentive to articulate an appeal that differentiates its supporters from other communities of opinion. Unlike the single-member district system in which parties were required to establish some regional basis for their support, this is not *necessary* in the proportional representation system unless separate regional slates are used in structuring at-large constituencies. This does not imply that regional factors are unimportant in organizing coalitions in these systems. Any party that can maintain support from a differentiable community of opinion based on any of a number of factors—language, religion, class, and so on—can be assured of success in proportion to the size of that community of opinion.

Even a small identifiable community of opinion assures success for some small fraction of representatives elected to a legislative assembly so long as their aggregate vote exceeds the minimum ratio for representation.

The structure of incentives in an at-large electoral system with proportional representation yields a party system that places greater emphasis upon ideological differentiation of diverse communities of opinion. The rules of the electoral structure do not require diverse interests to compromise or soften their particularized interests to gain seats in a representative body except to the extent that those interests surpass an established minimum. The appeal to diverse communities of opinion yields a tendency toward multiple parties. Governing, however, still requires the formation of larger groups that may soften the intensity of conflict in the at-large system. There exists a tension in proportional representation systems between the rules that apply to electoral contests and the rules of government formation (Katz 1980). In essence, electoral incentives promote greater ideological differentiation, while incentives of governance require room for compromise within the programs of the different parties.

The formation of coalitions that are capable of maintaining majority support typically occurs in negotiations among parties since any one party is not likely to command majority support from the larger national constituency. Bargaining among party leaders becomes an essential process in the formation of a government. Bargaining power is not proportioned directly to the votes received from the electorate but rather to the value of each party in forming a coalition. Representatives of the marginal party forming a governing coalition exercise greater weight in establishing a coalition's program than an equivalent number of representatives of the most numerous party in the coalition (Dodd 1976). Although the party with the largest number of representatives is most likely to gain the major governing positions, its share of positions will be proportionately less than those granted the marginal party or parties. The power exercised by each coalition member is a function of his or her contribution to the majority status of the coalition and the availability of acceptable alternative coalition partners. Parties may be constrained in the formation of coalition governments by ideological distance between potential coalition partners (Grofman 1982).

If no ideologically acceptable alternative coalition members exist, the marginal party will have an even greater degree of bargaining potential in the formation of the new government. Failure on the part of the marginal party to go along may result in the collapse of the government, requiring a new election. In this instance, the marginal party can demand a much higher proportion of the political advantage from other coalition members.

Success in future elections depends upon the maintenance of the continued support for distinguishable communities of opinion by the leadership of each party. When essential interests of any one party are threatened, it is capable of exercising a veto. Unless an oversize coalition exists, withdrawal of a party's support means the collapse of the majority coalition and the necessity to either form a new coalition or call a new election. Potential veto positions can be exercised by any party in a coalition, but they may be very costly to the party exercising the veto. If the veto attempt fails to result in a weakened position for the major parties in the coalition, the vetoing party may be excluded from future positions of government or be punished in the resulting electoral contest. Thus, we would expect coalition members to use their veto only when *essential* interests are in question.

Processes of government are bounded by combinations of votes and vetoes. The viability of small splinter parties representing extreme opinions exposes a political system relying upon an at-large electoral system with proportional representation to the formation of parties that use electoral processes and parliamentary processes of government to attack those institutions and to obstruct efforts to seek deliberative solutions to common problems. Although obstruction is possible by splinter parties, proportional representation systems maintain some level of consensual support by punishing parties that promote extreme conflict by excluding them as potential governing partners. A party that continually proposes extreme positions unacceptable to a majority of the population may by that strategy exclude itself as an unacceptable coalition partner. Without the potential for governing power, extreme parties may find it difficult to maintain their electoral position, particularly if voters base their choices on expected policy decisions as we assume.

The basic justification of the at-large electoral system with proportional representation is to provide an accurate reflection of public opinion in the councils of government. With regard to deliberations in the representative body this may be true. However, in terms of governmental decisions, the operation of such a system yields counterintuitive results. Extensive veto capabilities and significant bargaining power to those who represent relatively minor positions among the population creates the potential for underrepresentation of major interests in society. In essence, the potential for underrepresentation in the single-member-constituency system may simply be moved to the postelection governing stage in the proportional representation system.

Post-World-War-II European constitutions have significantly modified at-large electoral systems by disqualifying splinter parties receiving less than a specified fraction of the vote, the use of regional lists as a substitute for or as a supplement to national lists, and the use of voter preferences to establish priority among candidates on any one list. Some elements of the single-member-constituency electoral system have also been introduced as complements to at-large electoral systems with proportional representation to yield mixed electoral systems.

An even greater concern than the question of underrepresentation present in each of the systems is the problem associated with the indecisiveness of election outcomes for government formation in proportional representation systems. When the decision of government formation is made separate from the electoral contest, it is difficult for the individual voter to know how that vote will be translated into government action. A party with issue positions of great distaste to the voter may, in fact, be a part of the governing coalition their vote is used to support. At the time of voting, the voter may be unaware of what positions and issues will be traded to other parties in an effort to establish a majority coalition. Their vote for one party may have been based on the party platforms along the issue dimensions that are now largely controlled by ministers of an opposing party. Although voters may have some information regarding possible coalition partners prior to the election, the exact balance and distribution of positions and control will depend on the final vote totals and, thus, not be available at the time of the vote. This presents a serious problem in proportional representation systems. Guidance of government decisions by voters becomes difficult when they are unsure of how those votes will be translated into policy. The extent to which voters are able to exercise control over governmental decisions, in large part, rests on the ability of parties to practice a veto within the coalition over issues of importance to the voters. If parties are effective when essential interests are concerned, voters may exercise a voice with reference to essential interests.

5. Conclusion

No electoral system will permit the "Voice of the People" to prevail. Any electoral system permits people to exercise some voice in collective decisions. The nature of that voice does *not* turn upon the fact of election per se but upon the way that elections are linked to the diverse processes of government in a society. Whoever is elected to some office serves as a link between those who do the electing, those who assist in the election, and the decision-making structure in which the office is embedded. Meeting the requirements of these diverse structures is a limiting condition that bounds the strategic opportunities for anyone who aspires to an elective office.

Electoral processes provide participants with opportunities to speak and be heard. The nature of the communication, the matters that are addressed, the positions taken, and the nature of the response are significantly affected by the way that electoral processes are structured. The single-member constituency yields quite different results than at-large electoral arrangements with proportional representation.

Participating in electoral processes enables citizens to compare their circumstance with the assessments and proposals advanced by candidates. Candidates learn about the problems, preferences, and responses of constituents. But, the opportunities to act depend upon meeting the conditions of choice both in authorizing and implementing collective decisions. A multitude of others is involved. Limits to human knowledge, constraints upon opportunities to have a say, and self-serving distortions of information mean that collective choice is plagued by propensities for errors. The problem is how to limit those propensities to err and to provide human beings with constructive opportunities to develop their potentials and take account of one another.

In devising appropriate decision-making arrangements, no simple arrangement will suffice. If everyone were required to agree to any course of collective action, then each person would be in a position to veto collective action. Under those circumstances, each person would become the judge of his or her own interest in relation to the interest of others. Once a rule of unanimity is relaxed to allow for some form of majority or plurality decisions, then an opportunity is created for some to exploit others. Such possibilities can be constrained only as anyone is capable of interposing a veto whenever his or her essential interests or rights are jeopardized. This is the ground for establishing judicial remedies to interpose limits upon official discretion.

Authority to act always needs to be viewed in relation to correlative limits upon authority to act. But votes and vetoes are necessary complements to one another in establishing appropriate boundaries to the exercise of human discretion and to create a structure of incentives to search out mutually productive resolutions to joint problems. The way structures become linked together through mechanisms like elections has an important effect upon the rule-ruler-ruled relationship. Linkages can create bonds of reciprocity in human societies; or they can align elements in a society to war upon one another.

There is no perfect mode of election. Elections cannot provide mandates for action without regard for the interests of others. Elections always need to be viewed in the way that they link structures together to take account of diverse interests and diverse sets of consideration. They provide occasions for dialogue and deliberation among the elected about what is in the joint interest of those who elect.

The dialogue between candidates for public office and voters and the relationship of this dialogue to who wins and who loses serve as a mechanism for targeting legislative deliberations and collective action. Veto mechanisms serve to bound or constrain the realm of feasible collective action. By both targeting and bounding the realm of public discourse and collective action, votes and vetoes perform an important function in steering processes of collective action along channels that establish the scope of what is politically feasible.

REFERENCES

Arnold, R. D. (1979): *Congress and the Bureaucracy.* New Haven, Conn.: Yale University Press.

Beyme, K. von (1982): *Parteien in westlichen Demokratien.* München: Piper.

Black, D. (1958): *Theory of Committees and Elections.* Cambridge, Mass.: Cambridge University Press.

Crotty, W., and G. Jacobson (1980): *American Parties in Decline.* Boston: Little, Brown.

Dixon, R. (1968): *Democratic Representation.* New York: Oxford University Press.

Dodd, L. (1976): *Coalitions in Parliamentary Government.* Princeton, N.J.: Princeton University Press.

Downs, A. (1957): *An Economic Theory of Democracy.* New York: Harper and Brothers.

Duverger, M. (1954): *Political Parties.* New York: John Wiley and Sons.

Farquharson, R. (1969): *Theory of Voting.* New Haven, Conn.: Yale University Press.

Ginsberg, B. (1982): *The Consequences of Consent: Elections, Citizen Control and Popular Acquiescence.* Reading, Mass.: Addison-Wesley.

Grofman, B. (1982): "A Dynamic Model of Proto-Coalition Formation in Ideological N-Space." *Behavioral Science* 29/1: 77–90.

Katz, R. (1980): *A Theory of Parties and Electoral Systems.* Baltimore, Md.: Johns Hopkins University Press.

Lawson, K. (ed.) (1980): *Political Parties and Linkage: A Comparative Perspective.* New Haven, Conn.: Yale University Press.

Mayhew, D. (1974): *Congress: The Electoral Connection.* New Haven, Conn: Yale University Press.

Mazmanian, D. (1974): *Third Parties in Presidential Elections.* Washington, D.C.: Brookings Institution.

Michels, R. (1962): *Political Parties.* New York: The Free Press.

Milnor, A. J. (1969): *Elections and Political Stability.* Boston: Little, Brown.

Neimi, R., and H. Weisberg (eds.) (1976): *Controversies in Voting Behavior.* San Francisco: Freeman.

Ostrogorski, M. (1964): *Democracy and the Organization of Political Parties.* Garden City, N.J.: Doubleday.

Page, B., and R. Brody (1972): "Policy Voting and the Electoral Process: The Vietnam War Issue." *American Political Science Review* 66/3: 979–95.

Rae, D. (1967): *The Political Consequences of Electoral Laws.* New Haven, Conn.: Yale University Press.

Riker, W. (1962): *The Theory of Political Coalitions.* New Haven, Conn.: Yale University Press.

Riker, W. (1982): *Liberalism against Populism.* San Francisco: Freeman.

Rokkan, S. (1970): *Citizens, Elections, Parties.* New York: D. McKay.

Sartori, G. (1965): *Democratic Theory.* New York: Fr. A. Praeger.

Sorauf, F. (1980): *Party Politics in America.* Boston: Little, Brown.

Tufte, E. (1973): "Relationship Between Seats and Votes in Two Party Systems." *American Political Science Review* 62/2: 540–54.

CHAPTER 6

Negative Decision Powers and Institutional Equilibrium: Experiments on Blocking Coalitions

Rick K. Wilson and Roberta Herzberg

Over the past decade an important strand of political science research has turned to studying actors' behavioral strategies within political institutions. This "new institutionalism" has focused largely on legislative processes, with the approaches to that study varying from the historical to the analytical (see the discussion by Shepsle 1986). The principal question asked by these researchers is: In what ways do institutions affect the actions of political agents? The study carried out here relies on analytic methods and empirical testing to demonstrate one general institutional procedure that changes the behavior of political actors. We do not offer a general theory characterizing all institutional effects on behavior. Instead, we concentrate on "negative agenda powers" that impede political decision making. This focus provides insight into the effects of disparate negative agenda powers such as executive vetoes, bureaucratic intransigence, "pigeonholing" legislation by committees, and filibustering.

The study of agenda processes has dominated much of the analytic work on political institutions. Largely this is due to the importance of agendas for translating individual preferences into collective choices. Agenda procedures are not dictated by happenstance. They are endogenous to political institutions and are defined by detailed rules and precedents. For example, the House of Representatives adheres to strict procedures for the introduction, discussion, and passage of general legislation. Bills placed in the legislative hopper are typically routed to a committee of jurisdiction, legislation is returned to the committee-of-the-whole under well-defined rules, and the processes

Originally published in *Western Political Quarterly* 40, no. 4 (December 1987): 593–609. Reprinted by permission of the University of Utah, copyright holder and the authors.

Authors' note: The authors gratefully acknowledge the comments and criticisms of a number of colleagues, including David Brady, Melissa Collie, Joseph Cooper, Patricia Hurley, Elinor Ostrom, Charles Plott, Kenneth Shepsle, and the editor of *Western Political Science Quarterly*. Partial funding for this research was provided by the National Science Foundation Grant No. SES81–04472. A version of this essay was originally presented at the Annual Meeting of the Public Choice Society, Phoenix, Arizona, March 28–31, 1984.

governing voting on the legislation are well understood (see Oleszek 1984).

Two major points have emerged from the formal study of agenda processes. The first is that the structure of an agenda matters for political choices. Plott and Levine (1978) demonstrate, both theoretically and empirically, that the order of an agenda dictates the outcomes selected by decision makers. The second point is that different agenda procedures generate different strategies for political actors. Shepsle and Weingast (1984) pinpoint a variety of agenda procedures that yield very different outcomes, and Wilson (1986a) provides empirical support for their conjectures. Both points stress the fact that institutions can dramatically affect the choices of political actors.

Much of the institutional work on agendas has focused on legislative settings (cf., McKelvey 1976, 1986; Enelow and Koehler 1980; Denzau and MacKay 1983; Shepsle and Weingast 1984; Denzau, Riker, and Shepsle 1985). However, a body of literature has also emerged pointing to the importance of agendas within bureaus (Hammond 1986) and the role of bureaus for building budgets (Romer and Rosenthal 1978; Eavey and Miller 1984; Bendor and Moe 1985). Likewise, Kiewet and McCubbins (1985) elaborate the agenda powers exercised by the president vis-à-vis Congress. Above all, this scattered work illustrates the importance of agendas in a variety of institutional settings.

The bulk of the research on agenda processes points to the "positive" power of agenda setting. By this we mean the power to affect the elements that are added to an agenda or the order in which those elements are introduced. This has meant focusing on institutional rules awarding agenda power to a subset of decision makers (such as a committee in Congress) or identifying procedures specifying the order of introducing alternatives (such as amendments to bills under an open rule). However, a variety of institutional procedures also encompass negative agenda powers. These powers, which impede or block policy choices, have been less studied even though in most decision-making settings such powers are important for defending the status quo. In a legislative setting, Krehbiel (1985) has studied the "obstructive" powers of committees vis-à-vis the preferences of the committee-of-the-whole. Focusing on unidimensional settings, Krehbiel characterizes an agenda procedure whereby a subcommittee uses negative agenda powers to obstruct changes to a policy. Our interest in his study is similar to that of Krehbiel. We wish, however, to focus on generic negative agenda powers awarded to subsets of decision makers. We focus our attention on multidimensional results and seek to provide empirical evidence for the effect of this agenda power.

Contributions of Formal Theory

Social choice theory provides a useful tool for highlighting and dissecting the concern we have with blocking power. Social choice models begin with specifying a simplified set of decision rules used for aggregating individual preferences. These deductive models use this institutional context to lay out the strategic choices of purposive actors and to predict collective outcomes. Recent work in social choice points to two findings that have bearing on our concern with blocking power. The first is the general instability of simple majority rule institutions, while the second points to specific institutional mechanisms that take advantage of the instability of simple majority rule and redound to the benefit of those controlling the institutional mechanisms. We draw on these findings in order to highlight the effect of blocking rules.

Simple Majority Rule Models

Considerable formal work has focused on democratic decision processes involving simple majority rule. These processes are characteristic of pure participatory democracy in which no participant is advantaged when proposing policies for consideration and all participants have equal weight when deciding issues. The troubling aspect of such decision processes is that no decision rule (including simple majority rule) yields a consistent (equilibrium) collective choice (see Arrow 1963). Since collective choices should reflect the preferences of those charged with making choices, the same rules should yield the same choices when members' preferences remain unchanged. However, for simple majority rule processes this is rarely the case. Indeed, equilibrium outcomes will fail to occur unless very restrictive conditions are placed on the configuration of members' preferences (Plott 1967; Sloss 1973; Kramer 1973). The disturbing implication of this general disequilibrium result is that the decision process can wander anywhere, settling on any alternative (McKelvey 1976, 1979; Cohen 1979; Cohen and Matthews 1980; Schofield 1978).

In figure 6.1 we provide an example of this general instability finding. There we illustrate a five-person decision setting over a two-dimensional issue space. The points marked by roman numerals represent each person's ideal policy. Beginning from a status quo, x_0, we can determine whether or not there exist other alternatives preferred by a simple majority. In our example any small move into the shaded cones is supported by some majority, and thus x_0 is unstable. These cones are constructed by finding simple majority intersections of hyperplanes passing through x_0. Each hyperplane is orthogonal to a directional gradient originating from x_0 and pointing to a member's ideal policy (see Slutsky 1978, 1979; or Sloss 1973). This amounts to identifying feasible directions of movement from x_0, even though this movement might be very slight. In this example, any alternative in the policy space has a feasible direction of movement, and consequently no equilibrium exists.

A number of researchers have noted that choices under simple majority rule do not scatter throughout the feasible alternative space. In particular, laboratory experimental studies of simple decision settings by Fiorina and Plott (1978), McKelvey, Ordeshook, and Winer (1978), and Laing and Olmsted (1978) note persistent patterns to collective choices. This has given rise to a number of explanations for these outcomes. These include the Competitive solution (McKelvey and Ordeshook 1978), the Stochastic solution (Ferejohn, Fiorina, and Packel 1980), and the Copeland winner (Grofman and Owen 1985). Also, recent work by McKelvey (1986), exploring the bounds of the uncovered set, points to outcomes clustering around a centrally located subset of alternatives. Even though these models point to patterns among outcomes under simple majority rule, outcomes can still wander through a subset of the alternative space.

Structural Blocking Rules

In contrast to simple majority rule, we consider a form of negative agenda power centered in the institutional structure of the decision-making arrange-

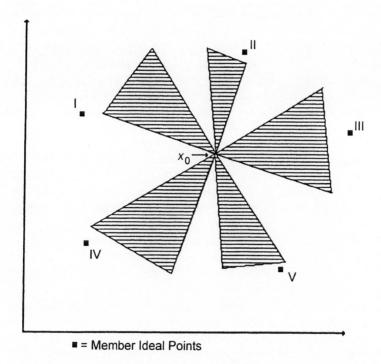

■ = Member Ideal Points

Fig. 6.1. Set of unblocked points to x_0 under simple majority rule

ment. This power allows specific coalitions to prevent changes in the decision. Such blocking powers are associated with the rules of procedures and division of labor schemes designed to organize complex decision processes. For instance, legislatures often employ a variety of agenda rules to structure debate and decision making. Each of these rules empowers a subset of the membership with the ability to propose and order policy alternatives to shape the decision process and speed decision making. In addition they also allow members to block majority-preferred moves by failing to propose or order alternatives (Plott and Levine 1978; McKelvey 1976; Shepsle and Weingast 1984). Similarly, institutional rules such as jurisdictional restrictions and germaneness criteria provide additional opportunities for specially designated actors to proven majority action. Under these procedures, blocking power does not simply slow movement, but it can be used to bias the process in favor of those assigned such power (Shepsle 1979; Denzau and MacKay 1983; Krehbiel 1985).

The blocking aspects of these institutional rules can be systematically understood using a social choice structure. Herzberg (1985, 1986) demonstrates that a variety of equilibrium conditions emerge as blocking powers assigned to specific coalitions of decision makers. In particular, if a single member is assigned blocking power, the only stable policy outcome is that member's ideal policy. In figure 6.2, for example, Player I is awarded blocking power over proposed changes to any status quo. Suppose the status quo was located at x_0. Player I would block any proposed changes moving into the shaded half space away from his ideal policy (e.g., a change to x_1). On the other hand, a movement to x_2 would be unblocked. Suppose x_2 then became the new status quo. Fewer changes preferred by Player I are now possible, which means that Player I will block even more changes than before. In the absence of transaction costs, the agenda will be finitely long until the ideal policy of Player I is selected, at which point every proposed change to x^1 will be blocked. Obviously such powers bias the process in favor of the designated member, as only moves that improve his position go unblocked.

This brief sketch of formal results is useful for pointing to differences that arise over outcomes in different institutional settings. From a comparative statics standpoint the two models are identical except for the introduction of negative agenda powers. Any deduced differences between the models, then, are a function of those agenda powers. Theoretically, outcomes under simple majority rule will differ from outcomes under structural blocking rules. The former have outcomes that vary across the alternative space, while the latter converge on an equilibrium defined by those endowed with blocking power. What is interesting is that one setting lacks any equilibrium solution over outcomes, while an equilibrium is present in the other setting. The equilibrium for institutions with structural blocking rules is a function of what Shepsle (1979) and others have characterized as a "structurally induced" equilibrium. In this case the equilibrium is induced by an institutional procedure awarding a specific agenda power to a subset of participants. We rely on these comparative statics results to generate empirical predictions for our experimental settings.

Empirical Method

We now turn to empirically investigating the effect of blocking on collective decision making. Since we wish to focus explicitly on blocking power, holding constant other institutional features, we turn to laboratory experimental methods. In our empirical discussion we consider two different experimental settings. The first represents a control setting operating under simple majority rule (SMR). The second treatment represents structural blocking in which a single individual is granted the power to block changes to a decision (SMB—single member blocking). These settings, as discussed below, mirror our theoretical observations.

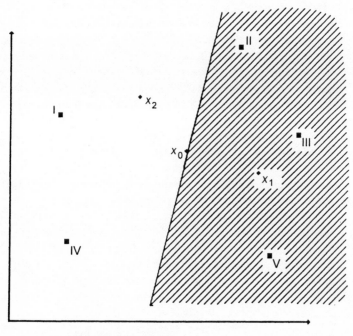

■ = Member Ideal Points

Fig. 6.2. Blocking by a single member

Control Setting

In the experiments reported below five participants are charged with selecting a single alternative from a two-dimensional policy space. The experimental structure employed here is almost identical to that used and described in Berl et al. (1976), McKelvey, Ordeshook, and Winer (1978), Fiorina and Plott (1978), Laing and Olmsted (1978), and Krehbiel (1986). The major difference is that our experiments required that all participant interaction take place via a computer system.[1] Participants were seated in the same room, although at separate computer terminals. Seating was arranged such that no participant could view the terminal screen of another. This provided for control over attributes of face-to-face interaction that are frequently uncontrolled in committee experiments. Furthermore, this structure ensured the experiment's social choice theoretic underpinnings through eliminating mechanisms by which participants could coordinate joint strategies.

Participants were introduced to the experiment and quizzed on its operation solely by a set of computer-displayed instructions, thus eliminating differences in administration of the experiments. This computer-controlled experiment also simplified the task complexity faced by participants. Values for proposals in the alternative space were automatically calculated for committee members based on their utility function. In addition, the computer displayed alternatives on the policy space, sketched representative indifference curves, and provided more explicit information as to who sent proposals.

Communication in these experiments was highly stylized. Players were only able to submit proposals for consideration by others and second other players' proposals. This procedure was imposed in order to minimize participant differences in these experiments. Individuals differ in their capacities to articulate alternatives or to convince others to consider those alternatives. Since the focus of this research is with the effect of specific institutional rules on behavior, we ensure that all participants begin from the same skill levels with respect to communication.

Each proposal provided participants information concerning an alternative's coordinates, who proposed the alternative, and the member's utility for the alternative. The information exchanged by participants was straightforward, and absorbing that information required little effort. Consequently this meant that the time spent in discussion could be limited and that individuals could participate in more than a single experiment at one sitting. In these experiments, participants engaged in three to four different "rounds." In addition, before beginning the experiment participants went through a short practice round with one another in which they were not compensated. This was designed to familiarize them with the equipment (Wilson 1986b).

At the outset of each round, members were randomly assigned a new identity and an ideal point from a specific preference configuration (the standard configuration is given by fig. 6.3). Assigning a new identity serves to eliminate the possibility that coalitions might form associating members with particular computer terminals. Randomization prevents participants from making choices based on knowledge of others in the experiment and contributes to participants pursuing their own induced valuation across alternatives since all members remain anonymous and payoffs are made in private. Participants' valuation for alternatives in the space were based on a monotonic decreasing function of distance from their ideal point with preferences being circular around each members' ideal point. In the SMB experiments, members' ideal points also were randomly assigned, and the configuration of ideal points was rotated in the space.[2]

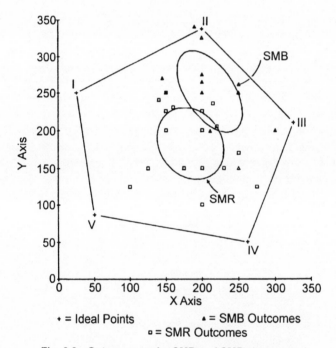

Fig. 6.3. Outcomes under SMR and SMB treatments

Proposals were made and compared using a forward-moving agenda process (see Wilson 1986a). This meant participants were presented with an initial status quo to which they could propose amendments. Each member was limited to offering 20 proposals during a single experimental round.[3] Using a binary voting procedure, members paired alternatives with the status quo. Since a number of alternatives could be on the floor at any time, alternatives required a "second" from another committee member in order to be considered an amendment to the status quo. Once an amendment was seconded all members voted. Simple majority rule decided the outcome, with the amendment supplanting the status quo if it received a majority vote and, if not, the status quo remaining unchanged. The agenda, in this instance, amounted to a process of continually amending the status quo. Members could choose to adjourn at any time, or the committee meeting was adjourned following 20 minutes of proposing and seconding alternatives. In both instances, the committee's outcome was the current status quo. In the discussion that follows, all outcomes are normalized to the same coordinate system.

Structural Blocking Setting

The treatment group retains the same general structure as that outlined above. Participants constructed an agenda of alternatives that led to the choice of a single point from a two-dimensional alternative space. However, in this treatment, a single individual was assigned blocking power. When an amendment to the status quo was proposed, all committee members voted either for the amendment or for the status quo. The exception was the member holding blocking power. That individual's choice was to *vote for the amendment* or to *veto the amendment.* If the latter choice was made, the status quo was retained even if a majority voted in favor of the amendment. Participants were informed that, although the amendment passed, it was vetoed. Except for the member holding blocking power, no one else was informed as to the identity of the blocker (though all were informed that someone held blocking power). As a consequence, if outcomes systematically favor the blocker this should be a function of the agenda structure, rather than an implicit recognition of power asymmetries by the other members. In these experiments blocking power was randomly assigned to a participant who held the (normalized) position of committee member II (see fig. 6.3).[4] This setting is similar to the designs by Isaac and Plott (1978) and Kormendi and Plott (1982) in which a single participant is assigned agenda power. The difference between our design and those others lies with our focus on negative agenda powers.

Expectations

Our expectations are quite simple. The pattern of outcomes should differ between the control and treatment groups. Our concern, then, is with predictions about how those two will differ. From the standpoint of formal theory we have no strong expectations as to the pattern of outcomes under our SMR control groups. Theoretically these outcomes can wander throughout the alternative space. However, from past empirical experience (McKelvey, Ordeshook, and Winer 1978; Fiorina and Plott 1978; Laing and Olmsted 1978) we recognize that outcomes from such a setting will tend to converge toward the center of the pareto optimal subspace. Here a variety of game theoretic solutions are suggestive—the Competitive solution, the Bargaining set, and the Copeland winner.

By contrast our model of blocking power yields two related predictions for our SMB treatment groups. The first is a *strong* prediction in which the equilibrium exerts strong attractive properties and is retentive. This means that an agenda will be constructed that converges on the equilibrium and once an outcome is at the equilibrium it will remain there. This predicts that SMB outcomes will fall at member II's ideal point. A second, *weak* prediction is also possible. The model of blocking sketched above and the existence of an equilibrium are dependent on zero transaction costs and a finitely long agenda. While there is no transaction cost for voting, there are opportunity costs for participating in these experiments. Consequently, assuming that participants will continue to build and vote on an agenda until the equilibrium is attained is too strong an assumption. Taking a weaker view of the attractive properties of the equilibrium, the second prediction holds that outcomes will be systematically biased toward the blocker—member II. At the same time, retaining the retentive properties of the equilibrium, we expect that blocking power will be exercised to prevent amendments that leave the blocker worse off.

From a comparative statics standpoint, then, outcomes between treatments will differ by the extent to which member II is advantaged by negative agenda power. Under a strong prediction SMB outcomes will be at member II's ideal point, while SMR outcomes will not. Under a weaker prediction SMB outcomes will converge to member II's ideal point while SMR outcomes will not. Under either the strong or weaker predictions, SMB outcomes will be closer to member II's ideal point than will be SMR outcomes. This reflects the "bias" introduced by a relatively simple institutional procedure.

Empirical Results

A total of 29 committee rounds was conducted, 18 using SMR and 11 using SMB treatment conditions. The outcomes under each treatment are plotted in fig. 6.3, and those outcomes are listed in table 6.1. Two points are highlighted by the results displayed in figure 6.3. First, the SMR and SMB outcomes are divergent—they differ as to their distributions in the alternative space. Second, SMB outcomes do not fall directly at the equilibrium. We address both of these points below.

TABLE 6.1. Transformed Experimental Outcomes

SMR		SMB	
Experiment: Round	Outcome	Experiment: Round	Outcome
D05:2	(125, 150)	G01:2	(250, 250)
D07:5	(150, 225)	G01:3	(200, 275)
D09:4	(140, 240)	G02:4	(150, 250)
D10:2	(200, 225)	G03:2	(250, 150)
D11:2	(100, 125)	G04:3	(211, 199)
D12:3	(200, 200)	G05:2	(145, 270)
D13:2	(275, 125)	G05:3	(200, 250)
D15:5	(220, 205)	G05:4	(190, 340)
D16:5	(200, 150)	G07:2	(200, 325)
D17:2	(250, 170)	G09:2	(300, 200)
D17:3	(175, 150)	G09:3	(200, 265)
D17:4	(230, 150)		
D17:5	(160, 230)		
D18:2	(150, 200)		
D20:2	(200, 100)		
D20:3	(150, 200)		
D21:2	(215, 235)		
D21:5	(150, 250)		

The striking point to these results is how SMB and SMR outcomes differ. Although the SMB outcomes do not fall precisely at the equilibrium, they appear to systematically favor the individual with blocking power. Meanwhile, under SMR, no single member appears to be advantaged. The apparent advantage enjoyed by member II under SMB was tested by calculating the Euclidean distance from member II's ideal point to each outcome. The relative ranking of distances under SMB and SMR committees was compared using a nonparametric estimator (Kruskal-Wallis). The results, listed in table 6.2, are striking. The differences in distances from member II's ideal point are significant ($p < .001$) and precisely in the expected direction. There is a consistent bias in outcomes that favors the individual endowed with blocking power.

To consider this point further, we can ask whether outcomes under these two treatments are part of the same distribution. In figure 6.3 we plot the one standard deviation confidence interval ellipsoids for both treatments. Here it is assumed that these outcomes are bivariate normally distributed in the alternative space (see Morrison 1976). While there is some overlap between these treatments, it is also apparent that the outcomes from each treatment come

TABLE 6.2. Comparison of SMR and SMB Committees

Variable	Treatment	Number of Cases	Mean	SD	Kruskal-Wallis (χ^2)	$P > \chi^2$
Distance from	SMR	18	161.5	46.6	11.22	.0008
Member II	SMB	11	93.9	57.4		
Time	SMR	18	888.1	321.9	1.89	.169
(seconds)	SMB	11	995.6	346.1		
Vote	SMR	18	10.8	4.5	0.32	.573
Total	SMB	11	11.7	6.1		
Proposal	SMR	18	32.9	14.8	0.55	.458
Total	SMB	11	33.9	12.9		

from very different distributions. In short, this specialized blocking power makes a real difference for collective choices.

It is interesting to note that the agenda processes characterizing these two different treatments are remarkably similar. It appears that blocking power interferes with the process of agenda construction only in that an individual holding power is able to prevent disadvantageous changes to the agenda. To see that this is the case, SMB and SMR committees are compared as to differences in the amount of time taken to reach agreement, the number of amendments voted on, and the total number of proposals offered during the round. Nonparametric comparisons over each of these measures indicate no significant differences between SMB and SMR committees (see table 6.2). Simply examining these variables might lead to a conclusion that both types of committee agendas were the same. However, this ignores the powerful effect of blocking power in shaping the agenda, since such power was employed in over a third of the votes cast.

Blocking power need not be passively utilized, since a blocker can combine such power with the clever use of proposal making. That is, not only can blockers prevent the adoption of disadvantageous amendments to the status quo, but such individuals can build an agenda that moves toward their own ideal point. One such example is indicated by figure 6.4. During this experiment (G5:2), the blocking power held by member II meant disadvantageous amendments to the status quo could be eliminated. Furthermore, member II either proposed or seconded alternatives 4, 5, and 7. This amounted to playing off committee members I and V against III and IV in order to move the status quo closer to her own ideal point. This shows that blocking power and skillful agenda manipulation can represent a powerful combination benefitting the blocker.

Strong and weak predictions were offered above concerning SMB out-

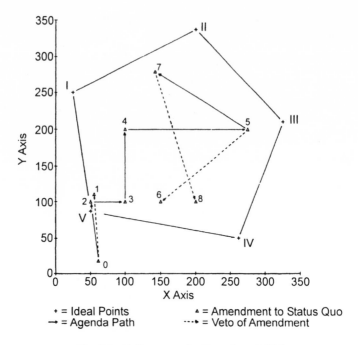

* = Ideal Points   ▲ = Amendment to Status Quo
→ = Agenda Path   ···• = Veto of Amendment

Fig. 6.4. Voting agenda: Experiment G5:2

comes. The strong prediction held that outcomes would converge to member II's ideal point. In the absence of transaction costs, once that point was selected as the new status quo it would be retained. However, only two SMB outcomes end up quite close to member II's ideal point. The remaining choices were scattered below that point. Two of 11 outcomes at the equilibrium hardly constitutes support for the strong prediction.

The weak prediction, on the other hand, held that outcomes would be biased in favor of the blocker. This was demonstrated above by using the SMR results as a comparative baseline. In part these findings indicate that the equilibrium is weakly attractive since outcomes are systematically biased toward member II's ideal point. "Attraction" in this sense means that changes to the status quo converge on and do not move away from the equilibrium. Of 68 amendment votes passed by a simple majority across all committees the blocker *never* failed to overturn amendments worsening his position. Only once did a blocker exercise negative agenda power, and this was to prevent a favorable amendment. This happened in experiment G5:2 discussed above, occurred at the first vote, and may only reflect the blocker experimenting with such power. Otherwise, blocking power was always appropriately used, and as a consequence the agenda always converged toward the equilibrium. These findings are consistent with our *weak* predictions concerning SMB procedures.

Conclusion

The experimental results reported here point to the effect of negative agenda power on outcomes. Quite simply granting an individual the power to obstruct a decision is to bias outcomes in favor of that blocker. This point is borne out by our experiments where blocking power is wielded by a single individual. Blocking power does not assure a specific outcome in the alternative space; it can only be used to protect the status quo. Clearly, such power can be paramount if the status quo is compatible with the interests of the blocker. This point is also true for more general forms of blocking where the power to obstruct is put into the hands of many blockers (see Herzberg 1986).

It is surprising that any decision-making body would employ blocking rules favoring a subset of individuals. Yet specialized blocking powers are often awarded in collective choice settings. In particular, committees in legislatures exercise these powers. Expertise and the division of labor are often cited as reasons for granting these specialized powers, even though blocking carries with it the possibility for abuse. Indeed the contemporary decentralization of the U.S. House of Representatives, coupled with respect for committee jurisdictional bounds, has enhanced the obstructive powers of committees. Even though legislation is now easier to force out of a committee, few bills are ever discharged. Committees, then, can (and often do) exercise tremendous blocking power. This point is well demonstrated by the findings from our experiments.

Negative agenda powers are not limited to a single individual or a committee. Such powers can be found in subcommittees in legislatures as well as with lower level policy planners in a bureaucratic hierarchy. The point is that whenever an individual or a group of individuals is endowed with the power to obstruct the agenda, those blockers are advantaged. Of course, these effects are mediated by the presence of other institutional rules affecting the agenda process. As we have shown by example and as Krehbiel (1985) argues, blocking in conjunction with proposal making makes negative agenda powers very powerful. This is consistent with empirical observations about the powers held by subcommittees and committees in modern legislatures. On the other hand, where many different groups are granted blocking power, then achieving policy change becomes more difficult, since the status quo becomes harder to amend. A contemporary example of many blockers is reflected in the increasing use of multiple referrals in the U.S. House of Representatives. More committees now are involved in shaping legislation.

In a general sense, generic negative agenda powers are indeed important. They bias outcomes toward those exercising such power. These powers
are a function of institutional rules that specify which subset of political actors is endowed with blocking powers. That such rules can make a difference
for the collective choice process is in line with recent work on "structurally
induced" equilibrium.

NOTES

1. For further information concerning participant instructions, see Wilson 1986b.
2. As with player identities, payoff functions were randomly assigned to participants
during each round. The payoff function and the associated parameters are given by

Payoff $= [B_i * \exp(A_i * \|X-p_i\|)]$			
			where X = any alternative
Participant (i)	B_i	A_i	p_i
1	$8.00	−.0116	(25,250)
2	$10.00	−.0120	(200,337)
3	$12.00	.0125	(325,210)
4	$13.00	−.0125	(262, 50)
5	$15.00	−.0140	(50, 87)

Three distinct sets of ideal points were used in the SMB experiments. They are given by

Participant	Set 1	Set 2	Set 3
1	(25,250)	(100, 25)	(325,100)
2	(200,337)	(13,200)	(150, 13)
3	(325,210)	(140,325)	(35,140)
4	(262, 50)	(300,262)	(88,300)
5	(50, 87)	(263, 50)	(300,263)

The coordinates of ideal points contained in Sets 2 and 3 are 90 and 180 degree rotations
(respectively) from the set of ideal points in Set 1. All data presented for analysis and
discussion have been transformed to reflect the coordinate system for Set 1.

3. Such a constraint did not appear to be too onerous for committee members. On
average each member sent only 7 proposals (with a standard deviation of 4.3). In no
instance did members send the full complement of 20 proposals.
4. Assigning the same relative position to each blocker should introduce no
systematic bias in the results. First, ideal points were rotated through the alternative space over each experiment. Consequently, roughly a third of all blocking
outcomes were selected in different areas of the alternative space. Second, the 11
different blocking outcomes were obtained from seven different committees. Con-

sequently, little bias should be introduced from committee members keying on the relative (but unknown) position of the blocker's ideal point relative to the ideal points of all others.

REFERENCES

Arrow, Kenneth. 1963. *Social Choice and Individual Values,* 2nd ed. New Haven: Yale University Press.

Bendor, Jonathan, and Terry Moe. 1985. "An Adaptive Model of Bureaucratic Politics." *American Political Science Review* 79: 755–74.

Berl, Janet, Richard McKelvey, Peter Ordeshook, and Mark Winer. "An Experimental Test of the Core in a Simple N-Person Cooperative Non-Sidepayment Game." *Journal of Conflict Resolution* 20: 3 (September): 543–79.

Cohen, Linda. 1979. "Cyclic Sets in Multidimensional Voting Models." *Journal of Economic Theory* 20 (February): 1–12.

Cohen, Linda, and Steven Matthews. 1980. "Constrained Plott Equilibrium, Directional Equilibrium and Global Cycling Sets." *Review of Economic Studies* 47: 5 (October): 975–86.

Denzau, Arthur, and Robert MacKay. 1983. "Gatekeeping and Monopoly Power of Committees: An Analysis of Sincere and Sophisticated Behavior." *American Journal of Political Science* 27: 4 (November): 740–61.

Denzau, Arthur, William Riker, and Kenneth Shepsle. 1985. "Farquharson and Fenno: Sophisticated Voting and Home Style." *American Political Science Review* 79: 1117–34.

Eavey, Cheryl, and Gary Miller. 1984. "Bureaucratic Agenda Control: Imposition or Bargaining." *American Political Science Review* 78: 719–33.

Enelow, James, and David Koehler. 1980. "The Amendment in Legislative Strategy: Sophisticated Voting in the U.S. Congress." *Journal of Politics* 42: 396–413.

Ferejohn, John, Morris Fiorina, and Edward Packel. 1980. "Non Equilibrium Solutions for Legislative Systems." *Behavioral Science* V. 25 (March): 140–48.

Fiorina, Morris, and Charles Plott. 1978. "Committee Decisions Under Majority Rule: An Experimental Study." *American Political Science Review* 72: 2 (June) 575–98.

Grofman, Bernard, and Guillermo Owen. 1985. "A New Look at an Old Solution Concept in the Spatial Context, and Some Evidence for its Predictive Power in Experimental Games." Paper presented at the Annual Meeting of the Public Choice Society, New Orleans, February 21–23.

Hammond, Thomas. 1986. "Agenda Control, Organizational Structure, and Bureaucratic Politics." *American Journal of Political Science* 30: 379–420.

Herzberg, Roberta. 1985. "Blocking Coalitions and Institutional Equilibrium." Ph.D. dissertation. Washington University.

_____. 1986. "Institutional Structure and the Power to Say No: Legislative Veto Powers." Manuscript.

Isaac, R. Mark, and Charles R. Plott. 1978. "Cooperative Game Models of the Influence of the Closed Rule in Three Person, Majority Rule Committees: Theory and Experiment." In Peter C. Ordeshook, ed. *Game Theory and Political Science*. New York: New York University Press.

Kiewiet, D. Roderick, and Matthew D. McCubbins. 1985. "Appropriations Decisions as a Bilateral Bargaining Game between President and Congress." *Legislative Studies Quarterly* 9: 181–202.

Kormendi, Roger C., and Charles R. Plott. 1982. "Committee Decisions under Alternative Procedural Rules." *Journal of Economic Behavior and Organization* 3 (June and September): 175–95.

Kramer, Gerald. 1973. "On a Class of Equilibrium Conditions for Majority Rule." *Econometrica* 41 (March): 285–97.

Krehbiel, Keith. 1985. "Obstruction and Representativeness in Legislatures." *American Journal of Political Science* 29: 3 (August): 643–59.

Krehbiel, Keith. 1986. "Sophisticated and Myopic Behavior in Legislative Committees: An Experimental Study." *American Journal of Political Science* 30: 3 (August): 542–61.

Laing, James D., and Scott Olmsted. 1978. "An Experimental and Game Theoretic Study of Committees." In Peter C. Ordeshook, ed. *Game Theory and Political Science*. New York: New York University Press.

McKelvey, Richard. 1976. "Intransitivities in Multidimensional Voting Models and Some Implications for Agenda Control." *Journal of Economic Theory* 12 (June): 472–84.

_____. 1979. "General Conditions for Global Intransitivities in Formal Voting Models." *Econometrica* 47: 1085–111.

_____. 1986 "Covering, Dominance, and Institution Free Properties of Social Choice." *American Journal of Political Science* 30: 283–314.

McKelvey, Richard, and Peter Ordeshook. 1978. "Competitive Coalition Theory." In Peter C. Ordeshook, ed. *Game Theory and Political Science*. New York: New York University Press.

McKelvey, Richard, Peter Ordeshook, and Mark Winer. 1978. "The Competitive Solution for N-Person Games Without Transferable Utility, with an Application to Committee Games." *American Political Science Review* 72: 3 (September): 599–615.

Morrison, Donald F. 1976. *Multivariate Statistical Models,* 2nd ed. New York: McGraw-Hill.

Oleszek, Walter. 1984. *Congressional Procedures and the Policy Process,* 2nd ed. Washington, D.C.: CQ Press.

Plott, Charles. 1967. "A Notion of Equilibrium and Its Possibility under Majority Rule." *American Economic Review* 57 (September): 787–806.

Plott, Charles, and Michael Levine. 1978. "A Model of Agenda Influence on Committee Decisions." *American Economic Review* 68 (March): 146–60.

Romer, Thomas, and Howard Rosenthal. 1978. "Political Resource Allocation, Controlled Agendas, and the Status Quo." *Public Choice* 33: 27–43.

Schofield, Norman. 1978. "Instability of Simple Dynamic Games." *Review of Economic Studies* 45: 3 (October): 575–94.

Shepsle, Kenneth A. 1979. "Institutional Arrangements and Equilibrium in Multidimensional Voting Models." *American Journal of Political Science* 23: 1 (February): 27–29.

_____. 1986. "The Positive Theory of Legislative Institutions: An Enrichment of Social Choice and Spatial Models." *Public Choice* 50: 135–78.

Shepsle, Kenneth A., and Barry Weingast. 1984. "Uncovered Sets and Sophisticated Voting with Implications for Agenda Institutions." *American Journal of Political Science* 28: 1 (February): 49–75.

Sloss, Judith. 1973. "Stable Outcomes in Majority Rule Voting Games." *Public Choice* 15: 19–48.

Slutsky, Stephen. 1978. "Characterization of Societies with Consistent Majority Decisions." *Review of Economic Studies* 44 (June): 211–25.

_____. 1979. "Equilibrium Under α-Majority Voting." *Econometrica* 47(5): 1111–25.

Wilson, Rick K. 1986a. "Forward and Backward Agenda Procedures: Committee Experiments on Structurally-Induced Equilibrium." *Journal of Politics* 48: 390–409.

_____. 1986b. "Results on the Condorcet Winner: A Committee Experiment on Time Constraints." *Simulation and Games* 17: 217–43.

CHAPTER 7

Policy Uncertainty in Two-Level Games: Examples of Correlated Equilibria

Michael D. McGinnis and John T. Williams

Nearly all game models in the international relations literature are based on the assumption that states can be modeled as if they are unitary rational actors maximizing some collective utility function. This assumption has proven to be a useful analytical device for tracing out the implications of fundamental assertions of realist and neorealist theories, but it remains inherently controversial (see Allison 1971; Young 1972; Krasner 1978; Waltz 1979; Bueno de Mesquita 1981; Gilpin 1981; Snidal 1985).

Recent research on two-level game models of international relations emphasizes fundamental connections between the domestic and foreign policies of states (Most and Starr 1984, 1989; Lake 1988; Putnam 1988; Mastanduno, Lake, and Ikenberry 1989; Bueno de Mesquita and Lalman 1990, 1992; Tsebelis 1990). This work represents a potentially significant advance over classic applications of game models to important patterns of interstate interactions, such as arms races (Rapoport 1960), crises (Snyder and Diesing 1977), nuclear deterrence (Schelling 1960; Powell 1990), wars (Bueno de Mesquita 1981), the balance of power (Niou, Ordeshook, and Rose 1989), and international cooperation (Axelrod 1984; Oye 1985). However, by modeling states as if they are unitary rational actors pursuing well-defined goals in both the domestic and international policy arenas, recent research on two-level games perpetuates the tendency of international relations game theorists to overlook fundamental dilemmas of social choice (Arrow 1963; Sen 1970; Plott 1976; Riker 1982).

Snidal (1985, 40–44) calls for development of a "theory of the payoffs" that would ground the policy preferences of states in the interests of domestic

Originally published in *International Studies Quarterly* 37 (1993): 29–54. Copyright © International Studies Association. Published by Blackwell Publishers. Reprinted by permission of Blackwell Publishers and the authors.

Authors' note: This research was supported by Grant No. SES–8810610 from the National Science Foundation. An earlier version was presented at the thirty-first Annual Convention of the International Studies Association, Washington, DC, April 10–14, 1990. We would like to thank Jeff Hart, Roberta Herzberg, Mark Lichbach, Cliff Morgan, Elinor Ostrom, and Rick Wilson for their helpful comments.

economic groups. However, social choice theorists have identified imperfections in preference (or interest) aggregation processes that can have important consequences for foreign policy. In particular, uncertainty about policy outcomes is a necessary consequence of the dynamic nature of coalition formation among domestic groups. Even if some coalition of groups "captures" the state (as in Snyder 1991), there is no guarantee that members of a governing coalition will continue to agree among themselves, especially with regard to newly arising issues. In the extreme case of majority voting cycles and other forms of social intransitivities, no consistent preference ordering can be defined at the collective level without violating basic conditions originally postulated in Arrow's theorem, the core result of social choice theory (to be discussed below).

We develop policy game models in which individuals compete to influence both domestic and foreign policy outcomes. Unlike previous two-level game models, our analysis does not require the existence of an autonomous state actor. Instead, the locus of strategic action remains at the individual (or group) level. Since strategic individuals use relevant information to form expectations about future outcomes and all other aspects of their games, we model individual policy actors as Bayesians. We argue, following Aumann (1987), that their expectations necessarily converge to a "correlated equilibrium." We demonstrate that this game theoretic solution concept implies that policy outcomes at the state and international levels follow regularized patterns, even in the presence of voting cycles and other sources of policy uncertainty. Furthermore, these patterns can be related in a meaningful way to the preferences of individual policy actors within one or more states.

In the first section we justify our assertion that the nature of states as collective entities requires the use of probabilistic strategies in game models of international relations. We then outline the behavior of Bayesian actors in generic policy games and use an example of a Chicken game (between unitary rational states) to illustrate the solution concept of correlated equilibrium. We then develop an example of political competition among three domestic groups in which voting cycles can occur, and we briefly discuss the implications of correlated equilibrium for the "meaning" of social choice. Our final example incorporates the domestic and foreign policies of two rival states into a single game involving groups within both states. We conclude with implications of correlated equilibrium for the dilemmas of social choice associated with Arrow's theorem and for the analysis of political institutions at all levels of analysis.

Probabilistic Strategies and Policy Uncertainty

Important insights about arms races and crisis behavior have been drawn from simple game models (see, e.g., Snyder 1971). However, we argue that some of the basic analytical tools of game theory are poorly suited to models of international relations. We are particularly concerned about the appropriateness of the game theoretic concept of "strategy," which specifies a rational actor's behavior in all possible sets of circumstances. Any assertion that states are fully strategic in this sense is problematic.

Analysts of bureaucratic politics emphasize that foreign policies are affected by the actions of multiple organizations that often pursue conflicting goals (Allison 1971, Model III). Also, any examination of organizational processes demonstrates that errors in policy implementation cannot be avoided (Allison 1971, Model II). Furthermore, major changes in foreign policy goals and behavior can result from leadership changes or from the changing perceptions of individual leaders (Jervis 1976). Finally, social choice theorists have demonstrated that policy outcomes can change dramatically as winning coalitions form and dissolve for a wide variety of reasons (Riker 1982). In short, policies that result from the outcome of competition among domestic groups are generally more difficult to interpret than are policies determined by a single rational actor.

Applications of game models in the international relations literature usually require a state to select a "pure strategy" that unambiguously determines its behavior in all circumstances, whether the game is played once or repeated over time. Even "mixed strategies" in which strategic players select specific actions from a well-defined probability distribution allow neither implementation errors nor the uncertain consequences of domestic competition over foreign policy. If states are incapable of playing a consistent strategy, then neither pure nor mixed strategies can suffice to model their foreign policy behavior. Instead, a stochastic error (or noise) component should be included as a fundamental component of game models of international relations, and this stochastic component should be directly related to processes of domestic competition over policy.

In recent models of international crises as games of incomplete information, states are uncertain about each other's preferences and send signals of their level of resolve in order to influence the outcome of the crisis (e.g., Powell 1990). However, the extent of uncertainty is tightly controlled in these signaling game models, since each preference "type" is assumed to be a unitary actor that can implement optimal strategies without error. We investigate a more fundamental source of policy uncertainty, namely, the nonunitary nature of states.

Similar issues have been discussed in more informal analyses. Putnam (1988) examines the relationship between domestic and international constraints on feasible agreements in "two-level games" in which negotiators must deal with each other and with the domestic groups needed to ratify any agreement. Mastanduno, Lake, and Ikenberry (1989) examine interactions between the domestic and foreign policy instruments that states use to pursue their goals in either policy arena. In their analysis of the substitutability of policy instruments, Most and Starr (1984, 1989) provide brief sketches of a formal model of a government's efforts to maintain its viability against domestic and external threats to its security. Finally, Tsebelis (1990) examines "nested games" in which players maximize the utility they receive from two overlapping games.

Any model of two-level or nested games necessarily raises questions about how individual preferences are aggregated into a collective preference structure. Social choice theory indicates that any mapping of individual preferences into a collective preference ordering is, at best, imperfect. The dilemmas of social choice are usually ignored in international relations theory, although there are a few notable exceptions. Bueno de Mesquita (1981, 12–18) carefully demonstrates that his calculations of the expected utility of a single, dominant actor are consistent with Arrow's Theorem, but this consistency is obtained at the cost of ignoring domestic politics. Bueno de Mesquita, Newman, and Rabushka (1985) apply a similar model to coalitions among domestic political groups, but this model does not directly address the problem of social intransitivities. In another rational model of domestic-international linkage, Bueno de Mesquita and Lalman (1990) emphasize the uncertain costs of domestic opposition that a (unitary) leader faces in decisions to go to war.

Two recent works deal directly with social choice dilemmas, but each is applicable only to special cases. Fogarty (1990) examines the uncertainties entailed in efforts to deter an adversary state in which domestic actors have different preferences over crisis outcomes. However, to ensure that a state's collective preferences are transitive and "deterrable" he imposes the restrictive condition that each individual's preferences are also deterrable. Achen (1988) demonstrates that any policy outcome can be represented as the winner selected by weighted majority vote based on some specific distribution of voting power among individual advisors (see also Pattanaik and Peleg 1986). However, the distribution of voting weights must be redefined to explain each outcome.

Finally, Morrow (1988) demonstrates that the international system aggregates the interests (preferences) of different states into policy outcomes at the international level (see also Oppenheimer 1979). Yet, Morrow's analysis remains an incomplete application of social choice theory because it leaves the unitary nature of state action uncontested. We develop a game theoretic solution concept that deals with social intransitivities and other sources of policy uncertainty in games within or between states, or both.

Bayesian Rationality and Policy Games

We base our game models on the fundamental premise of Allison's (1971) Model III: foreign policies are determined by some process of interaction among individuals with different personal interests and divergent perceptions of the national interest. We use *policy actor* as a generic term covering any individual or cohesive group having potential influence over a given state's policy in a given issue area and *policy game* to represent the overall structure of competition among policy actors seeking to influence a state's policy in some issue area. Participants in policy debates pursue both common and conflicting interests, and each individual seeks to influence policy outcomes by disseminating information, making proposals, forming coalitions, manipulating agendas, and other means.

Policy games are, in the technical sense, noncooperative games. Although policy actors communicate with each other on a regular basis, there is no way to ensure the enforcement of agreements. Nonetheless, policy actors regularly find ways to implicitly or explicitly coordinate their behavior. In particular, policy actors establish political institutions and develop norms of behavior that facilitate the convergence of their expectations with the expectations of other actors. Rational individuals establish institutions in order to facilitate communication, reduce uncertainty, and otherwise help coordinate the behavior of a large number of actors sharing some common interests (Keohane 1984; Bendor and Mookherjee 1987). Furthermore, institutional rules facilitate policy coordination by delimiting the range of individuals who are allowed (or have the right) to participate in policy debates, as well as by setting deadlines for decisions and procedures for conflict resolution and otherwise structuring interactions (E. Ostrom 1986).

A policy game is also a game of incomplete information, since no individual can be sure of the preferences of any other actor. We follow the standard practice of assuming that a rational actor assigns subjective probability assessments over relevant and uncertain aspects of the decision environment. An actor's evaluation of the likelihood of possible outcomes is defined as that actor's "beliefs," and we assume a rational actor chooses the option that maximizes expected utility given these beliefs.

Because information is costly, no rational actor will gather and evaluate all possible information before acting (see Hogarth and Reder 1986). Yet, Bayesian rational actors use information efficiently, by updating their prior beliefs (or expectations) according to Bayes' theorem.[1] The resulting posterior beliefs become the priors for later decisions and will be updated again whenever new information is observed. Although not intended as a detailed model of actual decision processes,[2] Bayesian rationality provides an abstract representation of the responsiveness of actors to new information. In short, Bayesian decision theory is a natural extension of the rational choice approach to cover dynamic and uncertain situations.

The behavior of other policy actors is a very important part of each policy actor's decision environment. We assume that each individual policy actor assigns subjective probability distributions over the feasible actions of other individuals, as well as over feasible policy outcomes. By doing so we depart from the standard approach to noncooperative games of incomplete information developed by Harsanyi (1967–68), in which individual actors assign beliefs over the other players' possible "types." Each type of actor is defined by different preferences or access to other private information.

Harsanyi asserts that game players cannot directly assign probabilities to the actions of other strategic agents, who must instead be assumed to be following some equilibrium strategy. Harsanyi players use Bayes' rule in a very specific way, namely, to update their beliefs concerning the preference types of other actors after observing these actors' behavior. Aumann (1987) argues that game players are Bayesian in a more fundamental sense; they use whatever information they find useful, from whatever source, in assigning subjective probabilities over whatever aspects of their environment they consider important, including the behavior of other actors. These expectations are informed by direct observation of other actors' past behavior, indirect inferences concerning the goals these other actors are pursuing, and any other potentially useful information. Aumann (1987, 2) justifies this conceptualization as follows.

> According to the Bayesian view, subjective probabilities should be assignable to every prospect, including that of players choosing certain strategies in certain games. Rather than playing an equilibrium, the players should simply choose strategies that maximize their utilities given their subjective distributions over the other players' strategy choices.

Aumann's conceptualization is particularly suited to policy games. Because competing policy advocates routinely provide evidence to support their proposals, participants in policy debates are confronted with an environment rich in information. In fact, policy actors are confronted with so much information distributed by so many different sources and interpreted in so many different ways that it is not unreasonable to say that exchange of information constitutes the very essence of politics. In short, policy actors interpret a vast amount of relevant information in a relatively effective manner (Wittman, 1989).

Correlated Equilibria in a Chicken Game

Aumann (1987) demonstrates that Bayesian rational actors in a noncooperative game of incomplete information will find some way to implicitly correlate their behavior, even if the result falls short of the perfect coordination allowed in models of cooperative games. He derives the following general result: if game players are Bayesian rational actors, then a nonempty set of correlated equilibria to that game exists (Aumann 1987, 7).

To introduce the concept of correlated equilibrium we present in figure 7.1*a* a Chicken game discussed by Aumann (1987, 5; 1974, 71–72). The CD and DC outcomes are pure strategy Nash equilibria, and in the mixed strategy equilibrium each actor plays C with probability 2/3 and D with probability 1/3. Mixed strategies are, by definition, independent; therefore, the joint probability distribution over outcomes is the product of the marginal probabilities of each player's choice, as shown in figure 7.1*b*.

Aumann argues that Bayesian actors have access to information relevant to the play of the game that cannot be summarized in the game matrix and that they will naturally find some way to use this information to coordinate their behavior. Formally, this information amounts to a "common random device" that suggests actions to each player. Aumann derives conditions for a "correlated equilibrium distribution" (CED) in which no player can benefit by deviating from these suggestions. Each player's strategy is a probability distribution over its available options, as in a mixed strategy, but the choices suggested to different players are correlated because they are drawn from the same underlying joint probability distribution. That is, strategies are correlated because different actors condition their behavior on related information.[3]

Aumann uses the probability distribution in which the DD outcome never occurs and the other three outcomes occur with equal probability as an example of a CED (fig. 7.1*c*). To illustrate the basic logic of correlated strategies, consider the following arrangement, taken from Kreps (1990, 411–12). The two players condition their behavior on the roll of a single die, as reported to them by some third party. B receives a suggestion to play C whenever one, two, three, or four appears, and A is suggested to play C for values of three, four, five, or six. The die plays the role of the common random device, and each player has only limited "private information" concerning its outcome. Players know their suggestion but not the actual number that appears on the die. However, each suggestion also implies a conditional probability distribution over the other player's behavior. If B gets the suggestion to play C, then B believes A has a .5 probability of playing C or D; if B is signaled to play D, B believes A will play C with certainty.

Players behave differently if they have complete information. For example, if player B receives a signal to play C but also knows that the die is a four, then B would know that A was signaled to play C, and B will deviate from the signal by playing D. It is the remaining uncertainty that makes this equilibrium self-enforcing in the sense that "no player can gain by deviating from the suggestions, given that the others obey them" (Aumann 1987, 4). Any coordinated action must remain in the self-interest of each player.

An extensive game formulation is implicit in Aumann's analysis. First,

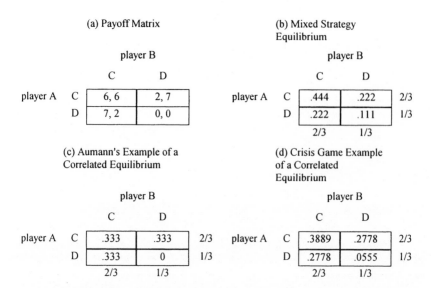

Note: In part (a) the entries in each cell represent the utility to each player, with payoffs to row player A listed first. In the remaining parts, the entry in each cell represents the probability of that outcome's occurrence under the appropriate equilibrium concept. The marginal distributions denote the overall probability distribution over a given player's choices. The payoff matrix and the example in part (c) is taken from Aumann (1987, 5), although Aumann does not label the players or their options.

Fig. 7.1. Payoffs and correlated equilibrium distributions in a Chicken game

each player receives a suggestion to play C or D by observing private information drawn from some common source (which need not literally be some third party). This signal enables each player to form conditional beliefs about the other player's likely behavior. Second, players select their moves (simultaneously) by using their beliefs about the other player's behavior to calculate their own expected utility of following or deviating from this suggestion. Given their beliefs, players select the option that maximizes their expected utility. In a correlated equilibrium distribution each player always follows either suggestion.

To demonstrate that the distribution in figure 7.1c is a CED, we define q_1 and q_2 as A's conditional beliefs that B receives a suggestion to play C given that A receives a suggestion to play C and D, respectively. If A receives a suggestion to play C, then A's expected utility of following this suggestion equals $6q_1 + 2(1 - q_1) = 4q_1 + 2$. By deviating to play D, A can expect to receive $7q_1 + 0(1 - q_1) = 7q_1$. Thus, player A should follow this suggestion if and only if $4q_1 + 2 \geq 7q_1$, which reduces to $q_1 \leq 2/3$. If A receives a suggestion to play D, then A should follow if and only if $7q_2 \geq 6q_2 + 2(1 - q_2)$, which reduces to $q_2 \geq 2/3$. In summary, $q_2 \geq 2/3 \geq q_1$. A similar condition can be derived for p_1, and p_2, denoting B's beliefs that A will play C conditional on B's suggestion of C and D, respectively. In Aumann's example $q_2 = p_2 = 1$ and $q_1 = p_1 = 1/2$.

The mixed strategy equilibrium is a special case of zero correlation. Independence of the player's strategies implies that $q_1 = q_2 = q$ and $p_1 = p_2 = p$, in which case the conditions reduce to $q = p = 2/3$. Each player receives an expected payoff of 5 from Aumann's example CED, compared to 4.67 from the mixed strategy equilibrium. Players could do still better by always playing CC, but this arrangement would require a high degree of coordination that may not be feasible in all situations. After all, if player A is certain that B will play C, then A will be tempted to play D. Instances of unilateral defection do occur in Aumann's correlated equilibrium distribution but not so often as to upset a mutually beneficial arrangement in which CC also occurs with some regularity.

This Chicken game can also be interpreted as a game between two unitary rational states. To illustrate, assume that two states are engaged in a long-standing rivalry involving a series of crises and other confrontations. In each time period each state selects a high (D) or a low (C) level of resoluteness. If both states are resolute at the same time (DD), then a crisis erupts that might escalate to war. If each state is willing to compromise, then CC occurs. If one state is more resolute than the other (CD or DC), then the state playing D wins that confrontation. If each state most prefers winning to compromise to losing and least prefers the risk of crisis escalation, then each state's payoffs match those of a Chicken game.

Aumann's example distribution (fig. 7.1c) is not credible in this game, since rival states cannot be expected to avoid crises entirely. But consider the correlated equilibrium distribution listed in figure 7.1d, in which the DD outcome occurs only half as often (.0555) as under the mixed strategy. Both states receive a higher expected utility (4.83) than in the mixed strategy equilibrium but less than if they could attain Aumann's example. In effect, both states make themselves better off by alternating their periods of resoluteness.

Rival states might realistically attain such a behavioral distribution during a series of repeated confrontations, provided they are able to find or establish some shared source of information that helps each determine at least some of the times when the other state is more likely to be resolute. Consider an example in which two rival great powers face a series of opportunities to intervene in various regional disputes. Each state knows whether or not a given situation threatens its vital interests, but each remains uncertain about whether the other state sees its own vital interests at stake. These two states see themselves as rivals, and the relationship that necessarily exists between their security interests may be the means by which they are able to correlate their behavior. In particular, they may be able to work out an informal understanding of their respective spheres of influence. That is, state B can be said to receive a suggestion to cooperate during crises in regions within state A's sphere of influence, whereas A would receive a signal to stand firm. If a particular region does not lie within either state's sphere of influence, then a compromise outcome could be easily worked out. However, instances of DD could still occur whenever a situation threatens the vital interests of both states.

Even if the two rival states manage to work out a clear demarcation of nonoverlapping spheres of influence, complications may arise from domestic political pressures. That is, policymakers facing a loss of domestic support may stand firm when their vital interests are not threatened or even when the confrontation occurs within the rival's recognized sphere of influence. In addition, uncertainty about the domestic pressures confronting the other state may make a state more willing to compromise on some confrontations occurring within its own sphere of influence. In general, neither state will be certain whether or not the other state will stand firm in a given situation, but each state will have access to a wide variety of informational clues. By operating on the basis of common information about their respective spheres of influence and informed guesses (based on private information) concerning the extent of domestic pressure in the other state, these rival states should be able to arrive at some arrangement that could be represented in the form of a correlated equilibrium distribution.

Communication, Information, and Correlated Equilibria

In subsequent sections we show that correlated strategies are ideal for modeling the logic of two-level games, but one of the strengths of correlated strategies is that they are also useful for modeling games of uncertainty between unitary actors. As suggested above, policymakers of rival states facing a series of confrontations will find it in their interests to arrive at some mutually beneficial arrangement. However, standard game models of pure or mixed strategy equilibria are inadequate to address the possibilities open to them. It has long been known that the concept of mixed strategies, the typical approach to rational probabilistic behavior in game models, is grossly inappropriate to the study of communication.

> A randomized strategy is dramatically anticommunicative; it is a deliberate means of destroying any possibility of communication, especially communication of intentions, inadvertent or otherwise. It is a means of expunging from the game all details except the mathematical structure of the payoff, and from the players all communicative relations. (Schelling 1960, 105)

Mixed strategies are a means of hiding information from other players, to make sure that one's opponents are unable to take advantage of any information about one's own choice. Political actors certainly hide some information from others, but they also devote considerable effort to exchanging information as well as to gathering information, from whatever source, about the likely behavior of other political actors. Mixed strategies require the very restrictive assumption that all these efforts to extract clues about possible future behavior of other players are doomed to failure, in the sense that these efforts cannot provide any useful information. In general, players' information-gathering efforts will be imperfectly successful: some information remains secret, but much information is commonly available.

Schelling (1960, 73) asserts that "any analysis of explicit bargaining must pay attention to what we might call the 'communication' that is inherent in the bargaining situation, the signals that the participants read in the inanimate details of the case." If game models are to be useful as descriptive models of political behavior they must allow policy actors to make efficient use of whatever information is available to them. Yet, it is generally impossible for analysts to include all relevant details in a specific model. Without access to this detailed information, analysts are unable to predict exactly which outcome actors will select, but the set of correlated equilibria includes all outcomes that can be sustained on the basis of individual self-interest. In other words, Aumann's correlated equilibrium enables analysts to implicitly incorporate the effects of all available information without explicitly representing these informational clues in detail.

Of course, if details about important common sources of information are known to the analyst, then it is worthwhile incorporating these details into the model. Aumann's common random device provides a general means to represent the implications rational actors draw from the details of their situation. Our interpretation of Aumann's Bayesian correlated equilibrium differs in fundamental ways from most applications of his earlier (1974) definition. Maskin and Tirole (1987) are typical in their use of sunspots as an exogenous source of common signals. We see no need to resort to such implausible devices. In his subsequent formulation, Aumann (1987) demonstrates that the existence of correlated equilibrium is a direct consequence of the assumption that it is common knowledge that all players are Bayesian.

The solution concept of correlated equilibrium has a fundamental connection to processes of communication among game players. As Myerson (1991, 249–58) demonstrates, correlated equilibria in a normal form game correspond to Nash equilibria in the extension of that game to a "communication game" in which players are allowed to freely communicate with each other. However, analysis of communication games would require a representation of all feasible messages and actions, and thus the corresponding set of correlated equilibria is considerably more convenient to analyze.

The common random device is best interpreted as an abstract representation of the information (from whatever source) that enables players (individuals or states) to converge to a mutually consistent set of expectations. As such, it is an essential component of the play of the game. In particular, political institutions, norms, and rules are crucial components of the "common random device" endogenous to the play of policy games. In our example above, an informal demarcation of spheres of influence contributes to the ability of rival states to coordinate their behavior, as does some understanding of domestic political pressures in both states. In effect, the correlated equilibrium distribution given in figure 7.1*d* balances the effects of these external and domestic signals.

To some extent the existence of correlation between players' strategies is a simple consequence of their common reaction to shared information. But even more important, players actively seek out information regarding each other's behavior. One of the major reasons political actors establish institutions is that they provide crucial clues as to the likely behavior of individuals occupying institutional positions. By monitoring compliance with these prescriptions, institutions also facilitate the updating of a Bayesian's beliefs. In short, Bayesian game players' efforts to learn more about the determinants of other players' behavior are a fundamental part of their strategic interaction.

In summary, we interpret player strategies in game models of political phenomena as inherently probabilistic and potentially correlated. For the case of game models of the policies of states, these probability distributions should reflect the outcome of competition among domestic groups.

Correlated Equilibria in a Voting Game

In this section we assess the ability of Bayesian actors to effect coordination of policy choice in the presence of social intransitivities. Although a full-fledged model would have to deal with aggregation issues at all levels of analysis, for simplicity we examine a two-level formulation in which three policy actors determine the domestic and foreign policies of state A. These actors can be interpreted as individual policy advisers or as cohesive groups. Whether policy actors are interpreted as policy advisers, bureaucratic organizations, political parties, or socioeconomic classes, if they are Bayesian then their beliefs will naturally result in correlated behavior.

Figure 7.2 depicts a simple spatial voting game with two policy dimensions, three voters (1, 2, 3), three alternatives (K, L, M), and a "deadlock" outcome N. The use of two-dimensional policy spaces is commonplace in formal models of elections or committee voting (see Ordeshook 1986), and similar models have been applied to international crises (Morgan 1984; Morrow 1986). For purposes of illustration we interpret the vertical dimension as a measure of the hostile activity (whether military or diplomatic) state A directs toward its rival, state B. The horizontal dimension represents A's level of activity with regard to some domestic issue, such as welfare policy. The three nondeadlock outcomes represent potential compromises on these two issues, with the set of feasible alternatives limited by (unspecified) institutional constraints.

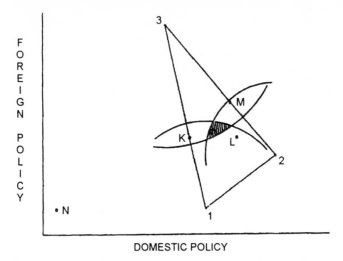

DOMESTIC POLICY

Note: Each player's ideal point is represented by a number, proposals by capital letters, and the status quo by the deadlock result N. The straight lines between each player are contract curves. Each arc represents a set of hypothetical proposals to which each voter is indifferent when compared with the correlated equilibrium distribution #1 described in Table 7.1. The hatched region represents hypothetical proposals which all voters prefer to this correlated equilibrium distribution.

Fig. 7.2. A spatial voting example

Of the three policy actors portrayed in figure 7.2, player 3 most prefers a higher level of hostility directed toward the rival than occurs in any of the feasible compromise outcomes. Similarly, player 1 most prefers a lower level (on that same dimension) than is feasible, and player 2 most prefers an unattainable level of activity on the other dimension. The deadlock outcome N represents a breakdown in the capacity of this state to act effectively in either policy area.[4]

For an alternative interpretation, assume that domestic groups prefer different combinations of policies concerning imperialistic expansion and domestic redistribution (see Snyder 1991). Higher values on the vertical dimension represent increased emphasis on imperialistic expansion, and the horizontal dimension designates alternative income redistribution policies that convey differential advantages to domestic groups. As drawn, group 3 is most and group 1 least interested in an expansionary foreign policy, and group 1 lies between the other groups on the domestic policy dimension.

We use this highly stylized example to highlight the implications of voting cycles, since situations in which voting cycles occur present a "worst case scenario" for anyone trying to base consistent state policy on the aggregation of domestic interests. These alternative interpretations should be seen as heuristic only, as this example is not intended to represent any specific instance of domestic-international policy linkage. Presentation of a detailed example of a particular empirical case would require identification of the most salient policy dimensions and the most influential domestic groups, as well as a specification of their ideal policy positions and institutional constraints on the range of feasible compromises. The controversies that would necessarily arise in these specifications would detract from making our more general point about the implications of the possible existence of social intransitivities in nonunitary models of foreign policy.

For the purpose of modeling domestic competition within one state, we (temporarily) treat the other state's policy as exogenous. For simplicity we assume that each actor shares the same level of expected threat, namely, that each player in A perceives that the policies of state B will constitute a high security threat with a probability of 1/3. (These beliefs are consistent with the equilibria defined below.)[5]

Using the standard assumption that each player has a quadratic-based utility function with circular indifference contours, players prefer outcomes closer to their ideal point in the spatial representation of figure 7.2.[6] As drawn, player 1 orders the outcomes K-L-M-N, player 2 L-M-K-N, and player 3 M-K-L-N. Deadlock is each voter's least preferred outcome, and a voting top cycle exists among all three alternatives to N, assuming simple majority rule.

In theory, cyclical outcomes in multidimensional space are common, but cyclical outcomes are observed only rarely.[7] Recent research focuses on the roles political institutions play in producing stability and equilibria in voting games. "Structurally induced" equilibria result from specific agenda procedures, rules of order, and jurisdictional restrictions on legislative committees (Shepsle 1979; Shepsle and Weingast 1981). Yet, institutions are not always exogenous, and they are subject to change by "frustrated majorities" (Riker 1982, 189).

It is standard practice to represent specific agendas in a game theoretic model in extensive form. Assuming complete information and sincere voting on the part of committee members, the individual controlling the rules or agenda can structure the game tree to have a unique (structure induced) pure strategy equilibrium. However, a large number of different extensive form games can represent institutional structures that induce outcomes. The choice of agendas is itself endogenous to the play of policy games (Austen-Smith 1987), so by assuming that a specific extensive form is exogenously determined we would ignore some of the most fundamental aspects of politics.[8]

To avoid imposing an exogenously determined agenda we analyze this policy game in normal form, as illustrated in figure 7.3a. Implicit in our analysis is an extensive form representation in which each player receives a suggestion to support one of the alternatives and then decides whether or not to follow that suggestion. For simplicity we assume that players support either their first or their second most preferred choice at the final stage of the policy process.[9] A simple majority (two voters) is necessary for a proposal to win, once an agenda plays itself out. Deadlock occurs when no alternative receives a majority.[10] This is essentially a game of coordination because each voter has an interest in avoiding deadlock.

There are three pure strategy Nash equilibria (denoted by *) in this game. In each of these equilibria two players form a winning coalition, one supporting his or her most preferred outcome and the other his or her second choice. Outcomes K and M result when player 3 forms a coalition with players 1 and 2, respectively, and L results when players 1 and 2 form the winning coalition.[11] No core exists, so all Nash equilibria are unstable and new coalitions can be formed to change the outcome.

Bayesian players use available information to make their strategic choice. However, they do so in the absence of certain knowledge about the behavior of their potential coalition partners and other coalitions. Their beliefs about the likely choices of other players are based on knowledge of coalitions and constituents, policy arguments and floor debates, past votes, expectations of membership in future coalitions, directions from leaders, and inside information. For example, committee members may use evidence of constituency pressure on one of their fellow members to determine that he or she cannot support outcomes that appear to violate constituent interests, thus limiting that member's capacity to vote strategically (see Denzau, Riker, and Shepsle 1985). Furthermore, committee chairs and party leadership send signals about the type of agenda and rules to use, and they may suggest who should vote for a less preferred choice in any particular circumstance.

Rather than attempt to model these probabilistic beliefs directly, we summarize actors' beliefs in a joint distribution of suggestions. Each voter receives a suggestion to support one of the three proposals, and this suggestion also conveys more detailed information in the form of a conditional probability distribution over the likely behavior of the other players. If no player can gain by deviating from this suggestion, then this collection of beliefs sustains a correlated equilibrium.[12]

(a) Outcomes and payoffs

Player 1			K				L	
Player 3		M		K		M		K
Player 2 L	N	4.65 0.0 1.68	K*	7.68 7.23 6.51	L*	7.53 8.28 5.64	L	7.53 8.28 5.64
M	M*	6.75 7.80 6.84	K	7.68 7.23 6.51	M	6.75 7.80 6.84	N	4.65 0.0 1.68

(b) Distribution of #1 (symmetric)

Player 1		K		L	
Player 3		M	K	M	K
Player 2 L		.000	.333	.333	.000
M		.333	.000	.000	.000

(c) Distribution of #2 (asymmetric)

Player 1		K		L	
Player 3		M	K	M	K
Player 2 L		.037	.277	.077	.038
M		.277	.138	.138	.018

Note: In part (a) each player's payoffs are given for each outcome, with player k's payoffs in the kth row. Each player's actions are arranged in order from 1st to 2nd preferred outcome; in this 2×2×2 version, players are restricted to choose between their 1st and 2nd most preferred outcome. Pure strategy Nash equilibria are denoted by asterisks. Each of the probability distributions in parts (b) and (c) is a correlated equilibrium distribution to this game.

Fig. 7.3. Examples of correlated equilibria in a spatial voting game

Two examples of correlated equilibria distributions for our voting game are listed in parts *b* and *c* of figure 7.3.[13] Table 7.1 summarizes the overall probabilities of outcomes and the expected payoffs associated with both examples.

In the first example (CED 1), players always manage to arrive at one of the three Nash equilibrium outcomes, each of which is equally probable. This example presumes a high degree of correlation among the players, since they are always able to avoid deadlock.

In the second example (CED 2), all possible signal combinations occur with nonzero probability, and outcome L occurs much less frequently than K or M. This asymmetric example represents a situation in which players 1 and 2 find it more difficult to form a coalition with each other than either does with player 3. As shown in table 7.1, player 2 supports his or her first choice less than half the time, reflecting that player's relatively poor bargaining position. Asymmetries of this type could result from many sources. Political

TABLE 7.1. Equilibrium Payoffs and Outcome and Signal Distributions
of Examples of Correlated Equilibrium Distributions in Spatial Voting Game

	Distribution #1 Symmetric	Distribution #2 Asymmetric
Outcome probabilities		
K	.333	.415
L	.333	.115
M	.333	.415
N (deadlock)	.000	.055
Probability of supporting most preferred outcome		
Player 1	.667	.728
Player 2	.667	.428
Player 3	.667	.528
Average Equilibrium Payoff		
Player 1	7.32	7.11
Player 2	7.77	7.19
Player 3	6.33	6.28

Note: The average equilibrium payoffs are computed by multiplying the probability of an alternative winning by the utility associated with that alternative for each player and summing these products over all possible outcomes.

institutions convey differential advantages to groups, and some groups may have a history of adversarial relations that is difficult to overcome. Conversely, other groups may see each other as natural allies.

Each player's expected payoff is larger for CED 1, since players avoid deadlock completely. Yet, the second distribution may be more realistic, given that deadlocks may occur if strategic individuals manage to convince others they are willing to accept deadlock. Although the threat in this case may not be credible in a single play (no rational player would vote for N at the last decision node in a specific agenda structure), a player may seek to cultivate a reputation of being an "irrational," or emotional, player. However, the nature of Bayesian updating of beliefs should reduce the probability of deadlock in the long run. We base all following examples on the first distribution.

Bayesian Beliefs and the Meaning of Social Choice

Another property of the symmetric example deserves mention. Each player's expected payoff lies between that player's payoffs for his or her second and third choices. Figure 7.2 includes the three indifference contours that represent hypothetical combinations of outcomes that give a player the same utility as that expected from CED 1. Each player would prefer any point within the hatched region to the correlated equilibrium, but any single outcome within that region could be defeated by many proposals, including K, L, or

M. If players have complete information about each other's preferences they can manipulate agendas and/or vote strategically in efforts to determine the final outcome (McKelvey 1976). Unless some specific agenda structure is imposed, the policy outcome remains indeterminate. In contrast, Bayesian players, facing incomplete information, manage to maintain relatively desirable expected payoffs.

Each player's expected payoffs must be given a long-run interpretation, because for each single play either K, L, M, or N wins. Whether interpreted as an alternation scheme in repeated plays of the same game or as a suggestion in a single play game, Bayesian correlated equilibrium grounds these outcomes in actors' beliefs about the nature of their strategic interaction.

Under a repeated play interpretation, the specific policy outcome selected varies over time, but the correlated equilibrium still imparts a regularity to state policy absent in many classical game models of strategic interaction. In the long run, multiple plays by the same actors serve to distribute benefits to all players, thus having the same effect as logrolling without explicit cooperation. Yet, the probability distribution over outcomes is derived from the players' beliefs, thus avoiding the artificiality of mixed strategy equilibria in voting games (see Ordeshook 1986, 180). For example, implicit in our signal distribution are factors such as constituency pressure that might lead all players to prefer some regular alternation of outcomes to the vagaries of less constrained strategic manipulation.

However, Bayesian actors also use their beliefs to inform their decisions in single play games. A correlated equilibrium represents a convergence of actor beliefs that implies a distribution over feasible outcomes. Given these beliefs, it remains in each player's self-interest to follow the suggestion implied in the correlated equilibrium distribution, for only then can successful coordination be achieved.

In our example we specify the probability values associated with each combination of signals, but in real games players arrive at these beliefs through observation of the many sources of information discussed above. As is usually the case, multiple equilibria exist, and we cannot model the detailed information actors use to select a particular equilibrium. Even so, we consider our results more plausible than the well-known "Folk theorem," which states that players in coordination games with multiple Nash equilibria can use "trigger strategies" to sustain any probabilistic mixture over these equilibria by credibly threatening to retaliate for any deviation from that agreement (see Aumann 1981; Fudenberg and Maskin 1986; Kreps 1990). Yet, trigger strategy equilibria require a high degree of coordination among players since even a single mistake by an otherwise cooperative player can trigger a long series of retaliations. In our policy game the threat of deadlock plays a role comparable to retaliation in trigger strategy equilibria, but occurrence of deadlock does not destroy the basis of cooperation. In fact, in some cases, as in CED 2, the occasional realization of this threat is a natural part of the equilibrium solution.[14]

We consider the imperfect cooperation implied by a correlated equilibrium to be a more realistic representation of political behavior than the unrelenting cooperation required in trigger strategy equilibria. Furthermore, since specific policy outcomes vary according to the information individuals observe at different times, a correlated equilibrium distribution connects policy changes to relevant developments in the domestic and international political environments. Therefore, even though policy outcomes are selected from some probability distribution, this outcome is given meaning through its responsiveness to the information contained in the changing strategic context of a particular game. Access to this same information can make it easier for outside observers to predict a state's policy outcomes. In the next section we extend this example to demonstrate that policy actors in different states can implicitly correlate their behavior, even in the absence of formal institutional arrangements for international policy coordination.

A Model of Domestic Political Competition and International Rivalry

Our analysis has two levels of implications for rational models of international relations. First, as shown above, correlated equilibrium can be applied to standard models of games between unitary rational states. Second, our conceptualization enables rational models to incorporate the effects of domestic competition over foreign policy, even if that competition results in voting cycles. In this section we combine our earlier examples into a model of two-level policy games.

To simplify our presentation, we assume that each state's policies are determined by a majority vote of three domestic groups (or policy advisers) with utility functions as specified in our earlier voting game example. Groups within state A select among policy outcomes K, L, and M. Groups in B select among analogously defined K, L, and M outcomes on policy dimensions distinct from those of state A.

The nine feasible combinations of policy outcomes of these two states can be grouped to form a 2 × 2 game at the international level, with policy outcomes K and L interpreted as a cooperative play of C in the international game and policy outcome M interpreted as a play of D (see fig. 7.4). In any one play of the game the utility actually obtained by each group depends on the rival's policy. We assume that groups in one state are indifferent between the K and L policy outcomes for the other state but that all feel threatened (and thus receive less utility) when the rival plays M.

An international rivalry should help domestic groups avoid deadlocks that cripple their ability to conduct foreign policy; therefore, we use example CED 1 to represent the outcome of policy competition within each state. In this case, each state is resolute (M) with probability 1/3 and irresolute (K or L) with probability 2/3. This distribution defines the expected level of security threat perceived by the other state's groups.

The specific utility values listed in fig. 7.4*a* were selected to integrate the numerical values used in our previous example. Given a uniform distribution over the other state's three policy outcomes (as defined by CED 1), each group's expected utilities for outcomes K, L, and M correspond to the values assigned in our spatial voting game example. Also, the difference in utility between the C and D outcomes of the international level game (within the same row for A's groups and the same column for B's groups) equals the difference between the utility values of Aumann's Chicken game example. Overall, each group orders the four outcomes at the international level as in a Chicken game, and each state's groups exhibit cyclic preferences over that state's K, L, and M policy outcomes.

In general, there is no reason to expect close relationships between the correlated equilibria of policy games in different states. But consider a pair of rival states in which most influential policy actors in both states perceive the other state to be the primary threat to national security. This fundamental security concern provides a strong incentive for policy actors to seek out information relevant to predictions of the rival's behavior. To the extent that participants in foreign policy debates in both states condition their behavior on similar information, the policy outcomes exhibited by rival states can be said to constitute a correlated equilibrium distribution of a dyadic "rivalry game."[15]

When a group in state A receives a suggestion to support K, L, or M, this player associates, as before, a conditional probability distribution over the suggestions received by the other players in A. Even if groups within one state do not have access to sufficient information to assign meaningful beliefs over the behavior of groups within the rival state, they should be able to form beliefs over the foreign policy of the rival state as a whole. As Allison (1971, 250–51) argues, it is often easier to base prediction of another state's behavior on its pursuit of a postulated national interest than it is to predict the outcome of domestic debates within other states.

The ability of groups within one state to obtain information relevant to predicting the rival's resoluteness will affect the set of feasible correlated equilibria. If each state's groups always assign a 1/3 probability to the rival's play of M, then the distributions over their policy outcomes are independent. Since CED 1 assigns a uniform distribution over K, L, and M, all nine combinations of policy outcomes occur with equal probability (fig. 7.4*b*). When

(a) Payoffs to groups

policy of state B

		K		L		M	
		K	L	M			
policy of state A	K	9.01 8.56 7.84	9.01 8.56 7.84	9.01 8.56 7.84	8.86 9.61 6.97	5.01 4.56 3.84	9.08 10.13 9.17 C
	L	8.86 9.61 6.97	9.01 8.56 7.84	8.86 9.61 6.97	8.86 9.61 6.97	4.86 5.61 2.97	9.08 10.13 9.17 D
	M	9.08 10.13 9.17	5.01 4.56 3.84	9.08 10.13 9.17	4.86 5.61 2.97	2.08 3.13 2.17	2.08 3.13 2.17

C   D

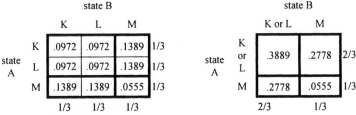

(b) CED #1 within each state, no correlation between states

state B

		K	L	M	
	K	.111	.111	.111	1/3
state A	L	.111	.111	.111	1/3
	M	.111	.111	.111	1/3
		1/3	1/3	1/3	

state B

		K or L	M	
	K or L	.444	.222	2/3
state A	M	.222	.111	1/3
		2/3	1/3	

(c) CED #1 within each state, imperfect correlation between states

state B

		K	L	M	
	K	.0972	.0972	.1389	1/3
state A	L	.0972	.0972	.1389	1/3
	M	.1389	.1389	.0555	1/3
		1/3	1/3	1/3	

state B

		K or L	M	
	K or L	.3889	.2778	2/3
state A	M	.2778	.0555	1/3
		2/3	1/3	

Note: In each cell of part (a) the payoffs to A's groups are listed in the first column and B's groups in the second column, with player k's payoffs in the kth row. The 3×3 matrices in parts (b) and (c) list the probabilities of each combination of K, L, M policies of states A and B. The 2×2 matrices on the right list the probabilities of outcomes at the international level (C or D).

Fig. 7.4. Examples of correlated equilibria in a two-level game

interpreted as plays of C or D in an international game, the distribution over the four outcomes corresponds to the mixed strategy equilibrium in Aumann's example (fig. 7.1*b*). This simple concatenation of correlated equilibria for the policy game within each state constitutes a correlated equilibrium to the overall game; groups have no more information than they did in our earlier discussion of the voting game. That is, each group's conditional beliefs remain exactly the same as before, except that their expected level of external threat is now explicitly included in the correlated equilibrium distribution.

Now consider what happens if domestic groups in each state use information observed in a particular situation to form conditional beliefs about the rival's resoluteness. Suppose, as before, that two rival great powers face repeated opportunities to intervene in regional conflicts. If groups within both states understand their respective spheres of influence, then a clear signal will be sent about which state is more likely to be resolute in a confrontation. Yet, in some situations, groups advocating a more aggressive or expansionary policy will have some temporary advantage in the domestic political competition (perhaps as the result of a recent election, or a new set of appointments, or a scandal that tarnishes the reputations of their competitors). In other words, these groups can be identified as the source of domestic pressure for a show of strength in otherwise inappropriate situations, as discussed in our earlier interpretation of a game between unitary rational states. As before, the equilibrium probability distribution we discuss below combines these domestic and international signals into a single joint probability distribution.

In this interpretation, groups within both states share common information concerning their respective spheres of influence, but some information concerning the effectiveness of domestic pressure for an aggressive or expansionary foreign policy will remain unobserved by groups within the other state. Yet, these groups will have access (through their own intelligence agencies as well as public sources) to much information relevant to making predictions about the other state's behavior. If groups within one state have even limited success in predicting situations in which groups advocating a hard-line policy will prevail in policy debates within the other state, then the probability of the DD (or MM) outcome can be significantly reduced. However, their ability to predict policy outcomes remains imperfect, so it is implausible that the possibility of direct confrontation could be entirely eliminated.[16]

Figure 7.4c lists a correlated equilibrium distribution to our two-level game in which the DD outcome occurs only half as often (.0555) as it does when the distributions over these two states' foreign policies remain independent. The probability values for all combinations of signals to groups within one state are listed in figure 7.5. The overall distribution is identical to the CED in figure 7.3b, but players now have sufficient information to assign asymmetric beliefs over the other state's policy outcomes.

The equivalent figure for the CED listed in figure 7.4b would assign a value of .1111 to all the nonzero entries in figure 7.5. The probabilities for the CED in figure 7.4c are higher than .1111 whenever exactly one state plays M, reflecting the groups' success at lowering the probability that both

Player 1		K			L	
Player 3		M	K		M	K
Player 2 — L	N 0		K	K .0972 / L .0972 / M .1389	L	K .0972 / L .0972 / M .1389	L 0
Player 2 — M	M	K .1389 / L .1389 / M .0555	K 0		M 0	N 0

Note: This figure lists the example CED from part (c) of Figure 3 in terms of the choices of groups within one of the two states. For each combination of group choices, the probabilities associated with the other states's policy outcome (K, L, M) are listed.

Fig. 7.5. Domestic representation of a correlated equilibrium in a two-level game

states will be resolute in the same crisis. That is, the probabilities of the DC and CD outcomes to the international-level Chicken game have been increased and those of CC and DD decreased.

Several properties of this correlated equilibrium distribution bear emphasis. First, each domestic group receives a higher expected utility than in the original voting game example.[17] As in our Chicken game example, players benefit by finding some way to alternate their periods of resoluteness or expansion, even if they cannot always avoid the dangers of crisis escalation. Each group can gain more by further reducing the occurrence of DD, but there is likely to be some limit to the extent of correlation that policy actors in rival states can achieve.

Second, by keeping each state's correlated equilibrium distribution fixed we assume that the overall threat level remains unchanged. In other words, this type of correlation requires adjustment in the timing of threatening actions, but it does not imply a major change in either state's overall foreign policy. Some groups could further increase their expected utility by lowering the overall level of threat, that is, by achieving a correlated equilibrium distribution in the international game based on an asymmetric distribution over domestic policy outcomes in which M occurs less often than K or L. However, group 3 most prefers outcome M and should be expected to resist efforts to decrease the relative occurrence of M. This result suggests that international rivalries can sometimes be maintained by informal complicity between the most "hawkish" groups in each state.[18]

Finally, at the level of international outcomes, the CED in Figure 7.4c is identical to the CED in Figure 7.1d for the two-player Chicken game. In general, there may be significant differences between the implications of models of domestic group interaction and unitary state actors, but in this particular example a voting game model provides results identical to those of a game between supposedly unitary rational states. Thus, at least some models of unitary state action are consistent with a rational model of policy competition among domestic groups in the absence of an autonomous state.[19] Whether interpreted as unitary rational states or domestic groups, the players can, by effective use of available information, sustain a relationship that reduces the probability of crises and increases their expected utilities.

Institutions and Strategic Action

Our examples demonstrate that the concept of correlated equilibrium enables rational models of international relations to relax the problematic assumption that unitary rational states can implement optimal strategies. Instead of positing a single "state" actor playing games in two overlapping policy arenas, we assume both foreign and domestic policies are grounded in processes of competition and cooperation among domestic groups.

The domestic and international policy arenas are intertwined in our model. Each policy actor's utility function incorporates domestic and international factors, and each draws signals from both arenas. Because foreign policy outcomes emerge from domestic competition, changes in each state's foreign policy will be driven by domestic political concerns as well as by the external environment. At the same time, external events significantly affect the methods and outcomes of domestic political competition.

Economic and other forms of interdependence inspire policy actors to improve their ability to correlate their actions with policy actors in different states. Any information relevant to the unfolding of one state's domestic political process (economic conditions, elections, constituency pressure, etc.) that groups in the other state can observe helps increase correlation between the foreign policies of related states. Thus, specific correlated equilibria selected by participants in these policy games reflect changes in both policy arenas.

In this way the policies of interdependent states are linked together through the information individual policy actors use to influence each other's beliefs and behavior. In Allison's (1971, 178) words, "The actions of one nation affect those of another to the degree that they result in advantages and disadvantages for players in the second nation." Like his Model III, our multilevel policy game model "directs attention to intra-national games, the overlap of which constitutes international relations" (Allison 1971, 149).

This juxtaposition of social choice theory and bureaucratic politics may seem unusual. By claiming that institutions pursue well-defined goals, resist change, and constrain the range of acceptable behavior and feasible policy outcomes, researchers in the bureaucratic politics tradition tend to reify and rigidify political institutions. Conversely, for social choice theorists, political institutions are part and parcel of the political process, in the sense that institutions are necessarily subject to strategic manipulation by individuals pursuing conflicting policy goals.

Despite these differences, Allison's original conceptualization is fully consistent with the strategic behavior of individuals emphasized by social choice theorists. For example, Allison and Halperin (1972, 43) admit, in a parenthetical remark, that the pulling and hauling characteristic of bureaucratic politics "by no means implies that individual players are not acting rationally, given their interests." In addition, Allison explicitly uses the term *game* to differentiate Model III from Model II. "The name of the game is politics: bargaining along regularized circuits among players positioned hierarchically within the government. Government behavior can thus be understood . . . as results of these bargaining games" (Allison 1971, 144). Although most work on bureaucratic politics emphasizes the implications of routinized "action channels," political competition remains intense at several levels.

> In most cases, however, there are several possible channels through which an issue could be resolved. Because action channels structure the game by preselecting the major players, determining the usual points of entrance into the game, and by distributing particular advantages for each game, players maneuver to get the issue into the channel they believe is most likely to yield the desired result. (Allison and Halperin 1972, 50)

Thus, the bureaucratic politics tradition requires that policy outcomes are not simply and completely determined by existing institutional arrangements. Instead, individuals are seen as relatively sophisticated in their use of institutional rules to better pursue their own interests. Information asymmetries between principals and agents produce high monitoring costs that prohibit principals from exercising complete control over bureaucratic agents (Moe 1984). Furthermore, uncertainty about who will win and who will lose future policy debates often results in the design of bureaucratic structures that are ill equipped to provide effective performance (Moe 1989). In short, bureaucratic agents, within reasonable limits, are free to undertake opportunistic behavior.

The extent of common information and the availability of monitoring and sanctioning institutions will affect the range of behavior that self-interested actors can sustain, but few (if any) political situations are so well specified that analysis of the institutional structure, no matter how detailed, can provide specific predictions about policy outcomes. Thus, our theoretical perspective reveals limitations inherent in efforts to develop precise models of specific situations. Practically speaking, it would not be possible to incorporate all the many sources of informational signals available to policy actors engaged in a given substantive situation, although one might categorize these sources in some useful manner. For example, one might differentiate between the normative principles and the monitoring and sanctioning procedures specified in formal treaties embodied in a given international regime. However, the massive amount of information available to policy actors and the fundamentally dynamic nature of coalition formation make point predictions of policy outcomes infeasible.

Correlated equilibrium is a useful solution concept primarily because it delimits the range of patterns of behavior that are consistent with the rational pursuit of self-interest. Analysts willing to specify a particular array of domestic group interests could use this solution concept to determine what range of distributions over policy outcomes would be sustainable in the presence of strategic action on the part of domestic groups. By combining models of the diverse domestic structures of interdependent states, it should become possible to delimit sustainable distributions of policy outcomes at the international level.

By ultimately grounding international relations in the strategic behavior of individuals pursuing common and conflicting interests, Bayesian social choice theory reveals a fundamental similarity between domestic political processes and international interactions. The policies of states have (limited) coherence because common information enables individual policy actors to implicitly or explicitly coordinate their behavior. At the systemic level, persistent patterns of international interactions are sustained through the effective use of information by policymakers in interdependent states. Correlation may generally be higher for domestic politics, but our examples demonstrate that, even in the presence of voting cycles, Bayesian correlated equilibrium imparts a considerable degree of coherence and meaning to policy outcomes at the national and international level. Political institutions and norms constitute important sources of the signals that facilitate the coordination of the behavior of multiple actors, but our analysis suggests that stable patterns of political order exist precisely because strategic action remains at the heart of politics, domestic and international.

The Relevance of Social Choice to International Relations Theory

Voting cycles occur in democratic voting systems, but they merely reflect the stronger claim of social choice theorists that any system of preference aggregation intended to satisfy fairness criteria may result in intransitive group preferences. In this section we discuss the relevance for international relations theory of this broader normative claim.

Arrow's theorem demonstrates the logical impossibility of aggregating

individual interests into a consistent social preference ordering that also satisfies basic criteria of fairness (Arrow 1963). Specifically, the following set
of assumptions is logically inconsistent.[20]

1. Nondictatorship: No one individual's preferences determine social
   choice;
2. Unrestricted domain: Any individual preference ordering over social
   outcomes is permissible;
3. Pareto condition: If all individuals prefer outcome x over y, then y is
   not selected by the collective choice mechanism;
4. Independence of irrelevant alternatives: No extraneous information
   affects collective choices between any two given outcomes;
5. Transitive social choice: A consistent ordering of social outcomes
   can be generated.

The first four conditions represent widely accepted, normative criteria of
"fairness," and Arrow demonstrated that no collective choice mechanism that
satisfies the consistency criterion (condition 5) can simultaneously satisfy
all four fairness conditions. If no transitive social choice ordering can be
generated, then particular collective choices will be determined by factors
unrelated to the distribution of individual preferences, such as the relative
manipulative skill of policy actors. Furthermore, actors have little incentive to reveal their actual preferences, and no institutions can ensure that voting will be sincere rather than strategic (Gibbard 1973; Satterthwaite 1975).
As long as individuals can affect policy outcomes by manipulating agendas
or misrepresenting their preferences, policy outcomes will often not mean-
ingfully reflect individual preferences (Riker 1982).

Virtually all applications of rational choice models to international
relations sidestep the dilemmas of social choice by relying on an analogy
between a state pursuing its national interest and a utility-maximizing individual.[21] Ironically, Arrow was inspired to complete the research project that
resulted in his famous theorem when a colleague at RAND asked him
whether the emerging practice of treating states as unitary rational actors in
nuclear deterrence models was acceptable on logical grounds (Arrow 1983,
3; Feiwel 1987, 193). Arrow's initial answer was yes, since welfare economists typically presume the existence of a social welfare function. Upon further analysis he concluded otherwise.

Arrow's theorem places limits on the conditions under which it is valid
to assume the existence of a well-defined collective utility function. It raises
fundamental doubts about the logical consistency of rational choice models
of the foreign policies of states, unless these states are, literally, unitary actors. Yet, international relations theorists pay scant attention to this theorem

because of a widespread belief that processes of foreign policy decision making systematically violate at least one of Arrow's criteria (see Bueno de Mesquita 1981, 12–18; Bueno de Mesquita and Lalman 1990, 30).

The nondictatorship condition is most commonly questioned. If a single individual controls foreign policy, then policy consistency is provided by his or her own preference ordering. However, if the identity of this dictator can change over time, then collective preferences will not necessarily remain consistent. Even the case of an enduring dictatorship is less clearcut than often presumed, because all leaders face limits set by the preferences of influential groups of supporters. Neither absolute monarchs nor modern dictators can fight a war or implement other foreign policy actions without extracting at least some resources from domestic groups. Rulers who fail to take into account the interests of key domestic groups are unlikely to remain in power, so the assumption that foreign policy usually serves one individual's interests is ultimately untenable.[22]

A second possibility is violation of the assumption of unrestricted domain, since policy actors within a given state should share a common interest in national security. However, violating the unrestricted domain assumption does not guarantee a consistent group preference ordering. Voting cycles occur in our example of a two-level game even though all policy actors within a state share exactly the same preference ordering over the outcomes of the international game.

Alternatively, consistency of state preferences could be attributed to the common interests of a governing coalition of groups. However, even if some coalition of groups "captures" the state, there is no guarantee that these groups will always agree among themselves. Newly arising issues are likely to produce some intragroup divisions. Social intransitivities occur even within a relatively stable ruling coalition, and any coalition may be overturned by a new governing coalition that also must manage social choice dilemmas.

Our application of Aumann's correlated equilibrium to policy games among Bayesian rational actors violates the assumption of the independence of irrelevant alternatives.[23] Individual policy actors take account of much more information than their own preferences, and it is this information about each other's likely behavior that allows them to attain a correlated equilibrium distribution. Correlation among players' strategies results from access to information shared in their common environment. This information helps each of them decide when it is advantageous to vote strategically or otherwise misrepresent their true preferences.

A correlated equilibrium neither implies nor requires a consistent preference ordering at the collective level, but a considerable degree of regularity and coherence is imparted to policy outcomes, at both the state and international levels, by the responsiveness of correlated equilibria to the information available to policy actors. Nonetheless, our examples demonstrate the persistent relevance of social intransitivities, even though the independence assumption is violated.

In summary, social choice dilemmas are crucial to the analysis of foreign policy and international relations whether or not Arrow's conditions are violated. Policy uncertainty is unavoidable as long as competing groups are responsible for foreign policy.

The relevance of social choice theory to the international system is likely to increase if recent trends toward democratization continue. The institution of popular elections ensures that policymakers of liberal democratic regimes pay close attention to their levels of domestic support. Even if the foreign policy of democratic states remains under the centralized control of executive leaders, this policy must comport with the interests of influential domestic groups. Finally, demonstrations that consistent policy outcomes can be explained as equilibria induced by particular institutional arrangements (Shepsle 1979) fall well short of proving that social intransitivities can be eliminated easily (Riker 1982; Dion 1992).

In an international system primarily composed of liberal democratic states, both the aggregation of domestic interests into foreign policies and the aggregation of these foreign policies into international outcomes are subject to the normative dilemmas of social choice. More generally, the fundamental logical dilemmas of social choice complicate any aggregation of state policies into systemic outcomes, no matter what the nature of the international system (Morrow 1988). Our interpretation of foreign policy and international order as related manifestations of correlated equilibria in multiple-level policy games provides a means to retain some "meaning" to social choice despite the continued existence of policy uncertainty. Unlike previous conceptualizations that maintain a sharp demarcation between domestic and international politics, our analysis suggests that similar processes of correlated social choice occur at all levels of political interaction.

NOTES

1. Bayes' theorem is covered in basic probability and statistics texts. The classic statement of the Bayesian viewpoint remains Savage (1954); see also Raiffa (1968).

2. We do not deny that human cognition often deviates from a literal interpretation of the subjective expected utility maximization model (Kahneman and Tversky 1979) nor that individuals often use simple decision aids or heuristics to cope with complex decision environments (Simon 1969, 1985). Nevertheless, we use a rational model to ensure the generality of our analysis. We are primarily concerned with collective outcomes, and those individuals most likely to be influential in policy debates can effec-

tively adapt to new situations by modifying their heuristics and learning from their previous mistakes.

3. Actors will not observe exactly the same information. However, Aumann's model, like all game models, requires the assumption that game players share the "common knowledge" that all actors are rational (see Binmore with Brandenburger 1990, 132–7).

4. Explicit budget constraints need not be imposed in these models, for at least two reasons. First, it is an empirical fact that democratic governments do not always recognize budgetary constraints and instead find indirect ways to fund increased governmental programs. Second, each policy proposal could be interpreted as including the necessary tax increases needed to fund it. In this case each actor's preferences for total expenditure and tax levels are implicitly incorporated in the utility functions imposed in this model.

5. If the level of expected threat changes, each actor's ideal point in this two-dimensional expected utility space would also change. Conditions for correlated equilibria would then have to be recalculated.

6. The utility values defined by a specific quadratic-based utility function are listed in figure 7.3a. Our analysis is easily generalized to noncircular indifference contours and additional alternatives.

7. Work by McKelvey (1986) indicates that in institution-free contexts, much like those we consider, outcomes are centrally located in an uncovered set (see also Ordeshook 1986, 184–7). However, this centrality of outcomes is much the same as a top-cycle, and the size of the set can be quite large. Furthermore, with a strong agenda controller and "loose" institutions, points outside of this space can be chosen (McKelvey 1976; Schofield 1978).

8. We realize that extensive game models of voting are useful for other purposes. Including institutions and agendas in an extensive form model of a CED requires representation of multiple signals because voters update beliefs as each step in the agenda unfolds. This is a promising direction for future research.

9. In a more general formulation, players could choose to support their third choice in order to decrease the probability of deadlock (Williams and McGinnis 1991).

10. The N outcome would be a status quo point if rules provide for this outcome in case of deadlock. However, our conceptualization is not tied to such an agenda, and N can more generally be viewed as what happens if a majority cannot act.

11. Groups 1 and 2 would prefer to select a compromise point below L, but such outcomes are precluded in this example. Points on the line connecting their ideal points would have values on the vertical dimension close to the deadlock position, which we arise to denote a breakdown in this state's ability to conduct an effective foreign policy.

12. We do not model dynamic processes of adjustment to equilibrium. Basically, if a player deviates from the signal, other players react by updating their beliefs and changing their behavior, until a new equilibrium distribution is reached.

13. To check that these distributions satisfy Aumann's conditions, the conditional probabilities for each signal for each player can be calculated from the probability matrices in figure 7.3b or c. Each player's expected utility for following or defeating

from a signal is calculated by multiplying these conditional probabilities by the utility values in the corresponding cell of the payoff matrix in figure 7.3a.

14. For a similar conclusion regarding imperfect cooperation in the correlated equilibria of iterated Prisoner's Dilemma games, see McGinnis and Williams (1991).

15. For evidence that data on U.S. and Soviet military expenditures and levels of hostility support the predictions of rational expectations models of superpower rivalry, see Williams and McGinnis (1988, 1992a, 1992b); McGinnis and Williams (1989).

16. We assume, as before, that after receiving other actors' suggestions, players face some remaining uncertainty about their behavior. In future research we could model more explicitly how specific pieces of information change situations as perceived by some of the actors, thus leading to different patterns of behavior. Suppose, for example, that each group in state A receives credible information that B will, with certainty, be irresolute in the current crisis. Then A's policy will be M, for, as shown in figure 7.4a, each group prefers M to either K or L under these circumstances. As noted in our discussion of Aumann's example, players may behave differently if they are provided additional information.

17. Specifically, players 1, 2, and 3 receive expected utilities of 7.48, 7.93, and 6.49, respectively. Because of the symmetries we impose, each group gains .16 compared to CED 1 in the domestic voting game.

18. A promising direction for future research is to build game models in which domestic groups have different preferences over international outcomes; see Fogarty (1990) for an example of different preferences over deterrence outcomes.

19. Additional research is needed to determine the conditions under which domestic group competition models can provide consistent microfoundations for models of games between unitary rational states.

20. For detailed specifications of alternative versions of these conditions see Arrow (1963); Sen (1970); Plott (1976); Riker (1982).

21. For example, some arms race models are based on the assumption that one state's utility decreases whenever its rival increases its military capability (e.g., Brito 1972).

22. The situation in which a single individual obtains exclusive control over foreign policy remains a special case of our more general model.

23. This assumption has been challenged in a wide variety of models; for reviews see Sen (1970); Plott (1976); Riker (1982).

REFERENCES

Achen, C. H. (1988) "A State with Bureaucratic Politics Is Representable as a Unitary Rational Actor." Paper presented at the Annual Meeting of the American Political Science Association, Washington, DC, August 31–September 3, 1988.

Allison, G. T. (1971) *Essence of Decision: Explaining the Cuban Missile Crisis.* Boston: Little, Brown.

Allison, G. T., and M. Halperin (1972) "Bureaucratic Politics: A Paradigm and Some Policy Implications." *World Politics* 24:40–79.

Arrow, K. J. (1963) *Social Choice and Individual Values.* 2d ed. New Haven, CT:

Yale University Press

Arrow, K. J. (1983) *Social Choice and Justice.* Cambridge, MA: Harvard University Press.

Aumann, R. J. (1974) "Subjectivity and Correlation in Randomized Strategies." *Journal of Mathematical Economics* 1:67–96.

Aumann, R. J. (1981) "Survey of Repeated Games." In *Essays in Game Theory and Mathematical Economics in Honor of Oscar Morgenstern,* edited by R. J. Aumann et al., 11–42. Mannheim: Bibliographisches Institut.

Aumann, R. J. (1987) "Correlated Equilibrium as an Expression of Bayesian Rationality." *Econometrica* 55:1–18.

Austen-Smith, D. (1987) "Sophisticated Sincerity: Voting over Endogenous Agendas." *American Political Science Review,* 81:1323–30.

Axelrod, R. (1984) *The Evolution of Cooperation.* New York: Basic Books.

Bendor, J., and D. Mookherjee (1987) "Institutional Structure and the Logic of Ongoing Collective Action." *American Political Science Review* 81:129–54.

Binmore, K., with A. Brandenburger (1990) "Common Knowledge and Game Theory." In *Essays on the Foundation of Game Theory,* edited by K. Binmore, 105–50. New York: Basil Blackwell.

Brito, D. L. (1972) "A Dynamic Model of an Armaments Race." *International Economic Review* 13:359–75.

Bueno de Mesqita, B. (1981) *The War Trap.* New Haven, CT: Yale University Press.

Bueno de Mesquita, B., and D. Lalman (1990) "Domestic Opposition and Foreign War." *American Political Science Review* 84:747–65.

Bueno de Mesquita, B., and D. Lalman (1992) *War and Reason: Domestic and International Imperatives.* New Haven, CT: Yale University Press.

Bueno de Mesquita, B., D. Newman, and A. Rabushka (1985) *Forecasting Political Events: Hong Kong's Future.* New Haven, CT: Yale University Press.

Denzau, A., W. Riker, and K. Shepsle (1985) "Farquharson and Fenno: Sophisticated Voting and Home Style." *American Political Science Review* 79:1117–34.

Dion, D. (1992) "The Robustness of the Structure-Induced Equilibrium." *American Journal of Political Science* 36:462–82.

Feiwel, G. R. (ed.) (1987) *Arrow and the Ascent of Modern Economic Theory.* Washington Square, NY. New York University Press.

Fogarty, T. (1990) "Deterrence Games and Social Choice: Asymmetry, Aggregation of Preferences, and a Conjecture about Uncertainty." *International Interactions* 15:203–26.

Fudenberg, D., and E. Maskin (1986) "The Folk Theorem in Repeated Games with Discounting or with Incomplete Information." *Econometrica* 54:533–54.

Gibbard, A. (1973) "Manipulation of Voting Schemes: A General Result." *Econometrica* 41:587–601.

Gilpin, R. (1981) *War and Change in World Politics.* Oxford: Cambridge University Press.

Harsanyi, J. C. (1967–68) "Games with Incomplete Information Played by 'Bayesian' Players, Parts I, II, III." *Management Science* 14:159–82, 320–34, 486–502.

Hogarth, R., and M. Reder (eds.) (1986) *Rational Choice: The Contrast between*

Economics and Psychology. Chicago: University of Chicago Press.

Jervis, R. (1976) *Perception and Misperception in International Politics.* Princeton, NJ: Princeton University Press.

Kahneman, D., and A. Tversky (1979) "Prospect Theory: An Analysis of Decision under Risk." *Econometrica* 47:263–91.

Keohane, R. O. (1984) *After Hegemony.* Princeton, N J: Princeton University Press.

Krasner, S. D. (1978) *Defending the National Interest.* Princeton, NJ: Princeton University Press.

Kreps, D. M. (1990) *A Course in Microeconomic Theory.* Princeton, NJ: Princeton University Press.

Lake, D. A. (1988) *Power, Protection, and Free Trade.* Ithaca, NY: Cornell University Press.

McGinnis, M. D., and J. T. Williams (1989) "Change and Stability in Superpower Rivalry." *American Political Science Review* 83:1101–23.

McGinnis, M. D., and, J. T. Williams (1991) "Configurations of Cooperation: Correlated Equilibria in Coordination and Iterated Prisoner's Dilemma Games." Paper presented at the Twenty-Fifth North American Meeting of the Peace Science Society (International), University of Michigan, Ann Arbor, Michigan, November 15–17, 1991.

McKelvey, R. D. (1976) "Intransitivities in Multi-dimensional Voting Models and Some Implications for Agenda Control." *Journal of Economic Theory* 12:472–82.

McKelvey, R. D. (1986) "Covering, Dominance, and Institution Free Properties of Social Choice." *American Journal of Political Science* 30:283–314.

Maskin, E., and J. Tirole (1987) "Correlated Equilibria and Sunspots." *Journal of Economic Theory* 43:364–73.

Mastanduno, M., D. A. Lake, and G. J. Ikenberry (1989) "Toward a Realist Theory of State Action." *International Studies Quarterly* 33:457–74.

Moe, T. M. (1984) "The New Economics of Organization." *American Journal of Political Science* 28:739–77.

Moe, T. M. (1989) "The Politics of Bureaucratic Structure." In *Can the Government Govern?,* edited by J. E. Chubb and P. E. Peterson, pp. 267–329. Washington, DC: Brookings Institution.

Morgan, T. C. (1984) "A Spatial Model of Crisis Bargaining." *International Studies Quarterly* 28:407–26.

Morrow, J. D. (1986) "A Spatial Model of International Conflict." *American Political Science Review* 80:1131–50.

Morrow, J. D. (1988) "Social Choice and System Structure in World Politics." *World Politics* 41:75–97.

Most, B. A., and H. Starr (1984) "International Relations Theory, Foreign Policy Substitutability, and 'Nice Laws.' " *World Politics* 36:383–406.

Most, B. A., and H. Starr (1989) *Inquiry, Logic and International Politics.* Columbia: University of South Carolina Press.

Myerson, R. B. (1991) *Game Theory: Analysis of Conflict.* Cambridge, MA: Harvard University Press.

Niou, E. M. S., P. C. Ordeshook, and G. F. Rose (1989) *The Balance of Power: Stability in International Systems.* Cambridge, MA: Cambridge University Press.

Oppenheimer, J. (1979) "Collective Goods and Alliances." *Journal of Conflict Resolution* 23:387–407.

Ordeshook, P. C. (1986) *Game Theory and Political Theory.* Cambridge, MA: Cambridge University Press.

Ostrom, E. (1986) "An Agenda for the Study of Institutions." *Public Choice* 48:3–25. (Reprinted as chapter 3 of this volume.)

Oye, K. A. (ed.) (1985) *Cooperation under Anarchy. World Politics* 38:1–254 (special issue). Reprinted by Princeton University Press, 1986.

Pattanaik, P. K, and B. Peleg (1986) "Distribution of Power under Stochastic Social Choice Rules." *Econometrica* 54:909–21.

Plott, C. (1976) "Axiomatic Social Choice Theory, an Overview and Interpretation." *American Journal of Political Science* 20:511–96.

Powell, R. (1990) *Nuclear Deterrence Theory.* Oxford: Cambridge University Press.

Putnam, R. D. (1988) "Diplomacy and Domestic Politics: The Logic of Two-Level Games." *International Organization* 42:427–60.

Raiffa, H. (1968) *Decision Analysis.* Reading, MA: Addison-Wesley.

Rapoport, A. (1960) *Fights, Games and Debates.* Ann Arbor: University of Michigan Press.

Riker, W. H. (1982) *Liberalism against Populism.* San Francisco: Freeman.

Satterthwaite, M. (1975) "Strategy Proofness and Arrow's Conditions." *Journal of Economic Theory* 10:187–217.

Savage, L. J. (1954) *The Foundations of Statistics.* New York: Wiley.

Schelling, T. C. (1960) *The Strategy of Conflict.* New York: Oxford University Press.

Schofield, N. (1978) "Instability of Simple Dynamic Games." *Review of Economic Studies* 45:575–94.

Sen, A. K (1970) *Collective Choice and Social Welfare.* San Francisco: Holder-Day.

Shepsle, K (1979) "Institutional Arrangements and Equilibrium in Multi-dimensional Voting Models." *American Journal of Political Science* 23:27–59.

Shepsle, K., and B. R. Weingast (1981) "Structure-Induced Equilibrium and Legislative Choice." *Public Choice* 37:503–19.

Simon, H. A. (1969) *The Sciences of the Artificial.* Cambridge, MA: MIT Press.

Simon, H. A. (1985) "Human Nature in Politics: The Dialogue of Psychology with Political Science." *American Political Science Review* 79:293–304.

Snidal, D. (1985) "The Game 'Theory' of International Politics." *World Politics* 38:25–57.

Snyder, G. H. (1971) "Prisoner's Dilemma and Chicken Models in International Politics." *International Studies Quarterly* 15:66–103.

Snyder, G. H., and P. Diesing (1977) *Conflict among Nations.* Princeton, NJ: Princeton University Press.

Snyder, J. (1991) *Myths of Empire: Domestic Politics and International Ambition.* Ithaca, NY: Cornell University Press.

Tsebelis, G. (1990) *Nested Games.* Berkeley: University of California Press.

Waltz, K N. (1979) *Theory of International Politics.* Reading, MA: Addison-Wesley.

Williams, J. T., and M. D. McGinnis (1988) "Sophisticated Reaction in the U.S.-Soviet Arms Race: Evidence of Rational Expectations." *American Journal of Political Science* 32:968–95.

Williams, J. T., and M. D. McGinnis (1991) "Bayesian Correlated Equilibria in Spatial Voting Games." Paper presented at the Annual Meeting of the American Political Science Association, Washington, DC, August 29–September 1, 1991.

Williams, J. T., and M. D. McGinnis (1992a) "The Dimension of Superpower Rivalry: A Dynamic Factor Analysis." *Journal of Conflict Resolution* 36:86–118.

Williams, J.T., and M. D. McGinnis (1992b) "Expectations and the Dynamics of U. S. Defense Budgets: A Critique of Organizational Reaction Models." In *The Political Economy of Military Spending in the United States,* edited by A. Mintz, pp. 282–304. London and New York: Routledge.

Wittman, D. (1989) "Why Democracies Produce Efficient Results." *Journal of Political Economy* 97:1395–424.

Young, O. R. (1972) "The Perils of Odysseus: On Constructing Theories of International Relations." *World Politics* 24 (Supplement): 179–203.

CHAPTER 8

Shepherds and Their Leaders among the *Raikas* of India: A Principal-Agent Perspective

Arun Agrawal

This essay focuses on the relationship between the leaders and ordinary shepherds among the *raikas,* nomadic pastoralists in western India.[1] Nomadic pastoralism continues to provide basic subsistence to millions of households that dwell in the semiarid regions of Asia and Africa. It also constitutes a dramatically different mode of production in comparison to settled agriculture (Kroeber 1948; Sadr 1991; Sandford 1983, 1–3). Knowledge and institutions of the pastoralists, therefore, remain pertinent for the practically as well as the theoretically inclined.[2]

To unravel problems stemming from environmental risks, the raikas employ a number of strategies: diversification, storage (on the hoof), exchange and mobility (Halstead and O'Shea 1989, 1–3). Of these, mobility is perhaps the most critical. Indeed, collective mobility in the face of environmental risks undergirds the survival of most nomadic pastoralists.[3] Collective mobility also requires that pastoralists carefully order interactions among leaders and followers. However, as Niamir points out in her comprehensive review of the African literature, few studies provide an in-depth account of relationships between leaders and followers among pastoralists (Niamir 1990).[4] Those studies that do exist tend to treat the actions of the ordinary pastoralist and the leader independently of each other.

This study, by using a principal-agent framework, attempts to achieve two objectives at once. It employs a significant tool of rational choice approaches—game theory—to represent the actions of shepherds and leaders and explain their observed interactions. In contrast to much theorizing about decision making among hunter-gatherers, it, thus, treats the actions of the raika shepherds and their leader—the *nambardar*—as mutually dependent.[5] Second, in using game theory, the study engages recent arguments that pathologize rational choice theory and suggest that its much-

Originally published in *Journal of Theoretical Politics* 9, no. 1 (1997): 235–63. Copyright © 1997 by Sage Publications Ltd. Reprinted by permission of Sage Publications Ltd. and the author.

heralded achievements are, in fact, deeply suspect (Green and Shapiro 1994). Green and Shapiro would further assert that fundamental rethinking is needed if rational choice theorists are to contribute to the understanding of politics. One of their most significant criticisms relates to the relative "thinness" of rational choice accounts and the lack of empirical explanatory power. This essay accepts Chong's (1995, 37–8) argument that the value of any theory has to be measured by its empirical power and aims at explaining observed facts regarding the relationship between shepherds and their leaders. In adopting a game-theoretical approach to explain these facts, the essay undermines those criticisms of rational choice by Green and Shapiro that are based on its empirical deficiencies. The analysis additionally seeks to demonstrate how rational choice analysis can elicit insights that would not otherwise be available.

The decision-making arrangements that exist among the raikas are best conceptualized as informal institutions.[6] They are informal because they are neither codified nor legally incumbent upon different groups. They are institutions because they structure, constrain and facilitate interactions and behavior (Bates 1989; North 1990). In this essay, I first describe the raikas and the three groups of decision makers among them. The next section discusses the major types of decisions. The third section presents an analysis of decision making by briefly examining why specific types of decisions are made by particular decision makers. These sections set the stage for examining the association between the shepherds and their leaders. In the following three sections, a simple game-theoretical model is used to uncover the logic behind the shepherd-leader relationship. While the substantive focus of the essay is on migrant shepherds, the analysis is couched in a generalizable principal-agent framework and thus possesses implications for other contexts where multiple principals hire an agent. The essay demonstrates the possibility of deriving useful insights even through easily accessible game-theoretical models in contrast to some recent literature on principal-agent relationships that increasingly deploys highly complex reasoning and models.[7] To the extent that managers of renewable resources in other parts of the world also periodically select new leaders and confront issues of controlling their leaders, the discussion holds wider relevance: for example, for the elections of rural leaders, in relation to attempts to co-manage natural resources in many countries in Latin America, Africa, and South Asia by creating partnerships between the state and the local communities, or in the context of rural thrift and credit associations that periodically select new leaders.

Raikas and Their Decision Makers

The raikas are the largest group of nomadic shepherds in India. Most raikas live in western India, in the states of Rajasthan and Gujarat, and are believed to have been dwelling in this part of the country for more than 500 years.[8] Their expertise in tending and herding camels made them an important part of the camel corps of many of the kingdoms in Rajputana in the eighteenth and nineteenth centuries, but with the development of more modern means of communication in these semiarid regions, and the increasing dominance of settled agriculturists, their political importance is far lower. They must rely on sheep, camels, and part-time agriculture to subsist. The shepherds own their animals but usually graze them on commonly owned or public pastures or in fields that belong to others.

The raikas reside in permanent dwellings for a quarter of the year, but they migrate with their sheep over distances spanning up to 1,200 miles annually. Their migration across state borders brings them in contact with farmers and government officials in far-flung places in Haryana, Uttar Pradesh, Madhya Pradesh and even Maharashtra. As a rule the shepherds migrate collectively, and move to a new camp location with their sheep almost daily. A mobile camp—the *dang*—embraces up to 18 sheep flocks. Since a flock often comprises 400–500 sheep, a camp can consist of 3,000–8,000 sheep, 20–100 camels, and, perhaps, 100 men, women and children. It resembles nothing as much as an entire village on the move.

Three major centers of decision making exist in shepherd camps. Of these, the nambardar is the most important. He is an influential shepherd, boasts wide-ranging contacts among other shepherds, farmers, wool and sheep merchants and, on occasion, even government officials. He is familiar with a variety of issues relating to migration routes, the movements of other shepherd camps, outsiders such as government officials and farmers, the purchase of supplies, and the sale of pastoral products. His past experience, relations with outsiders and access to multiple sources of information make him pivotal to the success of the migration. The second-in-command in the camp is called a *kamdar*. He assumes the duties of the camp leader when the nambardar is sick or away from the camp. Since both perform the same duties I will not treat them as different loci of decision making. Usually the second-in-command plays a role as a member of the *panchayat*, the council of elders, in the camp. The council comprises five of the older and more experienced persons in the camp. The five members of the council tend to represent the spectrum of different interest groups in the camp since they often are responsible for mobilizing different shepherds' flocks for the camp or might represent different villages. Collectively, they possess information and experience that none of the other decision makers can match. Finally, there is the *mukhiya*, the leader of the individual flocks that comprise the mobile camp.[9] He is intimately familiar with his flock, its dynamics, and the members of the flock.[10]

Major Types of Decisions

Members of the different camps I interviewed identified 60 issues as important for the functioning of the camp (see table 8.1). Using the classification the shepherds themselves suggested, I divide them into six categories: camp formation and dissolution, migration, flock management, camp management, market interactions, and interactions with the government and settled populations. Each class of activities and decisions has its own specific characteristics.

Camp Formation and Dissolution

This category contains two major decisions—selection of the camp leader before the beginning of the migration and the breakup of the camp at the end. To select a new leader, a few mukhiyas approach an individual they can trust. Once selected, the camp leader tends to continue for several annual migration cycles. In the normal course of events, the breakup of the camp occurs after the migration cycle is complete. On the return journey, flocks leave the camp at points closest to their villages. They seldom leave before the cycle is over and the nambardar takes special care to ensure that members do not get so dissatisfied that they are forced to leave in mid-migration.

Migration

The direction of travel, the timing of migration, the daily distance to be covered and the setting of the camp are the central migration issues. The raikas camp in a new location almost every day. Decisions related to setting the camp and the distance to be traveled each day must, therefore, be made repeatedly. For these decisions, the necessary information is not easily available to all shepherds. Familiarity with the migration route and information about the villagers are a prerequisite to decisions about where and when to set camp. Only the more experienced shepherds have such information.

TABLE 8.1. Aggregate Data on Decision Making, Classified by Issue Area

Issue Area	No. of Decisions	Decision-Making Unit			
		Flock-leaders	Nambardar	Council	Total
Camp formation/ dissolution	(2)	49	9	——	58
Migration	(11)	1	265	48	314
Flock management	(13)	322	55	——	377
Camp management	(16)	53	377	48	478
Market interactions	(10)	71	171	58	300
External relations	(8)	8	175	53	236
Total	(60)	504	1,052	207	1,763

At the same time, decisions on this subject can be considered routine because they are made often and the risks associated with a wrong decision are low. Thus a wrong decision is unlikely to impose huge costs on the shepherds because farmers are rarely hostile and welcome the manure that sheep deposit in their fields during the night.[11]

Flock Management

Two subclasses of decisions can be distinguished: household decisions (about cooking or loading and unloading camels) and decisions about managing the sheep (grazing, watering, accounting, and so forth). Women perform most of the housekeeping. Men carry out the sheep management tasks. Decision making depends on intimate familiarity with the affairs of the flock and an ability to direct member shepherds. Few of the tasks, however, require much direction from the decision maker. All the decision maker may need to do is ask that the evening meal be cooked, ask that camp be moved, or wake the shepherds to take the sheep for grazing.

Camp Management

Three issues affect camp management: the management of people, allocation of collective tasks, and camp security. The management of people includes issues such as arbitrating disputes, dividing responsibilities related to the camp and keeping track of shepherds who leave on various errands. Collective tasks include cooking on festive occasions, taking care of and interacting with visitors, purchasing medicines, and supervising expenses from the common fund. The decision-making unit that undertakes these tasks must be able to command. To make arrangements for better security and to ensure that these arrangements will be followed, the decision maker must also have contacts among settled farmers and must be familiar with the migration route.

Market Interactions

Shearing, sale of sheep, and wool sales lead to the major decisions needed to operate in markets. Most decisions on market interactions affect the entire camp. High stakes hinge on such decisions as the rate at which sheep should be sold. Other decisions are more routine—to whom should wool be sold? Some decisions demand substantial asset-specific knowledge—when are the sheep ready for shearing or sale? Decisions also entail possession of the latest knowledge about market prices for sheep and wool. Finally, since shearing takes place over a week, the decision maker needs to persuade farmers to furnish space during this time.

External Relations

This last issue area poses the greatest level of uncertainty. Decisions involve interactions with government, with the legal system, and with settled populations. In addition to low information availability on these issues, shepherds face an additional complication: the stakes are very high. While shepherds make their decisions only irregularly and infrequently, wrong choices can lead to losses of significant sums of money, expedite grave trouble and precipitate major fights with farmers. Right decisions, on the other hand, promise no benefits except that the daily business of the camp can continue as usual. Decisions thus involve high stakes and asymmetries between returns and losses.

Allocation of Decision Making

Table 8.1 presents the data collected on decision making in 30 raika camps. From column 2 it can be seen that two decision issues fall under the general category of "Camp Formation and Dissolution." Respondents in each camp were asked who made the decisions on each of these issues and there were 58 usable responses. Of these 58 responses, the owners of the flocks were mentioned 49 times and the camp leader was mentioned in nine instances. Of the total of 1,800 responses for 60 decision issues from 30 respondents, 37 (2 percent) were either invalid or unavailable; and thus the number 1,763 was the total number of responses.

The flock owners make most of the decisions about camp formation and flock management (85 percent). The nambardars have extensive powers to make decisions in all other areas. They make 72 percent of the decisions in the other general categories of decision making. The shepherds have delegated the responsibility for making most of the decisions in the areas of migration and camp management to their leaders. In addition, with the council of elders they cooperate to make some of the decisions where interactions with outsiders such as farmers and government officials are involved; and with the flock leaders they share authority for decisions related to market interactions. Even in these two areas they make 65 percent of all the decisions.[12]

Table 8.1 suggests that the responsibility for decisions is not distributed randomly in the shepherd camps. Under each category a particular decision-making unit seems chiefly responsible. For camp formation and dissolution the flock leaders are the primary decision makers. They select the camp leader and decide the time at which they will leave the camp. To understand the logic behind the allocation of these decisions recall how a new camp is formed. The flock leaders approach an individual whom they trust to guide them through the migration. In choosing their leader they abdicate responsibilities for making a large number of important decisions. By selecting a leader to make decisions on their behalf, they save time and energy. They must, however, be able to exercise some control over their leader, else he could easily exploit them. So although they give up the power to make many decisions, they retain the right to choose a new leader and the right to leave the camp if dissatisfied.

For decisions related to flock management, it is again the flock leaders who are responsible. They make 85 percent of the decisions. Their decisions affect only a small number of people, and there are seldom any economies of scale to be harvested. At the same time, the mukhiyas have more knowledge about their flock than anyone else in the camp. A mukhiya can gain no advantage by allowing another decision maker the right to manage his flock.

The basis on which the shepherds have allocated the rights to make decisions is summarized in figure 8.1.

The last part of figure 8.1 needs further analysis. How precisely do the shepherds ensure that their leader will not misuse the power that is delegated to him? The shepherds seem caught in a dilemma. On the one hand, they stand to gain substantial economic benefits by delegating to their leaders the power to make decisions in a wide variety of situations. Yet, at the same time, they face the risk that these leaders, especially the nambardar, may abuse their powers. In response the shepherds use a comprehensive set of preventive and corrective measures that help guard against opportunistic behavior. An analysis of the mechanisms they deploy helps us to understand

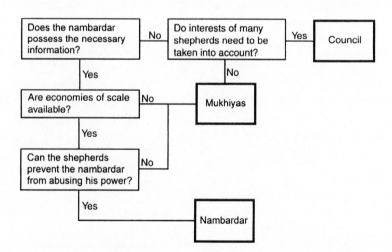

Fig. 8.1. A simplified model of decision making among the *Raikas*

better the leader-led relations among renewable resource users who periodically select new leaders.

The analysis is also useful in explicating the dynamics of a variety of other social situations in which individuals face the same type of problems as the raikas: how to control an agent who exercises responsibilities and powers on the behalf of a collective. Delegation, by permitting specialization, creates efficiency gains (Lupia and McCubbins 1994, 93). As Kiewiet and McCubbins (1991, 24) put it in their discussion of the relationship between the U.S. bureaucracy and the Congress:

> Delegation (of powers and responsibilities) is the key to the division of labor and development of specialization; tremendous gains accrue if tasks are delegated to those with the talent, training and inclination to do them. This…is what allows firms to profit, economies to grow, and governments to govern.[13]

But delegation also creates a range of opportunities for agents to shirk, subvert, or steal. The analysis in this case shows how a particular group of social actors resolves problems that arise because of delegation and how their strategy might differ from (or be similar to) that of other actors faced with the same problem.

A Neoinstitutional Perspective

With the recognition that transaction costs, information asymmetries, and property rights play exceedingly significant roles in determining economic outcomes, the neoinstitutional literature has burgeoned in the last two decades (Bromley 1989; Eggertsson 1990; North 1990; Williamson 1985). Many authors, in addition, have begun to apply insights from the neoinstitutional literature to rural contexts (Bates 1989; Berkes 1989; Cashdan 1990; Ensminger 1992; Ensminger and Rutten 1991; Ortiz and Lees 1992; E. Ostrom 1990; Wade 1988).[14] For most of these theorists, the notion of contracts among hierarchically located actors is fundamental to the analysis (Kreps 1990, 91–3).

The agency literature, which was launched by the seminal works of Grossman and Hart (1983), Holmstrom (1979), Hurvicz (1973), Jensen and Meckling (1976), and Ross (1973), focuses specifically on the outcomes that are a result of the divergence in the interests of principals and agents, and on the mechanisms that principals may use to tether agents to their will. While earlier theorists (Fama 1980; Fama and Jensen 1983; Jensen 1983) believed natural selection processes lead to the most efficient contracts between principals and agents, more recent mathematical literature examines mechanism design at great length to explore how principals may write efficient contracts ex ante (Biglaiser and Mezzetti 1993; Glover 1994; Laffont and Tirole 1987; Lewis and Sappington 1989). Few of these essays, however, consider the situation where multiple principals engage a single agent.[15]

We can consider the mobile shepherd camp as a congerie of relationships based on contracts between ordinary shepherds (principals) and the nambardar (their agent). According to this set of contracts, the shepherds select a leader for their migration so that they do not each have to divert their attention from the daily tasks related to shepherding. In exchange, the nambardar gains social prestige, cultivates political contacts and garners leadership status.[16] Even a cursory look at the range of issues over which the nambardar makes decisions reveals significant opportunities for malfeasance, graft, and moral hazard (see table 8.1). The nambardar decides upon the course of action in 60 percent of all situations mentioned by the shepherds as important. If we exclude the decision area of "flock management," which is related to issues internal to the management of each flock, the nambardar decides for the camp in nearly three-quarters of all decision situations.[17]

Opportunities for graft and exploitation arise from structural differences between the position of the shepherds and the nambardar. In turn, these structural differences can be traced to asymmetries of power and information. The nambardar possesses time-, place-, and asset-specific information that is unavailable to the shepherds. Indeed, this is the reason why they select him. But at the same time, the asymmetries of power and information place the nambardar in a position marked by the possibilities of adverse selection, moral hazard, shirking, and corrupt behavior.[18]

Each of these—adverse selection, moral hazard, shirking and corrupt behavior—can lead to major losses for migrating shepherds. These types of behavior may occur in any situation where the nambardar interacts with outsiders and sums of money or goods change hands. The leader can misrepresent both the true costs and the benefits of actions. Examples of the first type would include bribes negotiated to pay off government officials, payments to settle disputes with farmers, costs of medicines, and costs of supplies bought for feasts and festivals. Benefits might be misrepresented when farmers pay the camp for sheep folded in their fields or merchants pay for pastoral products such as wool and sheep. In response to the wrongdoings of their leader, the shepherds can try to deter him by sanctioning him in two possible ways: (1) by returning home in mid-migration; or (2) by declining to choose him as leader in future migration cycles.[19] The game in figure 8.2 represents the choices for the shepherds and their leaders.

SHEPHERD

LEADER	Sanction	Do Not Sanction
Cheat	$a_3 - a_1; - a_2 - (a_3/N)$	$a_3; - (a_3/N)$
Do Not Cheat	$- a_1; - a_2$	$0; 0$

Fig. 8.2. The relationship between cheating and sanctions

If the leader does not cheat, and the shepherds do not sanction, neither of them incurs any costs, and the payoffs to each are zero. If the leader cheats, and the shepherds do not sanction, the leader gains an amount a_3 and each shepherd in the camp loses a_3/N where N is the total number of shepherds in the camp. If the shepherds choose to sanction the leader, each of them incurs a_2 as the cost of sanctioning in addition to the cost of the leader's cheating. The cost on the leader is represented by a_1, and his payoff is reduced by that amount. If the leader does not cheat, and the shepherds sanction him, the leader suffers the cost of sanctions, a_1, and the shepherds incur the cost of sanctioning, a_2.

As mentioned earlier, the shepherds can impose two types of sanctions on their leader—leave the camp in mid-migration and return home; or decline to choose him as the leader in future migration cycles. Both strategies entail high costs: for the shepherds (principal), as well as for the nambardar (agent).[20] Leaving the camp in mid-migration is a high-cost/high-risk strategy. It defames the nambardar and carries a significant probability of ruining his reputation. The migration is risky enough for shepherds traveling in a group; for a lone shepherd returning home, it would be highly perilous.[21] A flock leader will follow this strategy only under extreme circumstances. The second possibility—to choose a different nambardar or to join a different camp after the current cycle concluded—would hurt the nambardar but only defer the current problem of the wayward agent. The costs of inappropriate performance by the agent will continue to be incurred.

The relationship among the payoffs for the leader and the shepherds are summarized in expressions 1 and 2, respectively. The relationships hold regardless of the form of sanction the shepherds choose.

For the leader

$$a_3 > 0 > a_3 - a_1 > -a_1 \tag{1}$$

For the shepherds

$$0 > - a_3/N > - a_2 > - a_2 - a_3/N \tag{2}$$

Under these payoffs there is an equilibrium in pure strategies, where the leader cheats and the shepherds do not sanction. This is clearly undesirable from the point of view of the shepherds. More importantly for analysis, it is also not the equilibrium observed during field research. The dominant strategy for the leader seemed to be seldom to cheat and, for the shepherds, to sanction rarely. How do we explain why the nambardar does not cheat?

The Restraint Exercised by the Nambardar

The key to understanding the equilibrium that prevailed in shepherd camps lies in the significance of information and the dynamics of monitoring in the camp. The shepherds, instead of waiting until their agent has successfully defrauded them, have created reliable mechanisms that monitor the nambardar and drastically reduce the possibility of hidden information and hidden actions. Before the nambardar's decision to cheat or not cheat and the shepherds' response to sanction or not sanction, thus, comes the choice of whether to monitor.[22] The nambardar also draws upon the expertise of elder shepherds in the camp (who are usually members of the council as well) to consult in cases where bribes or fines need to be paid to government officials or in disputes with farmers.

   To see how the shepherds monitor and use their information, I discuss four situations: the selection of camping sites, the purchase of medicines and supplies for feasts, the dealings with merchants, and finally, the payment of bribes to police, forest, and other government officials. These cases represent the greatest possibility of returns to an unscrupulous decision maker out of all the situations in which decisions are necessitated during the course of the migration. The interactions with outsiders over cash amounts can yield significant profits to a person possessing hidden information and undertaking hidden actions. The specific issues related to each of these situations and the distribution of responsibilities for making decisions over these issues are presented in table 8.2. This more detailed quantitative information on the distribution of decision-making responsibilities, together with the ensuing discussion about each specific situation, makes it clear that in most of the situations where opportunities for "hidden information and hidden action" are present, the interactions are primarily between the nambardar and the mukhiyas. Among all the listed decisions, the council of elders is involved mainly in determining when merchants and shearers should be called—situations where little opportunity for graft is present.[23]

Selection of Camp Sites

Raikas mostly camp on private fields. The nambardar negotiates with the farmers to select the fields in which the sheep will be folded for the night.

The amounts the nambardar receives on behalf of the shepherds are sizable, on the average Rs. 500.[24] for each flock over the year (see table 8.3).

The shepherds use a simple mechanism to ensure that the nambardar does not conceal information or siphon off returns from folding: the negotiations between the farmer and the nambardar often take place in the open where one or two shepherds may be present with the nambardar, and thus monitor the payment from the farmer.[25] Further, in as many as two-thirds of the cases, farmers pay in kind. The payments range between 50 and 100 pounds of grain. Because of its bulk it is difficult, if not impossible, for the leader to misrepresent the amount.

Purchase of Medicines and Supplies

The leader could purchase medicines from vendors known to him personally and skim a commission directly off the price, or he could misreport the price he paid. The shepherds guard against the latter possibility by having one or two members of the dang accompany the nambardar to monitor prices whenever he goes to buy medicines. Generally, on different occasions, different individuals go with the leader. This means that if the leader is successfully and systematically to report higher prices for the medicines he buys, he will have to collude with all the persons who go with him.

By patronizing a particular store and buying the medicine requirements of the entire camp in bulk, the nambardar could still make some gains; yet this opportunity for profit is limited by the fact that the prices he reports cannot exceed the prices in other stores. The shepherds use only a few types of western medicines sold in shops and can easily compare prices by random checks whenever they happen to be in a town. But because it is time consuming to visit a town regularly to buy medicines, it makes sense for the shepherds to delegate medicine purchases.

To prevent misreporting of prices of supplies for collective feasts, shepherds have created a similar mechanism: the nambardar appoints at least two shepherds to travel to a town to purchase supplies so that an individual cannot exploit the rest of the collective.

Sale of Pastoral Products

Three sets of individuals outside the camp are involved in the sale of pastoral products—the shearers, the wool merchants and the sheep buyers. The shepherds engage professional wool shearers called *lavas* to clip the wool. At the same time as the sheep are sheared, wool merchants receive invitations to come and purchase the wool. Individual sheep buyers frequently visit the camps to take advantage of the culling of sheep, which the shepherds are forced to do regularly owing to persistent cash flow problems.

TABLE 8.2. Distribution of Decision-Making Responsibilities for Specific Issues

Issue Area	Mukhiya	Nambardar	Council
Selection of campsite			
Choice of village	—	30	—
Choice of fields	—	29	01
Purchases			
Medicines	05	25	—
Supplies for feasts	—	28	02
Market interactions			
When to call wool merchants	—	17	13
When to call sheep merchants	—	10	20
When to call sheep shearers	—	12	18
Decision on rate for wool sales	—	27	03
Decision on rate for sheep sales	28	02	—
Decision on rate for shearing	—	28	02
Bribes			
Police officials	—	28	02
Forest officials	05	21	—
Total	38	257	74

TABLE 8.3. Income from Folding, and Expenses Incurred on Bribes and Fines: Data for 13 Flocks (in Rs.)

Flock Number	Income from Folding	Expenses on Fines/Bribes
1	264.00	107.00
2	335.00	134.00
3	227.00	100.00
4	396.00	150.00
5	396.00	150.00
6	335.00	134.00
7	396.00	154.00
8	396.00	160.00
9	791.00	300.00
10	791.00	300.00
11	791.00	382.00
12	396.00	161.00
13	791.00	350.00
Average	485.00	199.00

Two factors prevent wrong reports about the rates for shearing: first, the nambardar often makes the decision to invite the shearers in consultation with the council of elders. Since he does not have the discretion to call shearers known to him personally, it is difficult for him to engage in price fixing. Second, and this is not an intentional strategy, the entire party of 10–20 shearers stays with the camp for a week or more, which increases the chances of discovery if collusion did take place.

The rates for selling sheep are decided by flock leaders. This is a response to the possibility of large losses in case the nambardar colluded with the sheep buyers. Most flock owners are so vulnerable that if they did not get a good price in even one or two major sheep sales, they would be devastated. Also, the quality of sheep varies widely across flocks, and few flock owners trust anyone else to set prices for their sheep. The costs of not delegating remain low, since sheep buyers visit the camp regularly in the hope of buying sheep from cash-poor flock owners.[26]

It is in the case of wool sales that the nambardar finds the greatest opportunities for personal gains. He decides on the merchants to whom the wool should be sold, its quantity and the price. The wool merchants or their agents stay in the shepherds' camp very briefly, seldom more than a few hours at a time. Finally, since the price of wool fluctuates relatively rapidly in the urban markets, the ordinary shepherds would find it difficult to ascertain the best price for their wool at any given time. Misrepresentation of prices therefore is a simple matter.

However, the shepherds seem to exercise some control over the nambardar through ex post information collection. As they migrate, they constantly exchange information with shepherds in other camps on the rate at which nambardars sell wool. Sometimes new wool merchants visit shepherds' camps and in the process furnish price information. The vigilance of shepherds, and the nambardar's awareness of their vigilance, keeps within limits the commission that the nambardar can make on wool sales. Most shepherds believe only minor amounts to be at stake. Many of my interviewees felt that the nambardar deserved some returns for all the tasks that he performed for the collective. Using a common metaphor, a shepherd explained to me that the nambardar also had a "stomach to feed."[27] Clearly, as long as their ex-post price monitoring kept his hidden actions to a minimum, they were willing to allow him a little leeway,[28] and the efficiency losses from the nambardar's "corruption" seemed low enough for the shepherds to bear, if not grin while bearing.[29]

Bribes to Government Officials

Few members of the camp have much expertise in the fine art of bribing officials. The "rules of etiquette" that structure conversations with government officials in a rural context, where officials consider themselves superior, are beyond the ken of most shepherds. Even the nambardar possesses the necessary skills for negotiating bribes only to a small degree.[30] He, therefore, enjoys a near monopoly of power in this area of decision making. Yet, as table 8.3 shows, the nambardar does not gain much capital from it.

Table 8.3 indicates that on the average flock owners pay Rs. 200 each in bribes and fines during the entire migration cycle.[31] As a percentage of the total income, bribes and fines are no more than 3 percent for any of the 13 flocks surveyed. Even if the nambardar makes some money for himself in the payment of bribes, it can be only a small amount, a fraction of the money paid out as bribes. More importantly, the shepherds can do little, if anything, about it. They have to interact with government officials, and they need someone who can manage the necessary side payments.[32]

Explaining the Nambardar's Restraint

The preceding discussion makes it clear that shepherds actively seek information on the activities of their leader. In the various situations where moral hazard might arise—setting camp, purchasing supplies, selling pastoral products, or interacting with government officials—the shepherds monitor their leader closely or collect information after the fact. In most situations where the camp leader could make money on the side, institutionalized monitoring practices minimize large-scale exploitation. Where the nambardar enjoys some discretion in fixing prices and rates, the advantages that accrue to him are quite limited. One can conclude that overall, the shepherds monitor most of the time, and the leader cheats only infrequently. Figure 8.3 represents the discussion on the restraint exercised by the nambardar.

According to the figure, if the nambardar (leader) does not cheat, and the shepherds do not monitor, the payoffs to each are zero. If the leader cheats, and the shepherds do not monitor, the leader gains an amount a_3 and each shepherd loses a_3/N (see fig. 8.2). If the shepherds choose to monitor, and the leader does not cheat, each of the shepherds incurs a cost a_4, and the payoff to the leader is zero. The costs of monitoring are relatively low for the

SHEPHERD

LEADER	Monitor	Do Not Monitor	Probability
Cheat	$-a_5; -a_4$	$a_3; -a_3/N$	α
Do Not Cheat	$0; -a_4$	$0; 0$	$1-\alpha$
Probability	β	$1-\beta$	

Fig. 8.3. The relationship between cheating and monitoring

shepherds since they incorporate monitoring in their everyday interactions with the leader or in the daily routine of their livelihood.[33] If the shepherds monitor, and the agent cheats, then the shepherds prevent the loss of the amount a_3/N but still incur the low cost of monitoring. If shepherds know that their leader is a cheat, it imposes a cost on him equivalent to a_5. This cost is negligible in material terms but high in reputation and prestige. Given the fact that the shepherds monitor the nambardar closely, and live with him in daily contact, it should be obvious that systematic cheating would soon be discovered, and the ensuing gossip would affect the nambardar's reputation adversely.

The relationships among the payoffs for the leader and the shepherds are summarized in expressions 3 and 4 respectively.

For the leader

$$a_3 > 0 > -a_5 \tag{3}$$

For the shepherds

$$a_3/N > 0 > -a_4 \tag{4}$$

Under these payoffs, there is no equilibrium in pure strategies.[34] There is, however, a Nash equilibrium in mixed strategies, which can be calculated using the expected values of the payoffs for the leader and shepherds.[35] Assign the two strategies of the shepherds (monitor, do not monitor) the probabilities of β and $(1 - \beta)$, respectively. For the leader, similarly assign α and $(1 - \alpha)$ as the probabilities of cheating and not cheating. In equilibrium, the leader will be indifferent between cheating and not cheating because his payoff from the two strategies would be the same. Therefore,

$$[\beta(-a_5) + (1 - \beta)(a_3)] = 0 \tag{5}$$

where the left-hand side of the equation represents $E(U_1)$ from cheating and the right-hand side represents $E(U_1)$ from not cheating. Simplifying, the equilibrium value of β is

$$\beta^* = a_3/(a_5 + a_3). \tag{6}$$

Similarly, the shepherd will be indifferent between monitoring and not monitoring when his payoffs from the two strategies are the same.

$$[(1 - \alpha)(-a_4) + \alpha(-a_4)] = \alpha(-a_3/N) \tag{7}$$

where the left-hand side of the equation represents $E(U_s)$ from monitoring, and the right-hand side represents the $E(U_s)$ from not monitoring. The equilibrium value of α is, therefore,

$$\alpha^* = (a_4N)/a_3. \tag{8}$$

The derived values of α^* and β^* suggest some interesting interpretations. Two of these merit greater discussion. As a_3 (the payoff to the leader from cheating) increases, the equilibrium probability of cheating declines. The reason can be discovered as we examine the effect of a_3 on β^*, the equilibrium probability of monitoring. It arises, especially as it becomes high in relation to a_5, the cost that the shepherd imposes on the leader by monitoring if he is cheating. The rise in the probability of monitoring swamps the possible effect on the equilibrium probability of cheating. Further, as a_3 rises high in relation to a_5, the probability of monitoring gets closer to 1, but never reaches 1, unless $a_5 = 0$. Since equation 8 indicates that α^* will always be greater than zero, the game represented by figure 8.3 approximates closely the equilibrium observed among the shepherds where they monitored most of the time and their agent seldom cheated.[36]

Second, the value of N, the number of shepherds in the camp, plays an important role in determining the equilibrium α^*. As it rises, the value of α^* also rises. This confirms, perhaps, the "logic of collective action"; as group size rises, it becomes easier for the agent to cheat. Of course, α^* cannot rise above 1, and this imposes a limit on the value of N. Thus, $\alpha^* \leq 1$, if and only if, from expression 8,

$$N \leq a_3/a_4. \tag{9}$$

If N goes beyond this value, then the game will have an equilibrium in pure strategies where α^* will become 1, the leader will always cheat, and the shepherds will not monitor.[37] The gains from monitoring for each shepherd will be very low, and the losses from not monitoring for each shepherd will also be very low, enough to make him not wish to incur the costs of monitoring. The other implications of the equilibrium values of α^* and β^* are fairly straightforward and confirm earlier results of noncooperative games that have an equilibrium only in mixed strategies (see Tsebelis 1989). These include the results that as the cost of monitoring goes up, equilibrium probability of cheating goes up, and as monitoring becomes costless, the equilibrium probability of cheating declines to zero; the equilibrium probability of cheating is not affected by the size of penalty imposed on the wayward agent; and the equilibrium probability of monitoring is not affected by the size of monitoring costs.

$E(L^*)$ and $E(S^*)$, the expected returns to the shepherds and their leaders, are easy to determine. Since the leader is indifferent between cheating and not cheating in equilibrium, by expression 5, we can say that $E(L^*) = 0$. For the shepherds,

$$E(S^*) = \alpha^*(-a_3/N). \tag{10}$$

Substituting α^* in equation 10, $E(S^*) = -a_4$. Since $a_4 > 0$, the expected returns in equilibrium to the shepherds are negative and a function of the magnitude of their monitoring costs.

A Discussion by Way of Conclusion

Information, who possesses it, and how it is used emerge as crucial variables in determining whether the leader will cheat. At the same time, the leader's awareness that shepherds can impose sanctions upon discovery of cheating behavior plays a highly significant role in preventing him from mulcting his principals. The fact that the shepherds can choose a new leader for a new migration cycle is critical, then, to keeping the nambardar's actions within tolerable bounds. The essay, in showing the importance of direct monitoring by the agents, captures an important aspect of the environment in which shepherds operate. Lupia and McCubbins (1994), in their study of bureaucratic accountability, outline three possible methods of information collection for principals. Pointing to the costliness of direct monitoring, and the reluctance of agents to reveal information truthfully, they discuss whether third party information revelation might be efficient. For the shepherds, however, regular and close interactions with the agent, and the way in which monitoring is etched into the structure of their daily lives, mean that direct monitoring is the most effective means of information gathering.[38] This, perhaps, is a difference that marks the social context of many rural development and environmental projects in comparison to the prevailing situations in urban contexts. In most villages, the fact that the community is small and its residents engage in regular, overlapping, and multistranded interactions implies that they possess substantial direct monitoring capacities over their leaders. The achievement of the right to a voice in the selection and removal of leaders, together with the existing capacity for direct monitoring, might prove highly useful in enhancing the effectiveness of new rural development and environmental projects.

This analysis also presents some counterintuitive findings, in addition to confirming some of the principal results of two-person, noncooperative games without an equilibrium in pure strategies. The most important of these is that as gains from cheating increase for the agent, a higher probability of cheating behavior can be swamped by the increase in the probability of monitoring. The result is a consequence of several factors that may be common to community settings in a large proportion of the rural areas of the developing world.

1. The payoffs to the agent from cheating are related to the losses of the principals;[39]
2. The costs of monitoring, in relation to the losses from cheating, are low for the principals;
3. The loss the principals can impose on their agents by monitoring are relatively low when compared to the loss they suffer from cheating; but they still possess the capacity to impose significant costs through sanctions.

In light of these observations, the practical implications of the analysis are normatively attractive, if obvious. Policy provisions that stress greater accountability and permit higher monitoring levels of leaders by followers, especially when there are regular and frequent contacts among them, are likely to reduce a host of principal-agent problems. When unprincipled and self-interested actions of leaders directly reduce the benefits of their followers, extending greater authority to followers to hold leaders accountable is even more likely to reduce cheating behavior.

The second implication of the model, because of the explicit incorporation of the number of shepherds in the camp as a variable to address the existence of multiple principals, is the expected increase in the probability of cheating behavior with group size. The relationship can be interpreted in terms of the free-rider problem. A larger number of shepherds in the group weakens incentives to monitor because the losses owing to cheating are spread over more people. If the effect of a larger number of principals is to be overcome, institutional design must aim at lowering the costs of monitoring/collecting information.

One of the limiting assumptions throughout the foregoing analysis is that the shepherds are homogenous in their endowments. While the assumption is not very far fetched in the case of the raika camps, an obvious extension of the analysis to other social settings would require a consideration of the heterogenous nature of principals. The literature on common agency contains some directions for analysis on this issue.[40]

The formal model of shepherd-leader interactions I use is simple but captures some significant aspects of the rural social context in which principals interact with agents. These include the interdependence of payoffs (to agents) and losses (to principals) from cheating, the relationship between community and low monitoring costs, and relatively small group sizes. In outlining this model to theorize situations where multiple principals contract with a single agent, the essay formalizes the relationship between the shepherds and their agents and thereby permits lessons that may apply to other settings outside the context of nomadic shepherds' livelihoods. In so doing, it (1) demonstrates the possibility of developing new empirical insights even through simple game-theoretical models; and (2) contributes to a growing literature that attempts to formally analyze ethnographic/descriptive writings to understand rural politics.

Acknowledgments

The field research for this essay was conducted during the summers of 1990, 1992, and 1993. I interviewed individuals from 30 mobile camps of the shepherds, spreading the conversations over a one-year period. Apart from interviews with shepherds, I also met government officials responsible for development of migrant pastoralists. Financial assistance from the International Institute for Environment and Development greatly facilitated the fieldwork. I acknowledge the Center for Science and Environment, New Delhi, for the use of the maps of India and Rajasthan. Sri Dallaram Dewasi, Sri Bhopalaram Dewasi, Sri Achalaram Dewasi, Sri Arjundan Charan, Sri Gokulji Dewasi, Sri Sattaram Dewasi, and Sri Gairji Dewasi provided incalculable assistance in contacting shepherds and their camp leaders. I am grateful for their time and conversations. Government officials at the Sheep and Wool Development Board in Jaipur, India, clarified several puzzles. I wish to thank John Aldrich, Robert Bates, Sabine Engel, Kathryn Firmin, Clark Gibson, Michael Goldman, Anil Gupta, Paulette Higgins, D. L. Johnson, HeeMin Kim, Peter Lange, Margaret McKean, Elinor Ostrom, Harold Pollack, Amy Poteete, Michelle Taylor, and three anonymous reviewers for their comments, encouragement or criticisms in the different stages of writing this essay. I completed the final draft as a Fellow at the Program in Agrarian Studies at Yale University. Special thanks go to Venkatesh Bala for his help with the technical aspects.

NOTES

1. Nomadic pastoralists are defined as groups who "are economically reliant on their herds and who wander seasonally in search of pastures." This definition as well as a summary of the debate concerning the origins of pastoralism is available in Sadr (1991, 1–11). See also Khazanov (1984, 15–84).

2. As institutional and popular faith in top-down development policies erodes, studies valorizing local knowledge and institutions are gaining ground. An impressive body of research now stresses that the goal of securing an improved life for the

poor will be better served if we first appreciate their desires, knowledge, and institutions (Chambers, Saxena, and Shah 1989; Croll and Parkin 1992; Kaul 1996; Oldfield and Alcorn 1991; Posey and Balee 1989; Schmink and Wood 1984). In consequence of these emerging trends in "development science," new studies on local knowledge and institutions have become increasingly available (Brokensha, Warren, and Werner 1980; Warren 1991; Warren, Slikkerveer, and Brokensha 1991; McKean 1992; E. Ostrom 1990).

3. For a discussion of why nomadism/transhumance prevails in certain socioecological niches, see Nugent and Sanchez (1993).

4. Exceptions are available in the works of anthropologists who have produced several studies on decision making and organization among nomads and hunter-gatherers. See, for example, Abel and Blaikie (1990), Barth (1961), Mithen (1989), Weissleder (1978). See also Barlett (1980), Gladwin (1979), Ortiz (1967), and Quinn (1975) for anthropological studies of decision making. Few of these studies, however, focus on leader-led relations. The vast literature on decision making by game theorists and management scientists suggests possible future directions for research but few current applications. See, for example, Bell, Raiffa, and Tversky (1990) and March (1988).

5. Nambardar literally means "the holder of a number." The term dates back to the British period when the camp leaders were assigned individual numbers for ease in recording their movements.

6. The primary aim of this essay is to delineate the logic of shepherd-leader interactions and decision making in the camps, and so only limited information on raika society and economy and their relationship with their ecological context is provided. For the interested reader, general descriptions of the raikas are available in Agrawal (1993, 1994), Kohler-Rollefson (1992), Srivastava (1990, 1991), and Westphal-Hellbusch (1975).

7. The classic defense of the virtues of simplifying assumptions is, of course, available in Alchian (1950) and Friedman (1953). The radical simplification of reality embodied in the model I use, however, still incorporates the most significant features of the social context through the payoff structure of the games I use.

8. The literature on "nomadic" groups in India is far less extensive than what is available for the African or the West Asian context. Some general accounts of nomadic groups are available in Leshnik and Sontheimer (1975), Misra and Malhotra (1982), and Prasad (1994).

9. In the ensuing discussion, I will use the term *nambardar* interchangeably with the term *camp leader* and *mukhiya* with the term *flock leader*.

10. Rarely, if ever, do women hold positions of formal authority among the raikas. Thus, in none of the 30 dangs that I surveyed was a woman the leader of the camp. Nor was a woman ever the flock leader.

11. A number of writers have remarked on hostility between settled and migrant populations. See, for example, Nugent and Sanchez (1993). For the raikas, at least, such conflicts are the exception; mutual accommodation is the rule.

12. Despite the extensive authority, the nambardars were modest about their role in the migration, usually attributing success to God—*"Shivji ki kirpa."*

13. The ubiquity of principal-agent relationships is captured well in this grudging phrase from DiIulio (1994, 278): "Essentially, the so-called principal-agent problem exists whenever one individual depends on the action of another."

14. Among anthropologists, Bailey (1965, 1988, 1991), Barth (1966, 1967) and Colson (1974) are among the earliest exponents of approaches that are either compatible with, or presage the current institutionalist approaches to, the study of rural or "traditional" societies.

15. But, see Bernheim and Whinston (1985, 1986) and Li (1993). Several works on multiple agents and single principals are also now available (Demski and Sappington 1984; Mookherjee 1984; Nalebuff and Stiglitz 1983; and Tirole 1986). See also Krepps (1992).

16. Economic considerations seem to play only a small role in the desire to be the leader. The power to make decisions for a group of followers is, perhaps, the primary force impelling some individuals to seek status as a leader.

17. The council of elders plays a relatively minor role as a decision-making body in situations involving principal-agency problems owing to the rather limited opportunities it possesses to abuse authority. See the next section for a discussion of the situations where graft and malfeasance are possible, and the limited authority of the council in those situations. The council as a decision-making body plays a significant advisory role, however, as discussed earlier and depicted in figure 8.1.

18. Adverse selection, moral hazard, shirking, and corruption are discussed in Baiman (1982), E. Ostrom, Schroeder, and Wynne (1993), and Williamson (1985). See Kiser and Tong (1992) for a discussion of different types of corruption, and Shleifer and Vishny (1993) for the relation of corruption to efficiency.

19. A third alternative is logically possible—that the camp split into two or more parts as a result of dissension between the principals and the agent. I did not observe, nor could the shepherds recall, such a case.

20. None of the shepherds I interviewed felt that the nambardar could be removed during a migration cycle, even if he proved incompetent or corrupt.

21. Few shepherds leave a camp in the middle of the migration cycle, but the fact that they possess this choice remains an ultimate restraint on hasty, thoughtless, or arbitrary decisions. The extent to which the leader can impose his will is thus limited by the shepherds' freedom to express their preferences through exit and voice. The shepherds use "voice" not in complaining to the nambardar or protesting directly to him. Instead, they use voice to talk with other shepherds and thus affect the reputation of a nambardar and the possibility of his continued selection as nambardar (see Hirschman 1970). In a rare conversation shepherds remembered an instance where a flock leader, disgusted with his nambardar, left the camp in mid-migration. His nambardar was universally condemned by other camp leaders and shepherds as having behaved in a capricious and wilful fashion. "It is like a father deserting his son in the wilderness," was a common expression used by shepherds to describe the errant leader's behavior. When the shepherd left the camp, his leader lost status and respect.

22. One of the earliest theorists to stress the necessity of monitoring in principal-agent relationships was Max Weber (1992), who discussed how factors such as complexity and diversity of tasks, and ease of communication and transportation, affect

the possibility of effective monitoring.

23. The nambardar and the council of elders are each responsible roughly for about half of the cases requiring a decision on the time at which sheep shearers and wool merchants should be called. The most important factor governing who should make the decision seemed to be the size of the camp—in smaller camps, the nambardar was far more likely than the council to make this decision. The reason behind this distribution seems to relate to the increasing amount of information needed to make the decision about when the entire camp's sheep would be ready for shearing (as the camp grew in size).

24. At the time of this writing, US$1.00 equals Rs. 30.00.

25. As the raikas say, "Even your enemy does not cheat you while looking you in the eye."

26. For a detailed explanation of why sheep buyers often visit the shepherd camps, see Agrawal (1992).

27. On the other hand, the interviewed nambardars, unsurprisingly, did not acknowledge the possibility that they might earn a return without informing their camp members.

28. See Crémer (1995) and Frey (1993) for a discussion of how tighter monitoring may turn out to be counterproductive. Frey examines the role of trust and loyalty in principal-agent relationships, while Crémer discusses how more information might make threats of punishment less credible. Holmstrom (1979) provides the classical statement regarding the benefits of gaining and using all available information. Lewis and Sappington (1993) discuss how ignorance might introduce discontinuities in the performance of the agent.

29. See Alam (1989) and Shleifer and Vishny (1993).

30. The character of the asymmetrical relationship between subaltern groups such as the raikas, and those in power such as the police, assumes an almost timeless quality in the Indian context. See Arnold (1985) for a discussion of similar dynamics in South India but with the constabulary as the main character.

31. Usually the fines paid to settle disputes with villagers are negotiated openly. Thus all shepherds know the amount. The negotiations for bribes are carried out more covertly, and seldom is anyone other than the nambardar involved. An elder member of the camp may contribute his expertise, but most of the time the nambardar alone decides the amount of the bribe.

32. In a colorful phrase, a shepherd, evoking widespread laughter, explained to me that government officials absorb bribes just as the parched earth sucks in rain.

33. Collectively, the shepherds spend perhaps no more than 3—4 hours monitoring the nambardar each day since the rest of the nambardar's day is filled with innocuous tasks. Individually, each shepherd need spend no more than 10—15 minutes in the course of the daily routine on monitoring tasks. During the approximately 300 days of migration when shepherds work 12—15-hour days, they spend no more than 4—5 days on monitoring tasks. This implies, in monetary terms, an expenditure of about Rs. 100—50 per shepherd in labor time (see also note 34).

34. The cost of monitoring is very low for the shepherds because they constantly engage their agent in face-to-face interactions. If the cost of monitoring, a_4, were to

rise to a point where it is greater than a_3/N, the losses suffered by the agent's cheating, then there would be an equilibrium in pure strategies. The shepherds will never monitor, and the leader will always cheat.

35. I follow the standard procedure for calculating the equilibrium in mixed strategies. See Ordeshook (1986) and Rasmusen (1989).

36. See the exchange between Bianco, Ordeshook, and Tsebelis (1990) in the *American Political Science Review* for the relative merits and problems of using a mixed strategies game model.

37. Given the low cost of monitoring to the shepherds, and the fact that the size of the camp seldom rises above 20 flocks, this constraint seems never actually to be crossed. In other words, the number of flocks in the camp would have to rise far above 20 for the pure strategies equilibrium to emerge.

38. As North (1993) suggests, the process of direct monitoring and sanctioning, when feasible, avoids additional principal-agent problems that third-party involvement can introduce.

39. See Tsebelis (1989, 1990).

40. I am grateful to one of my anonymous reviewers for pointing this out to me. See Baron (1985) and Braverman and Stiglitz (1982).

REFERENCES

Abel, N. and P. Blaikie (1990) "Land Degradation, Stocking Rates and Conservation Policies in the Communal Rangelands of Botswana and Zimbabwe," *Pastoral Development Network,* Paper 29a, London: ODI.

Agrawal, A. (1992) "Risks, Resources and Politics," dissertation, Duke University, Durham, NC.

Agrawal, A. (1993) "Mobility and Cooperation among Nomadic Shepherds: The Case of the Raikas," *Human Ecology* 21(3): 261–79.

Agrawal, A. (1994) "I Don't Need It but You Can't Have It," *Pastoral Development Network,* Set 36a, 36–55. London: ODI.

Alam, M. S. (1989) "Anatomy of Corruption: An Approach to the Political Economy of Underdevelopment," *American Journal of Economics and Sociology* 48(4): 441–56.

Alchian, A. (1950) "Uncertainty, Evolution, and Economic Theory," *Journal of Political Economy* 58: 211–21.

Arnold, D. (1985) "Bureaucratic Recruitment and Subordination in Colonial India," in R. Guha (ed.) *Subaltern Studies IV: Writings on South Asian History and Society,* 1–53. New Delhi: Oxford University Press.

Bailey, F. (1965) "Decisions by Consensus in Councils and Committees with Special Reference to Village and Local Government in India," in M. Gluckman and F. Eggan (eds.) *Political Systems and the Distribution of Power.* New York: Praeger.

Bailey, F. (1988) *Humbuggery and Manipulation: The Art of Leadership.* Ithaca, NY: Cornell University Press.

Bailey, F. (1991) "Why Is Information Asymmetrical? Symbolic Behavior in Asymmetrical Organizations," *Rationality and Society* 3(4): 475–95.

Baiman, S. (1982) "Agency Research in Managerial Accounting: A Survey," *Journal of Accounting Literature* 1: 154–213.

Barlett, P. (ed.) (1980) *Agricultural Decision-making: Anthropological Contributions to Rural Development.* New York: Academic Press.

Baron, D. (1985) "Noncooperative Regulation of a Non-localized Externality," *Rand Journal of Economics* 16: 553–68.

Barth, F. (1961) *Nomads of South Persia.* Oslo: Oslo University Press.

Barth, F. (1966) "Models of Social Organization," Occasional Papers No. 23. London: Royal Anthropological Institute.

Barth, F. (1967) "On the Study of Social Change," *American Anthropologist* 69: 661–69.

Bates, R. (1989) *Beyond the Miracle of the Market: The Political Economy of Agrarian Development in Kenya.* Cambridge: Cambridge University Press.

Bell, D., H. Raiffa, and A. Tversky (eds.) (1990) *Descriptive, Normative and Prescriptive Interactions.* New York: Cambridge University Press.

Berkes, F. (ed.) (1989) *Common Property Resources: Ecology and Community Based Development.* London: Belhaven Press.

Bernheim, B. and M. Whinston (1985) "Common Marketing Agency as a Device for Facilitating Collusion," *Rand Journal of Economics* 16: 269–81.

Bernheim, B. and M. Whinston (1986) "Common Agency," *Econometrica* 54(4): 923–42.

Bianco, W. (1990) "Crime and Punishment: Are One-Shot, Two Person Games Enough?," *American Political Science Review* 84(2): 569–73.

Biglaiser, G. and C. Mezzetti (1993) "Principals Competing for an Agent in the Presence of Adverse Selection and Moral Hazard," *Journal of Economic Theory* 61: 302–30.

Braverman, A. and G. Stiglitz (1982) "Sharecropping and the Interlinking of Agrarian Markets," *American Economic Review* 72: 695–715.

Brokensha, D., D. Warren, and O. Werner (eds.) (1980) *Indigenous Knowledge Systems and Development.* Lanham, MD: University Press of America.

Bromley, D. (1989) *Economic Interests and Institutions: The Conceptual Foundations of Public Policy.* Oxford: Basil Blackwell.

Cashdan, E. (ed.) (1990) *Risk and Uncertainty in Tribal and Peasant Economies.* Boulder, CO: Westview Press.

Chambers, R., N. Saxena, and T. Shah (1989) *To the Hands of the Poor: Water and Trees.* London: Intermediate Technology Publications.

Chong, D. (1995) "Rational Choice Theory's Mysterious Rivals," Critical Review (Special Issue) *Rational Choice Theory and Politics* 9(1–2): 37–58.

Colson, E. (1974) *Tradition and Contract: The Problem of Order.* Chicago: Aldine.

Crémer, J. (1995) "Arm's Length Relationships," *Quarterly Journal of Economics* 110(2): 275–95.

Croll, E. and D. Parkin (eds.) (1992) *Bush Base, Forest Farm: Culture, Environment and Development.* London: Routledge.

Demski, J. and D. Sappington (1984) "Optimal Incentive Contracts with Multiple

Agents," *Journal of Economic Theory* 33: 152–71.

DiIulio, J. (1994) "Principled Agents: The Cultural Bases of Behavior in a Federal Government Bureaucracy," *Journal of Public Administration Research and Theory* 4(3): 277–318.

Dyson-Hudson, R. and N. Dyson-Hudson (1980) "Nomadic Pastoralism," *Annual Review of Anthropology* 9: 15–61.

Eggertsson, T. (1990) *Economic Behavior and Institutions: Principles of Neoinstitutional Economics.* Cambridge: Cambridge University Press.

Ensminger, J. (1992) *Making a Market: The Institutional Transformation of an African Society.* Cambridge: Cambridge University Press.

Ensminger, J. and A. Rutten (1991) "The Political Economy of Changing Property Rights: Dismantling a Pastoral Commons," *American Ethnologist,* 18(4): 683–99.

Fama, E. (1980) "Agency Problems and the Theory of the Firm," *Journal of Political Economy* 88: 288–307.

Fama, E. and M. Jensen (1983) "Separation of Ownership and Control," *Journal of Law and Economics* 26: 301–26.

Frey, B. (1993) "Does Monitoring Increase Work Effort? The Rivalry with Trust and Loyalty," *Economic Inquiry* 31 (October): 663–70.

Friedman, M. (1953) *Essays in Positive Economics.* Chicago: University of Chicago Press.

Gladwin, C. (1979) "Production Functions and Decision Models: Complementary Models," *American Ethnologist* 6(4): 653–74.

Glover, J. (1994) "A Simpler Mechanism that Stops Agents from Cheating," *Journal of Economic Theory* 62: 221–29.

Green, D. and I. Shapiro (1994) *Pathologies of Rational Choice Theory.* New Haven, CT: Yale University Press.

Green, D. and I. Shapiro (1995) "Pathologies Revisited: Reflections on Our Critics," Critical Review: (Special Issue) *Rational Choice Theory and Politics* 9(1–2): 235–76.

Grossman, S. and O. Hart (1983) "An Analysis of the Principal-Agent Problem," *Econometrica* 51(1): 7–45.

Halstead, P. and J. O'Shea (eds.) (1989) *Bad Year Economics: Cultural Responses to Risk and Uncertainty.* Cambridge: Cambridge University Press.

Hirschman, A. (1970) *Exit, Voice and Loyalty: Response to Decline in Firms, Organizations and States.* Cambridge, MA: Harvard University Press.

Holmstrom, B. (1979) "Moral Hazard and Observability," *Bell Journal of Economics* 10: 74–91.

Hurvicz, L. (1973) "The Design of Mechanisms for Resource Allocation," *American Economic Review* 63: 1–30.

Jensen, M. (1983) "Organization Theory and Methodology," *Accounting Review* 50: 319–39.

Jensen, M. and W. Meckling (1976) "Theory of the Firm: Managerial Behavior, Agency Costs, and Capital Structure," *Journal of Financial Economics* 3: 305–60.

Kaul, M. (1996) "Two Centuries on the Commons: The Punjab," draft manuscript.

Khazanov, A. (1984) *Nomads and the Outside World.* Cambridge: Cambridge University Press.

Kiewiet D. and M. McCubbins (1991) *The Logic of Delegation: Congressional Parties and the Appropriations Process.* Chicago: The University of Chicago Press.

Kiser, E. and X. Tong (1992) "Determinants of the Amount and Type of Corruption in State Fiscal Bureaucracies: An Analysis of Late Imperial China," *Comparative Political Studies* 25(3): 300–31.

Kohler-Rollefson, I. (1992) "The Raika Camel Pastoralists of Western India," *Research and Exploration* 8(1):117–19.

Krepps, M. (1992) "Bureaucrats and Indians: Principal-Agent Relations and Efficient Management of Tribal Forest Resources," Discussion Paper 1601. Cambridge, MA: Harvard Institute of Economic Research.

Kreps, D. (1990) "Corporate Culture and Economic Theory," in J. Alt and K. Shepsle (eds.) *Perspectives on Positive Political Economy,* 90–143. Cambridge: Cambridge University Press.

Kroeber, A. (1948) *Anthropology.* New York: Harcourt Brace.

Laffont, J. and J. Tirole (1987) "Using Cost Observation to Regulate Firms," *Journal of Political Economy* 94: 614–41.

Leshnik, L. and G. Sontheimer (1975) *Pastoralists and Nomads in South Asia.* Wiesbaden: Otto Harrassowitz.

Lewis, T. and D. Sappington (1989) "Countervailing Incentives in Agency Problems," *Journal of Economic Theory* 49: 294–313.

Lewis, T. and D. Sappington (1993) "Ignorance in Agency Problems," *Journal of Economic Theory* 61: 169–83.

Li, S. (1993) "Competitive Matching Equilibrium and Multiple Principal-Agent Models," Discussion Paper 267. Minneapolis, MN: Center for Economic Research, University of Minnesota.

Lupia, A. and M. McCubbins (1994) "Designing Bureaucratic Accountability," *Law and Contemporary Problems* 57(1): 91–126.

McKean, M. (1992) "Success on the Commons: A Comparative Examination of Institutions for Common Property Resource Management," *Journal of Theoretical Politics* 4(3): 247–81.

March, J. (1988) *Decisions and Organizations.* Oxford: Basil Blackwell.

Marten, G. (1986) *Traditional Agriculture in Southeast Asia: A Human Ecology Perspective.* Boulder, CO: Westview Press.

Misra, P. and K. Malhotra (1982) *Nomads in India: Proceedings of the National Seminar.* Calcutta: Anthropological Survey of India.

Mithen S. (1989) "Modeling Hunter-Gatherer Decision-Making: Complementing Optimal Foraging Theory," *Human Ecology* 17(1): 59–83.

Mookherjee, D. (1984) "Optimal Incentive Schemes with Many Agents," *Review of Economic Studies* 51: 433–56.

Nalebuff, B. and G. Stiglitz (1983) "Prizes and Incentives: Towards a General Theory of Compensation and Competition," *Bell Journal of Economics* 14: 21–43.

Niamir, M. (1990) "Herders' Decision Making in Natural Resource Management in Arid and Semi-Arid Africa," FAO Forestry Papers. Rome: FAO.

North, D. (1990) *Institutions, Institutional Change and Economic Performance.*

Cambridge: Cambridge University Press.

North, D. (1993) Personal Communication, during interview for the video, *Common Resources, Uncommon Grounds*, Workshop in Political Theory and Policy Analysis, Indiana University, Bloomington, IN.

Nugent, J. and N. Sanchez (1993) "Tribes, Chiefs, and Transhumance: A Comparative Institutional Analysis," *Economic Development and Cultural Change* 42 (1): 87–113.

Oldfield, M. and J. Alcorn (eds.) (1991) *Biodiversity: Culture, Conservation and Ecodevelopment*. Boulder, CO: Westview Press.

Ordeshook, P. (1986) *Game Theory and Political Theory*. Cambridge: Cambridge University Press.

Ordeshook, P. (1990) "Crime and Punishment: Are One-Shot, Two Person Games Enough?," *American Political Science Review* 84(2): 573–5.

Ortiz, S. (1967) "The Structure of Decision-Making among Indians of Columbia," in R. Firth, (ed.) *Themes in Economic Anthropology*. London: Tavistock.

Ortiz, S. and S. Lees (eds.) (1992) *Understanding Economic Process*. Monographs in Economic Anthropology, No. 10. Lanham: University Press of America.

Ostrom, E. (1990) *Governing the Commons: The Evolution of Institutions for Collective Action*. Cambridge: Cambridge University Press.

Ostrom, E., L. Schroeder and S. Wynne (1993) *Institutional Incentives and Sustainable Development. Infrastructure Policies in Perspective*. Boulder, CO: Westview Press.

Posey, D. and W. Balee (eds.) (1989) *Resource Management in Amazonia: Indigenous and Folk Strategies*. New York: The New York Botanical Garden.

Prasad, R. (1994) *Pastoral Nomadism in Arid Zones of India: Socio-Demographic and Ecological Aspects*. New Delhi: Discovery.

Quinn, N. (1975) "Decision Models of Social Structure," *American Ethnologist* 2(1): 19–46.

Rasmusen, E. (1989) *Games and Information: An Introduction to Game Theory*. Oxford: Basil Blackwell.

Ross, S. (1973) "The Economic Theory of Agency: The Principal's Problem," *American Economic Review* 63: 134–9.

Sadr, K. (1991) *The Development of Nomadism in Ancient Northeast Africa*. Philadelphia: University of Pennsylvania Press.

Sandford, S. (1983) *Management of Pastoral Development in the Third World*. Chichester: John Wiley & Sons.

Schmink, M. and C. Wood (eds.) (1984) *Frontier Expansion in Amazonia*. Gainesville: University of Florida Press.

Shleifer, A. and R. Vishny (1993) "Corruption," *The Quarterly Journal of Economics* 108(3): 599–617.

Srivastava, V. (1990) "In Search of Harmony between Life and Environment," *Journal of Human Ecology* 1(3): 291–300.

Srivastava, V. (1991) "Who are the Raikas/Rabaris?," *Man in India* 71(1, Special): 279–304.

Tirole, J. (1986) "Hierarchies and Bureaucracies: On the Role of Collusion in Organizations," *Journal of Law, Economics and Organization* 2: 181–213.

Tsebelis, G. (1989) "The Abuse of Probability in Political Analysis: The Robinson Crusoe Fallacy," *American Political Science Review* 83: 77–91.

Tsebelis, G. (1990) "Crime and Punishment: Are One-Shot, Two Person Games Enough?," *American Political Science Review* 84(2): 576–85.

Wade, R. (1988) *Village Republics: Economic Conditions for Collective Action in South India*. Cambridge: Cambridge University Press.

Warren, D. (1991) "Using Indigenous Knowledge in Agricultural Development," World Bank Discussion Paper 127. Washington D.C.: World Bank.

Warren, D., J. Slikkerveer and D. Brokensha (1991) Indigenous Knowledge Systems: *The Cultural Dimensions of Development*. London: Kegan Paul International.

Weber, M. (1922) *Economy and Society* (Rpt. 1968). New York: Free Press.

Weissleder, W. (ed.) (1978) *The Nomadic Alternative: Modes and Models of Interaction in the African-Asian Deserts and Steppes*. The Hague: Mouton.

Westphal-Hellbusch, S. (1975) "Changes in Meaning of Ethnic Names as Exemplified by the Jat, Rabari, Bharvad, and Charan in Northwestern India," in L. Leshnik and G. Sontheimer (eds.) *Pastoralists and Nomads in South Asia*, 117–38. Wiesbaden: Otto Harrasowitz.

Williamson, O. (1985) *The Economic Institutions of Capitalism: Firms, Markets, Relational Contracting*. London: The Free Press.

Part III
Rules, Regulations, and Resource Management

Introduction to Part III

This part combines a general defense of game theory with two models of the ways in which rules and regulations shape community efforts to manage natural resources.

Games and Models

In "Heterogeneous Players and Specialized Models" (chap. 9), Eric Rasmusen defends game theory as a method that introduces analytical discipline into the efforts of social scientists to understand specific situations. (This essay was originally included in a special issue of *Rationality and Society,* "The Use of Game Theory in the Social Sciences.") Game theory is often dismissed as being unable to handle situations in which the participants have heterogenous interests or capabilities. Rasmusen argues that only those overly simple game models that claim comprehensive coverage are guilty of this shortcoming. He argues that the real strength of game theory lies in its ability to trace out the implications of interactions among actors with divergent interests and capabilities, provided that each model is designed specifically to apply to particular empirical situations. With such detailed models an analyst can investigate the likely consequences of changes in the parameters of a situation.

Although this chapter's running example of teams choosing between offensive and defensive strategies in a football game may seem trivial, the general point is an important one. Rasmusen wrote this essay before he became closely associated with the Workshop, but his comments reveal an attitude very much in line with the characteristic Workshop emphasis on models and empirical analyses of narrowly defined empirical situations. It is only in the context of rich empirical settings that the implications of alternative institutional arrangements can be fully understood. The other two selections in this part of the volume present detailed models (formal or informal) of particular empirical contexts related to resource management.

Evaluating Alternative Legal Doctrines

In "Governing a Groundwater Commons: A Strategic and Laboratory Analysis of Western Water Law" (chap. 10), Roy Gardner, Michael Moore, and

James Walker use game models and laboratory experiments to investigate the differing implications of alternative legal doctrines. Their specific empirical referent is governance of groundwater resources in the western United States. They define four legal doctrines that are in place for different groundwater systems in these states. Each legal doctrine defines the rights that participants have to extract water from these systems. The prior appropriation doctrine, for example, asserts that those users who have already been using the water have established the right to continue to extract water in the future. For the state of California, this system was described by Vincent Ostrom (1967) as an after effect of the practices adopted by miners during the era of the gold rush in the mid–eighteenth century.

The authors interpret each of these legal doctrines in terms of the concepts used to model different common-pool resources. For example, the prior appropriation doctrine amounts to a restriction on entry. Thus, the crucial aspect of this legal doctrine is its value on the entry rule dimension of the rule configuration, as presented by Elinor Ostrom in chapter 3. The authors argue that this entry restriction should act to mitigate the negative consequences of strategic and stock externalities. (See also Schlager and E. Ostrom 1993; Schlager, Blomquist, and Tang 1994.)

The authors next formulate a model of the extraction of water by homogenous users. The "optimal" level of water that would be allocated by a "benevolent central planner" is used to define the benchmark of an "efficient" outcome. This outcome is unlikely to be achieved in real-world settings, because the participants are directly concerned with their own payoffs and do not take into account the external effects of their own actions on other players. The subgame perfect (Nash) equilibrium predicted by this noncooperative game version results in a less-efficient outcome, in the sense that more water is extracted from the system than can be sustained over the long term.

Finally, the authors use different experimental settings to represent the implications of alternative legal doctrines. For example, the effect of entry restrictions (as implicit in the prior appropriation doctrine) is implemented in the laboratory by comparing the levels of the common-pool resource appropriated by groups of 5 or 10 subjects. The other parameters of the game are adjusted to isolate a pure size effect. The procedure followed in these experiments is discussed more fully in E. Ostrom, Gardner, and Walker (1994) and in part V of this volume. The results of these experiments clearly demonstrate that institutions matter, in the sense that the systematically different outcomes are observed from the different experimental treatments representing different legal doctrines. In particular, subjects in the experimental treatment corresponding to limited entry obtained a higher level of efficiency than the unlimited entry treatment, even though in both cases the results were even worse than that predicted by the subgame perfect (Nash) equilibrium.

Stock quotas had a more demonstrable positive effect, but were still well short of the overall optimum. These results are quite similar to those produced by other experiments on common-pool resources. (See E. Ostrom, Gardner, and Walker (1994); and part V of this volume.)

Governance and Transaction Costs

The next chapter moves us to the private sector. In "Bottlenecks and Governance Structures: Open Access and Long-Term Contracting in Natural Gas" (chap. 11), Thomas P. Lyon and Steven C. Hackett build upon influential research by Oliver Williamson (1975, 1985, 1996) on the many ways in which firms strive to reduce the transaction costs of conducting business in sectors where long-term contracting is an appealing alternative to spot markets. One major solution is to engage in vertical integration. That is, by incorporating more aspects of the sequence by which raw materials are transformed into products made available to customers, a firm can lower its danger of being taken advantage of by suppliers or distributors at each step of this process. Williamson details the transition from long-term contracting to vertical integration, which incorporates both parties to these contracts within the same organization.

The authors examine a sector in which vertical integration has obvious advantages to the firm, namely, the production and delivery of natural gas. In particular, long-term contracts are very common in the operation of gas pipelines. However, the energy industry has also been an area of active intervention by the U.S. government as it seeks to limit the extent of vertical integration. Lyon and Hackett begin by stating that the basic problem confronting government regulators is how to keep this market as competitive as possible while still assuring the enforcement of the long-term contracts that are necessary for firms to invest in this inherently risky sector.

Firms (or governments) that have frequent interactions with each other may agree to designate each other as most-favored customers (or nations). In this way each is assured that it will not be hurt by any subsequent contracts (or treaties) signed with other corporations (or governments). As Besanko and Lyon (1993) demonstrate, such arrangements can sustain an oligopoly's domination of an industry. However, in an empirical analysis of the natural gas industry, Crocker and Lyon (1994) conclude that the reduction of transaction costs is the major reason behind most-favored-nation clauses in natural gas contracts.

Lyon and Hackett focus their attention on the conditions under which owners of natural gas pipelines implement a policy of open access, which in this context means that pipelines should be treated as common carriers rather than being dedicated to the use of a single firm. (In common-pool resource settings, open-access refers to the absence of restrictions on the number of users.) They argue that government regulators have given insufficient consideration to the transaction cost benefits of discriminatory policies (i.e., the oppositive of open-access rules). They investigate a series of hypotheses concerning the relative magnitude of spot markets and long-term contracting in this sector of the economy that should occur under different regulatory conditions. They conclude that although an increased reliance on open-access policies does reduce the threat of opportunism by those controlling the pipelines, these policies also increase the costs of transacting in this sector.

This analysis by two Workshop-affiliated scholars draws on the standard presumption that governments should act to reduce the costs of economic transaction and thereby contribute to the public good of economic growth. In practice, however, the rules and regulations enforced by governmental authorities may have exactly the opposite effect. For example, Eggertsson (1992, 1996) investigates an "equilibrium trap" that Icelanders experienced for an incredibly long time period. Despite the ready availability of fish in Icelandic waters, a sophisticated fishing industry never arose in Iceland until well after the establishment of comparable industries elsewhere. In the intervening years, fishers from many parts of Europe came thousands of miles to Icelandic waters while local fishers stayed closer to shore.

The title of Eggertsson's 1996 essay alludes to the phrase "great experiments and monumental disasters" that Vincent Ostrom (1991, 1997) uses to describe the experience of the great revolutions of the twentieth century. In countries such as the Soviet Union and China, revolutionary movements intended to create a more egalitarian system instead ended up increasing the misery of ordinary people. Ostrom uses this phrase to sound a cautionary note in response to repeated calls for wide-ranging and immediate reform. He argues that policy analysts, while remaining open to exploring the potential benefits of institutional change, should remain restrained in their ambitions. By trying to do too much all at once, would-be revolutionaries may make things worse.

Eggertsson uses his trademark gentle humor to remind us that the absence of reform can have equally detrimental consequences. The inability of Icelandic fishers to exploit readily available resources so close to home significantly stunted their economic development. He shows how difficult it was for Icelandic fishers to step outside their assigned role within the closed political system of governance in Iceland. External factors also contributed, especially the interest of the Danish crown in providing certain towns with monopoly rights to trade with Iceland. In effect, then, Iceland's equilibrium trap is an example of a "two-level game" (see chap. 7) in which the interests of internal and external actors reinforced each other to produce a stable policy outcome. Eventually, however, Iceland emerged as a major fishing nation.

The topic of the longer-term consequences of large-scale fishing in the North Atlantic lies outside the scope of this volume, but the chapters in this part clearly demonstrate the importance of understanding the institutional context of resource management. Similar concerns must be addressed whether the resource is managed primarily via private ownership (as in the natural gas industry) or by public agencies (as in the case of watershed management in the western United States). Depending on its nature, the existing structure of governance can facilitate more efficient management of resources, speed the destruction of a resource, or delay development indefinitely (as in Icelandic fisheries). As Rasmusen argues, game models of specific situations can help us understand the consequences of alternative institutional arrangements. Such an understanding is a crucial part of the process of institutional design and policy evaluation.

REFERENCES

Besanko, David, and Thomas P. Lyon. 1993. "Equilibrium Incentives for Most-Favored Customer Clauses in an Oligopolistic Industry." *International Journal of Industrial Organization* 11: 347–67.

Crocker, Keith J., and Thomas P. Lyon. 1994. "What do 'Facilitating Practices' Facilitate?: An Empirical Investigation of Most-Favored Nation Clauses in Natural Gas Contracts." *Journal of Law and Economics* 36: 297–322.

Eggertsson, Thráinn. 1992. "Analyzing Institutional Successes and Failures: A Millenium of Common Mountain Pastures in Iceland." *International Review of Law and Economics* 12: 423–37.

————. 1996. "No Experiments, Monumental Disasters: Why it Took a Thousand Years to Develop a Specialized Fishing Industry in Iceland." *Journal of Economic Behavior and Organization* 30: 1–23.

McGinnis, Michael D., ed. 1999a. *Polycentric Governance and Development.* Ann Arbor: University of Michigan Press.

————, ed. 1999b. *Polycentricity and Local Public Economies.* Ann Arbor: University of Michigan Press.

Ostrom, Elinor, Roy Gardner, and James Walker, with Arun Agrawal, William Blomquist, Edella Schlager, and Shui-Yan Tang. 1994. *Rules, Games, and Common-Pool Resources.* Ann Arbor: University of Michigan Press.

Ostrom, Vincent. 1967. "Water and Politics California Style." *Arts and Architecture* 84 (July/August): 14–16, 32. Reprinted in McGinnis 1999a.

————. 1991. *The Meaning of American Federalism: Constituting a Self-Governing Society.* San Francisco: ICS Press.

————. 1997. *The Meaning of Democracy and the Vulnerability of Democracies: A Response to Tocqueville's Challenge.* Ann Arbor: University of Michigan Press.

Schlager, Edella, William Blomquist, and Shui Yan Tang. 1994. "Mobile Flows,

Storage, and Self-Organized Institutions for Governing Common-Pool Resources." *Land Economics* 70(3) (Aug.): 294–317. Reprinted in McGinnis 1999a.

Schlager, Edella, and Elinor Ostrom. 1993. "Property-Rights Regimes and Coastal Fisheries: An Empirical Analysis." In *The Political Economy of Customs and Culture: Informal Solutions to the Commons Problem,* ed. Terry L. Anderson and Randy T. Simmons, 13–41. Lanham, MD: Rowman & Littlefield. Reprinted in McGinnis 1999a.

Williamson, Oliver E. 1975. *Markets and Hierarchies: Analysis and Antitrust Implications.* New York: Free Press.

———. 1985. *The Economic Institutions of Capitalism.* New York: Free Press.

———. 1996. *The Mechanisms of Governance.* New York: Oxford University Press.

CHAPTER 9

Heterogeneous Players and Specialized Models

Eric Rasmusen

A theme common to the criticisms by Wildavsky (1992), Tullock (1992), and Hechter (1992) is that game theory is a constricted theory, relying too heavily on narrow assumptions to be useful in explaining how the world works. Small changes in assumptions produce big changes in conclusions, and, in particular, game theory has trouble with the heterogeneity of human beings—their differences in tastes, information, and culture.

The game theorist's reply must be, I think, that game theory's building blocks do apply to a wide variety of situations and are, in fact, particularly well suited to heterogeneity, but that the situations being modeled do not lend themselves to general theories. To the extent that heterogeneity is important, models ought to be narrow, for a general model is effectively saying that situations are homogeneous. If situations apparently similar are actually different, then tailoring the game to the facts is better than relying on a single game for all facts. In a good theory, at least some seemingly minor details make a difference, for how else are we to discover that anything but the obvious is important?

To make this concrete, let us follow Tullock's (1992) good example and analyze a particular game. Tullock uses the matrix in figure 9.1. In its stark form, this matrix conjures up players in a laboratory, whose behavior, while analyzable, has no intrinsic interest for us. But to call the properties of a bare mathematical matrix unrealistic is not a well-posed statement; to build a model, something must exist to be modeled, something whose essence is to be captured. Starting with just the matrix, we are in the position of the RAND game theorists in the early 1950s who were perplexed by a certain 2 × 2 matrix that generated perverse results. Albert Tucker, on being asked to give a talk on game theory to the Stanford psychology department, decided to attach a story to the numbers. The result was the Prisoner's Dilemma and a deeper understanding than the mathematics alone could give (Straffin 1980).

Originally published in *Rationality and Society* 4, no. 1 (1992): 83–94. Copyright © 1992 by Sage Publications Ltd. Reprinted by permission of Sage Publications Ltd. and the author.

	A γ	B $1-\gamma$
C θ	−5 5	5 −5
D $1-\theta$	1 −1	−1 1

Fig. 9.1. The Tullock Game

The Tullock matrix is not a Prisoner's Dilemma, but a story can nonetheless be attached to it. Let us try discussing it as a model of the conflict between offense and defense in a football game.[1] The offensive team is trying to decide between passing the ball and running with it, and the opposing team must decide whether to set up a defense against passing or running. The offensive team would like to do the unexpected, since that is the way to advance the ball toward the goal. But more is at stake with passing than with running—five yards instead of one, if the numbers in figure 9.1 are retained.[2] Let us call this the Football Game. Its Nash equilibrium is in mixed (random) strategies. The offense passes with probability 1/6, and the defense uses a running defense with probability 1/2.[3]

After discussing the characteristics of the original matrix, Tullock objects that the players should be, like most people, risk averse. Gaining five yards may not please as much as losing five yards displeases—at least if one's team is currently ahead in the game. This means that the numbers in figure 9.2 are no longer valid, for they no longer represent the payoffs-utility—but an instrumental means to those payoffs-yards. The obvious solution is to change the numbers to fit risk-averse payoffs. But how can this be done? It requires knowledge of how much each player values winning the overall game and how much the gains and losses of yardage affect winning.

		Defense	
		Running Defense γ	Passing Defense $1-\gamma$
Offense	Pass θ	−5 5	5 −5
	Run $1-\theta$	1 −1	−1 1

Fig. 9.2. The Football Game

The difficulty of measuring possible payoffs is a common criticism of game theory, but empirical work in any subject runs into measurement problems, and game theory presents no special difficulties—only the standard, hard ones. The same problem faces the economist who is asked how many more oranges John will buy if the price of apples rises. About both fruit and football, he can make an informed guess using general knowledge and 10 minutes' thought, or he can use a government grant and three years' work to come up with a somewhat better answer. In the absence of a grant (and three years), let us make a guess, based on the assumption that the offensive player is already ahead on points, and replace our earlier matrix with figure 9.3 (I have also added variable c, which will be explained shortly).

Defense

		Running Defense γ	Passing Defense $1-\gamma$
Offense	Pass θ	$3-c$ ⟋ -3	$-8-c$ ⟋ 8
	Run $1-\theta$	-3 ⟋ 3	1 ⟋ -1

Fig. 9.3. The Football Game with risk aversion and jealousy

The game is no longer zero sum, but the Nash equilibrium can still be calculated. It will depend on the term c, which is added to answer another of Tullock's questions: What if the Row player does not like playing his upper action? For the Football Game, the question can be made more concrete: What if the offensive team captain becomes jealous when his pass receivers get attention from the crowd and suffers disutility pangs of c when he uses the passing strategy? Jealousy is easy to incorporate; that is what c depicts. The mixed-strategy equilibrium is for the offense to pass with probability $\theta = 4/15$ and the defense to use a running defense with probability $\gamma = c/15 + 3/5$.[4] This yields the interesting prediction that if the offensive captain comes to hate his receivers more (c increases), he will not pass any the less, but the defense will use the passing defense less often, knowing that the offense is reluctant to take advantage of the opportunity. Thus not only can risk aversion and preferences for certain actions be incorporated into the game, but the model can make predictions about how behavior varies depending on those parameters.

I have already mentioned that measurement is difficult for the modeler, but it is also difficult for the players, a distinctly different objection. So far, the Football Game has assumed that they know each other's utility functions. Is this justified? On a general level, it seems that individuals do act as if they know something about each other's utility functions. When a businessman opens up a shoe factory, it is under the belief that he knows pretty well how much other people value shoes—enough so that in tight competition, he can still make a normal profit. He may be mistaken, but he is willing to bet on his knowledge, and since the information is very important to him, he has invested some effort in acquiring it (even more, perhaps, than a scholar writing an article on the subject). So, the question is really what happens if the players know each other's payoff functions, but imperfectly.

A complete model requires specification of what it is that the players know and do not know. In the Football Game, let us see what happens if the offensive captain knows how jealous he is, but the defensive captain does not. Assume that (1) the defensive captain does not know the exact value of c and attaches equal probabilities to $c = 0$ and $c = 1$; (2) the offensive captain knows that the defensive captain has those beliefs; and (3) in actual fact, $c = 1$.[5] In equilibrium, the offensive player will pass with probability $\theta_{c=1} = 0$ and the defensive player uses a running defense with probability $\gamma = 3/5$. The equilibrium also specifies the behavior that the offensive player would adopt if $c = 0$; it is to pass with probability $\theta_{c=0} = 8/15$.[6] From the point of view of the defensive player, who does not know the value of c, the probability of the offensive player passing is $\theta = .5(0) + .5(8/15) = 4/15$. The change in assumptions has changed the offensive player's behavior, from $\theta = 4/15$ to $\theta_{c=1} = 0$. But this is what we should expect: The offensive player should pass less often when facing an opponent who overestimates his incentive to pass. Thus game theory easily adapts to differences in tastes and knowledge, but by the very fact that it does so, predicting different outcomes under different circumstances, it tells us that a perfectly general theory is impossible.

Even if one accepts that game theory can accommodate different sorts of utility functions and beliefs, however, one might still quarrel with the very idea of the mixed-strategy equilibrium—as indeed Tullock does. Not only do mixed strategies involve "carefully random" behavior, but the equilibrium is weak in the sense that each player is indifferent between at least two of his pure strategies or he would not be willing to mix between them. Yet if the equilibrium is to exist, it seems this indifferent player must carefully pick just the right probabilities for each action.

A mixed-strategy equilibrium does not, however, actually require any randomization. What it requires is that the mixing player's actions seem random to the other players, whether this results from literal randomization or not. A football captain might not throw dice in the huddle, but he surely wishes to take actions unpredictable to the other team. If the captain chooses his plays based on a deterministic device such as whether the time remaining in the game is an odd or even number, the game theorist is justified in modeling the choice as random if it seems random to the defense. The captain may even decide in advance to pass on the first play of the game with probability 1, but if the other team believes that in general only proportion θ of captains pass on the first play, the result is the mixed-strategy equilibrium.

The Harsanyi (1973) explanation for mixed strategies cited by Tullock (1992) is similar in flavor but based on player heterogeneity. Harsanyi suggests that there are always small features of the situation that would push the deciding player to one or the other pure strategy but that cannot be observed by the other players. The Football Game with unknown c hints at this, because what to the defensive player seems randomization with a 4/15 probability is actually probability 8/15 for one type of offensive player ($c = 0$) and probability 0 for the other type ($c = 1$). If there were a continuum of types from $c = 0$ to $c = 2$, practically all types would be using pure strategies, but to the defensive player, who cannot observe c, it would seem that the offensive player was randomizing. The differences in types need not even be large; there could be a continuum of types from $c = .98$ to $c = 1.02$, and while they would all be choosing one pure strategy or the other, they would appear to the defensive player (and the modeler) to be randomizing.

It should be kept in mind, too, that a mixed-strategy equilibrium is still a Nash equilibrium: No player can profitably deviate from his assigned strategy, even though he may be indifferent about it. This is illustrated by Tullock's other example, the Hillman-Samet (1987) rent-seeking game. In that game, two players simultaneously offer bribes to an official who will grant $100 to the highest bidder but keep both bribes. In the mixed-strategy equilibrium, the two players randomize their choices of bribes, between $0 and $100. Tullock asks what happens if the game is repeated and a third player enters and bids $90 each time. This new strategy will do very badly. If the original two players fail to react, they too will do badly, but that does not make the "bid 90" strategy rational unless the entrant is malicious. If the original players do react, they can achieve higher payoffs than the entrant by mixing between bids of $0 and bids on the interval between $90 and $100.[7] They will not be driven from the game; often each will bid $0, sometimes one will bid high and win, and sometimes both will bid high and one will lose his bid without reward. The example only shows that rational players should adapt their behavior to the behavior of irrational ones, not that irrational ones do better.[8]

I have spent so much time on the Football Game because it illustrates the sort of sensitivity analysis that is useful for honing intuition. Where changing the assumptions makes a difference, it should make a difference and shows that apparently minor assumptions are not so minor. This approach places the burden on the modeler to describe the game carefully, but that burden is inescapable under any method of analysis; a car that moves along fine on a solid highway will spin its wheels uselessly when you try to drive it on sand.

Let me turn now to Wildavsky (1992), who says that game theory ignores culture and that culture is important, because "without a supportive cultural context, no strategy makes sense" (10). Thus the rational man of game theory is not the reasonable man of law; he is a sociopath, without preexisting values or relations to defend. In particular, Prisoner's Dilemma is not nearly as useful as has been claimed, since most people are actually not sociopaths. People in different cultures play the Prisoner's Dilemma differently, and a given situation will be a Prisoner's Dilemma in one culture but not in another.

The issue comes down to what is similar about humans and what is different. The position of game theory is that everyone, whatever his or her culture, is best analyzed as a rational maximizer, but what is rational depends on the particular preferences and constraints available to the individual. To quote O'Rourke (1988), "A Japanese raised in Riyadh would be an Arab. A Zulu raised in New Rochelle would be a dentist" (4). Japanese, Arabs, Zulus, and dentists are all rational actors, even if we who are outside their particular cultures do not share their tastes and beliefs.

As Wildavsky (1992) says, one physical situation might call for different models for different cultures, or even within one culture, if preferences differ. The Football Game's variants—the basic game and the variants with risk aversion, jealousy, and asymmetric information—were all based on the same physical situation. Game theory does treat people as if they had preexisting values; the payoffs are literally values, and the model takes them as preexisting. Wildavsky hits the point precisely when he says the modeler's position should be "Tell me how individuals understand their situation in terms of their preferences, and I will then model the strategic aspects of this situation" (12). Once provided with the rules of the game—players, payoffs, actions, and initial beliefs—game theory predicts the outcome. If the rules of the game are those of the Prisoner's Dilemma, the outcome is that both players will confess, whatever their culture may be. But if people in the culture enjoy prison or prefer confession to lying, then the payoffs are different and the game is no longer a Prisoner's Dilemma. This is not a matter of how the game is analyzed but of how its rules are specified.

Culture theorists may object to the entire idea that people respond to incentives, believing instead that their responses are preconditioned. This certainly makes for generality, since it leads to predictions independent of the parameters of the particular situation. But it seems much like an extreme form of the Tullock (1992) objection that a player may not enjoy playing one of his actions, which, as seen earlier, is easily handled by game theory. The objection must then become that the theory is unimportant because the answer is determined by the empirical measurement of the disutility of the action. That is unobjectionable; theory and empirical investigation each have their place.

In some contexts, however, one must wonder whether "culture" really just refers to situations with different payoff matrices. Wildavsky's (1992) cultural groups—hierarchists who trust authorities, individualists who trust market exchange, egalitarians who trust voluntary groups, and fatalists who trust no one—might be identical human beings facing different incentives. Consider, for example, a game in which Smith must decide whether to buy a computer from Jones and Jones must decide whether to sell him a working computer or a broken one. Smith will be a fatalist and not buy the computer if there is no enforcement mechanism penalizing Jones for selling a broken one. Smith will be a hierarchist and buy if Jones has a boss who will punish him for selling broken equipment. Smith will be an individualist and buy if warranty law would force Jones to replace a defective computer. Smith will be an egalitarian and buy if he and Jones are members of the same university department and Jones's reputation will be ruined if he defrauds Smith. Although these are not all Prisoner's Dilemmas, and each story changes the game, Prisoner's Dilemma is the basic building block for them all. One must only remember that a house requires more than the basic concrete blocks.

Finally, I turn to Michael Hechter's (1992) discussion of the repeated Prisoner's Dilemma. Essentially, he has two objections: that the game has multiple equilibria and that cooperation is difficult to achieve under incomplete information. The first objection is quite valid—and often forgotten. The folk theorem tells us that a huge variety of outcomes can occur in equilibria of the infinitely repeated Prisoner's Dilemma, including cooperate each period and fink each period. Arguments have been put forward for why cooperation might be more likely, but the debate is still very much alive.[9]

That cooperation is more difficult when information is incomplete, however, is dubious. Incomplete information, in fact, is the most widely accepted explanation for why cooperation might ensue. Kreps et al. (1982) show that even if Prisoner's Dilemma is only finitely repeated (and a fortiori if it is infinitely repeated), the unique equilibrium outcome can be cooperation until close to the end of the game. The extra assumption they add is a small probability that one player (let us call him Smith) is not rational—Smith plays the tit-for-tat strategy whether it helps him or not. The rational player knows of this possibility, and so he cooperates until near the end of the game. If Smith is indeed irrational, the other player need not fear an unprovoked confession, but even if Smith is rational, it is to his advantage to pretend to be irrational until near the end of the game. The argument is subtle, but the conclusion is simple: Uncertainty over the player's payoffs can make cooperation more likely, not less.[10] Under complete information, cooperation is difficult, so muddying the waters can hardly hurt.

Hechter and Wildavsky would probably agree with Tullock (1992) when he says, "My objection to game theory is as a formal body of mathematics allegedly applying to human action, not as a heuristic which makes it easier to think about certain problems" (24). Game theory should indeed be a branch of storytelling, not of mathematics—storytelling with the i's dotted and the t's crossed. One can tell stories that are consistent and stories that are not, and formalism helps spot the inconsistencies by forcing the storyteller to tell one story at a time. When Tullock says that "the problem is that parties with no knowledge of formal game theory are likely to go through the same process as the trained game theorist" (31), he pays a compliment to formal game theory, for economists have learned only slowly that participants in the marketplace are often wiser than scholars. Long experience, inherited tradition, and the careful deliberation that self-interest motivates often lead to behavior whose usefulness is not apparent to the outsider. But theory has an advantage similar to that of the factory worker over the craftsman: The theorist can make do without experience, tradition, and self-interest and use superior capital to produce a product that is cheaper and more uniform, if perhaps not so reliable. Let us not be Luddites; without theoretical tools, only the born craftsman is able to produce decent scholarship.

A criticism common to all three critics is that game theory relies too heavily on situation-specific assumptions. We all would prefer generalizing theory to exemplifying theory, a model of what must happen instead of what can happen, as Franklin Fisher (1989) puts it in his own critique of game theory. The danger in this is that the modeler may try to force-fit a model to situations for which it is unsuited, as Wildavsky (1992) says is done with Prisoner's Dilemma. But sociology, political science, and anthropology are in a wonderful position to escape this danger. These disciplines have been heavily data driven, with many descriptive studies of particular situations, which is the empirical analog of the exemplifying style of game theory. Game theory can explain not only what happened in a case study but what might have happened had conditions been different and what will happen if the parameters change. If game theory is used in this way, I think that the critics will be happier.

NOTES

1. If your response is "He's using a game to model a game!" mentally substitute a military conflict between offense and defense where the attack can come on the right or left flank.

2. These numbers are not quite realistic, although their ordinal rank fits football. Partly this is a matter of the units of measurement. If the numbers in the matrix are doubled and one yard is added to each entry, they become more realistic. Such a change will not affect the optimal strategies in the slightest—it is just a change of measurement units.

3. To calculate these, use the fact that in a mixed-strategy equilibrium, a player is indifferent between his pure strategies. The expected payoffs for the offense from his two pure strategies are $\pi_o(\text{pass}) = 5\gamma - 5(1 - \gamma)$ and $\pi_o(\text{run}) = -1\gamma + (1)(1 - \gamma)$. Equating these yields $\gamma = .5$. Similarly, equating $\pi_d(\text{running}) = -5\theta + (1)(1 - \theta)$ and π_d (passing) $= 5\theta + (-1)(1 - \theta)$ yields $\theta = 1/6$.

4. To calculate this, use the fact that in a mixed-strategy equilibrium, a player is indifferent between his pure strategies. The expected payoffs for the offensive player from his two pure strategies are $\pi_o(\text{pass}) = (3 - c)\gamma + (-8 - c)(1 - \gamma)$ and $\pi_o(\text{run}) = -3\gamma + (1)$ $(1 - \gamma)$. Equating these yields $\gamma = c/15 + 3/5$. Similarly, equating $\pi_d(\text{running defense})$ $= -3\theta + 3(1 - \theta)$ and $\pi_d(\text{passing defense}) = 8\theta + (-1)(1 - \theta)$ yields $\theta = 4/15$.

5. If the offensive captain is uncertain about the defensive captain's beliefs, that too can be incorporated into the model, at the cost of extra complexity. If he is certain but wrong, it is even simpler to incorporate.

6. The equilibrium is calculated as follows: Whatever value of γ is picked, it cannot be the case that the offensive player would mix for both $c = 0$ and $c = 1$; in one case or the other he will use a pure strategy. In computing the equilibrium with known c, it was established that if the value of θ was greater than 4/15, then $\gamma = 1$. Hence, it cannot be that either $\theta_{c=0} = 1$ or $\theta_{c=1} = 1$, or the average would exceed 1/2 and we would have $\gamma = 1$. Since it is the player with $c = 1$ who is more reluctant to pass, it must be that $\theta_{c=1} = 0$, and $\theta_{c=1} \in (0, 1)$. If γ is to be between zero and one, it must be that the average value of θ equals 4/15, so $4/15 = 5(\theta_{c=1}) + .5(0)$, which yields $\theta_{c=1} = 8/15$. Since γ must be chosen to make the player with $c = 0$ willing to mix, it is chosen using the formula $\gamma = c/15 + 3/5$ found in the earlier equilibrium, which since $c = 0$ gives $\gamma = 3.5$.

7. Let $F(x)$ be the cumulative probability that a player bids up to x. In equilibrium, the payoffs from the pure strategies are equal, and the pure strategies are postulated to be 0 and the (90, 100) interval. The payoff from bidding 0 is 0, and the probability of winning with a bid of $x > 90$ is the probability $F(x)$ that the other rational player has bid less than x, so $\pi(0) = 0 = \pi(x) = -x + 100F(x)$, and $F(x) = x/100$. This implies that a player bids 0 with probability .9 and spreads the rest of his probability over the interval (90, 100). The "bid 90" player will win with probability $.9^2$, so his expected payoff is $.81(100) - 90 = -8.9$.

8. In other contexts, however, such as bargaining, irrationality can be a positive advantage (see Rasmusen 1989, chap. 10).

9. Technical notes, however: (1) The strategy of tit-for-tat is *not* a credible (perfect) equilibrium strategy because there is insufficient incentive to punish a *Confess* deviation with *Confess* the next period (see Kalai, Samet, and Stanford 1988, or Rasmusen 1992). (2) It is claimed that for large but not too large discount rates there is a unique, efficient equilibrium. In fact, (always *Confess*) remains an equilibrium outcome for any discount rate for which (always *Cooperate*) is an equilibrium out-

come. (3) The repeated Prisoner's Dilemma is not a game of perfect information. It includes simultaneous moves, so information sets are not singletons (see Rasmusen 1989, chap. 2). (4) Discounting hinders cooperation, rather than aiding it.

10. For elaboration, see chapter 5 of Rasmusen (1989).

REFERENCES

Fisher, Franklin. 1989. Games economists play: A noncooperative view. *RAND Journal of Economics* 20:113–24.

Harsanyi, John. 1973. Games with randomly distributed payoffs: A new rationale for mixed-strategy equilibrium points. *International Journal of Game Theory* 2:1–23.

Hechter, M. 1992. The insufficiency of game theory for the resolution of real-world collective action problems. *Rationality and Society* 4:33–40.

Hillman, A., and D. Samet. 1987. Dissipation of contestable rents by small numbers of contenders. *Public Choice* 54:63–82.

Kalai, Ehud, Dov Samet, and William Stanford. 1988. Note on reactive equilibria in the discounted prisoner's dilemma and associated games. *International Journal of Game Theory* 17:177–86.

Kreps, David, Paul Milgrom, John Roberts, and Robert Wilson. 1982. Rational cooperation in the finitely repeated prisoners' dilemma. *Journal of Economic Theory* 27:245–52.

O'Rourke, P.J. 1988. *Holidays in hell.* New York: Atlantic Monthly Press.

Rasmusen, Eric. 1989. *Games and information.* Oxford: Blackwell.

———. 1992. Folk theorems for the observable implications of repeated games. *Theory and Decision* 32 (March): 147–64.

Straffin, Philip. 1980. The prisoner's dilemma. *UMAP Journal* 1:101–3.

Tullock, G. 1992. Games and preference. *Rationality and Society* 4:24–32.

Wildavsky, A. 1992. Indispensable framework or just another ideology? Prisoner's Dilemma as an antihierarchical game. *Rationality and Society* 4:8–23.

CHAPTER 10

Governing a Groundwater Commons: A Strategic and Laboratory Analysis of Western Water Law

Roy Gardner, Michael Moore, and James Walker

I. Introduction

Between the poles of rent maximization and complete rent dissipation, wide latitude exists for institutions to manage or allocate common pool resources (CPRs) with reasonable economic performance. Two topics addressed in previous research are salient. One concerns the role of limiting entry by users into a commons. In the seminal article on the economics of CPRs, Gordon (1954) described how monopolist ownership would internalize CPR externalities, thereby creating incentives for rent maximization. Eswaran and Lewis (1984), applying a model of a CPR as a time-dependent repeated game, derived a related analytical result that the degree of rent accrual depends inversely on the number of users depleting the resource. In the context of groundwater, Brown (1974) and Gisser (1983) reasoned that existing laws restricting entry into groundwater CPRs would improve rent accrual. Empirical experience with more than five users, however, reached pessimistic conclusions in two cases. Libecap and Wiggins (1984) found that cooperative behavior in oil pool extraction occurred only with fewer than five firms. Otherwise, state law was required to coerce cooperation with roughly 10–12 firms. Indeed, with hundreds of firms operating in the East Texas oil fields there was no cooperation and, apparently, complete rent dissipation. Walker, Gardner, and E. Ostrom (1990) and Walker and Gardner (1992) reached a similar conclusion in analysis of data from laboratory experiments on noncooperative game CPRs. A high degree of rent dissipation or a high probability of resource destruction occurred even with access limited to eight users.[1]

The second topic concerns the ability of additional regulations or property rights, other than entry restrictions, to mitigate CPR externalities in light

Originally published in *Economic Inquiry* 35 (April 1997): 218–34. Reprinted by permission of Western Economic Association International and the authors.

Authors' note: Research support from USDA Cooperative Agreement No. 43-3AEM 1-80078 and National Science Foundation Grant No. SBR-9319835 is gratefully acknowledged.

of noncooperative behavior. Forms of property rights, such as firm-specific fishing rights or quotas, are widely recognized as reducing or removing the incentive for a race to exploit a CPR, as in Levhari, Michener, and Mirman (1981). Specific to groundwater, Smith (1977) recommended that rights to a share of the groundwater stock should replace Arizona's then-existing rule-of-capture, while Gisser (1983) noted that New Mexico's individual rights to annual water quantities, combined with a guaranteed time period of depletion, effectively define a share right in the stock. Both reasoned that this form of property right—stock quotas—would go far toward achieving optimal groundwater depletion.

State governance of groundwater resources in the western United States provides an institutional setting to study the effect of property rights and regulations on rent appropriation. Sax and Abrams (1986) and Smith (1989) write that, in the early to mid-1900s, independent state authority over groundwater resulted in adoption of four distinct legal doctrines governing groundwater use in the 17 western states. Each doctrine established a set of principles directing entry and allocation rules. Further, concern about the pace of groundwater mining has spawned major legal reforms in five states within the last 25 years.[2] The reforms primarily involved adopting specific regulations that either limit entry into groundwater basins to the set of existing groundwater pumpers or define permit systems setting quotas on individuals' pumping levels, or both. The variety across states of general doctrinal principles and specific regulations creates a diverse set of groundwater property-right systems in the West.

This essay develops and empirically applies a modeling framework of governing a groundwater CPR. Section II qualitatively describes the groundwater property-right systems in the West in terms of externalities present in a groundwater commons. Following the literature on CPRs as dynamic games originating in Levhari and Mirman (1980), Eswaran and Lewis (1984), and Reinganum and Stokey (1985), section III develops a formal model in which depletion from a fixed stock is modeled as a noncooperative game. Solving the model for its optimal solution and subgame perfect equilibrium provides benchmarks for behavior observed in laboratory experiments. Section IV describes an experimental design that implements the modeling framework. The design involves three experimental treatments, all of which depict legal doctrines governing groundwater depletion. Section V presents evidence from laboratory experiments that apply the experimental design. Performance is judged by an efficiency measure, the ratio of rent earned to maximum possible rent. Given the high cost and imprecise measurement that confront collection of field data, laboratory experiments offer a unique method for assessing the performance of various groundwater property rights and the applicability of game theory to behavior in such systems.

II. Groundwater Externalities and Water Law:
An Analytical Framework

This section develops an analytical framework to guide subsequent model development and empirical analysis.[3] It adopts the perspective that western water law developed as a response to the externality problems of a groundwater CPR. The framework isolates the key features of the major groundwater laws applied throughout the West, rather than replicating groundwater law in any particular state.

CPR Externalities

As described in Eswaran and Lewis (1984), Gardner, E. Ostrom, and Walker (1990), Negri (1990), and Reinganum and Stokey (1985), users depleting a CPR typically face three appropriation externalities: a strategic externality, a stock externality, and a congestion externality.[4] These externalities induce inefficiently rapid depletion or destruction of CPRs, commonly described by the adage "tragedy of the commons."

Negri (1989) and Provencher and Burt (1993) show that groundwater depletion for irrigated agriculture creates the potential for all three CPR externalities.[5] Individual agricultural producers invest in deep wells drilled into aquifer formations and pump groundwater from the wells for application in crop production. The strategic externality occurs because, under some legal doctrines governing groundwater depletion, water use offers the only vehicle to establish ownership. Ownership through use creates a depletion game. The stock externality occurs because, with groundwater pumping costs, individual water depletion reduces the aquifer's water-table level, thereby increasing pumping costs for all producers. The congestion externality occurs by spacing wells too closely together, with a subsequent direct loss in pumping efficiency. Thus, one producer's current effort can reduce the current output of another producer. The congestion externality, however, is not a focus of this study.[6]

Groundwater Law

A state's groundwater property-rights system consists of a general legal doctrine in combination with distinctive regulations adopted by the state when implementing the doctrine. In the authoritative source on water law, Sax and Abrams (1986) define the four legal doctrines applied to groundwater in the West.

Absolute Ownership Doctrine: The "absolute ownership rule was that the landowner overlying an aquifer had an absolute right to extract the water situated beneath the parcel. No consideration was given to the fact that the groundwater extracted from one parcel might have flowed to that location from beneath a neighbor's property..." (787).

Reasonable Use Doctrine: As a minor modification of the absolute ownership rule, the "reasonable use rule may have curtailed some whimsical uses of groundwater that harmed neighbors, but it continued the basic thrust of the absolute ownership rule that treated groundwater as an incident of ownership of the overlying tract" (792).

Correlative Rights Doctrine: "The central tenets of the doctrine...are [that:] (1) the right to use groundwater stored in an aquifer is shared by all of the owners of land overlying the aquifer, (2) uses must be made on the overlying tract and must be reasonable in relation to the uses of other overlying owners and the characteristics of the aquifer, and (3) the groundwater user's property right is usufructuary" (795).

Prior Appropriation Doctrine: "As with surface streams, states that follow prior appropriation doctrine in regard to groundwater protect pumpers on the basis of priority in time....Most jurisdictions which employ the prior appropriation doctrine to groundwater protect only 'reasonable pumping levels' of senior appropriators" (794). Further, again adopting a principle of the surface water appropriation doctrine, an appropriative right is established by demonstrating use of the water rather than being incidental to landownership.

Of the 17 western states, 12 apply the prior appropriation doctrine to establish basic principles of groundwater rights.[7] Texas is the only state to continue with the absolute ownership doctrine, the common-law doctrine adopted from English law. Nebraska (beginning 1982) and Oklahoma (beginning 1972) utilize general principles of the correlative rights doctrine. Arizona, a state that applied the reasonable use doctrine until recently, replaced existing law with the 1980 Arizona Groundwater Management Act. The Act primarily uses principles from the correlative rights doctrine because water scarcity is shared "equitably" among landowners. In California, groundwater management occurs at the local level, rather than at the state level. There, several local basins—including the region of the state reliant on groundwater for irrigated agriculture—operate without a legal structure to govern use.

In addition to the general doctrinal principles, most western states created permit systems to administer groundwater law. Aiken (1980) and Jensen (1979) note that to implement the sharing rule of correlative rights, Nebraska and Oklahoma set annual permit levels based on an individual's share of the land overlying the aquifer. In the case of the prior appropriation doctrine, states set annual permit levels based on the pumper's historical use of water. Groundwater permits typically define an individual's maximum annual use rather than specifying a fixed level of use.

Several states with permit systems also define a planning horizon that specifies a minimum time period before exhaustion could occur. With information on the stock of water in an aquifer, individual permits can be specified to guarantee a minimum depletion period, that is, a year through which water in the aquifer is guaranteed. For example, New Mexico designated a minimum 40-year life for some aquifers, while Oklahoma set a minimum 20-year period for its groundwater.[8]

Analytical Elements

CPR externalities in groundwater depletion lead to the problem of creating property rights that provide incentives for more efficient intertemporal depletion of stocks. As one conceivable property right, annual quotas could be assigned to users in a way that reproduces the optimal depletion path. This approach, however, is a planning solution; it requires perfect information on the part of a central planner to implement the optimal path. In contrast to the optimal program, features of existing groundwater law may partially remedy the externality problem.[9] A framework with three elements develops from the key features of groundwater law in the West (table 10.1). It is these three features that motivate the experimental design developed in section IV.

Common-Pool Depletion
The absolute ownership doctrine establishes a baseline for studying groundwater property rights. As applied in its pure form in Texas—and used implicitly in regions of California—the doctrine imposes no constraints on groundwater depletion by overlying landowners. This creates an environment for the rule-of-capture to prevail, providing depletion incentives to a fixed number of users. Rent dissipation is most likely to occur under the absolute ownership doctrine.

TABLE 10.1. An Analytical Framework for Groundwater Law in the U.S. West

Legal Doctrine	Analytical Element	States
Absolute ownership	Common-pool depletion: fixed number of agents in commons	Texas; regions of California
Prior appropriation	Entry restriction: reduced number of agents in commons	Colorado, Idaho, Kansas, Montana, Nevada, New Mexico, North Dakota, Oregon, South Dakota, Utah, Washington, Wyoming
Correlative rights	Stock quota: property right to share of groundwater stock	Nebraska, Oklahoma

Entry Restriction

The key feature of the widely used prior appropriation doctrine is restricted entry of users into a groundwater commons. The doctrine gives chronologically senior pumpers security in the maintenance of "reasonable" depths-to-water. To effect this provision, in prior appropriation states, administrative agencies commonly close groundwater basins to additional entrants. Moreover, groundwater users have successfully sued under the doctrine to block entry.[10] In contrast, the other legal doctrines grant entry to a groundwater CPR based solely on ownership of overlying land. Since the concept of monopolistic ownership or unitary behavior does not apply to groundwater, limited entry to the commons primarily should mitigate, as opposed to remove, the strategic and stock externalities.[11]

Stock Quota

The key feature of the correlative rights doctrine is land-based apportionment of an aquifer, that is, a user's share of the overlying land determines the share of the groundwater stock. We label this a *stock quota:* a water right that assigns an ownership share in the stock without specifying intertemporal use.[12] In practice, the states applying this doctrine—Nebraska and Oklahoma—also specify annual depletion permits. However, Smith (1989) cautions that these permits likely impose nonbinding constraints on annual use because they are based on historic use.

In terms of externalities creating CPR inefficiency, a stock quota removes the strategic externality but ignores the stock externality. That is, it ends the strategic race to capture a share of the stock but continues the incentive to capture a *cheap* share. Nevertheless, Smith (1977)[13] and Anderson, Burt, and Fractor (1983) speculate that, by removing incentives given by a rule-of-capture, a stock quota would significantly reduce the magnitude of CPR externalities in groundwater depletion.[14] This, of course, is an empirical question—one that this research addresses directly.

III. A Noncooperative Game Model of CPR Depletion

In the following CPR model, we will refer to the CPR as a groundwater aquifer. Other interpretations are available, however, such as appropriation activities in forests, fisheries, and irrigation systems.

Consider an aquifer described by the state variable depth to water at time t, d_t. There are n users of the water, indexed by i. User i withdraws an amount of water x_{it} in period t. The depth to water evolves according to the following discrete time equation:

$$d_{t+1} = d_t + k \, \Sigma_i x_{it} - h. \tag{1}$$

The parameter k depends on the size and configuration of the aquifer; the parameter h represents a constant recharge rate. Here we examine the special case where $h = 0$.

We assume that water pumped to the surface is used in agricultural production. The instantaneous benefit accruing to user i at time t, B_{it}, is quadratic:

$$B_{it}(x_{it}) = ax_{it} - bx_{it}^2 \tag{2}$$

where a and b are positive constants. This implies diminishing returns to production at the surface, an assumption that accords with production experience from aquifers like the Ogallala (Kim et al. 1989). Users are assumed to be homogeneous, so that equation (2) applies to each. Notice also that since the parameters a and b are time independent, so is the benefit function.

The cost for user i to pump water to the surface at time t, C_{it}, depends on both water pumped to the surface and depth to water. For our purposes we use the following transformation of physical units into monetary units, measured in cents:

$$C_{it}(x_{it}, X_t, d_t) = [(d_t + AX_t + B)x_{it}], \tag{3}$$

where A and B are positive constants and X_t is the sum of all users' withdrawals from the aquifer at time t. Cost is proportional to water pumped to the surface. Cost is increasing in depth to water, and in total water pumped in a given period. The latter effect is due to the fact that depth to water increases within a period, as a function of current pumping. Given the common pool nature of groundwater, each user has an incentive to pump the relatively cheap water near the surface before others do.

Solve the depletion problem in equations (1) through (3) for its optimal solution. An authority with total control over pumping maximizes net benefits from groundwater depletion over a planning horizon of length T by solving the following optimization problem:

$$\text{maximize } \Sigma_i \Sigma_t [B_{it}(x_{it}) - C_{it}(x_{it}, X_t, d_t)]$$

subject to (1), (2), (3), the initial condition d_1, and the terminal time T. Notice that in this maximization, there is no discounting of future benefits. The solution can be easily amended if discounting is desired.

Solve this optimization by dynamic programming. Let $V_t(d_t)$ denote the optimal value of the resource at time t, given that the depth to water is d_t. The recursive equation defining the value function is given by

$$V_t(d_t) = \max \Sigma_i [B_{it}(x_{it}) - C_{it}(x_{it}, X_t, d_t)] + V_{t+1}(d_{t+1}). \tag{4}$$

The transversality condition for this problem is that the value of the resource after the terminal period is zero, regardless of the depth to water:

$$V_{T+1} = 0. \tag{5}$$

By varying the transversality condition (5), one can map out a variety of optimal paths.

In order for the resource to have a positive optimal value, it is necessary that the following condition on the parameters of the net benefit function (measured in cents) be satisfied:

$$a - d_T - B > 0. \tag{6}$$

It remains to find the form of the optimal value function $V_t(d_t)$. Consider the last period T. One can show, differentiating (4) and using (5), that the optimal decision in the last period is given by

$$\Sigma_i x_{iT} = (a - d_T - B)/(2b/n + 2A). \tag{7}$$

Further, the optimal value function (in cents) for the last period is given by

$$V_T(d_T) = 0.5(a - d_T - B)^2/(2b/n + 2A). \tag{8}$$

One can show by mathematical induction that for any time t, the optimal decision function takes the form

$$\Sigma_i x_{it} = L_t(a - d_t - B), \tag{9}$$

and the optimal value function takes the form

$$V_t(d_t) = K_t(a - d_t - B)^2. \tag{10}$$

The proportionality factors L_t and K_t in equations (9) and (10) are given by the nonlinear recursive equations:

$$L_t = (1 - 2kK_{t+1})/(2b/n + 2A - 2k^2 K_{t+1}) \tag{11}$$

and

$$K_t = L_t - (b/n + A)L_t^2 + K_{t+1}(1 - kL_t)^2. \tag{12}$$

One derives the optimal solution by starting the recursion with (5), substitut-

TABLE 10.2. Backward Recursion and Optimal Solution *n*=10

t	K_t	L_t	x_{it}	c_t	Cumulative Earnings
10	1/4	1/2	2.00	179.6	$219
9	2/6	1/3	2.00	159.6	$215
8	3/8	1/4	2.00	139.7	$207
7	4/10	1/5	2.00	119.7	$195
6	5/12	1/6	2.00	99.7	$179
5	6/14	1/7	2.00	79.8	$159
4	7/16	1/8	2.00	59.8	$136
3	8/18	1/9	2.00	39.9	$108
2	9/20	1/10	2.00	19.9	$ 76
1	10/22	1/11	2.00	0.0	$ 40

ing into (11) to get L_t, substituting into (12) to get K_T, and working back from there to the beginning, $t = 1$. Equations (7) and (8) represent the first two steps of the solution process. For all values of the eight-dimensional parameter space $(a, b, n, A, B, k, d_1, T)$ satisfying inequality (6), one can show that the optimal solution path has each user withdrawing water at a uniform rate. This rate is such that the last unit of water withdrawn in the terminal period has zero net benefit.

For illustration, consider the parameter values chosen for our baseline design $(a, b, n, A, B, k, d_1, T) = (220, 5, 10, 0.5, 0.5, 1, 0, 10)$. For these parameters, table 10.2 gives the backward recursion solution for the series L_t and K_t. The optimal aggregate withdrawal in the first period is given by

$$\Sigma x_{i1} = (1/11)(220 - 0.5) = 19.95, \tag{13}$$

whence the optimal withdrawal by each individual user is 19.95/10, or 1.995. The optimal value in cents of the entire resource, $V_1(d_1)$, from table 10.2, is

$$V_1(d_1) = (10/22)(220 - 0.5)^2 = 21900. \tag{14}$$

Any other withdrawal path will have a lower value. The coefficient of resource utilization, or CRU (Debreu 1951) measures how efficiently a resource is being used. The CRU, which lies between 0 percent and 100 percent, can be expressed as the ratio of the value of the resource from any other withdrawal path to its optimal value.

Depletion patterns associated with game equilibria are important to establish benchmarks for behavior observed in the laboratory experiments. In a noncooperative game, each user maximizes his own net benefit without regard to the effect of this behavior on other users. This is the basis for the externality created when a rule-of-capture defines resource ownership. Analyze the game played by users in extensive form and characterize its symmetric subgame perfect equilibrium. A strategy for user i, x_i, is a complete plan for the play of the game, given the history available to the player when he has to make a decision. At the beginning of the game, player i's decision, x_{i1}, is based on no history. Recall that X_t is the sum of all users' withdrawals at time t:

$$X_t = \Sigma x_{it}. \tag{15}$$

In the same period, user i's decision x_{i2} depends on depth to water d_2, which in turn depends on the previous period's water withdrawal. Write this dependence as $x_{i2}(X_1)$. Proceeding inductively, write a complete plan of play as

$$x_i = [x_{i1}, x_{i2}(X_1), \ldots, x_{iT}(X_1, \ldots, X_{T-1})]. \tag{16}$$

Now solve the depletion game whose net benefit functions and transition equations are given by (1) through (3) for its symmetric subgame perfect equilibrium. Since the game is symmetric, it has such an equilibrium. User i chooses his strategy x_i to maximize net benefits from groundwater depletion over a planning horizon of length T by solving the following optimization problem:

$$\text{maximize } \Sigma_t B_{it}(x_{it}) - C_{it}(x_{it}, X_t, d_t)$$

subject to (1), (2), (3), the initial depth to water d_1, and the terminal time T.

Solve this optimization problem by dynamic programming. Let $V_{it}(d_t)$ denote the optimal value of the resource to user i at time t, given that the depth to water is d_t. The recursive equation defining the value function is given by

$$V_{it}(d_t) = \max B_{it}(x_{it}) - C_{it}(x_{it}, X_t, d_t) + V_{it+1}(d_{t+1}). \tag{17}$$

The transversality condition for this problem is that the value of the resource to user i after time T is zero, regardless of the depth to water:

$$V_{iT+1} = 0. \tag{18}$$

It remains to find the form of the optimal value function $V_{it}(d_t)$. Consider the last period T. One can show, differentiating (17), and using (18), that the optimal decision in the last period is given by

$$x_{iT} = (a - d_T - B)/[2b + (n + 1)A)]. \tag{19}$$

Further, the optimal value function for the last period is given by

$$V_{iT}(d_T) = 0.5(2b + 2A)(a - d_T - B)^2/[2b + (n + 1)A]^2. \tag{20}$$

One can show by mathematical induction that in each period, the equilibrium decision function takes the form

$$x_{it} = L_{it}(a - d_t - B), \tag{21}$$

and the equilibrium value function takes the form

$$V_{it}(d_t) = K_{it}(a - d_t - B)^2. \tag{22}$$

The proportionality factors L_{it} and K_{it} in equations (21) and (22) are given by the nonlinear recursive equations

$$L_{it} = (1 - 2kK_{it+1})/[2b + (n + 1)A - 2k^2nK_{it+1}] \tag{23}$$

and

$$K_{it} = L_{it} - (b + nA)L_{it}^2 + K_{it+1}(1 - knL_{it})^2. \tag{24}$$

One derives the symmetric subgame perfect equilibrium by starting the recursion with (18), substituting (18) into (23) to get L_{iT}, substituting L_{iT} into (24) to get K_{iT}, and working back from there to the beginning, $t = 1$. Equations (20) and (21) represent the first two steps of the solution process.

Since this is a symmetric equilibrium, the solution for user i is the same for all users. Note that the recursive equations (23) and (24) are different from those defining the optimal solution. Thus, the subgame perfect equilibrium is not an optimum. Suppose that the program is one period long ($T = 1$). Then the equilibrium and the optimum both start at the initial depth to water d_1. Comparing (11) and (23) yields

$$nL_{i1} = 1/\{2b/n + [(n + 1)/n]A\} > 1/(2b/n + 2A) = L_t. \tag{25}$$

The subgame perfect equilibrium withdraws too much water. This continues to hold true more generally: the subgame perfect equilibrium path withdraws too much water in the first period regardless of the length of the game. Table 10.3 shows the subgame perfect equilibrium path using the same parameters as for table 10.2. The subgame perfect path is virtually exponential, thus differing markedly from the optimal path's constant depletion rate. The first two periods have high depletion rates, while later periods have almost no depletion. At this equilibrium, each user has the incentive to deplete the relatively cheap water at the top of the aquifer before other users capture it. This equilibrium naturally produces a lower payoff from the water resource. In particular (from table 10.3), the aggregate value in cents at the subgame perfect equilibrium is

$$nK_{iT}(a - d_1 - B)^2 = 10(0.0269)(219.5)^2 = 12,960.$$

Compared to the optimum, the subgame perfect equilibrium has an efficiency of $12,960/21,900 = 59\%$.

TABLE 10.3. Backward Recursion and Symmetric Subgame Perfect Equilibrium $n=10$

t	K_t	L_t	x_{it}	c_t	Cumulative Earnings
10	0.0229	0.0645	0.00	219.5	$130
9	0.0263	0.0634	0.00	219.4	$130
8	0.0268	0.0633	0.01	219.3	$130
7	0.0269	0.0632	0.04	218.9	$130
6	0.0269	0.0632	0.09	218.0	$130
5	0.0269	0.0632	0.25	215.5	$130
4	0.0269	0.0632	0.70	208.5	$130
3	0.0269	0.0632	1.90	189.5	$130
2	0.0269	0.0632	5.07	138.8	$127
1	0.0269	0.0632	13.88	0.0	$112

IV. Experimental Design and Decision Setting

The experimental design focuses on three conditions: (1) a baseline with no restrictions on individual levels of appropriation, group size equal to 10 and $T = 10$; (2) a treatment with no restrictions on individual levels of appropriation, but group size restricted to $n = 5$ with the terminal round extended to $T = 20$; and (3) a treatment imposing a stock quota restriction on each individual's total level of appropriation (see table 10.4). The three conditions depict, respectively, common-pool depletion under the absolute ownership doctrine, an entry restriction under the prior appropriation doctrine, and a stock quota under the correlative rights doctrine.[15]

Subject i makes a decision x_{it} in each round t. The decision x_{it} is itself integer valued with a lower bound of zero and an upper bound, if any, given by the institutions. The units of the decision are called "tokens." Payoffs according to the net benefit function are evaluated at integer values of the arguments of that function.[16]

TABLE 10.4. Parameterization of Laboratory Experiments

	Number of Players	Number of Decision Periods	Water Use Quantity Constraints
Baseline ($n = 10$)	10	10	∞
Entry Restriction ($n = 5$)	5	20	∞
Stock Quota Rule[a]	10	10	$\sum x_{it} \leq 25$

[a]The quantity constraint for the stock quota states that accumulated multi-period water use cannot exceed a specified quantity.

All experiments satisfy the following net benefit function parameterizations, measured in cents:

$$a = 220, b = 5, A = .5, B = .5, d_1 = 0.$$

As discussed above, with the additional parameter $k = 1$ governing the depth to water transition equation (1), the optimal solution for the case $n = 10$ and $T = 10$ is

$$V_1(d_1) = \$219, \quad x_{it} = 2.$$

As shown in table 10.5, the treatment with $n = 5$ and $T = 20$ gives the same optimal value and individual withdrawal rate. The exhaustion condition is reached by half as many appropriators withdrawing the same amount of water per period for twice as many periods. Thus, holding the value of the resource constant, this parameterization allows us to investigate a pure "number of appropriators" effect.[17]

In contrast to the optimal value, the valuations generated by the subgame perfect equilibria are lower. As discussed above, for $n = 10$ and $T = 10$ the subgame perfect equilibrium reaches its maximum cumulative earnings, $130, by the fourth period, for an efficiency of 59 percent. For $n = 5$ and $T = 20$ the subgame perfect equilibrium reaches its maximum cumulative earnings, $136, by the sixth period, as shown in table 10.6, with an efficiency of 62 percent. Thus, according to subgame perfection, restricting group size from ten to five players increases efficiency by only 3 percent.

For our parameterizations ($d_1 = 0$, $k = 1$), $d_{T+1} - d_1 = d_{T+1}$ represents the amount of groundwater ultimately pumped from the aquifer. A stock quota places an upper bound on the water an individual player can withdraw over the life of the resource. This type of quota mitigates the impact of especially high individual withdrawal paths.[18] In our experiments, the stock quota was 25 tokens per individual.[19] Note, this quota does not act as a constraint to subgame perfect equilibrium behavior, which requires only 22 tokens per individual.

TABLE 10.5. Backward Recursion and Optimal Solution $n=5$

t	K_t	L_t	x_{it}	c_t	Cumulative Earnings
20	1/3	1/6	2.00	189.6	$219
19	1/4	2/8	2.00	179.6	$218
18	1/5	3/10	2.00	169.7	$216
17	1/6	4/12	2.00	159.7	$212
16	1/7	5/14	2.00	149.7	$207
15	1/8	6/16	2.00	139.7	$202
14	1/9	7/18	2.00	129.7	$195
13	1/10	8/20	2.00	114.7	$188
12	1/11	9/22	2.00	109.8	$179
11	1/12	10/24	2.00	99.8	$170
10	1/13	11/26	2.00	89.8	$160
9	1/14	12/28	2.00	79.8	$148
8	1/15	13/30	2.00	69.8	$136
7	1/16	14/32	2.00	59.9	$122
6	1/17	15/34	2.00	49.9	$108
5	1/18	16/36	2.00	39.9	$ 92
4	1/19	17/38	2.00	30.0	$ 76
3	1/20	18/40	2.00	20.0	$ 58
2	1/21	19/42	2.00	10.0	$ 40
1	1/22	20/44	2.00	0.0	$ 20

Placing the stock quota at a level below 22 tokens per person would artificially lead to improvements in efficiency. Our purpose was to investigate the role of a stock quota on behavior without disturbing potential equilibrium behavior.

All experiments were conducted at Indiana University. Volunteers were recruited from graduate and advanced undergraduate economics courses. These subjects were paid in cash in private at the end of the experiment. Subjects privately went through a series of instructions and had the opportunity to ask the experimenter a question at any time during the experiment. The decision problem faced by the subjects can be summarized as follows.

Each subject had a single decision to make each round, namely how many tokens to order. Each knew his/her own benefit function (expressed in equation and tabular form), and that every subject faced the same benefit function. A base token cost of $0.01 was stipulated for round 1. The instructions explained that token cost increased by $0.01 for each token ordered by the group and token cost for an individual in a given round would be the average token cost for that round times the number of tokens the individual ordered in that round. The base cost for the next round was computed by adding one to the aggregate number of tokens ordered in previous rounds and then multiplying this total by $0.01. All subjects made purchasing decisions simultaneously. Subjects were explicitly informed of the maximum number of rounds in the experiment. After each decision round, subjects were informed of the total number of tokens ordered by the group, the cost per token for that round, the new base cost for tokens purchased in the next round, and their profits for that round. Subjects were also told if the base token cost ever reached a level where there was no possibility of earning positive returns to buying tokens, the experiment would end.[20]

V. Laboratory Results and Discussion

The experimental results are drawn from nine experiments conducted over the three design conditions: (1) the baseline condition where $n = 10$ and $T = 10$; (2) the entry restriction condition where $n = 5$ and $T = 20$; and (3) the stock quota condition where $n = 10$, $T = 10$, and the stock quota is 25. In each condition, we examine results from two experiments using subjects inexperi-

TABLE 10.6. Backward Recursion and Symmetric Subgame Perfect Equilibrium $n=5$

t	K_t	L_t	x_{it}	c_t	Cumulative Earnings
20	0.0325	0.0769	0.00	219.5	$136
19	0.0446	0.0738	0.00	219.4	$136
18	0.0512	0.0725	0.01	219.4	$136
17	0.0541	0.0719	0.01	219.3	$136
16	0.0555	0.0714	0.02	219.2	$136
15	0.0561	0.0714	0.03	219.0	$136
14	0.0564	0.0713	0.05	218.8	$136
13	0.0565	0.0713	0.08	218.4	$136
12	0.0566	0.0713	0.12	217.8	$136
11	0.0566	0.0713	0.19	216.8	$136
10	0.0566	0.0713	0.30	215.4	$136
9	0.0566	0.0713	0.46	213.0	$136
8	0.0566	0.0713	0.71	209.5	$136
7	0.0566	0.0713	1.11	203.9	$136
6	0.0566	0.0713	1.73	195.3	$136
5	0.0566	0.0713	2.68	181.9	$135
4	0.0566	0.0713	4.17	161.0	$132
3	0.0566	0.0713	6.48	128.6	$127
2	0.0566	0.0713	10.07	78.2	$113
1	0.0566	0.0713	15.65	0.0	$ 80

TABLE 10.7. Overview of All Experiments

Case	Aggregate Net Benefits	Efficiency	Periods to Exhaustion
Optimum	$219.00	100%	10 ($n = 10$) or 20 ($n = 5$)
Baseline ($n = 10$)			
Base 1	$ 88.50	40%	3
Base 2	$ 38.80	18%	2
Base-experienced	$ 69.00	32%	4
Entry restriction ($n = 5$)			
Entry 1	$ 83.00	38%	6
Entry 2	$ 93.10	42%	5
Entry-experienced	$116.30	53%	8
Stock quota			
Stock quota 1	$125.30	57%	7
Stock quota 2	$128.60	59%	4
Stock quota-Experienced	$ 98.30	45%	3

enced in the decision environment and from one experiment using experienced subjects randomly recruited from the subject pool of the inexperienced runs.

An overview of our experimental results is presented in table 10.7. For each experiment, aggregate payoffs, experimental efficiency, and duration of the experiment are displayed. The set of baseline and entry restriction experiments reflect an environment in which resource use is the only way to establish ownership. As expected, paths with later exhaustion periods are typically associated with higher efficiencies. With $n = 10$, the average exhaustion round was 3; with $n = 5$ the average increased to 6.33. In the stock quota experiments, the average increased to 4.67. While increasing the life of the resource is not an economic goal per se, it does help explain the increase in average efficiency across experimental settings.

SUMMARY RESULT 1: *In each of the three baseline experiments, efficiencies were well below the efficiency level generated by the optimum and even below that generated by the subgame perfect equilibrium.*
Table 10.8 reports detailed results for the three experiments with $n = 10$ and $T = 10$, including the actual appropriation levels by decision round and summary statistics. In the first round of these experiments, subjects ordered on average 164 tokens, implying an average second round base cost of $1.65. This compares to an optimal order of two tokens per subject for a total order

TABLE 10.8. Summary Results: Baseline *n*=10 Experiment

		1	2	3	4	5	6	7	8	9	10	Base Cost	Average Cost	Total Order
						Token Order by Subject Number								
Experiment: Base 1 Overall Efficiency = 40.4%														
Round	1	10	22	15	6	17	22	15	10	22	15	.01	.77	154
	2	6	0	12	3	4	7	10	3	2	5	1.55	1.81	52
	3	0	0	0	0	0	0	1	2	2	0	2.07	2.09	5
Experiment: Base 2 Overall Efficiency = 17.7%														
Round	1	22	20	3	20	20	22	13	7	22	11	.01	.80	160
	2	0	6	10	10	2	14	13	10	22	3	1.61	2.00	80
Experiment: Base-Experienced 1 Overall Efficiency = 31.5%														
Round	1	10	22	16	23	15	22	17	18	14	22	.01	.90	179
	2	1	3	1	2	2	0	3	2	2	3	1.80	1.89	19
	3	1	1	1	1	1	0	1	1	1	1	1.99	2.04	9
	4	1	0	0	1	1	0	1	1	1	0	2.09	2.11	6

of 20 tokens in the first round and a second round base cost of $0.21. The subgame perfect equilibrium predicts an order of 14 tokens per subject for a total order of 140. This explosive appropriation of cheap tokens in the first round guarantees very low efficiencies. Efficiencies averaged only 30 percent of optimum.

SUMMARY RESULT 2: *In each of the three experiments that limit entry to five players, efficiencies again were below levels generated in both the optimum and the subgame perfect equilibrium. However, the average efficiency generated by this treatment was distinctly higher than that of the baseline experiments.*

Table 10.9 reports detailed results for the three experiments with $n = 5$ and $T = 20$. In the first round of these experiments, subjects ordered an average of 86 tokens, implying an average second round base cost of $0.87. This compares to an optimal order of 10 tokens in the first round and a second round base cost of $0.11. The subgame perfect path predicts an order of 16 tokens per subject for a total order of 80. Efficiencies averaged 44 percent of the optimum.

The set of three experiments using a stock quota rule is summarized in table 10.10. These experiments were conducted in a manner identical to the baseline experiments where $n = 10$ and $T = 10$, except that each subject was constrained to order no more than 25 tokens over the course of the experiment. This treatment variable was announced in public.

TABLE 10.9. Summary Results: Entry Restriction *n* = 5 Experiments

		Token Order by Subject Number					Base Cost	Average Cost	Total Order
		1	2	3	4	5			
Experiment: Entry 1 Overall Efficiency = 37.96%									
Round	1	15	22	5	5	22	0.01	0.35	69
	2	10	22	8	15	22	0.70	1.08	77
	3	0	3	10	18	3	1.47	1.64	34
	4	5	2	4	3	2	1.81	1.89	16
	5	0	1	2	6	1	1.97	2.02	10
	6	0	1	1	1	1	2.07	2.09	4
Experiment: Entry 2 Overall Efficiency = 42.5%									
Round	1	22	22	15	14	22	0.01	0.48	95
	2	14	7	20	5	8	0.96	1.23	54
	3	6	4	10	4	5	1.50	1.64	29
	4	2	2	9	3	4	1.79	1.89	20
	5	1	2	4	1	2	1.99	2.04	10
Experiment: Entry-Experienced 1 Overall Efficiency = 53.1%									
Round	1	22	20	14	16	22	0.01	0.47	94
	2	11	9	8	8	8	0.95	1.17	44
	3	5	5	6	4	6	1.39	1.52	26
	4	3	4	5	4	5	1.65	1.75	21
	5	2	2	3	2	2	1.86	1.91	11
	6	1	1	2	1	2	1.97	2.00	7
	7	1	1	1	1	2	2.04	2.07	6
	8	1	1	1	0	1	2.10	2.12	4

SUMMARY RESULT 3: *In each of the three experiments using the stock quota rule, efficiencies increased markedly relative to baseline, but remained well below the optimum. Efficiencies averaged 54 percent of the optimum.*

In the first round of these experiments, subjects ordered on average 125 tokens, implying an average second round base cost of $1.26. Thus, the upper bound on orders slowed down, but did not eliminate, the race to cheap water. These results call into question the optimistic conjectures made in previous research (e.g., Anderson, Burt, and Fractor 1983) about the ability of stock quotas to capture most of a groundwater CPR's scarcity rent.

Note that group behavior most closely resembles the subgame perfect equilibrium in the stock quota experiments. Efficiencies in experiments 1 and 2 (57 percent and 59 percent) are in line with subgame perfect equilibrium efficiency (59 percent). In these two experiments, first-round orders averaged 11.8 tokens per subject, lower than the equilibrium prediction of 14; second-round orders averaged 5.2 tokens per subject, slightly higher than the equilibrium prediction of 5. Interestingly, it is the experienced run in the stock quota

TABLE 10.10. Summary Results: Stock Quota Rule Experiments

		1	2	3	4	5	6	7	8	9	10	Base Cost	Average Cost	Total Order
		Token Order by Subject Number												

		1	2	3	4	5	6	7	8	9	10	Base Cost	Average Cost	Total Order
Experiment: Stock Quota 1 Overall Efficiency = 57.2%														
Round	1	15	15	10	14	13	22	8	20	4	10	0.01	0.66	131
	2	2	0	4	6	6	3	2	5	5	0	1.32	1.48	33
	3	2	4	10	2	3	0	3	0	2	3	1.65	1.79	29
	4	0	0	1	2	2	0	1	0	2	0	1.94	1.98	8
	5	2	0	0	1	1	0	1	0	1	0	2.02	2.04	6
	6	0	0	0	0	0	0	1	0	0	1	2.08	2.09	2
	7	0	0	0	0	0	0	1	0	0	0	2.10	2.10	1
Experiment: Stock Quota 2 Overall Efficiency = 58.7%														
Round	1	18	15	3	10	6	13	10	6	20	5	0.01	0.54	106
	2	7	10	7	11	9	5	8	1	2	10	1.07	1.42	70
	3	0	0	4	3	4	2	3	2	3	4	1.77	1.89	25
	4	0	0	3	1	1	5	4	4	0	1	2.02	2.11	19
Experiment: Stock Quota-Experienced 1 Overall Efficiency = 44.9%														
Round	1	14	14	17	14	19	14	22	3	12	10	0.01	0.70	139
	2	6	8	8	6	5	11	3	7	6	3	1.40	1.71	63
	3	1	1	0	1	1	0	0	15	1	0	2.03	2.13	20

design that resulted in the poorest performance, generating an efficiency 14 percent below that predicted by subgame perfection. More generally, this experiment demonstrates a point that holds true across all of our experiments. Individual behavior is quite diverse. As in experiments reported by E. Ostrom, Gardner, and Walker (1994) and Herr, Gardner, and Walker (1997), average behavior across groups often follows a path similar to that predicted by noncooperative game theory. At the individual level, however, there is too much variation to argue strong support for the theory.

VI. Conclusions

This essay considers the depletion of a groundwater CPR within a setting of state governance of groundwater resources in the western United States. A benchmark model is constructed with a fixed stock of groundwater and fixed exhaustion time. The optimal solution and subgame perfect equilibrium provide benchmarks for efficiencies observed in laboratory experiments. Although the model and experiments are couched in terms of groundwater CPRs, the research is also informative to dilemmas encountered in other CPRs, such as forests, fisheries, and cooperative irrigation systems.

The laboratory experiments examine the effect on individual strategic behavior of three legal rules for governing groundwater depletion in the West. The experiments show the relative performance of the rules given the study parameters. Average efficiency equals only 30 percent in the baseline experiments, with a group size of ten players under common-pool depletion.

Common-pool depletion mimics the absolute ownership doctrine, in which property rights in land also convey a right to deplete groundwater. Restricting entry to five participants, while still operating under common-pool depletion, increases average efficiency to 44 percent. The prior appropriation doctrine—the prevalent doctrine in use—uses entry restrictions as its main mechanism to reduce rent dissipation. A stock quota, as a replacement for common-pool depletion, increased efficiency to 54 percent with group size held at ten. The correlative rights doctrine effectively imposes stock quotas on landowners overlying aquifers. Although entry restrictions and stock quotas distinctly improve performance, a substantial amount of rent remains unappropriated.

NOTES

1. The result that fewer than five firms are necessary for cooperation has received theoretical support from Selten (1971).

2. The states are Arizona, Colorado, Kansas, Nebraska, and Oklahoma.

3. Several previous studies also address issues related to the performance of groundwater institutions. The costliness of collecting data on groundwater use and the difficulty of applying game-theoretic models explain the overwhelming reliance in that research on analytical results (Dixon 1988; Negri 1989; Provencher and Burt 1993), simulation methods (Dixon 1988), or reasoned institutional arguments concerning the desirable properties of specific groundwater property-right systems (Anderson, Burt, and Fractor 1983; Gisser 1983; Smith 1977). For a more empirical approach, see Blomquist (1992) for an insightful investigation of groundwater institutions in southern California.

4. Provencher and Burt (1993) also identified a risk externality that pertains to the case of agricultural irrigation using groundwater in conjunction with stochastic surface water supply. Study of the risk externality is beyond the scope of this essay.

5. The model is developed for the case of irrigated agriculture because agriculture is the dominant water-consuming sector in the 17 western states. The sector commonly consumes 85 percent to 90 percent of total water consumption in those states. Groundwater provides roughly 37 percent of water withdrawn for irrigation, with surface water supplying the remainder (U.S. Department of the Interior 1993). Groundwater pumping distances vary substantially depending on aquifer conditions. Over the Ogallala Aquifer in the Great Plains region, for example, average depth-to-water in the Great Plains states in 1988 ranged from 70 to 154 feet (U.S. Department of Commerce 1990).

6. Virtually every western state has a well-spacing statute to avoid this externality. Further, Negri (1989) notes that well spacing is less interesting in a modeling context because it does not require a dynamic model.

7. The 12 states are Colorado, Idaho, Kansas, Montana, Nevada, New Mexico, North Dakota, Oregon, South Dakota, Utah, Washington, and Wyoming. See Grant (1981).

8. For details see Gisser (1983), Nunn (1985), and Jensen (1979).

9. This essay does not address the role of water markets in achieving optimal groundwater allocation. Since we assume homogeneous users with stationary benefit

functions, markets (or other forms of transaction) are not a necessary component of achieving optimality. Other research, such as Gisser (1983) and Smith (1977), emphasizes the importance of markets for groundwater rights.

10. See Bagley (1961), Grant (1981), and Nunn (1985) for further discussion of these issues.

11. With groundwater, irrigation development proceeded via settlement of arable cropland by individual farm families. The conceptual artifice of sole ownership thus lacks sufficient realism to be incorporated into this groundwater model except as a benchmark. Further, unlike oil or natural gas, the economic value of water in agriculture cannot support transportation of groundwater to distant markets. This feature, together with the high cost of negotiation relative to resource value, removes the incentive for unitization of aquifers developed for agriculture. In contrast, Libecap and Wiggins (1984) found that unitization is an incentive that operates successfully in many cases for oil fields.

12. A second configuration of legal rules also resembles a stock quota. A permit specifying an annual limit on depletion, along with a guaranteed time horizon for use of the permit, combine to produce a stock quota. In literal terms, however, the stock quota is binding only if the annual permit is binding over the time period. Nonetheless, this configuration reinforces the need to analyze a stock quota.

13. Smith (1977) recommended three elements to solve Arizona's groundwater mining problem: a stock quota to define a right to the groundwater stock; an annual quota to define a right to annual groundwater recharge; and water markets in which the rights could be transferred freely. The analysis here focuses on the first element. Notably, the direct connection between the correlative rights doctrine and a stock quota has not been made in the literature.

14. A strength of a system of stock quotas is that annual depletion rates would not be specified; individual and aggregate intertemporal depletion paths would be determined endogenously. Thus, such a system would economize on an agency's information requirements relative to selecting the optimal depletion path.

15. The model and experiments contain a number of restrictive assumptions, including no resource recharge, no discounting, and a known finite horizon. These restrictions were made to make the model solvable and the experiment less complex. The simplicity of the design allows subjects to focus on the strategic and stock externalities without the further complexities associated with field settings. Relaxing the restrictions would allow for a richer, yet more complex, decision setting.

16. It would have been preferable to have parameterized an experimental design with the subgame perfect equilibrium path and the optimal path each taking on integer values at each point at time. Given the complexity of this decision problem, meeting each of these criteria was impossible.

17. Alternatively, we could have held $T = 10$ and merely reduced n to 5. This parameterization would yield an arbitrary reduction in the value of the resource. Thus, the design we investigate examines the impact of reducing the number of users in a situation where maximal resource value is held constant.

18. Alternatively, one could investigate the impact of placing flow quotas on an individual user's per period withdrawals. Further, one could investigate an even more

complex environment where flow or stock quotas are marketable. In fact, we intend to pursue these types of settings in future research.

19. A stock quota of 20 would allow the players to follow the optimal path; 22 would allow players to follow the subgame perfect equilibrium path. In baseline experiments, subjects often ordered tokens in the last round that went beyond the economically valuable range. For comparisons, we chose a stock quota of 25 to allow this type of behavior in our stock quota design.

20. A complete set of instructions is available from the authors on request.

REFERENCES

Aiken, J. David. "Nebraska Ground Water Law and Administration." *Nebraska Law Review,* October 1980, 917–1000.

Anderson, Terry L., Oscar R. Burt, and David T. Fractor. "Privatizing Groundwater Basins: A Model and Its Application," in *Water Rights,* edited by Terry L. Anderson. Cambridge: Ballinger Publishing, 1983, 223–48.

Bagley, Edgar S. "Water Rights Law and Public Policies Relating to Ground Water 'Mining' in the Southwestern States." *Journal of Law and Economics,* October 1961, 144–74.

Blomquist, William. *Dividing the Waters: Governing Groundwater in Southern California.* San Francisco: ICS Press, 1992.

Brown, Gardner, Jr. "An Optimal Program for Managing Common Property Resources with Congestion Externalities." *Journal of Political Economy,* January/February 1974, 163–73.

Debreu, Gerard. "The Coefficient of Resource Utilization." *Econometrica,* July 1951, 273–92.

Dixon, Lloyd S. "Models of Ground-Water Extraction with an Examination of Agricultural Water Use in Kern County, California." Ph.D. dissertation, Department of Economics, University of California, Berkeley, 1988.

Eswaran, Mukesh, and Tracy Lewis. "Appropriability and the Extraction of a Common Property Resource." *Economica,* November 1984, 393–400.

Gardner, Roy, Elinor Ostrom, and James Walker. "The Nature of Common-Pool Resource Problems." *Rationality and Society,* July 1990, 338–58.

Gisser, Micha. "Groundwater: Focusing on the Real Issue." *Journal of Political Economy,* December 1983, 1001–27.

Gordon, H. Scott. "The Economic Theory of a Common Property Resource: The Fishery." *Journal of Political Economy,* April 1954, 124–42.

Grant, Douglas L. "Reasonable Groundwater Pumping Levels under the Appropriation Doctrine: The Law and Underlying Economic Goals." *Natural Resources Journal,* January 1981, 1–36.

Herr, Andrew, Roy Gardner, and James Walker. "An Experimental Study of Time-Independent and Time-Dependent Externalities in the Commons." *Games and Economic Behavior,* 19, 1977, 77–96.

Jensen, Eric B. "The Allocation of Percolating Water under the Oklahoma Ground

Water Law of 1972." *Tulsa Law Journal,* July 1979, 437–76.

Kim, C. S., Michael R. Moore, John J. Hanchar, and Michael Nieswiadomy. "A Dynamic Model of Adaptation to Resource Depletion: Theory and an Application to Groundwater Mining." *Journal of Environmental Economics and Management,* July 1989, 66–82.

Levhari, David, Ron Michener, and Leonard J. Mirman. "Dynamic Programming Models of Fishing: Competition." *American Economic Review,* September 1981, 649–61.

Levhari, David, and Leonard J. Mirman. "The Great Fish War: An Example Using a Dynamic Cournot-Nash Solution." *Bell Journal of Economics,* Spring 1980, 322–34.

Libecap, Gary D., and Steven N. Wiggins. "Contractual Responses to the Common Pool: Prorationing of Crude Oil Production." *American Economic Review,* March 1984, 87–98.

Negri, Donald H. "The Common Property Aquifer as a Differential Game." *Water Resources Research,* 25(1), 1989, 9–15.

Negri, Donald H. " 'Stragedy' of the Commons." *Natural Resource Modeling,* 4(4), 1990, 521–37.

Nunn, Susan C. "The Political Economy of Institutional Change: A Distribution Criterion for Acceptance of Groundwater Rules." *Natural Resources Journal,* October 1985, 867–92.

Ostrom, Elinor, Roy Gardner, and James Walker. *Rules, Games, and Common-Pool Resources.* Ann Arbor: The University of Michigan Press, 1994.

Provencher, Bill, and Oscar Burt. "The Externalities Associated with the Common Property Exploitation of Groundwater." *Journal of Environmental Economics and Management,* 24(2), 1993, 139–58.

Reinganum, Jennifer F., and Nancy L. Stokey. "Oligopoly Extraction of a Common Property Natural Resource: The Importance of the Period of Commitment in Dynamic Games." *International Economic Review,* February 1985, 161–73.

Sax, Joseph L., and Robert H. Abrams. *Legal Control of Water Resources.* St. Paul: West Publishing Company, 1986.

Selten, Reinhard. "A Model of Imperfect Competition Where 4 Are Few and 6 Are Many." *International Journal of Game Theory,* July 1973, 141–201.

Smith, Vernon L. "Water Deeds: A Proposed Solution to the Water Valuation Problem." *Arizona Review,* January 1977, 7–10.

Smith, Zachary A. *Groundwater in the West.* San Diego: Academic Press, 1989.

U.S. Department of Commerce, Bureau of the Census. "Farm and Ranch Irrigation Survey (1988)." AC87RS–1, 1987 Census of Agriculture, Volume 3, Part 1. Washington, D.C., May 1990.

U.S. Department of the Interior, U.S. Geological Survey. "Estimated Use of Water in the United States in 1990." USGS Circular 1081. U.S. Government Printing Office, 1993.

Walker, James M., and Roy Gardner. "Probabilistic Destruction of Common-Pool Resources: Experimental Evidence." *The Economic Journal,* September 1992, 1149–61.

Walker, James M., Roy Gardner, and Elinor Ostrom. "Rent Dissipation in Limited Access Common Pool Resource Environments: Experimental Evidence." *Journal of Environmental Economics and Management,* 19(3), 1990, 203–11.

CHAPTER 11

Bottlenecks and Governance Structures: Open Access and Long-Term Contracting in Natural Gas

Thomas P. Lyon and Steven C. Hackett

1. Introduction

In a number of regulated industries, market evolution has made one (or more) vertical stages of production increasingly competitive.[1] The challenge for public policy is to deregulate these largely competitive activities, while maintaining adequate antitrust and/or regulatory safeguards over remaining "bottleneck" facilities such as local telephone switches, pipelines, and electric transmission lines.[2] These industries historically provided a tied or bundled[3] product, combining upstream production with transportation to downstream markets. However, recent policy has generally involved unbundling upstream supply from transportation through the bottleneck and requiring bottleneck facilities to operate as common carriers providing transportation services on a nondiscriminatory basis.[4] We use the shorthand "open-access policy" to refer to policy of this type and "transporter" to designate the operator of a bottleneck facility.

This essay offers an initial transaction-cost analysis of open-access policy and its effects on the institutions used to govern exchange through bottleneck facilities. While much of the existing empirical work on contracting involves transactions in regulated industries, there has been relatively little investigation of the effects of regulation on vertical integration or long-term contracting, and what there is focuses mainly on the effects of price regulation rather than access policy.[5] The only prior article of which

Originally published in *Journal of Law, Economics, and Organization* 9, no. 2 (October 1993): 381–98. Reprinted by permission of Oxford University Press and the authors.

Authors' note: We have benefited from the comments of Keith Crocker, Roy Gardner, Curt Hribernik, Ron Johnson, Paul Joskow, Ron Lafferty, Scott Masten, John Morris, Dick O'Neill, Benjamin Schlesinger, Michael Toman, Oliver Williamson, an anonymous referee, seminar participants at the Rutgers University Twelfth Annual Conference of the Advanced Workshop in Regulation and Public Utility Economics, and the sixty-eighth Annual Western Economic Association Meetings.

we are aware that integrates open-access regulatory policy with an assessment of institutional change is Doane and Spulber's (1991) study of the natural gas spot market. They show that in the wake of open-access policy for natural gas pipelines, price movements in regional spot markets have become closely correlated. Their findings suggest the natural gas pipeline network has developed to the point that interregional arbitrage opportunities have been largely eliminated.

The novelty of our approach is twofold. First, we identify a set of transactional and regulatory attributes common to most bottleneck facilities. We then use these attributes to structure a general framework that addresses the timing and welfare effects of open-access policy (sec. 2). As in the antitrust analysis of mergers, there is a trade-off between enhancing competition and exploiting potential cost efficiencies (see Williamson 1968). Perhaps the most important of the latter are economies that flow from network coordination.[6] We also investigate the implications of open-access policy for long-term governance structures in upstream supply and identify a number of motives for long-term contracting in an open-access environment (sec. 3).

Our second contribution is to apply our analytic framework to bottleneck facilities in the natural gas industry (sec. 4). Gas pipelines historically provided delivered gas service through long-term contracts. The Federal Energy Regulatory Commission (FERC) has recently mandated full unbundling of gas supply and transportation. We discuss the costs and benefits of this policy, arguing that FERC appears to have been oblivious to the transaction-cost motives for bundled pipeline service. We then assess the role of long-term contracts after open access, focusing on the fact that gas buyers are heterogeneous in their valuation of supply reliability. We develop testable hypotheses based on the contracting motives identified in section 3 and evaluate them in light of the available evidence on recent contracting practices. The dominant motive for long-term contracting after open access appears to be economizing on the transaction costs of providing reliable supply. Concluding comments are made in section 5.

2. Bottlenecks and Governance Structures: An Overview

The electricity, natural gas, long-distance telecommunications, oil, and rail freight industries feature bottleneck transportation facilities with relatively large economies of scale and substantial asset specificity. Moreover, these facilities initially linked otherwise isolated buyers and sellers. These circumstances have led to similar institutional adaptations and regulatory control.

First, because transportation links feature extremely large quasi rents, they typically require protection via some form of long-term governance structure. For example, electric transmission lines historically were vertically integrated with both generation and distribution. Similarly, oil pipelines were vertically integrated with refineries.[7]

Second, exchange relationships associated with bottleneck facilities have been governed by regulatory policies that protect both investors and customers. Markets for these infant industries were relatively small, and investment in site-specific transportation capital was risky. Specific investment was protected by limiting entry through elaborate certification procedures. Yet protecting a transporter's specific investment implied that many downstream customers and upstream suppliers would be served by monopoly transporters. To protect consumers, regulators also imposed price and profit controls.

A third important characteristic of bottleneck facilities is that they began as providers of a service that tied upstream supply with transportation of that supply—for example, *delivered* natural gas, electricity, oil, freight, or station-to-station communications. In contrast, after service unbundling, bottleneck facilities must provide a transportation access service, and the remainder of the delivered service bundle may be provided by other firms. Thus, after the breakup of the Bell system, the regional Bell operating companies allow customers to use rival long-distance firms. Likewise, as a condition for a merger, railroads may be required to provide trackage access and so allow customers to ship freight on other firms' trains.[8]

Transporters may choose to tie upstream supply with transportation to enhance efficiency by exploiting vertical economies of scope. These occur when it is cheaper to coordinate upstream and downstream production in a single firm than through arms-length contracts or markets. For example, Kaserman and Mayo (1991, 499) estimate that arms-length contracting between electricity generators and distributors raises costs by 11.95 percent relative to vertically integrated production. One important vertical economy results from the internalization of network externalities. Such externalities occur when the actions of individual buyers and sellers impose external congestion or reliability costs on others. For example, if markets for wellhead gas supply and transportation are separated, individual wellhead producers lack incentives to maintain the systemwide pressure required to provide reliable service to downstream customers.[9]

Transporters may also choose to tie to exploit market power opportunities created by perverse regulatory incentives.[10] If the bottleneck is regulated, backward integration may allow for cost inflation upstream and the collection of rents through an unregulated upstream affiliate. For example, AT&T was accused of using its manufacturing affiliate, Western Electric, for this purpose prior to divestiture, and the Bell operating companies were excluded from the equipment business for the same reason (Noll and Owen 1989).[11] If a vertically integrated monopolist's prices (but not its quantities) are regulated, it also might restrict output to increase profits in upstream markets.

A fourth and final bottleneck characteristic is that networks tend to grow in scope and complexity over time. This creates the potential for greater competition upstream of the bottleneck facility (e.g., at the gas wellhead or electric power generation stages).[12] As we discuss more fully below, the natural gas pipeline network grew threefold between 1945 and 1970. The pro-competitive implications of this development are suggested in Doane and Spulber's (1991) finding that regional gas spot markets have merged into a unified national market. On the other hand, more numerous network transactions and greater network complexity imply that the vertical economies associated with network coordination may also increase.

The benefits of open-access policy follow from enhancing competition in upstream markets, while the costs derive from the (potential) loss of vertical economies of scope.[13] Greater upstream competition has several beneficial effects. Perhaps the most important is that allocative efficiency is enhanced as market prices gradually supplant regulated prices. For example, open access in natural gas alleviated persistent shortages by eliminating long-term contracts with formula-driven prices exceeding market-clearing levels. Greater upstream productive efficiency should also follow from open-access policy; for example, independent electric power producers often bid below the cost a regulated utility would incur if it constructed its own new capacity.[14] In addition, removal of rate-of-return regulation allows for a reduction in the resource costs of regulatory oversight. These savings include reduced regulatory personnel, less use of lawyers and expert witnesses in regulatory proceedings, and reduced use of court time in the appeals process. Enhanced upstream competition may also produce spillover efficiencies in related markets. For example, entry into long-distance telecommunications, made possible by open-access policy, undermined the traditional cross-subsidy from long-distance to local service (see Noll 1985). Finally, the presence of a verifiable market price reduces information-gathering costs for those services that must still be regulated.

On the other hand, open access may undermine Coasian efficiencies of coordination. First, as mentioned above, bottleneck facilities commonly feature network externalities that can be internalized through bundling. Second, coordinating interrelated markets by prices rather than bundling may require sequential commitment to transportation and upstream supply. Laboratory experiments find that overall efficiency is reduced substantially when buyers must contract sequentially, committing to one component of delivered service (either transportation or upstream supply) before the other.[15] Finally, valuable transaction-specific assets (e.g., specialized knowledge) may be lost, which reduces the credibility of forming enforceable long-term contracts in the future.

3. Governance Structures in an Open-Access Setting

Previous authors have suggested there is little role for long-term governance structures in an open-access environment.[16] However, we identify a number of motives for the use of long-term governance structures in transactions with upstream suppliers.

1. If product quality differs across suppliers but is difficult to verify prior to purchase (an experience good), and if buyers value quality (e.g., uninterrupted delivery of gas or electricity), then long-term relationships may emerge to protect costly information about supplier quality. To the extent that information is relationship specific, long-term relational contracts will arise between buyers who value reliability and suppliers found to provide it.[17]
2. Unless spot prices are free to adjust instantaneously, demand or supply shocks will periodically cause markets to fail to clear. Reliability-sensitive buyers fear even temporary shortages and may prefer long-term contracts or bundled service to spot purchases.
3. Long-term relationships may be desired, as Coase argued, because they reduce the cost of repeated spot market purchases.
4. Long-term contracts may be used to reduce price risk if a complete set of markets for hedging does not exist. Even where futures markets exist, they may not extend far enough into the future to meet all hedging needs.
5. Imperfections in the access market may create quasi rents in the upstream market that require protection under long-term contract or integration. Because the bottleneck facility remains a natural monopoly, the terms of access typically will remain regulated. The rigidities that thus arise are similar to those under traditional price regulation. Furthermore, the comparability of access across different users may not be exact; a particular concern is that the bottleneck operator will offer preferable terms to its own upstream affiliate.

4. Open Access and Governance Structures in the Natural Gas Industry

In the remainder of this essay, we apply the general arguments made above to the specific case of the natural gas industry. We begin with a brief history of the industry through the 1980s, highlighting the coevolution of technology, regulation, and vertical structure. FERC policy over the last decade has consistently moved toward open access, in the process placing restrictions on contract forms. Our assessment of open-access policy turns on two key industry characteristics: (i) buyers vary in their valuation of supply reliability; and (ii) enhanced supply reliability is costly. Two organizational issues are associated with the provision of reliability. First, vertical economies of scope may be associated with bundled service, making FERC's policy of mandatory unbundling (promulgated in Order 636) a policy with questionable economic merit.[18] Second, given an open-access setting, the transaction costs of providing reliable service may be minimized through the use of long-term contracts, as argued above. Building on the motives for long-term contracting identified in section 3, we develop a number of testable hypotheses and assess them using the available evidence on contracting in the open-access environment.

4.1. Historical Background

The governance of vertical relationships in the natural gas industry has changed greatly over time.[19] In the early twentieth century, gas supplies were predominantly from the Appalachian area, and relatively short pipelines were adequate to move the supplies to markets in the northeast. Most producers had multiple pipelines, and—consistent with predictions from transaction-cost reasoning—short-term contacts were the typical mode of exchange between wellhead producer and pipeline. In the 1920s, new supplies in the southwest and improved pipeline technology led to the building of long-distance pipelines to end-use markets in the midwest and northeast. Many producing fields were served by a single pipeline, taking advantage of the scale economies in pipeline gas transportation. Vertical integration became common. Many pipelines and producing fields were owned jointly by several local distribution companies downstream, thereby countering the increased potential for opportunism arising from highly specific assets. Vertical relationships were transformed again in the 1930s by the Public Utility Holding Company Act (PUHCA) of 1935 and the Natural Gas Act (NGA) of 1938. The PUHCA separated most distribution companies from the upstream pipeline/production entities. The NGA subjected the pipelines to public-utility-style regulation by the Federal Power Commission (FPC). FPC accounting practices had the effect of reducing the profitability of pipelines owning gas production sites, and, according to Mulherin (1986, 533), "interstate pipelines constructed after the Natural Gas Act was implemented produced no gas of consequence themselves but instead purchased their supplies from unaffiliated producers."

While the originally proposed text of the NGA called for pipelines to provide unbundled transportation services on a nondiscriminatory basis, pipeline protestations eliminated this requirement, and pipelines were allowed to provide bundled service to downstream ("citygate") markets. Bundled service increased the scope of pipeline opportunism by allowing pipelines to credibly threaten to cut off a producer's market access. Producers thus had incentives to obtain long-term contracts with minimum purchase guarantees ("take-or-pay clauses"). Pipelines in turn passed these purchase obligations downstream via contracts with local distribution companies (LDCs) that included minimum bills.

Contracts between producer and pipeline typically had a variety of price adjustment provisions, the most common of which were "two-party most-favored-nation (MFN) clauses that indexed contract price to the price paid by the purchaser to other suppliers in a stipulated area and renegotiation provisions, typically calling for open-ended price discussions every four or five years" (Crocker and Masten 1991, 80). Contracts between pipeline and local distribution company generally set price at the pipeline's weighted average cost of gas (WACOG).

In 1954 the Supreme Court's decision in *Phillips Petroleum v. Wisconsin,* 347 U.S. 672 (1954), required the FPC to regulate prices at the wellhead. By the 1970s, price regulation was causing serious shortages in the interstate gas market, and in 1978 the U.S. Congress attempted to remedy this problem by passing the Natural Gas Policy Act (NGPA). The NGPA divided the wellhead market into a complex set of categories with differing price ceilings that escalated over time. Price ceilings for categories such as "high cost" gas were high enough to induce substantial new exploration and development, and new contracts often indexed the highest applicable NGPA price ceiling.

During the 1950s and 1960s, the pipeline network expanded rapidly, growing from a total of 82,000 miles in 1945 to 252,000 miles in 1970. Growth rates dropped precipitously after 1970, indicating the maturation of the transmission network. However, the higher wellhead prices allowed under the NGPA brought forth substantial growth in gathering lines connecting pipelines to wells (American Gas Association 1948; U.S. Bureau of the Census 1970, 1991). Thus, by 1985 technological improvement and network expansion had created a system in which open access stood to significantly enhance competition at the wellhead.

FERC policy from the mid-1980s into the 1990s has moved consistently toward enhanced competition in the wellhead market and greater access to pipeline transportation.[20] The first major regulatory step toward open access was Order 436, promulgated in 1985. Under Order 436, a pipeline was free to choose whether to adopt open-access status, but if it wanted to provide any unbundled transportation services at all, it had to offer these services on a nondiscriminatory basis to all customers. Eventually, all major pipelines chose to become open-access transporters.

Open access fostered the growth of a successful market in interruptible gas supply. However, increased sales by non-pipeline suppliers came at the expense of reduced pipeline sales, and pipelines were left with contractual obligations to "take or pay" for gas at prices well above market-clearing levels. The D.C. Court of Appeals eventually remanded Order 436 to the FERC for reconsideration, largely because of its "insouciance" on the take-or-pay issue. FERC's response, Order 500, attempted to (among other things) lay the groundwork for an industry structure in which a take-or-pay crisis could never again emerge. One part of its program was the establishment of "gas inventory charges" (GICs) that would allow pipelines to recover the costs of assuring reliable gas supply but would be more flexible than the old take-or-pay contracts.

While FERC's open-access program was largely successful for interruptible supply, a growing chorus of complaints eventually made clear to FERC that the provision of firm supply and transportation was a much more complex task. One key concern was that pipelines were unduly favoring their own gas supply affiliates in developing transportation arrangements. However, pipelines argued that they were forced to hold substantial capacity in reserve for peak periods, in order to meet their ongoing obligations to provide reliable delivered supply to LDCs. Consistent with both of the above arguments, bundled pipeline service in 1991 accounted for 52.5 percent of peak day deliveries but only 18.8 percent of total annual throughput (FERC 1992, 13273).

FERC Order 636 was designed to solve a number of problems in the provision of reliable gas deliveries. FERC Order 636 sought to assure competitive access to firm transportation by requiring pipelines to unbundle all services (e.g., gas supply, transportation, storage, etc.) and offer them individually on a nondiscriminatory basis. At the same time, pipelines were no longer required to petition FERC for permission to abandon service to buyers whose contracts had expired. Our assessment of Order 636 turns on several important transactional characteristics, which we briefly describe below.

4.2. Transactional Characteristics of the Natural Gas Industry

An important feature of the natural gas industry is the heterogeneity of consumers. Residential customers can neither store gas nor readily switch their heating and cooking appliances from gas to other fuels in response to short-run price variations or delivery interruptions. High short-run switching costs make reliability particularly important for these customers, who comprised roughly 26 percent of sales volume in 1990. In contrast, many industrial electric utility customers have dual-fuel boiler technology that allows them to switch instantaneously to on-site storable fuels in the event of gas price increases or delivery interruptions. Low short-run switching costs make reliability a minor concern for these customers, who comprised roughly 32 percent of sales in 1990.[21]

In natural gas, reliable service takes two slightly different forms. One, called "instantaneous" service, allows a buyer to take gas out of the pipeline downstream as soon as the buyer has gas put into the pipeline upstream. The pipeline typically requires a producer selling under an instantaneous price contract to nominate its demand level a few days in advance of actual delivery. The pipeline then monitors whether the designated supplier has delivered on a timely basis and, if it has not, can interrupt the service of the buyer whose supply has failed. A second type of service, called "no notice" peak service, allows a buyer to take as much gas as needed without giving advance warning to the pipeline or putting equivalent amounts of gas into the pipeline upstream.

Reliability costs can be divided into two components. First are inventory costs of holding excess gas deliverability at the wellhead or in downstream storage facilities. Second are pipeline costs such as holding excess transportation capacity, maintaining extra system pressure (called "linepack"), and monitoring suppliers. When gas supply and transportation are unbundled, there are potential network externalities in maintaining system pressure. If a producer takes costly actions to maintain system pressure, the benefits are shared by all users of the system, thus creating a standard free-rider problem. When the pipeline provides bundled service, its rights to control operation of the system are more complete than under decentralized contracting in an unbundled environment. These residual rights of control are similar to those identified by Grossman and Hart (1986) as central to vertically integrated production. A maintained hypothesis in the contracting literature is that residual rights of control cannot be separated from asset ownership. Because bundled service allows the pipeline to internalize network externalities, an unbundled system that achieves the same level of reliability is expected to generate greater costs.

4.3. Reliable Gas Delivery: A Transaction-Cost Critique of Order 636

In an issue of the *Journal of Law, Economics, and Organization,* Joskow (1991) argues that transaction-cost economics merits an increased role in public utility policymaking, particularly with respect to issues of vertical restructuring. Recent refinements of open-access policy in natural gas demonstrate well the value of transaction-cost reasoning in the policy process. We focus on the transactional complexities involved in the organization of a market for "firm" (reliable) transportation services. This issue has been of particular difficulty for the FERC and was addressed in the recent FERC Order 636, which mandates the unbundling of gas supply from pipeline transportation. Keeping in mind the costs and benefits of open access discussed in section 2, we argue that FERC has paid insufficient attention to the transaction costs of mandatory unbundling.

The FERC is clearly sensitive to the benefits of unbundling but appears less aware of the potential costs. This is illustrated in Order 636, where the FERC states that after full unbundling, a customer will be able to "receive its natural gas supplies in a fashion as reliable as the customer had been receiving under a bundled, citygate service, with the added advantage of providing greater opportunities to purchase that supply at competitive prices. Hence, the customer will have the 'best of both worlds'—reliable service and competitively priced gas" (FERC 1992, 13288). While FERC clearly recognizes—but rejects—the possibility of degraded service reliability, it appears to assume that organizational form has no impact on transportation *costs.*

Throughout Order 636, the FERC discusses in surprising detail the operational consequences of full unbundling for pipelines, illustrating in the process the complexity of the transactions involved. For example, FERC has been quite explicit regarding unbundled pipeline tariffs, which will clearly be very complex contingent contracts. FERC requires that pipelines set forth in their tariffs

> their methods for allocating aggregate receipt point capacity, individual receipt point capacity, mainline segment capacity, storage capacity, and delivery point capacity [;]...provisions governing shipper flexibility, in changing receipt and delivery points [;]...provisions concerning supply and capacity curtailments, the scheduling of gas injections into the mainline and into storage, the scheduling of gas deliveries from storage and from the mainline, the setting and charging of penalties, balancing rights, and the instantaneous receipt and delivery of gas [;]...[and] provisions under which they will provide the "no notice" transportation service required by this rule. (FERC 1992, 13282)

FERC also recognizes the importance of network externalities and the usefulness of vertical command-and-control governance structures.

> Because the pipeline is operating as a transporter, its ability to effectively manage its system will depend in part on its shippers injecting gas into the mainline (packing the line) and into storage at the right place and time. While the pipeline and its shippers (or their suppliers or agents) may be able to achieve what is needed through communication, cooperation, coordination, and compromise, it may be necessary for the pipeline to retain compulsory powers where it dictates to shippers where and when to act by, for example, operational flow orders. All shippers must recognize that the action or nonaction by a single shipper may affect a pipeline's ability to serve all other shippers. (FERC 1992, 13287)

While the exact contractual meaning of "compulsory powers" is unclear, it suggests that contracts must be incomplete, and assigns residual rights of control to the pipeline independent of ownership of gas in the system.

The one place in Order 636 where the FERC explicitly recognizes the potential transaction costs of unbundling comes when it states that pipelines and customers need to consider

> the need for appropriate equipment to accurately monitor and measure injections into the system on a timely basis. It may not be cost-effective on some pipeline systems to install the necessary equipment. In those cases, the pipelines should consider allowing other gas merchants to provide the pipeline with pre-determined allocation plans for the merchant's gas. (FERC 1992, 13290)

The above quotes show that the provision of reliable supply in a fully unbundled marketplace will involve the coordination of many individual transactions; the use of detailed contingent contacts between suppliers, buyers, and the pipeline; extensive monitoring of supplier quality; and a considerable amount of residual command-and-control authority on the part of the pipeline. These are exactly the sort of transactional characteristics that we identified above as likely to lead to bundled service.

Concern over the transaction costs of unbundling was manifested during FERC deliberations over Order 636. In their recorded comments to the Notice of Proposed Rule (NOPR) for Order 636, a number of pipeline companies argued that bundling was *necessary* for reliable no-notice service. A later technical conference sponsored by FERC, however, determined that such service can be provided as long as pipelines are allowed to "retain adequate operational control of the use of their facilities" (FERC 1992, 13278). This policy represents an experiment in divorcing residual rights of control from asset (gas) ownership. Even if we accept the apparent industry consensus that system reliability can be maintained, we predict that the cost of internalizing externalities in the unbundled system will be greater than in the bundled system.

4.4. Governance Structures in an Open-Access Setting: Transaction-Cost Predictions and Evidence

In this section we combine our general framework for analyzing the governance of bottleneck transactions with the microanalytic detail of natural gas institutions and regulations from the preceding section to arrive at some potentially testable hypotheses. We begin by identifying the hypotheses that follow from each of the motives for contracting discussed in section 3. We then evaluate these hypotheses using the available evidence on the scope, pricing, and structure of long-term contracts in the open-access environment.[22]

4.4.1. Summary of Hypotheses

Although we have identified five motives for long-term contracting after open access, their relative importance is an empirical question. As we discussed above, customers are heterogeneous with respect to their demand for supply reliability, and reliability implies added costs. Recognizing these characteristics, we develop below a number of testable hypotheses regarding the role of long-term contracts in the open-access environment.

First, suppliers vary in their ability to deliver reliable service, so if supplier quality information is important and costly, then we should observe the coexistence of spot transactions and long-term contracts. Duann, Burns, and Nagler (1989, 51) note that "[l]iterally thousands of potential suppliers and many gas transporters exist. All have diverse sizes, financial conditions, gas field locations, ownership affiliations, and years of experience in the gas industry. Task such as finding suppliers, evaluating their reliability, and arranging transportation require an intense effort in collecting data and analyzing alternatives." Roughly 32 percent of gas sales are to dual-fuel customers who assign a small value to gas supply reliability. These customers will not pay to assess the quality of a supplier and have no incentive to lock in particular suppliers via long-term contract. In contrast, residential customers comprise roughly 25 percent of total sales and assign a high value to service reliability. This characteristic generates a motive for assessing supplier quality. Assessment costs are minimized by locking in quality suppliers, once they are found, to long-term contracts. In the resulting separating equilibrium, spot sales should be bounded between 32 percent and 75 percent. The remainder of sales occur through contracts with reliability-sensitive buyers, who pay a price premium that reflects the added resource costs of providing premium reliable service. Furthermore, contracts should have considerable quantity flexibility, since reliability-sensitive buyers have demand that is strongly affected by temperature.

Second, price stickiness also leads to a mix of spot purchases and long-term contracting but for slightly different reasons. Because gas is a flow good, the Coasian transaction costs of continuous real-time pricing are high; instead, prices remain fixed for at least some minimal length of time. (The standard "spot" contract in the industry lasts for 30 days.) Supply or demand shocks that occur during this period thus can lead to temporary shortages. Reliability-sensitive buyers are willing to bypass the market and enter a contract with a supplier at a price reflecting the added cost of reliability. Again, quantity flexibility should be observed in such contracts.

Third, if the transaction costs of repeated spot exchange dominate other motives for contractual choice, we should observe only long-term contracts. We would expect these contracts to allow considerable quantity flexibility, allowing buyers to adjust to temperature-induced shifts in demand. Conversely, if the costs of repeated exchange are small, "spot" (or short-term contract) exchange should be adequate for all transactions, regardless of the degree of reliability involved. In either case, we would expect a price premium for reliable supply that reflects the added costs of providing reliability.

Fourth, if some buyers and sellers are averse to income variance, we should observe long-term contracts with relatively rigid prices and quantities coexisting with a spot market. As shown by Hubbard and Weiner (1992), the size of a contract price premium and the scope of the spot market depend upon the distribution of risk preference among buyers and sellers and the relative importance of demand and supply shocks in the industry. In the gas industry, we would expect demand shocks to be more important, since uncertainty over both future weather conditions and the price of alternate fuels is critical. Given this, the Hubbard and Weiner (1992) model predicts that relatively rigid contract prices should be *below* expected spot prices.[23] Furthermore, an increase in demand uncertainty should increase reliance on contracting. One can argue that gas demand uncertainty increased over the 1980s, as the number of dual-fuel customers grew and interfuel competition intensified. Thus, the scope of the spot market should have narrowed.

Fifth, if there are regulatory rigidities in the transportation market or imperfect comparability of access, sellers lack perfect access to alternative buyers. This lowers the value of their product in the next best use, creating quasi rents appropriable by the pipeline. We expect to see such quasi rents protected under long-term contracts with minimum quantity constraints as occurred before open access.

4.4.2. Evidence

Extent of the Spot Market. The spot market in natural gas grew steadily from 5 percent of total sales in 1983 to 75 percent in 1988. Since then it has declined to 65 percent in 1989, 50 percent in 1990, and 45 percent in 1991 and was estimated to be 35–40 percent in 1992.[24] These sales are solely for undifferentiated interruptible gas supply.

Since 1986, the percentage of trade in the spot market has been in the range predicted by the reliability-based explanations for contracting. As the data show, for several years the scope of the spot market trended near the top of the predicted range. This high reliance on the spot market was driven in large part by regulatory pricing distortions. As we discussed in section 4.1, the NGPA created incentives for massive sunk investment in new wellhead supplies. While this investment was taking place, falling oil prices enticed many gas buyers to switch fuels, ultimately leading to a collapse in natural gas prices. At the same time, pipelines retained an obligation to stand ready to provide no-notice reliable peak service to LDCs. Thus, LDCs (whose customers highly value reliability) could shop for bargains in the spot market, knowing the pipeline was obliged to be available as a backup supplier on peak days. As market conditions have begun to equilibrate, however, LDCs have reduced their use of spot supplies. This decline in use of the spot market by reliability-sensitive buyers is unlikely to be reversed, since Order 636 eliminated the pipeline's role as backstop supplier.

The very existence of a spot market shows that pure Coasian transaction costs are not prohibitively high; given this, however, a notion more subtle than simple repeat purchase costs is required to explain the fact that spot sales do not account for 100 percent of gas trade. The risk-aversion hypothesis is not supported by the evidence, since spot sales grew rather than shrank as oil price shocks increased demand uncertainty in the early to mid-1980s. The emergence of a robust spot market does suggest that open access has reduced the potential for opportunistic holdup by pipelines, though the threat of holdup apparently has not been eliminated.

Relative Prices in Long-Term Contracts and Spot Purchases. An analysis of long-term contract and spot price data from Louisiana for the period from November 28, 1988, to February 5, 1990, reveals that the average spot price was $1.68 while the average contract price was $1.83, reflecting a price premium of $.15. The standard deviation of spot prices was .25 while that for contract prices was only .09.[25] Similar findings are reported by Henderson et al. (1988). They collected and analyzed data on a group of 28 long-term contracts that were signed or modified between January 1985 and July 1987; "each contract...was matched with a corresponding spot price at the time the long-term contract was signed and at the distributor's location" (Henderson et al. 1988, 58). The estimated relation between contract price P_c and spot price P_s is $P_c = .363 + .929P_s$. The R^2 for the equation was .917, and the t-ratios for the constant and the coefficient were 2.887 and 16.947, respectively. From 1984 through 1991, spot prices ranged from approximately $.90 to $3.50, and in the period for which the regression was estimated, they ranged from approximately $1.20 to $3.35 (see Doane and Spulber 1991, fig. 4). Using the broader range of prices from 1984 to 1991, the resulting contract price premium estimates range from $.30 to $.11.[26]

The existence of a price premium is consistent with the notion that contracts are used to reduce the transaction costs of providing reliable supply. As mentioned above, however, the coexistence of spot sales and long-term contracts is inconsistent with the hypothesis that the costs of repeated exchange drive contracting in this industry. Again, the evidence lends little support to the risk-aversion hypothesis, suggesting that reliability costs dominate risk avoidance in the determination of contract prices. Access imperfections do not generate clear predictions regarding price premiums.

Predictions Regarding Contract Forms. The contracts that are emerging in the open-access environment exhibit a variety of different structures. In addition to a maximum purchase level, some contain a minimum purchase threshold, below which the buyer is obligated to pay a "deficiency charge" for each untaken unit.[27] This structure initially emerged in applications by pipeline companies to the FERC for permission to include a "gas inventory charge" (GIC) in rates; the GIC was designed to allow the pipeline to cover its costs of providing long-term, reliable supply (see FERC 1987). Interestingly, the same contract structures are also appearing in contracts between wellhead producers and downstream customers. For example, Southern California Gas Company recently asked the California Public Utilities Commission (CPUC) to approve a set of contracts with minimum takes of 50 percent to 90 percent of the maximum level and deficiency charges of $.25/MMBtu (see CPUC 1991).

On the other hand, some contracts forgo minimum take provisions. For example, Indiana Gas Company recently signed a set of 13 long-term contracts whose total volume represents about 25 percent of the company's annual purchases and 60 percent of purchases on a peak day. There is no minimum take level in the contracts, and performance is guaranteed by the seller. The contracts use a two-part pricing structure. Commodity prices are indexed to spot price reports from *Inside FERC's Gas Market Report.* The demand charge pays for first place in the allocation queue during peak demand periods.[28]

Either of these contract forms is consistent with the hypothesis that contracts are used for supply reliability. A deficiency charge may reflect a producer's opportunity cost of holding excess deliverability in reserve for a given buyer. In a two-part tariff, each part is an instrument that addresses a particular contractual problem. The fixed portion of the contract reflects the value of a high-priority position for the allocation of supplies if demand shocks exceed production capacity. As before, Coasian costs of repeated exchange have little implication for the structure of contracts. If contracting is driven by risk aversion, contracts should have rigid price and quantity terms, which are observed in neither of the contract forms described above. Again the evidence does not support the risk-aversion hypothesis.

Finally, the differences in contract forms are consistent with the notion that the access problems faced by producers vary. When there are potential holdup problems associated with the gas pipeline bottleneck after open access, we would expect contracts with quantity and perhaps price assurance to prevent ex post opportunism. The use of minimum take thresholds with deficiency charges may be appropriate for producers who face restricted alternatives for the sale of their gas and are unable to sell untaken quantities on the spot market. Such restrictions may occur if a new gas field is drilled and has limited access to downstream markets, or they may be due to rigidities in transportation markets. On the other hand, even before Order 636's mandatory unbundling, some pipelines had chosen to abandon the merchant function and provide only transportation service. Under these circumstances, pipelines have no incentive to hold up producer sales, and sellers do not need quantity protection in contracts.

Contract price flexibility has the benefit of allowing for efficient adaptation to supply and demand shocks but thwarts the price stability desired by risk-averse traders. One way to achieve price flexibility is by indexing contract prices to spot prices. Evidence suggests indexed contracts are being used in the gas industry. For example, the regression by Henderson et al. (1988) shows contract and spot prices to be closely correlated. In fact, the hypothesis that the correlation is perfect cannot be rejected at standard significance levels. Moreover, consultant Schlesinger (1991, 5) claims that "[i]ncreasingly, the pricing provisions contained in new long-term gas purchase contracts are being written to reference prices reflective of gas spot markets." Such contracts provide little insurance against price fluctuations, thus further undercutting the risk-aversion explanation for long-term contracts.

The success of indexing here may be surprising to some. Goldberg (1985, 542) remarks that "[i]f a unique, easily observable market price existed, the [price] adjustment problem would not be difficult. However, the conditions that make entering into a contract desirable in the first instance make it unlikely that such prices exist....The problem is that by entering into the long-term contract at least one of the parties deliberately isolated itself from the external market. The relationship between the external price and the opportunity costs of the parties need not be very close." In the case of natural gas, however, the homogeneity of methane molecules combines with the heterogeneity of buyers to make indexing possible. The spot market, made feasible by customers with low fuel-switching costs, provides an external benefit to customers who value reliability more highly, allowing them to write more efficient, indexed contracts.

Summary of Evidence. The above evidence is entirely consistent with the notion that long-term contracts are used to reduce the transaction costs of providing reliable gas supply. The extent of the spot market is consistent with the mix of downstream customer types; the price premium for long-term contracts reflects the added cost of reliability; and the structures used in new long-term contracts are consistent with an attempt to reflect the cost of reliability in an efficient fashion, shaped by potential access restrictions. The hypothesis that contracts are used only to reduce the transaction costs of repeated spot exchange cannot explain the emergence of a sizable spot market and gives little insight into the prices and structures of the contracts that have been signed. Nor does the risk-sharing hypothesis find much support in the available evidence. Neither the growth of spot sales nor the observed price premium in contracts are predicted by an income-smoothing model with increasingly important demand shocks. Furthermore, contracts lack the price and quantity rigidity predicted by the model. Finally, the growth of a spot market and reduced quantity rigidity in remaining contracts suggest that open access has successfully given producers access to alternative buyers and so reduced the threat of pipeline opportunism.

5. Concluding Comments

Open-access requirements are an increasingly common remedy for problems of market power in network industries such as natural gas, electricity, and telephones. The benefits of such policy—greater competition upstream of bottleneck facilities—must be traded off against the cost of potential loss of vertical economies. The transactional characteristics of each particular industry largely determine the importance of this cost and also determine the nature of the long-term governance structures that emerge in the wake of open access. We have discussed the key transactional characteristics of these network industries, laid out a framework for assessing costs and benefits, and identified motives for long-term contractual relationships in an open-access setting. We then applied these general considerations to a case study involving the natural gas industry.

Given the difficulties inherent in measuring the costs and benefits of unbundling, it seems to us that prudent public policy should leave as much room as possible for efficient governance structures to emerge. FERC's prohibition in Order 636 on bundled service forecloses this potentially efficient option. Furthermore, regulatory costs under Order 636 may increase, since FERC requirements for nondiscrimination will now extend to a vast array of pipeline interactions with buyers and sellers. If FERC later reverses its policy prohibiting bundling, this should be taken as evidence that residual rights of control (of gas flow) cannot easily be separated from ownership.

NOTES

1. For example, new telecommunication technologies have allowed entry into long-distance telephone service; extensive development of the natural gas pipeline network has facilitated competition at the wellhead; reduced transmission losses on

new electric power lines have made the wheeling of electricity less costly and facilitated competition in power generation.

2. The antitrust issues are discussed by Werden (1987), Tye (1987), and Owen (1990). Key antitrust cases include *MCI Communication Corporation v. AT&T*, 708 F.2nd 1081, *7th Cir., cert. den.*, 464 U.S. 891 (1983); *United States v. Terminal Railroad Association*, 224 U.S. 383 (1912); and *Otter Tail Power Co. v. United States*, 410 U.S. 366, 372 (1973).

3. We use these terms interchangeably below.

4. For example, in telecommunications, open access was implemented by vertical divestiture resulting from antitrust proceedings; the regional Bell operating communities (RBOCs) remain regulated. In electricity, access requirements have evolved on a case-by-case basis under antitrust principles.

5. For example, Goldberg and Erickson (1987) and Crocker and Masten (1988) have shown that price regulation tends to shorten contractual length. For other empirical work on contracts, see Joskow (1987), Hubbard and Weiner (1991).

6. As pointed out by Coase (1937), if control over a complex system is shifted from a single firm to a decentralized price system, transaction costs ("marketing costs" in Coase's terminology) of various types are increased.

7. In addition, long-distance telecommunications lines were vertically integrated with the local switch. The Public Utility Holding Company Act (PUHCA) of 1935 forced vertical disintegration of the natural gas industry.

8. Open-access pipelines allow buyers to purchase oil or gas from alternative upstream suppliers. Electricity distribution companies can procure electricity from independent electricity producers by "wheeling" power through open-access transmission facilities.

9. Similarly, a franchise has an incentive to use substandard inputs when the reputational costs are shared throughout the chain. Tying contracts can attenuate this external cost; see Klein and Saft (1985).

10. With a structurally competitive upstream sector, an *unregulated* bottleneck facility would generally have no incentive to foreclose the upstream market. It would, of course, have incentives to charge supercompetitive downstream prices.

11. In a similar vein, the Nevada Public Service Commission recently disallowed $6.2 million from Nevada Power's $18 million coal cost request, arguing that the company's relations with its supposedly unaffiliated coal-production subsidiary "raised major questions about the company's relationship with its subsidiary and its coal contracting practices in general" (*Wall Street Journal*, November 12, 1991).

12. Similarly, extensive development of the oil pipeline network has facilitated competition in the wellhead market; new telecommunication technologies have lowered the cost of long-distance phone calls, and allowed entry into long-distance telephone service.

13. Open access can substantially increase the competitiveness of markets upstream of the bottleneck, as evidenced in the markets for long-distance telecommunications (upstream of the local switch) and wellhead sales of natural gas (upstream of the pipeline). Greater competition would presumably arise in electric power generation (upstream of transmission lines) if mandatory wheeling were enacted. The bene-

fits and costs we identify here are similar to those considered in the antitrust literature on "essential facilities"; see, for example, Tye (1987). However, we view things
from a regulatory, as opposed to an antitrust, perspective, and place more emphasis
on transaction costs and economic institutions.

14. Competition in the market for long-distance telephone service forced AT&T to
cut overhead expenses such as underwriting the *Bell Journal of Economics* and spon-
soring pure science at Bell Labs; this presumably increased productive efficiency.

15. The Federal Energy Regulatory Commission has sponsored two sets of laboratory experiments in this area, the results of which are summarized in Alger and Toman (1990, 274).

16. Hubbard and Weiner (1991, 65) point out: "It is often argued, for example, a
switch from private to common carriage [of natural gas] would allow spot markets
and short-term marketing agreements to displace long-term contracts." Note that their
distinction between private and common carriage parallels ours between bundled
service and unbundled service with nondiscriminatory access.

17. Experiments on relational contracting by Hackett, Wiggins, and Battalio
(1993) find that when supplier quality information is costly and relationship-specific,
long-term exchange relationships form when buyers successfully find quality suppliers.

18. While we would also like to offer some numerical estimates of the welfare
effects of open-access policy, data availability precludes such analysis at this time.
It will be some years yet before an organizational equilibrium can be reached. A
particular difficulty is that FERC Order 636 imposes mandatory unbundling of
transportation from gas supply. This precludes the observation of the bundled
service alternative, even if it were the most efficient organizational form.

19. Our discussion of industry history closely follows Mulherin (1986).

20. Because the evolution of recent gas policy has been well documented in several other papers, here we limit ourselves to a very broad overview of recent policy
developments. See, for example, Pierce (1988).

21. Industrial and electric utility customers comprised 58 percent of gas sales in
1990. According to the *Wall Street Journal* (October 12, 1992, A1), 55 percent of
these customers have dual-fuel capability. Commercial customers account for 16 percent of total sales. Data on sales by customer class are from *Natural Gas Monthly*
(U.S. Dept. of Energy 1991), 13.

22. Unfortunately, the study of contracting within the gas industry has become
more difficult since FERC ceased requiring the filing of contract information for
transactions in which the pipeline serves only as a transporter.

23. If there is no long-term contracting, a sudden outward shift of the downstream
demand curve will increase sales and raise spot prices. Producer profits rise, and consumer surplus also increases, at least for additive demand shocks. Hence, spot prices
covary positively with income for both producers and consumers. Now suppose market participants consider signing a forward contract at a fixed price, a move akin to
speculating on the futures market. For buyers, speculative profits increase with the
spot price, adding to net variance. Risk-averse buyers will require a contract discount
to offset the increase in income variance. For sellers, speculative profits decrease
with the spot price, reducing net variance. Risk-averse sellers find this reduction in

variance desirable, so competitive pressures lead to a contract price discount.

24. Source: Benjamin Schlesinger and Associates (1992). The convention in the industry is to refer to contracts of 30 days or less as "spot" sales.

25. Source: *Natural Gas Intelligence,* Louisiana region, November 28, 1988–February 5, 1990. The data include all contracts of one year or more in duration, the standard for long-term duration in the gas industry, and are taken from "new contracts, renegotiations, posted prices, and market out actions." Price reports are given as a range from low to high each week.

26. The existence of a contract price premium is also reported by Schlesinger in CPUC (1992) based on his firm's confidential analysis of 40 long-term contracts for reliable supply.

27. Under a traditional take-or-pay contract, the buyer pays a lump sum for any quantity below a minimum threshold.

28. Personal communication from Curt Hribernik, Manager of Gas Supply, Indiana Gas Company, March 12, 1992.

REFERENCES

Alger, Dan, and Michael Toman. 1990. "Market-Based Regulation of Natural Gas Pipelines," 2(3) *Journal of Regulatory Economics* 263–80.

American Gas Association. 1948. *Gas Facts. New York: American Gas Association.*

Benjamin Schlesinger and Associates/Energy Futures Group. 1992. *Multi-Client Analysis of Natural Gas Markets: Blending Risk Management Tools for the 1990s.* Bethesda, Maryland: Benjamin Schlesinger and Associates.

California Public Utilities Commission. 1991. "Proposed Decision of ALJ Barnes on Application of Southern California Gas Company for Expedited Approval of Five Long-Term Supply Agreements," A. 91–04–038, (April 26, 1991).

California Public Utilities Commission. 1992. *Comments Submitted on En Banc Session on Natural Gas Procurement* (February 17, 1992).

Coase, Ronald H. 1937. "The Nature of the Firm," 4 *Economica* 386–405.

Crocker, Keith J., and Scott E. Masten. 1988. "Mitigating Contractual Hazards: Unilateral Options and Contract Length," 19(3) *RAND Journal of Economics* 327–43.

———. 1991. "Pretia ex Machina? Prices and Process in Long-Term Contracts," 34 *Journal of Law and Economics* 69–100.

Doane, Michael, and Daniel Spulber. 1991. "Open Access and the Evolution of the U.S. Spot Market for Gas," GMRCSM Discussion Paper 91–48, Northwestern University.

Duann, Daniel, Robert Burns, and Peter Nagler. 1989. *Direct Gas Purchases by Gas Distribution Companies: Supply Reliability and Cost Implications.* Columbus, Ohio: NRRI.

Federal Energy Regulatory Commission. 1985. *Order 436. Final Rule and Notice Requesting Supplemental Comments.* "Regulation of Natural Gas Pipelines after Partial Wellhead Decontrol." Docket No. RM85–1–000 (Washington, D.C., May 30, 1985).

———. 1987. *Order 500. Interim Rule and Statement of Policy.* "Regulation of Natural Gas Pipelines after Partial Wellhead Decontrol." Docket No. RM87–34–000 (Washington, D.C., August 7, 1987).

———. 1992. *Order 636. Final Rule.* "Pipeline Service Obligations and Revisions to Regulations Governing Self-Implementing Transportation; and Regulation of Natural Gas Pipelines after Partial Wellhead Decontrol." Docket Nos. RM91–11–000; RM87–34–065 (Washington, D.C., April 8, 1992).

Goldberg, Victor P. 1985. "Price Adjustment in Long-Term Contracts," *Wisconsin Law Review* 527–43.

———, and John R. Erickson. 1987. "Quantity and Price Adjustment in Long-Term Contracts: A Case Study of Petroleum Coke," 30 *Journal of Law and Economics* 369–98.

Grossman, Sanford, and Oliver Hart. 1986. "The Costs and Benefits of Ownership: A Theory of Lateral and Vertical Integration," 94 *Journal of Political Economy* 691–719.

Hackett, Steven, Steve Wiggins, and Ray Battalio. 1993. "The Endogenous Choice between Contracts and Firms: An Experimental Study of Institutional Choice," mimeo, Indiana University.

Henderson, J. Stephen et al. 1988. *Natural Gas Producer-Distributor Contracts: State Regulatory Issues and Approaches,* National Regulatory Research Institute Report NRR1 87–12 (January 1988).

Hubbard, R. Glenn, and Robert Weiner. 1991. "Efficient Contracting and Market Power: Evidence from the U.S. Natural Gas Industry," 34 *Journal of Law and Economics* 25–68.

———. 1992. "Long-Term Contracting and Multiple-Price Systems," 65 *Journal of Business* 177–98.

Joskow, Paul L. 1987. "Contract Duration and Relationship-Specific Investments: Empirical Evidence from Coal Markets," 77(1) *American Economic Review* 168–85.

———. 1991. "The Role of Transaction Cost Economics in Antitrust and Public Utility Regulatory Policies," 7 *Journal of Law, Economics, and Organization* 53–83.

Kaserman, David L., and John W. Mayo. 1991. "The Measurement of Vertical Economies and the Efficient Structure of the Electric Utility Business," 39(5) *Journal of Industrial Economics* 483–502.

Klein, Benjamin, and Lester F. Saft. 1985. "The Law and Economics of Franchise Tying Contracts," 28(2) *Journal of Law and Economics* 345–62.

Mulherin, John H. 1986. "Specialized Assets, Governmental Regulation, and Organizational Structure in the Natural Gas Industry," 142 *Journal of Institutional and Theoretical Economics* 528–41.

Natural Gas Intelligence. 1988–90. Great Falls, Virginia.

Noll, Roger G. 1985. " 'Let Them Make Toll Calls': A State Regulator's Lament," 75 *American Economic Review Papers and Proceedings* (May 1985) 52–56.

Noll, Roger G., and Bruce M. Owen. 1989. "The Anticompetitive Use of Regulation: *United States v. AT&T* (1982)," in J. E. Kwoka, Jr., and L. J. White, eds., *The*

Antitrust Revolution. Glenview, Illinois: Scott, Foresman and Company.

Owen, Bruce M. 1990. "Determining Optimal Access to Regulated Essential Facilities," 58 *Antitrust Law Journal* 887–94.

Pierce, Richard J., Jr. 1988. "Reconstituting the Natural Gas Industry from Wellhead to Burnertip," 9(1) *Energy Law Journal* 1–57.

Schlesinger, Benjamin. 1991. "Spot Trading, Futures, and Long-Term Contracts: Gas Market Mechanisms in the 1990s," *Energy in the News,* 2–6.

Tye, William B. 1987. "Competitive Access: A Comparative Industry Approach to the Essential Facility Doctrine," 8 *Energy Law Journal* 337–79.

United States Bureau of the Census. 1970, 1991. *Statistical Abstract of the United States.* Washington, D.C.: U.S. GPO.

United States Department of Energy, Energy Information Administration. 1991. *Natural Gas Monthly,* DOE/EIA–0130 (91/06), June 1991.

Wall Street Journal. 1991. "Nevada Power's Rate Rise Is 40% of Total Requested," November 17, 1991:A10.

Wall Street Journal. 1991. "Scary Prosperity: Natural-Gas Prices Surge and Producers Aren't Happy at All," October 12, 1992:A1–A6.

Werden, Gregory J. 1987. "The Law and Economics of the Essential Facility Doctrine," 32 *Saint Louis University Law Journal* 433–80.

Williamson, Oliver E. 1968. "Economies as an Antitrust Defense: The Welfare Tradeoffs," 58 *American Economic Review* 18–36.

Part IV
Models of Monitoring and Sanctioning

Introduction to Part IV

One of the major points made by Elinor Ostrom in chapter 3 is that institutional analysts need to incorporate the particular roles filled by different actors in empirical situations. Eric Rasmusen's argument in chapter 9 leads to a similar conclusion. Workshop scholars have demonstrated that monitoring and sanctioning are crucial to the success of diverse forms of governance, ranging all the way from microlevel irrigation systems to issues of global environmental change (McGinnis 1999). The essays included in part IV model the process of monitoring and enforcement in different ways and in different contexts.

Models of Rule Enforcement

In "Transforming Rural Hunters into Conservationists: An Assessment of Community-Based Wildlife Management Programs in Africa" (chap. 12), Clark C. Gibson and Stuart A. Marks use extensive form game models to represent the fundamental dilemmas of wildlife management in Africa. This essay is especially useful as an illustration of the interactions among the diverse factors discussed throughout this volume of readings. Their model highlights, for example, the central question of whether actors have the requisite incentives and capabilities to carry out the monitoring and sanctioning activities needed for a successful wildlife preservation program. The advantages of having locally managed preservation schemes come through loud and clear. Yet local self-governance is by no means automatic, for local farmers have multiple incentives, many of which are in conflict with the goal of wildlife management.

The next selection develops an abstract representation of an irrigation system to explore the implications of different forms of monitoring behavior. In "Irrigation Institutions and the Games Irrigators Play: Rule Enforcement on Government- and Farmer-Managed Systems" (chap. 13), Franz Weissing and Elinor Ostrom build on a previous model (Weissing and Ostrom 1991). In that model, no separate roles for monitors or sanctioners are included. The implications of that analysis serve as a benchmark for comparison with models including two different types of monitors. In particular, the authors identify conditions under which farmers should be expected to have the appropriate incentives to carry out monitoring activities. As is typical in any reasonably complex model, different equilibrium conditions apply for different configurations of parameters. Without going through all the equilibria conditions here, one implication deserving emphasis is that rational actors can monitor and sanction each other, at least under some conditions. In short, this model serves as a formal demonstration that self-governance is indeed possible.

At the same time, this model predicts a nonzero level of stealing in equilibrium. That is, some participants will, at least some of the time, violate the rules governing levels of water to be drawn from an irrigation system. Clearly, some form of monitoring is required.

Weissing and Ostrom introduce distinctions between two different types of monitors: "integrated" guards whose payoffs are directly related to the physical success of the local farmers and "disassociated guards" whose payoffs are determined by some measure of violations detected. Integrated guards represent monitors who are themselves part of the community managing the irrigation system, whereas disassociated guards represent the incentives facing monitors who, because they are appointed by the central government or some other authority, are not concerned with the long-term viability of this irrigation system.

By comparing the equilibria conditions for models of irrigation systems in which monitoring and sanctioning tasks are carried out by integrated or disassociated guards, Weissing and Ostrom demonstrate that under a reasonable range of conditions, the farmer-managed systems are more efficient and sustainable than are systems managed by governmental agents. This result comports nicely with the observation of Lam, Lee, and E. Ostrom (1997) of exactly this difference between effective farmer-managed and ineffective government-managed irrigation systems in Nepal. Thus, Weissing and Ostrom have provided a formal game theoretic representation of results observed in field studies of actual irrigation systems.

Rule Enforcement in Asymmetric Games

The first two essays in this part demonstrate that it is possible to build tractable game models in which the parameters of the model bear a clear connection to important aspects of the related real-world phenomenon. The Workshop research program on common-pool resources has been truly unique in its ability to draw close and complementary interconnections among formal game models, laboratory experiments, and field research.

In "Coping with Asymmetries in the Commons: Self-Governing Irrigation Systems Can Work" (chap. 14), Elinor Ostrom and Roy Gardner combine a simple model with the results of field research to demonstrate that farmer-managed irrigation systems can be sustained even when the community includes two groups of farmers with conflicting interests. In this case, farmers located at the head or tail end of an irrigation system have different interests. Specifically, "headenders" have an inherent advantage, because they can extract their needed water before the downstream "tailenders" get a chance. But in the situations modeled here, the headenders are themselves unable to shoulder the burden of maintenance of the entire irrigation system. Thus, headenders must find a way to encourage tailenders to contribute to the system's maintenance, which can only happen if the headenders can restrain themselves from extracting too much water. This is a very powerful form of asymmetry, one that is often found in real-world irrigation systems.

That particular model is less elaborate in a technical sense than the one in the preceding essay, but it extends that model's implications in one important way. Even when faced with a fundamental asymmetry in interests, there are conditions under which we can expect rational individuals of both actor types to cooperate in the maintenance of this irrigation system. Too often formal models or laboratory experiments assume or impose a condition of symmetry on participants, and this essay is a good illustration of the potential benefits of an alternative mode of analysis in which more of the most important aspects of actual situations can be incorporated into formal models or experimental studies. In addition, this essay demonstrates that formal models and field research can be cross integrated in productive ways. Similar results have been observed in experimental studies, which is the topic of the concluding part of this volume.

REFERENCES

Lam, Wai Fung, Myungsuk Lee, and Elinor Ostrom. 1997. "The Institutional Analysis and Development Framework: Application to Irrigation Policy in Nepal." In *Policy Studies and Developing Nations: An Institutional and Implementation Focus,* ed. Derick W. Brinkerhoff, vol. 5, 53–85. Greenwich, CT: JAI Press.

McGinnis, Michael D., ed. 1999. *Polycentric Governance and Development.* Ann Arbor: University of Michigan Press.

Weissing, Franz J., and Elinor Ostrom. 1991. "Irrigation Institutions and the Games Irrigators Play: Rule Enforcement without Guards." In *Game Equilibrium Models II: Methods, Morals, and Markets,* ed. Reinhard Selten, 188–262. Berlin: Springer-Verlag.

CHAPTER 12

Transforming Rural Hunters into Conservationists: An Assessment of Community-Based Wildlife Management Programs in Africa

Clark A. Gibson and Stuart A. Marks

1. Introduction

Hunters decimated African wildlife in the 1970s and 1980s. Decreasing commodity export prices, increasing international demand for wildlife products, and inflation led to startling decreases in the populations of many species, especially elephant and rhinoceros. Despite conservation policies that restricted access to firearms, established hunting quotas, designated conservation areas, issued hunting permits and used military-trained scouts to enforce the law, the underfunded wildlife departments of Eastern and Southern African countries failed to stem the tide of poaching.[1]

Given the apparent inadequacy of conventional wildlife conservation policies—inherited from the colonial era—to deal with the upsurge in illegal hunting, conservationists, international conservation organizations and African wildlife departments searched for new approaches. Their exploration coincided with a more general trend in development studies to include local communities in the planning and management of natural resources to promote economic growth (World Conservation Union 1980).

Currently, a growing number of specialists in both conservation and development consider the inclusion of local communities in wildlife management indispensable for successful conservation. These experts charge that conventional wildlife policies exclude rural residents from most legal uses of wildlife. While paying the costs for conservation in the form of damaged crops and even human lives, rural communities receive few legal benefits from wildlife. Consequently, such exclusionary wildlife policies provide few incentives for the sustainable use of wild animals; rural and urban residents

Originally published in *World Development* 23 (1995): 941–57. Reprinted by permission of Elsevier Science and the authors.

consistently choose to kill wildlife despite the restrictive legal codes.

Reflecting the recent rapprochement between development and conservation, planners now seek to return some revenues derived from wildlife to rural communities (Brandon and Wells 1992). By ensuring that "local communities participate in the benefits to be gained from the presence or utilization of wildlife" (Kiss 1990, 10), new initiatives hope to induce individuals away from their past practices, particularly unlicensed hunting, and toward behaviors that conserve wild animals. By providing economic goods such as employment and development projects, the new programs aspire to transform the would-be poacher into an individual with a sense of proprietorship over wildlife. Optimally, rather than continue their historic antagonism with wildlife scouts, rural residents would assist government efforts in monitoring and protecting the resource.

Yet these wildlife initiatives have been less successful than hoped. While the hunting of some of the larger mammals has decreased, in many areas rural residents continue to kill and to use wildlife illegally. Some residents are also openly hostile to scouts. Although arrests for illegal offenses have increased, the rate of illegal offtakes and game meat consumption has not diminished in some program areas.[2] Rather than resolve the dilemmas faced by rural communities, these wildlife conservation programs have added an additional layer of bureaucracy on local communities and further alienated them with increased enforcement.[3]

This essay seeks to explain why current wildlife policies have failed to achieve their goal of conserving wild animals through community-level conservation programs. Applying concepts from economics and game theory to a conservation initiative in Zambia, we demonstrate that program designers misunderstand the incentives that motivate rural residents—particularly hunters, the primary users—and the resources essential to sustain livelihoods in marginal environments. These programs generally fail to appreciate the social significance of local hunting. They also attempt to change individuals' behavior from wildlife consumption to wildlife conservation by offering goods that mimic public goods. As a result, rural residents continue to hunt and to trade wildlife. The programs' increased level of enforcement does, however, alter locals' tactics and prey selection. We suggest that community-based wildlife initiatives in other African countries present locals with comparable incentive structures and therefore result in analogous outcomes.

This essay has five sections. In section 2, we briefly review the assumptions and outcomes of conventional wildlife management policy in Africa. African governments' diminished enforcement ability and the increasing value of wildlife to rural residents helped make wildlife a de facto open-access resource. To illustrate these consequences, we model the interactions between the wildlife scouts and local hunters using noncooperative game theory. In section 3, we discuss the assumptions of the new "participatory" wildlife schemes and again use game theory to depict their intended results. These initiatives sought to end illegal hunting by transforming residents' incentives to kill into a desire to conserve. Section 4 examines the actual impact of a community-based wildlife program in Zambia, the Administrative Management Design for Game Management Areas (ADMADE). Section 5 discusses the reasons behind the weak performance of ADMADE and similar programs. An important finding is that by giving quasi-public goods to communities to abstain from hunting, programs fail to reward individual behavior. Consequently, programs create a free-rider problem in which individuals continue to hunt while receiving the benefits of community-level projects. In the last section we argue that in many important respects, the "new" conservation policies continue conventional practices. We claim that conservation will be more successful at the local level when rural residents possess significant legal claims over wildlife resources and their management.

2. The Incentives of Conventional Wildlife Policy

Since the day-to-day practice of wildlife management involves the enforcement of rules in field settings, we model the effects of wildlife policies by characterizing the choices confronting wildlife scouts and rural residents. Because the choices of local residents and wildlife scouts regarding hunting

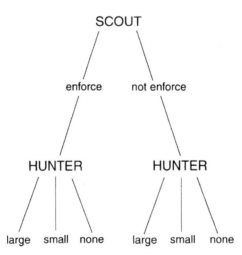

Fig. 12.1.   Choices confronting wildlife scouts and local hunters. Scouts choose to enforce or not enforce the law; hunters choose to hunt large, small or no wild animals.

and enforcement are strategic, that is, each depends on the choices of the other, we present a simple framework for their interaction in figure 12.1.

The scout faces two choices in this characterization, to enforce or not to enforce wildlife regulations. The rural resident has three possible actions: hunt large game (e.g., elephant, rhino, and hippopotamus), hunt small game (e.g., buffalo, most antelope species and warthog), or not to hunt at all.[4]

Conventional wildlife conservation policy accepts that wildlife is valuable and that enforcement is necessary to prevent illegal hunting. Such a policy, therefore, depends crucially on two assumptions: that scouts will choose to enforce wildlife law and that scouts' enforcement activities raise the costs of an individual's hunting to a level higher than its benefits (Leader-Williams and Milner-Gulland 1993). These two assumptions push the interaction in figure 12.1 to the Enforce/None outcome: scouts enforce the law, which deters locals from killing animals.

In the 1970s and 1980s, however, several factors belied those two important assumptions. The severe contraction of the budgets of most African governments led to underpaid and understaffed wildlife departments.[5] Decreasing commodity export prices, inflation, and a general decline in living standards made wildlife a more valuable commodity to rural residents.[6]

Such factors had a strong impact on the choices of wildlife scouts and rural hunters. In figure 12.2, we use the framework of figure 12.1 to construct a noncooperative game with complete information.[7] Figure 12.2 depicts this game between scouts and residents with payoffs resulting from the events of the 1970s and 1980s. We assume that each player calculates the payoffs of each outcome.[8] The numbers one through six represent each player's ranking of outcomes.[9]

Rural residents had always valued wildlife. The meat of wild animals augments diets, especially in drought-prone areas where crops are staples (Marks 1984; Murombedzi 1992). People exchange game meat and skins for the cash needed to purchase consumer goods and for school fees. Locals also use their skills and knowledge to assist strangers in their poaching, receiving income and other material benefits in return. In addition to economic goods, hunters gain social status from providing their lineage dependents with meat, money, crop protection and consumer goods (Marks 1976, 1979, 1984).

Given the increased relative value of wildlife to the rural resident during this period, and the decreased probability of getting caught, locals chose to hunt.[10] In addition, since killing larger species yields more meat and income than smaller species, hunters preferred to hunt bigger animals.[11] Thus, under conventional wildlife policy, the hunter had a dominant strategy to Hunt Large Game, in the sense that he is always better off choosing this strategy no matter what the scout chooses (Kreps 1990).

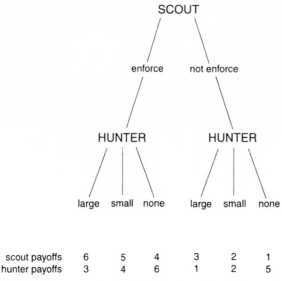

Note: Numbers represent rank order for the players (i.e., 1 is the most preferred outcome).

Fig. 12.2. Extensive form game between wildlife scouts and local hunters under conventional wildlife policies in the 1970s and 1980s

The scout, meanwhile, had little to gain by enforcing wildlife regulations during this time.[12] Enforcement meant going on patrols, enduring difficult conditions, and possibly facing well-armed poachers.[13] Upholding wildlife law also generated hostility from local communities, on whom scouts depend for social activities and petty trade. Scouts usually lacked adequate firearms and ammunition to protect themselves from hunters armed with modern weapons, and these shortages were made even more severe by departmental budget cuts. Further, scouts knew that their departments possessed little capacity to monitor their behavior; without sufficient funds and staff for oversight, most scouts avoided supervision.[14] While scouts preferred that residents did not hunt—rampant poaching in their areas could mean transfer—scouts preferred not to enforce the law even more.[15] Under conventional policy at this time, Not Enforce was the scout's dominant strategy.

Given these preferences of resident and scout, the game portrayed in figure 12.2 results in the outcome Hunt Large Game/Not Enforce.[16] This game helps explain the observed rapid increase in illegal hunting—especially of the larger mammals—across Africa at this time (Bonner 1993).[17] While the state owned and protected wildlife de jure, residents of local communities helped themselves to a de facto open-access resource.

3. The Shift to Community-Based Programs

The rapid escalation of illegal hunting throughout much of Africa in the 1970s and 1980s led scholars and practitioners to reevaluate the institutions of conventional wildlife management (Prescott-Allen and Prescott-Allen 1982; Eltringham 1984; Child and Child 1987; MacKinnon et al. 1986; Bell and McShane-Caluzi 1984; Western 1984; Marks 1984; Kiss 1990). They realized that previous wildlife policies perpetuated a pattern of costs and benefits that significantly disadvantaged rural residents in many ways.

Land gazetted in national parks and game reserves bars rural communities from using resources on these lands that had once been theirs to exploit. Government provides little, if any, compensation for these resource losses. In areas designated for hunting, quotas and licenses favor the wealthy and urban citizen over the poor and rural people. Both tourism and safari enterprises can generate significant revenues from wildlife, but most of these proceeds go to operators, owners, and government. Additionally, wild animals—ranging from elephants to insects—destroy a significant amount of crops grown by rural residents each year. Some species, such as elephants and crocodiles, are a constant threat to human lives as well.

Attributing economic rationality to the rural resident, scholars and practitioners suggested new management policies to close what had become a wildlife commons (Murombedzi 1992). Experts believed communities would actively seek ways to preserve wildlife over the long term if they possessed a legal and valuable stake in the resource (Western 1982; Motshwari 1981; Martin 1986; Lewis and Kaweche 1985). To this end, new policies emerged in Zimbabwe, Zambia, and Botswana that linked jobs and services to communities located in or near wildlife conservation areas.[18]

Despite their variety of institutional structures, the new initiatives share a key assumption: conservation policies will work only if local communities receive sufficient benefits to change their behavior from taking wildlife to conserving it. The benefits offered by the programs to induce behavioral changes fall into three broad categories. The first type is linked directly to wildlife. Locals receive jobs as scouts to protect and monitor the resource, or as wage earners with tourist and safari companies whose activities depend on viewing or consuming it. Local communities may also benefit from the meat of safari-killed game or through the sales of meat from supervised culling operations.[19] A few programs directly give households a share of the income earned from renting the hunting rights in their areas (Peterson 1991).

The second type of benefit is linked indirectly to wildlife and consists of the standard goods dispensed by development projects. These items include health clinics, schools, maize grinding mills, teachers' houses, water wells, famine relief, roads, and bridges. Usually, the government finances these development projects out of the revenue it earns by taxing safari and tourist operations.

The last category of benefit relates to the empowerment of rural residents. Many programs have created institutions that allow a greater degree of local participation in decision making than previous policies. Participatory structures range from district-level committees whose members include traditional and political authorities to community-level committees comprised of local elites, committees that oversee culling operations, and women's groups. The powers of these new committees vary as well and include controlling the distribution of hunting licenses in their area, discussing land-use issues with government officials, selecting local candidates for training as wildlife scouts, advising government agencies on their most-preferred development projects, and disciplining project employees.

Because commercial wildlife businesses (tourism and safari hunting) and the programs' benefits depend on sizable animal populations, every program increases the level of enforcement for existing wildlife law. Escalating enforcement includes expanding the number of scouts in a particular area, ensuring scouts have adequate equipment and provisions, increasing supervision of scouts in the field, and providing incentives for scouts to perform their assigned duties (e.g., cash bonuses for arrests, patrols, and confiscated firearms). Since most African wildlife departments cannot afford additional civil servants on the payroll, they hire rural residents as daily employees. As well as being an inexpensive way to enlarge the scout force, these "village scouts" offer another direct advantage: they can be dismissed for failure to perform their duties much more easily than civil servants. Wildlife departments also assume that village scouts possess detailed knowledge about their area's wildlife and those who pursue it. Further, since the new programs try to recruit local youths as scouts, they assume that the income and status of being a government employee will terminate some of the residents' hunting activities.

The new programs' array of development projects, employment opportunities, and increased enforcement intends to change the incentives of the rural hunter and wildlife scout into the pattern depicted by the noncooperative game in figure 12.3.

In this game, the incentives linked to the scout's enforcement efforts have changed his ranking of outcomes. Cash bonuses and greater supervision induce the scout to prefer to enforce wildlife law in every case. The scout places his highest priority on preventing the killing of elephant and rhinoceros, since his supervisors usually consider the protection of these large mammals the department's most important duty. Moreover, because such kills are the most easily observed, the scout's failure is more easily detected.[20] If the new wildlife conservation program works as intended, the scout's new dominant strategy becomes Enforce.

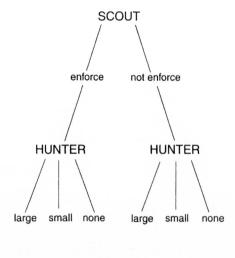

| scout payoffs | 2 | 3 | 1 | 6 | 5 | 4 |
| hunter payoffs | 5 | 3 | 1 | 6 | 4 | 2 |

Note: Numbers represent rank order for the players (i.e., 1 is the most preferred outcome).

Fig. 12.3. Extensive form game between wildlife scouts and local hunters intended by ADMADE

The new programs also change the preferences of the rural resident. The programs' package of benefits should make the strategy of not hunting superior to all other options. Given a new sense of ownership over wild animals, the resident should prefer that scouts enforce wildlife laws to protect these resources from any illegal offtakes, since the monies that drive his benefits derive from tourism and safari hunting, which require adequate animal populations. If the rural resident is forced to hunt by extraordinary circumstances, the new programs induce him to hunt smaller, more numerous animals rather than the larger mammals. The latter generate the greatest amount of income for the community. If these new programs' assumptions are appropriate, Not Hunting becomes the rural resident's dominant strategy.

The game in figure 12.3 produces the outcome most desired by the new programs, Not Hunt/Enforce.[21] If the new package of incentives works, local residents would choose not to hunt, enjoying instead new infrastructure, purchases of legally culled meat, jobs, and income offered by the new programs. Simultaneously, scouts would strictly enforce the laws and receive rewards for their successes. Such a scenario would boast gains for all interested groups. Conservationists and wildlife departments would achieve their goal of preserving wild animals. Communities living in these conservation areas would gain revenue for development. In addition, governments could shift the responsibilities for management and development to local communities while sharing safari hunting and tourism revenues.

As the following case study demonstrates, however, such an outcome did not occur in Zambia. Improved conditions of service under the AD-MADE program did significantly change the behavior of wildlife scouts, leading to an increased number of arrests and the likely decrease in the killing of larger mammals.[22] The program did not, however, induce rural residents to forgo hunting altogether. Rather, local hunters have changed their tactics and prey, and the taking of smaller species continues unabated.

4. Zambia's ADMADE Program

(a) ADMADE's Origins and Institutions

The Administrative Management Design for Game Management Areas (ADMADE) program evolved from a workshop held in the Luangwa Valley in 1983 (Dalal-Clayton 1987). The workshop's members included representatives from international donor agencies, conservation organizations, Zambian government departments, and the National Parks and Wildlife Services (hereafter referred to as the Wildlife Department). The participants agreed that wildlife was a resource critical for rural development and that successful wildlife management required participation by local communities. The workshop gave rise to a small experiment, the Lupande Development Project, that the Wildlife Department later expanded into the ADMADE program with the financial assistance of the World Wildlife Fund (USA) and the United States Agency for International Development (USAID).

ADMADE is an approach for conserving wildlife based upon the sustainable offtake of wild animals, with the proceeds of these harvests shared with local residents. ADMADE explicitly acknowledges the need to shift the colonial command and control philosophy of management to a more community-based approach in which local authorities are partners in wildlife decisions (Lewis, Kaweche, and Mwenya 1988; Mwenya, Lewis, and Kaweche 1990).[23]

ADMADE's institutional structures comprise three levels. An ADMADE Directorate, composed of senior Wildlife Department officers, manages the planning and daily operations of the program. Each Game Management Area (GMA) participating in ADMADE has a Wildlife Management Authority, whose members include district and local government officials, and is chaired by the District Executive Officer. A warden from the Wildlife Department serves as the Authority's executive secretary. The Authority's terms of reference include monitoring the offtakes of wildlife, approving the Department's allocation of hunting quotas, overseeing the revenues and expenditures resulting from the exploitation of wildlife, and choosing development projects to be financed by ADMADE revenues. ADMADE also creates Wildlife Management Sub-Authorities, whose area usually corresponds to chiefdoms. The chief chairs Sub-Authority meetings, which are organized and convened by the ADMADE Unit Leader, the Wildlife Department's staff member in charge of the local program. Sub-Authority duties include monitoring and solving wildlife problems at the local level, identifying community development projects for funding by the Authority, and facilitating the implementation of plans and projects.

ADMADE generates revenues from safari concession fees, hunting licenses, donor contributions, and those profits produced by its own activities, such as culling operations. The Wildlife Department holds these revenues in a revolving fund. Local communities are supposed to receive 35 percent of these revenues to spend on community development (Mwenya, Lewis, and Kaweche 1990). The rest of the funds support the Department's costs of administering ADMADE and other conservation endeavors.

Initially, ADMADE seemed to yield substantial benefits for both local communities and the Wildlife Department. On July 14, 1989, at a ceremony at State House, President Kaunda presented checks worth US$230,000 to chiefs representing ADMADE's first 10 Wildlife Management Authorities (Mwenya, Lewis, and Kaweche 1990). These revenues came from the 35 percent share of ADMADE revenues allocated to local community improvement. The Wildlife Department reported that illegal hunting had dropped 90 percent in the program's pilot area (Lewis, Kaweche, and Mwenya 1988). By 1990 ADMADE employed over 300 village scouts and had plans to train another 1,700. Chiefs actively cooperated with Department staff to implement ADMADE. A training center for village scouts—complete with dormitory, classrooms, and a library—was constructed in the Luangwa Valley. Officials from wildlife departments in other African countries visited Zambia to learn about the ADMADE program. Despite these apparent successes, the ADMADE program experienced serious difficulties that reflected its fundamental assumptions and design.

(b) ADMADE's Distribution of Benefits at the Community Level

(i) Chiefs

From the beginning, ADMADE identified chiefs, the "traditional rulers" in GMAs, as its link with rural communities and gave practical and historic reasons for this choice. Gaining the support of chiefs, in the Department's opinion, would overcome the alienation between Wildlife Department staff and rural residents over rights to wildlife. The Department's strategy was to reinvest chiefs with some of their former symbolic functions and distributive powers over wildlife (Lewis, Kaweche, and Mwenya 1988).[24]

ADMADE gave two important powers to the chief. First, the chief chaired the Wildlife Management Sub-Authority and appointed some of its members. Among the Sub-Authority's duties was the oversight of local wildlife and development projects that resulted from the community's share of revenue. Members of this committee also received a sitting allowance. Second, the chief selected the individuals who were to be trained and employed as village scouts, the backbone of the new enforcement strategy.

Despite the promise of these powers, the initial discussions with chiefs had to overcome deep suspicion as changes in wildlife policies had historically meant greater restrictions. Chiefs quickly realized the potential for their position in the new scheme, and those initially not involved in ADMADE later clamored to join. Although ADMADE and its consultants regarded this enthusiasm as acceptance for locally oriented conservation practices, another perspective suggests chiefs were acting out of self-interest. Chiefs used these initiatives to secure more power and resources for themselves rather than to facilitate local participation or wildlife conservation.[25] Because ADMADE policy did not stipulate clearly the composition or operation of the Sub-Authority, chiefs generally controlled its agenda and membership. Chiefs' ideas dominated the list of development projects, which were often situated within or near chiefs' compounds. Chiefs' relatives and loyalists obtained many of the new salaried positions, resulting in charges of nepotism. As the person responsible for selecting village scouts, the chief became the gatekeeper for access to a valuable commodity—a salaried position with law-enforcement powers. Given the value of employment in rural areas, chiefs vastly increased their potential for surveillance over the activities of others; chiefs were not timid in using these scouts when disputes arose. Some observers claimed the scouts felt more loyalty to their patron chief than to their de jure employers, the Wildlife Department. Predictably, villagers accused chiefs of favoritism in their selection of scouts and in the enforcement of wildlife laws.

Chiefs contended with ADMADE on several levels, but their concern with their own position and resources dominated their complaints.[26] They protested the Authority's control over the community portion of the ADMADE bank account, its ability to veto the Sub-Authority's choices for development projects, and its time-consuming routines for project approval. Chiefs asked for a higher proportion of ADMADE revenues to be allocated to their communities, while questioning the Wildlife Department's delays in delivering funds and lack of financial accountability. On the local level, some chiefs used their powers to enhance the fortunes of their kin, to dismiss some arrests, and to organize their own illegal hunting gangs.[27]

(ii) Rural Residents

ADMADE planned to provide the GMA resident both economic and political goods. The program's largest economic contribution to communities was through its employment, especially the village scout program. Initially, an average of 20 residents per GMA received jobs as village scouts. While low when compared to the pay of a regular Department scout, their salaries significantly boosted the local economy (USAID/Zambia 1993). Other residents received jobs as grinding mill operators or builders of the village scouts' temporary residences. ADMADE revenues also funded community projects. During 1989–92, the community's 35 percent share paid for 60 projects, including houses for teachers (23), classrooms (14), maize grinding mills (9) and rural health centers (7) (USAID/Zambia 1993). Finally, some ADMADE areas benefited from culling operations intended to provide local residents with a supply of legally acquired and inexpensive game meat.

ADMADE's designers also lauded the program's political goals, especially the decentralization of control over wildlife resources (Mwenya, Lewis, and Kaweche 1990). Decision-making powers over wildlife, heretofore monopolized by centralized governments, were to be returned to the local community. When local communities demonstrated sufficient skill in administering ADMADE, the Department aimed to hand over responsibility for wildlife management to the Wildlife Management Authority, with the Department providing only technical advice.

Most rural residents, however, experienced little direct economic or political benefit from ADMADE. Only about 2 percent of the gross profits from sport hunting reached rural communities (DeGeorges 1992). In some cases even this small amount never made it: one year the revolving fund accountant used an area's entire 35 percent community share to pay village scout salaries rather than fund development projects (USAID/Zambia 1993). The paucity of ADMADE income received by the Sub-Authority meant that local communities acquired relatively few projects: each Sub-Authority started an average of less than three projects in ADMADE's first three years. Many of these projects were only partially completed because of a continuing lack of funds and high inflation. ADMADE's own community development officers heard the same complaint "over and over again": ADMADE monies were just too small to make much of a difference (DeGeorges 1992, 30).

Villagers also resented the skewed distribution of ADMADE's economic goods. Most of the projects clustered around chiefs' residences. The chiefs' families and friends were the most likely recipients of jobs (USAID/Zambia 1993). In addition, most culling operations failed to sell the subsidized meat away from the ADMADE unit headquarters; the bulk of the sales went to denizens of nearby towns, not the countryside (Balakrishnan and Ndhlovu 1992; DeGeorges 1992).[28]

Further, GMA residents came to doubt ADMADE's promises of local participation.[29] The Wildlife Department had designated the Sub-Authority as the vehicle by which locals' voices could be heard. But not only did chiefs exert great control over the Sub-Authority, the institution had few stipulated powers: the Sub-Authority—and thus community members—had no say about hunting regulations, quotas, licenses, or fees (Mwenya, Lewis, and Kaweche 1990). The Sub-Authorities neither received information about the revenues their area contributed to the revolving fund, nor could they actually select the projects paid for by the fund. Moreover, residents did not participate in either the negotiation or the approval process for safari concession leases, even though the safari companies would hunt on the community's land and generate the bulk of its ADMADE income.

While ADMADE's structure allowed local communities through the Sub-Authority to administer aspects of local development projects—such as selecting workers and ordering materials—the program gave almost no power to community members over managing wildlife. Many residents of ADMADE areas perceived the program as a government sponsored employment program benefiting some rural residents at the expense of others (USAID/Zambia 1993).

(iii) Wildlife Scouts

ADMADE did expand formal participation in wildlife management through the village scout program. ADMADE intended village scouts to pursue three objectives: to eliminate all illegal hunting, to monitor wildlife populations, and to develop links with the community to foster a more positive attitude toward wildlife conservation (Mwenya, Lewis, and Kaweche 1990). ADMADE's designers hoped that local residents would trust village scouts—who had been selected from local communities—more than regular Department forces.

Despite the intentions of ADMADE to make the relationship between residents and Department staff more congenial, village scouts became unpopular in many communities. In those areas without effective unit leaders, village scouts' commitment waned, and residents indicted scouts for their poaching, stealing, fighting, witchcraft, and drunkenness. Where unit leaders closely supervised the scouts, zealous enforcement of wildlife regulations quickly estranged the village scout from his or her community. The invigorated pursuit of poachers led to an increase in villager complaints about scouts' harassment, just as enforcement activities had done for decades.[30]

(c) The ADMADE Game

The incentives offered by ADMADE did not change the behavior of scouts and rural residents in the intended manner. Figure 12.4 presents the outcome

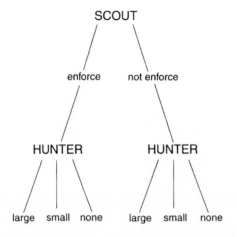

| scout payoffs | 3 | 2 | 1 | 6 | 5 | 4 |
| hunter payoffs | 4 | 3 | 6 | 1 | 2 | 5 |

Note: Numbers represent rank order for the players (i.e., 1 is the most preferred outcome).

Fig. 12.4. Extensive form game between wildlife scouts and local hunters after ADMADE implementation

of the noncooperative hunting game under ADMADE's actual incentives.

Despite ADMADE's benefits, the resident still ranks both hunting options higher than the Not Hunting strategy. The scout has, however, changed his behavior. The emoluments and supervision instituted by ADMADE have made enforcement the scout's most-preferred choice. Given that the resident knows that scouts will now enforce the law, he can no longer afford to hunt large game with impunity. Rather, hunting small game is now the best choice for the resident. Consequently, the result of the game in figure 12.3 is the equilibrium Hunt Small Game/Enforce.[31]

If the program's benefits changed the behavior for both resident and scout, as indicated by the payoffs in figure 12.3, we would expect certain characteristics or trends in data on the interaction between the two actors. First, residents' offtakes from wildlife should decrease as they enjoy AD-MADE's community benefits, especially the meat they purchase from culling stations. Ultimately, ADMADE would induce locals to forgo unlicensed hunting completely, resulting in no illegal offtake. Second, the number of arrests made by scouts should increase during the initial period of enhanced enforcement.[32] As residents accept ADMADE's new distribution of benefits and new regulations, however, the level of arrests should fall.

Evidence from the field suggests that these expected trends are not occurring. Data from the Munyamadzi GMA in the central Luangwa Valley, where ADMADE has been in effect for the past six years, reveal that local hunters have not reduced their offtakes (see table 12.1).[33] Besides noticeable drops during 1990 and 1992, caused by droughts that affected both human and animal behavior, offtakes returned to levels substantially the same or greater than before ADMADE's operation. Figure 12.5 displays the offtakes for hunters A, B, and C, for which there are yearly entries. Figure 12.5 clearly indicates that individual and total offtakes rebounded from 1990.

TABLE 12.1. Kilograms[a] of Meat Taken by Five Local Hunters (A–E), Munyamadzi Game Management Area, 1988–93

Year	A	B	C	D	E
1988	1,432	2,586	210	3,996	105
1989	1,362	2,249	225	2,580	315
1990	603	555	155	1,932[b]	60[c]
1991	1,170	1,787	260	—[d]	—
1992	1,120	603	200	10,395	—
1993	1,264[b]	3,930	239†	924	—

Source: Records of and interviews with local hunters (Marks, unpublished notes).
[a]Kilograms are calculated using estimated carcass yields for species and sex (50–60% of live weight). For example the carcass yield for a bull buffalo is estimated at 366 kg, a female impala at 27 kg, a male warthog at 53 kg.
[b]Results from partial year only.
[c]Hunter E left the area mid-1990.
[d]No data

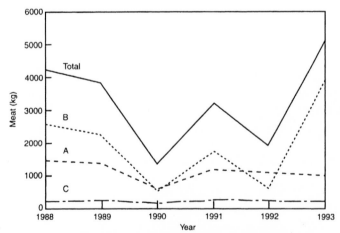

Fig. 12.5. Kilograms of meat taken by hunters A, B, and C

Given that the data from hunters A and C are incomplete, the figures presented may underestimate the actual extent of their hunting in 1993.[34]

These data, while limited, indicate that ADMADE has not prevented local hunting. Local hunters are able to maintain offtake levels by changing their routines and by switching to killing smaller mammals such as warthog, impala, and occasionally buffalo.[35] In addition, the data do not take into account the increase in snaring, a more universal and discreet practice whose offtakes are much more difficult to measure. Both Wildlife Department officials and conservationists contend that snaring has increased rapidly in AD-MADE areas.[36] These data support the outcome presented in figure 12.4: killing smaller animals by less-detectable methods is the most-preferred choice of the rural resident.

ADMADE has succeeded in increasing the number of arrests made by the Wildlife Department. The Department registered a record number of arrests and of firearms confiscated during the second year of ADMADE's implementation (Mwenya, Lewis, and Kaweche 1990). This trend is reflected both in the Munyamadzi GMA (table 12.2), and in the entire Luangwa area (table 12.3).[37] These numbers represent the commitment to putting more scouts in the field and rewarding them for enforcing wildlife laws, as intended by ADMADE from the outset.

TABLE 12.2. Law Enforcement by Wildlife Scouts, Munyamadzi Wildlife Subauthority, 1988–93

Year	Arrests	Convictions	Pending	Dismissed	Escaped
1988	4	3	0	0	1
1989	5	5	0	0	0
1990	8	7	0	0	1
1991	11	9[a]	0	0	2
1992	21	10	0	9	2
1993	19	4[b]	12	1	2

Source: Local records.
[a]Includes two punishments handed down by the chief.
[b]Includes three punishments handed down by the chief.

TABLE 12.3. Law Enforcement by the Zambian Wildlife Department, Luangwa Command, 1985–90

Year	Arrests	Convictions	Acquittals	Pending
1985	112	97	8	5
1986	149	116	5	28
1987	130	119	4	7
1988	155	137	2	16
1989	165	136	8	21
1990	197	183	7	7

Source: Zambian National Parks and Wildlife Department.

The number of arrests has not, however, diminished appreciably over time.[38] Although the Wildlife Department's record of arrests has not been published regularly over the past five years, officials indicate that the numbers have remained fairly static after their initial surge.[39] If ADMADE had succeeded in wooing locals out of illegal hunting, this fixed level of arrests must result from continuous pressure of hunters coming from outside the ADMADE areas (Lewis, Mwenya, and Kaweche 1991).[40] But Department officials, consultants, and rural residents all claim that outsider hunting has dropped off considerably, and locals still make up the vast majority of those arrested.[41] The fact that arrests have not declined over time implies that residents, despite ADMADE, continue to hunt.[42] Scouts admit that it does little good to put local hunters in jail since they "return to hunting as soon as they are released" (DeGeorges 1992, 54).

5. Discussion

While ADMADE's benefits and increased enforcement have changed locals' tactics and prey, the program has not convinced members of the rural community to conserve animals; neither has it stopped their illegal hunting practices. Although enhanced enforcement increases the chances for individuals to be apprehended, several factors sustain the rural residents' strong incentives to hunt.

First, case materials clearly demonstrate that the vast majority of residents receive little from the ADMADE program. Community projects and income-generating jobs are few, and little opportunity exists to exert any influence over the legal uses of wildlife resources. Individual returns from hunting far outweigh a resident's share of the goods that ADMADE delivers.

Second, not all residents respond to purely economic incentives.[43] The role of hunter is especially valued within lineage groups who live in areas where the presence of tsetse fly precludes the keeping of domestic stock and where wildlife remains relatively abundant. Descent groups sanction the hunter's duties, which include the protection of their property and lives in addition to providing meat and other goods.[44] A hunter seeks to distinguish himself, to build a following, and to become a leader, both materially and spiritually. For these hunters, killing animals becomes an identity, a way of becoming a man of local consequence and power. His goals are a lifetime quest emphasizing personal bravery, skill, and spiritual merit. The structure of benefits and opportunities of ADMADE clearly fails to meet these social goals of the lineage hunter.[45]

Third, ADMADE also fails to acknowledge the different values that individuals place on development projects. The logic of economic incentives assumes that individuals will choose that action that provides the most valuable payoff. Yet people may not secure equal levels of utility from the same project. A new maize grinding mill, for example, requires a cash payment to be used; cash may be too scarce to allow most rural residents to take advantage of the mill.[46] Further, the location and timing of a development project may also undermine its value. A health clinic, school, well, or grinding mill may provide less benefit to someone who does not live near it, even though the program may consider that individual part of a "target community." A delay in the delivery of a project also reduces its value and weakens the cause/effect linkage that ADMADE hopes to instill in community members. Thus, a benefit perceived by an outsider as a critical motivator may have little or no significant impact on local lives.

Even a more accurate targeting of higher value projects fails to resolve the incentive problem for ADMADE. The problem lies in the public nature of the goods it provides. Public goods are characterized by their nonexcludability, that is, their benefits are available to a group whether or not its members contribute to the provision of the good (Buchanan 1968; E. Ostrom, Schroeder, and Wynne 1993). The health clinics, schools, grinding mills, and most other projects offered by the new wildlife programs mimic public goods since their users are not regulated. If no one is excluded from consuming the good, then all rural residents—whether they poach or conserve wild animals—stand to gain from the project. Rational individuals would choose to receive the benefits from hunting activities while simultaneously enjoying the advantages offered by the public good. Since ADMADE fails to link its benefits to individual behavior within the village, the cessation of hunting is unlikely.

For the wildlife scout, however, ADMADE does connect reward with individual action. The program links cash, promotions, and job retention with actions such as arrests made, firearms confiscated, and days spent on patrol. Consequently, ADMADE successfully altered the incentive structure for the scout. The stepped-up enforcement of the wildlife scouts forced locals to change their preferred tactics and choices of prey.

Of course our models are simplifications of the realities of complex community-level interactions; scouts and residents confront a larger array of choices than the ones presented. One important alternative is collusion, a phenomenon endemic to the scout/resident relationship. Confronted with the social problems resulting from arresting a neighbor (e.g., exclusion from community life), and faced with the uncertainties of government support (ADMADE scouts in Zambia have waited up to six months for their pay and rations), some scouts choose to work with local hunters in their illegal efforts. Residents caught in the act of hunting try to bribe scouts, and scouts may encourage residents to change their stories to cover up collusion. Scouts and residents can gain from collaboration rather than conflict.

We also downplay the social tensions within the local community that can be generated by wildlife programs. One important effect of hiring village scouts is the generational conflict it foments. ADMADE recruits their scouts from among younger men. These men use their government job, which provides an income, a firearm, some education, and the power of arrest, for ascendancy within their lineages. Village scouts, therefore, present a challenge to their elders and to the institutions of traditional authority within local communities.

Neither do we consider the roles played by other groups important in the formulation and implementation of wildlife policy. These include international donors, owners of safari and tourist businesses, international and domestic conservation organizations, and professional hunters.

The models and evidence presented above do, however, help to explain why ADMADE and other community-based wildlife programs in Africa have not achieved their goals. These programs suffer from more than the lack of skilled personnel or poor implementation. Rather, their designers' assumptions and the institutions they created lead to weak outcomes for both conservation and community participation. Such outcomes are not limited to ADMADE: The Luangwa Integrated Resource Development Project (LIRDP) reduced elephant poaching by increasing the number of wildlife scouts and was popular with a committee of local authorities with which it shared revenue, but locals continued to take game (Marks, Mutelete, and Samson 1989). Zimbabwe's Communal Areas Management Program for Indigenous Resources (CAMPFIRE) seeks to include locals and to devolve ownership of natural resources to the rural community (Martin 1986). It also provides community-level benefits, offers few opportunities for local participation, and increases enforcement through scouts. After three years of operating in the Nyaminyami region, CAMPFIRE still contends with considerable levels of local poaching. Some reports claimed illegal hunting has increased after CAMPFIRE's introduction (Murombedzi 1992).[47] The NRMP project in Botswana based its "participatory" wildlife management on nonexistent community-level institutions and hoped to use existing nongovernment organizations (NGOs) to mobilize rural residents for its projects (Tropical Research and Development, Inc. 1993).

Rural residents rarely enter negotiations about wildlife resources as equal stakeholders. Local hunters are rarely considered primary users of wildlife; neither have they been the clients for whom wildlife is "managed." Therefore, as scouts boost enforcement, resident hunters continue to take wildlife and become increasingly wary of scouts when hunting. ADMADE and other new initiatives essentially ask rural hunters to trade direct access to wildlife—access which for them and their dependents means survival in marginal environments—for limited access to community-level infrastructure and minimal participation. Generally, rural residents find this exchange unappealing.

6. Conclusion

In this essay, we offer some reasons why community-based wildlife management schemes have failed to realize their intended goals. We argue that these programs fail to understand the significance of wildlife to rural residents and do not offer sufficient incentives for them to stop hunting. Wildlife conservation organizations and international donors currently provide substantial funds to community-based wildlife management initiatives, and, despite their weaknesses, this trend will likely continue. We conclude our analysis by raising issues we believe crucial to the success of effective and sustainable wildlife policy.

Many of the new conservation initiatives look remarkably similar to their colonial predecessors (Marks 1984; Murombedzi 1992). Ultimate ownership of wildlife remains in the hands of the state, whose agencies control de jure access to the animals.[48] Government employees make the most important decisions about wildlife-related revenues and quotas. Moreover, despite wildlife departments' calls for giving proprietorship over wildlife to the locals, few government officials push for this transfer of authority. Notwithstanding the material benefits that new wildlife programs deliver, locals remain disenfranchised from wildlife resources.

Africa's rural dwellers are accustomed to this exclusion. They have experienced nearly a century of separation from legal access to wildlife through regulation and enforcement by the state (Anderson and Grove 1987). They have been omitted from the negotiations regarding the use and fate of the resources with which they live. It is not surprising, therefore, that villagers are reluctant to accept the structure of incentives offered by these new programs.

The new initiatives have produced some successes. The poaching of the larger mammals, especially elephants, appears to have been checked in those areas with a heavy presence of scouts. Fresh streams of money and projects have flowed into some very impoverished areas. But closer examination reveals the fragility of these approaches: most are not financially self-supporting and require considerable and continuing support from donors. Thus, the goal of sustainability remains elusive. Given the fickle nature of aid money, and the immense pressure on African governments to provide social services, financial and political support for these conservation programs may quickly erode.

To their credit, the designers of these initiatives reintroduced the importance of the local community into a debate long stifled by quasi-military, hierarchical relationships. But to achieve the goals of conservation and participation, policies must include rural residents in more meaningful ways. As a first step, detailed studies must be undertaken to clarify the importance of wild animals to rural residents. Specifically, little is known about how rural residents make decisions regarding wildlife resources. The fact that ADMADE and other programs target their rewards at the level of community may reveal an inappropriate bias toward conceptualizing rural Africa in corporate terms. Different groups subsumed in the category "community" interact with wild animals in different ways: hunters will possess different incentive structures about wildlife resources from nonhunters, women, or resident civil servants. If the new schemes benefit safari hunters, scouts, lodge owners, and conservation consultants as individuals, perhaps such logic should be applied to rural residents as well. Currently, local hunters are individually targeted only for punishment.

One prospective difficulty is to incorporate the information from community-level studies into the state structures currently responsible for wildlife management. After decades of adversarial behavior, many African wildlife managers and conservationists believe that merely allowing villagers to become scouts and to share in the proceeds of wildlife are revolutionary undertakings. Yet these only begin to invest rural inhabitants with authority over wild animals.[49] More effective programs will entail a redistribution of power over wildlife, a topic that Government departments and conservation organizations generally find difficult to discuss, let alone relinquish (Murombedzi 1992).

The industrialization and urbanization that created lifestyle and land-use patterns conducive to tourism and recreational hunting in the West are unlikely to occur in the short term in rural Africa.[50] It is equally unlikely that African governments can afford the high costs required by conventional wildlife management to achieve successful conservation (Bell and McShane-Caluzi 1984). Since most wildlife in Africa still exists in nonprotected, inhabited areas, efficacious wildlife management continues to need local-level institutions that can effectively monitor and sanction behavior.

NOTES

1. We refer to the former British colonies and protectorates of Kenya, Tanzania, Zambia, Zimbabwe, and Botswana.

2. Much of the numerical data presented in this essay come from one geographical area, the Munyamadzi game management area in Zambia. Extension of this work to other parts of Zambia can be found in Marks, Mutelele, and Samson (1989); and Gibson (1999). Comparisons to Zimbabwe are found in Murombedzi (1992). Few published, in-depth analyses of rural residents' hunting practices and game consumption exist, despite the fact that "participatory programs" are meant to compensate locals for the losses they suffer by forgoing hunting. Unfortunately, most information emanates from the programs' managers or sponsors.

3. Marshall Murphee calls attention to the dangers of wildlife policy that use the modern rhetoric of including locals: " 'Participation' and 'involvement' turn out to

mean the co-optation of local elites and leadership for exogenously-derived pro-grammes: 'decentralization' turns out to mean simply the addition of another ob-structive administrative layer to the bureaucratic hierarchy which governs wildlife management" (Murphee 1990).

4. These choices are not necessarily mutually exclusive. Hunters may kill smaller or larger animals while seeking their opposite. We claim, however, that these catego-ries do account for general trends in species selection and hunting behavior.

5. For a detailed account of the effects of reduced budgets on the National Parks and Wildlife Department, see Gibson (1999).

6. The most valuable products during this period were ivory, rhino horn, and dried or smoked game meat.

7. Game theory allows researchers to model strategic action, i.e., action that is dependent on another's action. While we are aware that games simplify real-life situations, characterizing strategic interaction with games can lead to insights that descriptive analysis alone fails to highlight.

Noncooperation assumes that communication between the parties is irrelevant, impossible, or forbidden; complete information assumes that all parties know the full structure of the game and the payoffs associated with each outcome. We also present a one-shot rather than a repeated game. We defend this choice on two grounds. First, rural residents of developing countries generally have short time horizons. Second, because rural residents have experienced several policy changes or reversals over time, they know that they may be playing a new game tomorrow. For the fundamentals of nonco-operative game theory, an excellent introduction can be found in Kreps (1990).

8. Payoffs are individuals' calculations of expected utility for each outcome.

9. We base the ranking of choices by scouts and rural residents on observations of their behavior and interviews during fieldwork in Zambia conducted in 1966–67, 1973, 1988–89, 1991–92, and 1993 and in Botswana from 1991–92. Some of these observations are published in Marks (1976, 1984). These observations were extended by intensive interviews with hunters elsewhere in Zambia's Luangwa Valley (Marks, Mutelele, and Samson 1989).

10. Such strong incentives encouraged gun owners to lend their guns to others (particularly relatives) to provide meat, consumer goods, and crop protection for others.

11. Elephant, rhino, and hippo yield ivories or horns that could be readily sold on the black market in addition to meat.

12. Not enforcing the law meant overlooking illegal practices and colluding with others.

13. Hunters from outside rural areas usually boast better arms (semi- or fully auto-matic) than locals. In the Luangwa Valley, many local hunters still use muzzle-loading guns. Outside hunters are also more likely to shoot at scouts.

14. Of course, scouts often did more than just fail to uphold the law—they were often poachers in their own right. Allocating a small number of ammunition rounds to scouts was one way wardens tried to limit the scouts' tendencies to hunt illegally, given the difficulty of effective supervision in the field.

15. In a transfer to a new location, scouts lose the valuable social capital they might have built up while in their former area.

16. The game in figure 12.1 is dominance solvable. The scout will always choose to not enforce; the hunter will always choose to hunt large game. These strategies result in the Not Enforce/Hunt Large Game outcome.

17. Of course, numerous nonlocal commercial hunters and civil servants were also an important part of this wave of illegal hunting.

18. Among these were the CAMPFIRE (Communal Areas Management Program for Indigenous Resources) program in Zimbabwe, LIRDP (Luangwa Integrated Resources Development Project), and ADMADE (Administrative Management Design for Game Management Areas) in Zambia, and NRMP (Natural Resources Management Project) in Botswana.

19. "A cull" is a wildlife management euphemism for a legal, selective kill of wild animals, usually carried out by wildlife department staff.

20. Population trends for the larger animals, such as elephant, rhino, and hippo, have been documented for years in wildlife departments. Unlike the smaller animals, larger species' dark colors and great size make them easy to count by air, and wildlife managers use aerial transects to estimate population change over time. Moreover, the death of the bigger animals attracts a host of vultures and other scavengers (whose feasting lasts for days) and produces considerable skeletal material.

21. This game is also dominance solvable: the scout always prefers to enforce, and the resident always prefers not to hunt. The result of these two strategies is the outcome Not Hunt/Enforce.

22. Circumstances confound any attempt to provide definitive evidence. First, as elephants and other large mammals have become scarcer, the surviving individuals and groups are increasingly more difficult to find and approach. Second, scouts' stepped-up enforcement activities have conditioned local hunters to be reticent to share information on the location and movements of animals. Some programs, including ADMADE, base their claims of reduced poaching on the number of carcasses or new skeletons found, rather than counting living animals within statistically meaningful frames. Criticism of animal census techniques in Africa can be found in Adams and McShane (1992).

23. The Project Agreement signed by the governments of the United States and Zambia states, "the (ADMADE) project design strategy recognizes that when resident communities can realize a direct share in the benefits arising from legal natural resource exploitation, an incentive for conserving that resource base is created. It also recognizes that communities must be able to control access to the resource in order to receive future benefits from short-term conservation actions." See USAID, *Project Grant Agreement* (1990), p. 2 of Annex 1.

24. Despite ADMADE's claims, the Northern Rhodesia government had a long history of catering to chiefs under indirect rule. For wildlife management in particular, Pitman (1934) suggested chiefs as the important players in his promotion of new game policies; these same ideas were in a memorandum written by Vaughan-Jones (1938) that led to the establishment of the Department of Game and Tsetse Control in 1940. The idea behind the Controlled Area concept, which later became the Game Management Areas in Zambia, was that the proceeds from wildlife within them would be under the "Native Authorities" who would manage it as a meat supply for

their subjects (Vaughan-Jones 1948, 40). During the 1950s, several chiefs had their own game reserves that catered to tourists. Chiefs kept all the proceeds as well as some from safari concessions (Government of Northern Rhodesia 1951). Although the large game drives in the past commanded by important chiefs were well publicized events in western journals, lesser chiefs sought to extract tribute mainly from the larger mammals killed within their territories. The delivery of these commodities to the chiefs depended upon his/her capacity to enforce them. Chiefs rarely sought tribute from the smaller, more abundant wildlife as there were a host of lineage norms and customs governing its distribution (Marks 1976).

25. Such behavior by chiefs is not new. Chiefly power derives from managing a complex web of patron-client relationships. To maintain their authority over this network, chiefs act to secure their control over local resources and opportunities. See *inter alia* Hall (1976), Nawa (1990), Gluckman (1968), Gann (1958), Prins (1980), and Chipungu 1988). Bratton (1980) provides a more contemporary account of how chiefs attempt to manipulate development projects.

26. Some observers would claim that this contention is a healthy sign, that is, that some locals have become empowered by the program.

27. NPWS officers knew that chiefs would try to exploit ADMADE. Few local level institutions, however, exist in Zambia's rural areas upon which NPWS officers could build a wildlife program. NPWS officers believed that the benefits of incorporating chiefs within ADMADE outweighed their potential manipulation of AD-MADE resources.

28. Most local residents, who are not on salary, do not purchase meat even at comparatively low prices and instead depend on relatives to procure it from the bush for them. Most meat produced by culling stations in the Luangwa is consumed not by local residents but by civil servants and others on salary.

29. For a similar perspective from Zimbabwe's rural residents, see Murombedzi (1992).

30. The following story illustrates the interpretive problems in the scout/resident relationship.

Nkumbula heard an elephant feeding in his sorghum field during the night and left his house to chase it away. Instead of fleeing, the elephant charged the unarmed Nkumbula and followed him back into the village where it began to eat the thatched roof from the house where the young unmarried girls were sleeping. Nkumbula entered the house of the headman, away to consult with an African doctor, loaded a muzzle-loading gun and fired at the elephant, which had remained in the village. It fell, and the next morning, Nkumbula and the rest of the villagers found it still breathing. After killing it with another shot. Nkumbula reported the incident to the scouts some distance away. The scouts came, inspected the dead animal, discovered it was tuskless and had been wounded by three shots a month previously. They ordered the villagers to butcher the carcass and deliver all the meat to their camp. The scouts arrested Nkumbula, charging him with killing an elephant and for using a weapon registered in someone else's name. In Magistrate's Court, Nkumbula was sentenced to five years' hard labor, but the following year, he was released with his case before the Appeals Court. "Many people say that the game guards are not good. How can they imprison a young man who saved the young girls from the elephant?" (Munyamadzi notebook)

31. This game is also solved by using the concept of dominance. The scout has the dominant strategy of Enforce. Knowing this, the resident chooses his best option given the scout will enforce the law. The resident chooses to Hunt Small Game.

32. Such a decrease in arrests could also result from a change of local hunting tactics, making it more difficult for scouts to detect these activities, such as a switch from hunting with guns to snares.

33. The Munyamadzi GMA is one of Zambia's prime safari sites and a large revenue earner for ADMADE. This 3,000 sq. km. territory is surrounded on three sides by national parks and on the fourth by a steep escarpment.

34. Further, since these hunters operate in the vicinity of a culling station, we expect others' offtakes to be higher in areas further from such concentrations of scouts.

35. Records kept by local hunters show the steady changes in their choices of prey and search tactics due to the patrolling efforts of scouts.

36. From interviews in 1991 with *inter alia* Dale Lewis, ADMADE Technical Advisor; Ackim Mwenya, NPWS Director; Gilson Kaweche, NPWS Deputy Director; Clive Kelly, Chinzombo Lodge; Phil Berry, Chinzombo Lodge; Norman Carr, Kapani Lodge; and the authors' personal observations.

37. Like other areas, Munyamadzi has suffered from a succession of weak ADMADE Unit Leaders, the program's chief executive officer at the Subauthority level. The first four appointees to this post were dismissed in as many years for corruption, loss of funds, public drunkenness, abuse of local residents, and failure to implement the program. Such difficulties likely influenced the comparatively few arrests trade during the initial years of ADMADE. During 1991–92, arrests increased—due in part to the expiration of an amnesty for possession of unregistered firearms instituted by the chief.

38. Interview with Crispin Siachibuye, Chief Prosecutor NPWS, Chilanga, Zambia, November 12, 1991.

39. See note 36.

40. Arrests would also not decline if rural residents were uninformed about the ADMADE program. Given locals' generally rapid response to economic incentives (e.g., agriculture) it is unlikely that lack of information is the cause for the sustained level of arrests made by the Wildlife Department.

41. See note 36.

42. The following account illustrates how some locals thwart arrests.

Village scouts were suspicious of Chibeza, believing he had killed more than the single buffalo he had licensed. They asked a woman, an unrelated neighbor of Chibeza, to serve as an informer and report whenever Chibeza was seen with meat. When she saw him returning home with meat (an impala) she informed the scouts that he had killed another buffalo. The scouts entered his home and found fresh impala together with dried warthog meat in a basin. The scouts took the basin and confronted Chibeza, who was drinking beer at their camp, with the evidence. Chibeza accepted the evidence, claiming the impala had been killed by his father on license and the warthog by his son, who, in conjunction with the scouts' sons, had acquired the warthog with dogs several days previously. He had a license to kill a warthog. Despite his statements, Chibeza was arrested, handcuffed, and brought with his gun into camp. Chibeza's father, retired from work in the copper mines, showed the scouts his properly endorsed license for the impala, offered to reveal its kill site, and was prepared to testify in court for his son. Scouts figuring they would lose the case, released Chibeza after six days in custody. They fixed the hammer on his muzzle-loading gun when Chibeza noted evidence of their tampering with it. The woman, who turned Chibeza in, complained to the scouts that she was afraid now for her life. Chibeza's relatives had turned against her, and her house was destroyed by a bush fire three days after his arrest. Chibeza responded that he bore no grudges. Following his release, he married another wife, a cross-cousin, living in a village more removed from the scrutiny of the village scouts. (Munyamadzi notebook)

43. See for example Richard Bell (1987, 95). For the social significance of hunting in Zimbabwe, see Murombedzi (1992); for Zambia, see Marks (1976, 1984); for Botswana, see Lee (1979), and Biesele (1993).

44. To fill this important role, older generations choose and promote youths. For many in the Luangwa Valley, the process by which a boy becomes a lineage hunter begins with a dream interpreted by an elder as a summons from an ancestor who had served in this capacity. The ancestor becomes the youth's lifetime guardian (Marks 1976, 1979).

45. For example, when asked what would happen if local hunting was eliminated completely, a lineage hunter responded in the following way.

[Explication begins with shortness of breath, look of puzzlement, expression of disbelief]

I won't feel alright. I always dream of killing animals and of making my kinfolk happy. If I have to stop hunting, I will become worried, perhaps mad. My deceased grandfather always tells me through dreams that I should continue hunting. (Marks, Mutelele, and Samson 1989)

46. Cash is a scarce resource in many rural areas. Participation in or use of projects that require cash payment is a sure way of reducing the number of people willing to take part.

47. While CAMPFIRE is a national policy, the institutions through which it is implemented at the local level vary considerably. In areas where locals receive direct benefits and actively participate in decision making about wildlife, CAMPFIRE has achieved much better results.

48. One exception is found in Zimbabwe, where due to the 1975 Parks and Wild-

life Act, commercial farmers were given the right to manage the wildlife on their private land. See Martin (1994).

49. Different institutional arrangements and incentives would allow a more extensive role for local hunters in wildlife management. See Marks (1994a,b).

50. See Dunlap (1988) for a discussion of how the relationships between wild animals and people changed with urbanization in the United States.

REFERENCES

Abel, N. and P. Blaikie, "Elephants, people, parks and development: The case of the Luangwa Valley, Zambia," *Environmental Management,* Vol. 10, No. 6 (1986), 735–51.

Adams, J. S. and T. O. McShane, *The Myth of Wild Africa* (New York: Norton, 1992).

Alpert, P. and P. A. DeGeorges, "Midterm evaluation of the Zambia Natural Resources Management Project." Mimeo (July 1, 1992).

Anderson, D. and R. Grove, "The scramble for Eden: Past, present and future in African conservation," in D. Anderson and R. Grove (Eds.), *Conservation in Africa: People, Policies and Practice* (Cambridge: Cambridge University Press, 1987).

Balakrishnan, M. and D. Ndhiovu, "Wildlife utilization and local people: A case-study in Upper Lupande Game Management Area, Zambia," *Environmental Conservation,* Vol. 19, No. 2 (1992), 135–144.

Bell, R. H. V., "Conservation with a human face: Conflict and reconciliation in African land use planning," in D. Anderson and R. Grove (Eds.), *Conservation in Africa: People Policies and Practice* (Cambridge: Cambridge University Press, 1987).

Bell, R. H. V. and F. B. Lungu, "The Luangwa Integrated Resource Development Project. Progress of Phase I and Proposals for Phase II." Mimeo (Chipata, Zambia: LIRDP, August 11, 1986).

Bell, R. H. V. and E. McShane Caluzi (Eds.), *Conservation and Wildlife Management in Africa* (Washington, DC: U.S. Peace Corps, 1984).

Biesele, M., *Women Like Meat* (Bloomington, IN: Indiana University Press, 1993).

Bonner, R., *At the Hand of Man* (New York: Alfred A. Knopf, 1993).

Brandon, K. E. and M. Wells, "Planning for people and parks: Design dilemmas," *World Development,* Vol. 20, No. 4 (1992), 557–70.

Bratton, M., *The Local Politics of Rural Development: Peasant and Party-State in Zambia* (Hanover, NH: University Press of New England, 1980).

Buchanan, J. M., *The Demand and Supply of Public Goods* (Chicago: Rand McNally, 1968).

Child, G. and B. Child, "Economic characteristics of the wildlife resources," in *Proceedings of the International Symposium and Conference on Wildlife Management in Sub-Saharan Africa* (Harare: IGF/CIC, 1987).

Chipungu, S. N., *The State, Technology and Peasant Differentiation in Zambia: A Case Study of the Southern Province, 1930–1986* (Lusaka, Zambia: Historical

Association of Zambia, 1988).

Dalal-Clayton, D. B. (Ed.), *Proceedings of the Lupande Development Workshop* (Chipata, Zambia: Government of the Republic of Zambia, 1987).

DeGeorges, P. A., "ADMADE: An evaluation today and the future policy issues and direction." Mimeo (Chilanga, Zambia: Report prepared for USAID and NPWS, 1992).

Development Alternatives, Inc., "Regional National Resources Management Project." Mimeo (690–0251) (Washington, DC: USAID, 1989).

Dunlap, T. R., *Saving America's Wildlife* (Princeton, NJ: Princeton University Press, 1988).

Eltringham, S. K., *Wildlife Resources and Economic Development* (New York: John Wiley and Sons, 1984).

Gann, L. H., *The Birth of a Plural Society* (Manchester: Manchester University Press, 1958).

Gibson, Clark. *Politicians and Poachers: The Political Economy of Wildlife Policy in Africa.* (Cambridge: Cambridge University Press, 1999).

Gluckman, Max, *Essay on Lozi Land and Royal Property*, Rhodes-Livingstone Institute Papers No. 10 (Manchester: Manchester University Press, 1968).

Government of Northern Rhodesia, *Game and Tsetse Control Report 1951* (Livingstone: Government Printer, 1952).

Hall, Richard, *Zambia 1890–1964: The Colonial Period* (London: Longman, 1976).

Kiss, A. (Ed.), *Living with Wildlife: Wildlife Resource Management with Local Participation in Africa* (Washington, DC: The World Bank, 1990).

Kreps, D. M., *A Course in Microeconomic Theory* (Princeton, NJ: Princeton University Press, 1990).

Leader-Williams, N. and S. D. Albon, "Allocation of resources for conservation," *Nature,* Vol. 366, No. 6199 (December 8, 1988), 533–35.

Leader-Williams, N. and E. S. Milner-Gulland, "Policies for the enforcement of wildlife laws: The balance between detection and penalties in Luangwa Valley, Zambia." *Conservation Biology,* Vol. 7 (September, 1993), 611–17.

Lee, R. B., *The Kung San: Men, Women and Work in a Foraging Society* (New York: Cambridge University Press, 1979).

Lewis, D. M. and G. B. Kaweche, "The Luangwa Valley of Zambia: Preparing its future by integrated management," *Ambio,* Vol. 14 (1985), 362–65.

Lewis, D. M., G. B. Kaweche and Ackim N. Mwenya, "Wildlife conservation outside protected areas, lessons from an experiment in Zambia" (Nyamaluma Conservation Camp: National Parks and Wildlife Services, May 1988).

Lewis, D. M., A. Mwenya and G. B. Kaweche, "African solutions to wildlife problems in Africa: Insights from a community-based project in Zambia," *International Journal on Nature Conservation,* Vol. 7, No. 1 (1991), 10–23.

Luangwa Integrated Resources Development Project, "Proposals for Phase Two." Mimeo, Document 3 (Chipata, Zambia: LIRDP, 1987).

Lunga, F. B. and R. H. V. Bell, "Luangwa Integrated Resource Development Project, A presentation to the National Assembly." Mimeo (Chipata, Zambia: LIRDP, December 1990).

MacKinnon, J., K. MacKinnon, G. Child and J. Thorsell, *Managing Protected Areas in the Tropics* (Gland, Switzerland: IUCN/UNEP, 1986).

Marks, S. A., "Managerial ecology and lineage husbandry: Environmental dilemmas in Zambia's Luangwa Valley," in M. Hufford (Ed.), *Conserving Culture: New Discourse on Heritage* (Champaign, IL: University of Illinois Press, 1994a), 111–21.

Marks, S. A., "Local hunters and wildlife surveys: A design to enhance participation," *Journal of African Ecology*, Vol. 32 (1994b), 233–54.

Marks, S. A., *The Imperial Lion: Human Dimensions of Wildlife Management in Central Africa* (Boulder, CO: Westview Press, 1984).

Marks, S. A., "Profile and Process: Subsistence hunters in a Zambian community," *Africa*, Vol. 49 (1979), 53–67.

Marks, S. A., *Large Mammals and a Brave People: Subsistence Hunters in Zambia* (Seattle, WA: University of Washington Press, 1976).

Marks, S. A., H. Mutelele and K. Samson, "Interviews with local hunters, Lupande Game Management Area." Mimeo Consultancy Report (Chipata, Zambia: LIRDP, August 1989).

Martin, R. B., *Communal Areas Management Programme for Indigenous Resources (CAMPFIRE)* (Harare: Department of National Parks and Wild Life Management, 1986).

Martin, R. B., "The influence of governance on conservation and wildlife utilisation." Mimeo (Harare, Zimbabwe: Department of National Parks and Wild Life Management, 1994).

Motshwari Game Limited, "Feasibility study on the utilization of wildlife resources." Mimeo (Lusaka, Zambia: Government of the Republic of Zambia, 1981).

Munyamadzi Game Management Sub-Authority, "Minutes" (1988–93).

Murombedzi, J. C., "Decentralization or recentralization? Implementing CAMPFIRE in the Omay Communal Lands of the Nyaminyami District," Working Paper No. 2 (Harare: University of Zimbabwe, Centre for Applied Social Science, 1992).

Mwenya, A. N., G. B. Kaweche and D. M. Lewis, "Administrative Management Design for Game Management Areas (ADMADE)" (Chilanga, Zambia: National Parks and Wildlife Service, January 1988).

Mwenya, A. N., D. M. Lewis and G. B. Kaweche, "ADMADE: Policy, background and future" (Lusaka, Zambia: National Parks and Wildlife Service, 1990).

Murphee, M. W., "Communities and institutions for resource management," Paper presented to the National Conference on Environment and Development (Maputo, Mozambique: October 7–11, 1991).

Murphee, M. W., "Community conservation and wildlife management outside protected areas—southern African experiences and perspectives: Implications for policy formation. A report to the African Wildlife Foundation." Mimeo (August 22, 1990).

National Parks and Wildlife Services, "Proceedings of the First ADMADE Planning Workshop," Mimeo (Lusaka, Zambia: National Parks and Wildlife Services, December 1988).

Nawa, N., "The role of traditional authority in the conservation of natural resources in the Western Province of Zambia, 1878–1989." MA thesis (Lusaka, Zambia: University of Zambia, 1990).

Ostrom, E., L. Schroeder and S. Wynne, *Institutional Incentives and Sustainable Development* (Boulder, CO: Westview Press, 1993).

Peluso, N. L., "Coercing conservation: The politics of state resource control," in R. Lipshutz and K. Conca (Eds.), *The State and Social Power in Global Environmental Politics* (New York: Columbia University Press, 1993), 46–70.

Peterson, J. H. Jr., "CAMPFIRE: A Zimbabwean approach to sustainable development and community empowerment through wildlife utilization." Mimeo (Harare: University of Zimbabwe, Centre for Applied Social Sciences, 1991).

Pitman, C. R. S., *A Report on a Faunal Survey of Northern Rhodesia with Especial Reference to Game, Elephant Control, and National Parks* (Livingstone: The Government Printer, 1934).

Prescott-Allen, R. and C. Prescott-Allen, *What's Wildlife Worth?* (London: Earthscan Publications, 1982).

Prins, G., *The Hidden Hippopotamus: Reappraisal in African History: The Early Colonial Experience in Western Zambia* (London: Cambridge University Press, 1980).

Tropical Research and Development, Inc., "Midterm evaluation of the Botswana Natural Resources Management Project (690–0251)." Mimeo (Washington, DC: USAID, July 1993).

United States Agency for International Development/Zambia, "Project Paper Supplement for the Natural Resources Management Project" (including Technical and Social Analyses). Mimeo (Washington, DC: USAID, 1993).

United States Agency for International Development/Zambia, "Project Grant Agreement between the Republic of Zambia and the United States of America for Natural Resources Management (USAID Project Number 690–0251.11)." Mimeo (Washington, DC: USAID, January 16, 1990).

Vaughan-Jones, T. G. C., "A short survey of the aims and functions of the Game and Tsetse Control Department of Northern Rhodesia," *Rhodes-Livingstone Institute Journal,* Vol. 6 (1948), 37–47.

Vaughan-Jones, T. G. C., *Memorandum on Policy Concerning the Foundation of a Game Department and the Conservation of Fauna in Northern Rhodesia* (Lusaka, Zambian National Archives, file SEC1/994, 1938).

Western, D., "Amboseli National Park: Human values and the conservation of a savanna ecosystem," in J. A. McNeely and K. R. Miller (Eds.), *National Parks, Conservation and Development: The Role of Protected Areas in Sustaining Society: Proceedings of the World Congress on National Parks* (Washington, DC: Smithsonian Institution Press, 1984).

Western, D., "Amboseli National Park: Enlisting landowners to conserve migratory wildlife," *Ambio,* Vol. 11, No. 5 (1982), 302–8.

The World Conservation Union (IUCN), "The world conservation strategy" (Gland, Switzerland: The World Conservation Union, 1980).

Young, O., *Resource Regimes: Natural Resources and Social Institutions* (Berkeley, CA: University of California Press, 1982).

CHAPTER 13

Irrigation Institutions and the Games Irrigators Play: Rule Enforcement on Government- and Farmer-Managed Systems

Franz J. Weissing and Elinor Ostrom

Introduction

A central puzzle addressed here is how can theorists capture important aspects of the complexity found in field settings without losing the rigor of formal analysis. A second organizing theme is the analysis of nested structures or smaller games that are played within larger institutional structures. Two types of larger structures are identified for special study: (1) hierarchical structures and (2) networks. In an earlier essay, we analyzed the stealing and monitoring behavior of participants in an irrigation game that had elements common to many "crime and punishment" or "cheating and deterrence" games (Weissing and E. Ostrom 1991a). In this effort we embed this core game into two hierarchical contexts in order to examine some of the key themes of this volume.

First, we present the core game in an isolated setting where the only individuals involved are the farmers themselves who rotate into positions that give them the opportunity to (1) withdraw an authorized amount of water or to steal and (2) to monitor what is happening or not to monitor. Then we introduce specialists: irrigators who monitor at a different rate. Next, we introduce an official guard. We distinguish two different types of official guards. Then we investigate how motivating the guard affects strategies and outcomes.

A key question, then, is how to represent different types of hierarchical structures. We have chosen to represent different structures by the constraints we have placed on the relationships among payoff parameters. We overtly examine the difference between "integrated" and "disassociated" guards. In a farmer-managed system using integrated guards, the hierarchy is "inverted," with the farmers as principals on the higher level and the guards as agents on

Originally published in *Games in Hierarchies and Networks: Analytical and Empirical Approaches to the Study of Governance Institutions*, ed. Fritz W. Scharpf (Frankfurt: Campus Verlag; Boulder, CO: Westview Press, 1993): 387–428. Reprinted by permission of Campus Verlag and the authors.

the lower level. In a government system using disassociated guards, the hierarchy contains three levels where the principal is a government, the guards are hired as agents by that government, and the farmers are on the lowest level. In the last section we will focus on the empirical relevance of our models.

1. The Core Game: Irrigation Systems without Guards[1]

For all of our models, we assume there is a population of farmers who share a single irrigation system that is the main source of water supply for their crops. The authorization to take water is rotated among the farmers according to a predetermined formula. At each point in time, only one farmer is in the position of a turntaker (TT) who is authorized to take a certain quantity of water. The other N farmers wait for their turn to take water and are called turnwaiters (TW). A potential conflict of interest exists between the turntaker and the turnwaiters. The turntaker is in a position to extract more than the authorized amount of water from the system, while the turnwaiters are negatively affected by a stealing event.

The strategic decision facing the turntaker is whether to steal additional water from an irrigation canal (S) or to take the authorized amount of water ($\neg S$). Stealing is defined as taking actions that are not allowed by the rotation system in use. We assume that the turntaker always has a positive incentive to steal water. The only thing that deters a turntaker from stealing is the possibility that the stealing event might be detected by others and punishment imposed. The strategic choices available to a turnwaiter are to monitor (M) what the turntaker is doing or to refrain from monitoring ($\neg M$). We assume that monitoring is the only way that a stealing event can be detected and that the effort to steal, if detected, can be prevented effectively. The turnwaiters cannot, however, coordinate their behavior. In fact, each player (including the turntaker) has to take his decision without knowing what the other players are planning to do.

In addition to the pure strategies (S and $\neg S$ for the turntaker and M and $\neg M$ for the turnwaiters), we also consider some intermediate form of behavior that is typically represented by a real number $\sigma (0 \leq \sigma \leq 1)$ for the turntaker or a real number μ ($0 \leq \mu \leq 1$) for a turnwaiter. σ and μ are either interpreted as mixed strategies or—equivalently—as behavioral tendencies. According to the first interpretation, σ corresponds to the probability that the pure strategy S is chosen by the turntaker. Alternatively, σ may be viewed as a stealing rate, that is, as a level of stealing that is intermediate between the two extreme options of not stealing at all ($\sigma = 0$) or stealing at the maximal rate ($\sigma = 1$). Similarly, μ is either interpreted as the probability that a turnwaiter chooses M or as a monitoring rate that is intermediate between not monitoring and monitoring with maximal effort.

Considering intermediate levels of stealing and monitoring introduces a probabilistic element into our models. Another probabilistic element is related to the fact that monitoring is not assumed to be perfect. Even if the turntaker takes more water than authorized, a monitoring turnwaiter will only detect stealing with a certain probability, the detection probability α. For a turnwaiter monitoring at an intermediate level μ, a stealing effort will be detected with probability $\alpha\mu$ and it will remain undetected with probability $1 - \alpha\mu$. Since the turnwaiters monitor independently from one another, detection probabilities are combined in a multiplicative way. If, for example, v of the N turnwaiters monitor at a rate μ whereas the other turnwaiters do not monitor at all, a stealing event will not be detected with probability $(1 - \alpha\mu)^v$, and it will be detected (by at least one of the v turnwaiters) with probability $1 - (1 - \alpha\mu)^v$.

Each strategy combination of the $N + 1$ players introduces a probability distribution over the different outcomes of the conflict. We assume that the preferences of the players concerning the outcomes may be represented by payoff parameters in such a way that the expected value of payoffs is a good representation of the individual evaluation of a probability distribution over the basic outcomes (i.e., payoffs have the properties of von Neumann and Morgenstern's utility concept). Under this assumption, it is possible to calculate for each player the expected payoff of a strategy combination. When we talk about behavioral tendencies (instead of mixed strategies), identical payoff functions are obtained if we assume that the expected payoff of a player is a linear function of the strategic parameter of any participant.

In the core game, three different outcomes are relevant for the turntaker. If he does not steal, he gets the status quo amount of water, which is represented by a payoff of zero. If he steals and stealing remains undetected, he gets the benefits of an additional amount of water which are characterized by the positive payoff B. If, however, a stealing effort is detected, the turntaker cannot extract any extra benefit but incurs some impairment. This punishment received when detected is represented by the negative payoff $-P_M$.

The turnwaiters have four different outcomes. Again, the status quo outcome corresponds to a payoff of zero. Losing water due to stealing is worse than the status quo; therefore it is represented by a negative number $-L_M$. No water is lost if the stealing event is detected. Instead, there will be a "reward" (see Weissing and E. Ostrom 1991a) for the turnwaiters that is characterized by a number R_M. The costs of monitoring are represented by a negative number $-C_M$. In order to keep the number of parameters as small as possible, we assume that $-C_M$ corresponds to fixed costs that may be combined additively with the other payoff parameters 0, $-L_M$, and R_M.

The resulting payoff configurations for the turntaker and one of the turn-waiters are summarized in Table 13.1. The expected payoffs for pure and mixed strategies of player X will only be derived for the special case that all turnwaiters not identical to X choose the same strategy μ. Let us first focus on the turntaker. If he does not steal, he gets a payoff of zero. If he tries to steal water, the stealing effort will not be detected with probability $(1 - \alpha\mu)^N$ and it will be detected with probability $1 - (1 - \alpha\mu)^N$. Accordingly, the expected payoffs for the two pure strategies of the turntaker are given by

$$F_{TT}(\neg S, \mu) = 0, \tag{1a}$$

$$F_{TT}(S, \mu) = (1-\alpha\mu)^N \cdot B - [1-(1-\alpha\mu)^N] \cdot P_M. \tag{1b}$$

S is a better reply against μ than $\neg S$ ($S \succ \neg S$) if and only if $F_{TT}(S,\mu)$ is larger than $F_{TT}(\neg S, \mu)$ or, equivalently:

$$S \succ \neg S \;\leftrightarrow\; (1-\alpha\mu)^N > P_{rel}, \tag{2}$$

where
$$P_{rel} := \frac{P_M}{B + P_M} \tag{3}$$

denotes the *relative* cost of a detected stealing event for the turntaker.

TABLE 13.1. Payoff Configurations in the Core Game[a]

Decision of TW	Event	Probability of event	Decision of TT	
			S	$\neg S$
M	Somebody detects	$1 - (1-\alpha)(1-\alpha\mu)^{N-1}$	$R_M - C_M, -P_M$	$-C_M, 0$
	Nobody detects	$(1-\alpha)(1-\alpha\mu)^{N-1}$	$-L_M - C_M, B$	$-C_M, 0$
$\neg M$	Somebody detects	$1 - (1-\alpha\mu)^{N-1}$	$R_M, -P_M$	$0, 0$
	Nobody detects	$(1-\alpha\mu)^{N-1}$	$-L_M, B$	$0, 0$

[a]Payoff configurations for a turnwaiter (listed first) who has to decide between M and $\neg M$ and for the turntaker (listed second) who has to decide between S and $\neg S$. The probabilities of the basic outcomes depend on the turnwaiter's detection probability α and on the probability $(1 - \alpha\mu)^{N-1}$ that the other $N - 1$ turnwaiters who are all monitoring at rate μ do not detect a stealing attempt.
  In contrast to our earlier essay (Weissing and E. Ostrom 1991a), we concentrate on irrigation games where all turnwaiters are treated symmetrically. In addition, the analysis in the present essay is simplified considerably by not distinguishing between the rewards for the turnwaiter who detected a stealing event and the rewards for the other turnwaiters. The notation has also slightly been changed: $-L_M$ replaces $-L_i$ and R_M replaces $R_0 = R_i$.

The expected payoffs for the two pure strategies of a turnwaiter depend on the stealing rate σ of the turntaker and on the probability $(1-\alpha\mu)^{N-1}$ that a stealing effort is not detected by any of the other $N-1$ turnwaiters. Table 13.1 shows that the expected payoffs are of the form

$$F_{TW}(\neg M, \sigma, \mu) = \sigma \cdot [R_M - (1 - \alpha\mu)^{N-1}(R_M + L_M)], \tag{4a}$$

$$F_{TW}(M, \sigma, \mu) = \sigma \cdot [R_M - (1 - \alpha)(1 - \alpha\mu)^{N-1}(R_M + L_M)] - C_M. \tag{4b}$$

It is easy to see when M is a better reply against α and μ than $\neg M$:

$$M \succ \neg M \;\Leftrightarrow\; \sigma \cdot (1 - \alpha\mu)^{N-1} > C_{rel}, \tag{5}$$

where $\qquad\qquad C_{rel} := \dfrac{C_M}{\alpha(R_M + L_M)}, \tag{6}$

the ratio between the absolute cost of monitoring and the maximal benefit of monitoring, is a measure for the *relative* cost of monitoring.

The analysis of our games is based on the most fundamental solution concept of noncooperative game theory, the concept of a Nash equilibrium point. For simplicity, we shall usually assume that all N turnwaiters are equivalent with respect to their role in the game. By this we mean that the turnwaiters not only have the same strategies, detection probabilities, and payoff parameters but also similar starting conditions, a similar knowledge about the system, and a similar norm of reaction (see Weissing and E. Ostrom 1991a, sec. 3.5). These assumptions imply that we can concentrate on "symmetric" Nash equilibrium points where all N turnwaiters monitor at the same rate, μ^*.

The mixed-strategy components of a Nash equilibrium point are characterized by the fact that both pure strategies are payoff-equivalent at equilibrium. For the core game, $0 < \sigma^* < 1$ and $0 < \mu^* < 1$ require

$$(1 - \alpha\mu^*)^N = P_{rel} \tag{7}$$

and

$$\sigma^* \cdot (1 - \alpha\mu^*)^{N-1} = C_{rel} \tag{8}$$

respectively. The more general equilibrium analysis in Weissing and Ostrom (1991a, sec. 4.4) is directly applicable to the simplified core game studied here. The most important results are summarized in the following theorem and illustrated in figure 13.1.

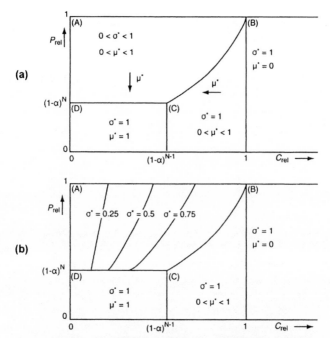

Fig. 13.1. Equilibrium diagrams illustrating the four equilibrium regimes and the comparative statistics of the core game (Theorem 1): (a) In regions (A) and (C), the monitoring rate increases for parameter changes in the direction of the arrows; (b) "isostealing" curves.

THEOREM 1: Nash Equilibrium Points of the Core Game

For each parameter constellation, the core game has exactly one[2] symmetric Nash equilibrium point that will be denoted by (σ^*, μ^*). The stealing rate σ^* is always positive at equilibrium. The symmetric equilibrium points of the core game can be classified as follows.

(A) The stealing rate remains at a moderate level (i.e., $\sigma^* < 1$) if the relative cost of stealing, P_{rel}, is larger than $(1 - \alpha)^N$ (a measure for the inefficiency of monitoring) and $(C_{rel})^{N/(N-1)}$ (a measure for the relative cost of monitoring). In this case, a mixed-strategy equilibrium is obtained which is given by

$$\sigma^* = \frac{C_{rel}}{(P_{rel})^{(N-1)/N}}, \quad \mu^* = \frac{1}{\alpha} \cdot [1 - (P_{rel})^{1/N}]. \tag{9}$$

(B) The stealing rate is maximal ($\sigma^* = 1$) and the monitoring rate is minimal ($\mu^* = 0$) at equilibrium if C_{rel} is larger than one.

(C) The stealing rate is maximal ($\sigma^* = 1$) and the monitoring rate is intermediate ($0 < \mu^* < 1$) at equilibrium if C_{rel} is smaller than one but larger than $(P_{rel})^{(N-1)/N}$ and $(1 - \alpha\mu)^{N-1}$. In this case, μ^* is given by

$$\mu^* = \frac{1}{\alpha} \cdot [(1 - C_{rel})^{(1/(N-1))}]. \tag{10}$$

(D) The stealing rate and the monitoring rate are both maximal at equilibrium ($\sigma^* = 1, \mu^* = 1$) if C_{rel} is smaller than $(1 - \alpha)^{N-1}$ and P_{rel} is smaller than $(1 - \alpha)^N$.

Figure 13.1 illustrates the four equilibrium regimes as well as the comparative statics of the core model. In parameter region (C), μ^* does not depend on P_{rel}, and it is negatively correlated with C_{rel}. In region (A), μ^* does not depend on C_{rel}, and it is negatively correlated with P_{rel}. In this region, σ^* is directly proportional to C_{rel} and negatively correlated with P_{rel}. The "isostealing curves" that connect parameter constellations leading to identical stealing rates at equilibrium are sections of convex curves through the origin in the (C_{rel}, P_{rel}) parameter space.

We chose to examine the core game first for analytical simplicity. Once one understands the pattern of stealing and monitoring in such a simple situation, it is possible to begin to embed the core game in more complex settings. Even though we chose our starting point for analytical rather than empirical reasons, many irrigation systems exist where there are no guards hired by the irrigators or by an external agency such as a Department of Irrigation. Of 51 self-organized irrigation systems in the Philippines, for example, 33 (65 percent) do not employ any guards while the remainder have hired their own guards responsible to the farmers (de los Reyes 1980). In the Philippines, the presence of guards is strongly related to the size of a communal irrigation system. Eighty percent (24 out of 30) of the small systems (less than 50 hectares) did not hire their own guards while only 30 percent (3 out of 10) of the large systems (more than 100 hectares) lacked guards. A survey with individual farmers served by these systems reveals the residual importance of farmers monitoring one another. Farmers indicated that water theft was an important factor affecting the level of observance of rotation schedules.

Schedules can only be enforced when farmers who have the right to draw water personally watch the irrigation of their fields. When they are absent, other farmers appropriate the water for their own use. Farmers concede this practice as theft, but they also argue that a farmer who neglects to oversee the irrigation of his field will not mind if he does not receive his due share (de los Reyes 1980, 53).

Consequently, 67 percent of the farmers on the communal systems included in this survey (412 farmers out of 617 responded to this question) reported that they personally patrolled the canals serving their farms (de los Reyes 1980, 53).

2. Specialization: A First Step toward a Guard's Position

Thus, one puzzle that we attempt to address in this essay is why some self-organized irrigation institutions continue for long periods of time without creating a specialized position while many others create a position that we call a guard but that is variously called a "water divisor," a "water tender," a "water guard," and so on. In other words, farmers create a specialized position with its own associated duties and payoffs and then fill that position with one or more persons from their own group. Each of the farmers may rotate into that position or hold that position for a set period of time, subject to being removed from it by their fellow farmers. If the farmers themselves are going to rotate into this position, what advantage is there to having a position itself as contrasted to allowing some specialization of monitoring behavior to evolve without the need of adding structural complexity to the institutional arrangement? In order to address this question, we relax the symmetry assumptions of our first model and ask what types of equilibria are produced if one specialized turnwaiter were to emerge. To begin with, we assume that this "distinguished" turnwaiter does not differ from the other turnwaiters in terms of payoffs and detection probabilities but only in his norm of behavior (he simply monitors at a different rate than the other turnwaiters).

Let us therefore again consider the core game with N turnwaiters who are equivalent in the sense that they all have the same detection probabilities and the same payoff parameters. Let $(\sigma^*_{sym}, \mu^*_{sym})$ denote the symmetric equilibrium which corresponds to this parameter combination and that is characterized by Theorem 1. Consider an alternative, asymmetric equilibrium $(\sigma^*, \mu^*, \gamma^*)$ where $N - 1$ turnwaiters monitor at the common rate μ^*, whereas one "specialized" turnwaiter monitors at a different rate γ^*. According to Theorem 1, the stealing rate σ^* is positive at equilibrium. Let us concentrate on those equilibrium points where σ^* is at an intermediate level. In such a case, the turntaker is indifferent between S and $\neg S$ at equilibrium. It is easy to see that this will only happen if

$$(1 - \alpha\gamma^*) \cdot (1 - \alpha\mu^*)^{N-1} = P_{rel}. \tag{11}$$

A comparison of (11) and (7) shows that μ^*_{sym}, the monitoring rate at the symmetric equilibrium, is always intermediate between μ^* and γ^*: a higher or lower monitoring activity of the $N-1$ "symmetric" turnwaiters is compensated by the monitoring activity of the distinguished turnwaiter. It is also easy to see that, generically, the strategies μ^* and γ^* will not both be pure or both be mixed. Thus, there are four possible types of asymmetric equilibrium with an intermediate stealing rate:

$$(S_{\mu 0}):\quad 0 < \sigma^* < 1,\quad \gamma^* = 0,\quad \mu^*_{sym} < \mu^* < 1; \tag{12a}$$

$$(S_{\mu 1}):\quad 0 < \sigma^* < 1,\quad \gamma^* = 1,\quad 0 < \mu^* < \mu^*_{sym}; \tag{12b}$$

$$(S_{0\gamma}):\quad 0 < \sigma^* < 1,\quad \mu^* = 0,\quad \mu^*_{sym} < \gamma^* < 1; \tag{12c}$$

$$(S_{1\gamma}):\quad 0 < \sigma^* < 1,\quad \mu^* = 1,\quad 0 < \gamma^* < \mu^*_{sym}. \tag{12d}$$

A characterization of these equilibrium regimes and the corresponding equilibrium points is rather straightforward and will not be given here. The shape and location of parameter regimes $(S_{\mu 0})$, $(S_{0\gamma})$, and $(S_{1\gamma})$ are illustrated in figure 13.2. In contrast to the symmetric case, several points may coexist for the same combination of payoff parameters. Notice that specialization is not always efficient: there are parameter constellations in region (A) of figure 13.1 which do not admit an asymmetric equilibrium with intermediate stealing rate. On the other hand, the equilibrium regions $(S_{\mu 0})$ and $(S_{0\gamma})$ overlap with region (C) of figure 13.1. Accordingly, specialization may reduce the stealing rate from the maximal to an intermediate level.

The following theorem shows that, quite generally, the equilibria corresponding to $(S_{\mu 0})$ and $(S_{0\gamma})$ are more efficient, whereas those corresponding to $(S_{\mu 1})$ and $(S_{1\gamma})$ are less efficient than the symmetric equilibrium.

THEOREM 2: Asymmetric Nash Equilibrium Points
Consider a parameter combination in region (A) of the core game where the symmetric equilibrium $(\sigma^*_{sym}, \mu^*_{sym})$ coexists with an asymmetric equilibrium $(\sigma^*, \mu^*, \gamma^*)$ corresponding to the specialization of one distinguished turnwaiter.

(1) If $\mu^* = 0$ or $\gamma^* = 0$, the asymmetric equilibrium is more efficient than the symmetric equilibrium in the sense that σ^* is smaller than σ^*_{sym}. In addition, the mean rate of monitoring is also smaller than μ^*_{sym}.
(2) If $\mu^* = 1$ or $\gamma^* = 1$, the asymmetric equilibrium is less efficient than the symmetric equilibrium in the sense that σ^* is larger than σ^*_{sym}. It depends on the specific parameter constellation whether the mean rate of monitoring is larger or smaller than μ^*_{sym}.

The results of this section show that specialization is possible even if the turnwaiters have identical payoffs and detection probabilities. Specialization is, however, not always efficient. Quite surprisingly, the stealing rate increases if the "specialized" turnwaiter monitors at a maximal rate whereas the other turnwaiters monitor at a reduced rate. The same is true if the $N - 1$ unspecialized turnwaiters increase their rate of monitoring from an intermediate to the maximal level. On the other hand, specialization is highly efficient if most of the turnwaiters do not monitor at all and thus rely completely on the monitoring efforts of their "specialized" colleague. Notice that the corresponding asymmetric equilibrium is only feasible if the relative cost of punishment is quite high for the turntaker (see figure 13.2).

One might argue that—in the context of specialization—it is quite artificial to assume that a different norm of behavior is the main distinction between a specialized turnwaiter and other turnwaiters. Probably, specialization will soon lead to differentiation with respect to payoffs and detection probabilities. This is, however, only a minor problem since all the results derived in this section are quite robust and continue to hold true if payoff asymmetries are taken into account (see Weissing and E. Ostrom 1991a, sec. 4.6).

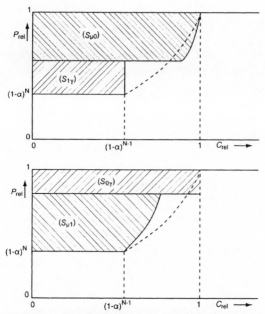

Fig. 13.2. Parameter regimes corresponding to equilibria with intermediate rate of stealing in a core game with distinguished turnwaiter. See text for explanation.

3. Irrigation Systems with a Motivated Guard

In the last section, we showed that the introduction of a specialist will sometimes, but not always, transform an irrigation system into a more efficient state where both the stealing rate and the monitoring rates are reduced. In this section, we investigate how the situation is changed if a formal position of a guard is created. This player has the strategic options not to guard at all ($\neg G$) or to guard as well as possible (G). Again, we shall also consider mixed strategies that are characterized by a parameter γ ($0 \leq \gamma \leq 1$) that corresponds to the probability that the guard is properly doing his job. Alternatively, γ will also be interpreted as a "guarding rate," that is, as a measure of the guard's willingness to fulfill his duties. Under the assumptions made in this essay, both interpretations are equivalent. We assume that guarding is not perfect. The detection probability of a guard who is guarding at a maximal rate (G) is limited by a parameter δ ($0 < \delta < 1$), and it increases proportionally with the guarding rate. Correspondingly, a guard guarding at rate γ will detect a stealing event with probability $\delta\gamma$, and he will not detect it with probability $1 - \delta\gamma$.

We assume that detection of a stealing effort by the guard may have different payoff consequences for the turntaker and turnwaiters than detection by a turnwaiter. For the turntaker, it may well make a difference whether his act of stealing was detected by one of his fellow farmers or by a guard who has an official legitimation to have an eye upon him. Accordingly, we consider the possibility that there are two levels of punishment, $-P_M$ and $-P_G$, which are imposed on a cheating turntaker depending on whether the stealing event was detected by a turnwaiter or by the guard. (Of course, in a special case, $-P_M$ and $-P_G$ may not differ from one another.) For the turnwaiters, it is mainly the "sociopsychological" component of payoffs that is affected by the introduction of a guard position. Formally, we distinguish between two "rewards," R_M and R_G, that is, the payoffs to a turnwaiter after a stealing effort has been detected by a turnwaiter or a guard, respectively. The payoff parameter R_G comprises the attitude of the turnwaiters toward the guard. It strongly depends on the system, and it might even be negative if the farmers do not like others meddling into their affairs. The possible payoff configurations for the turntaker and one of the turnwaiters are represented in table 13.2.

It is essential for the analysis to understand how the guard is motivated to do his job properly. Let us define the "status quo" point of the guard as the situation in which he does not guard but nevertheless no stealing is going on. This is our standard of comparison for the payoffs of the guard, and it will be assigned the numerical value 0. We assume that guarding involves a cost that is proportional to the effort allocated to this activity. Guarding at a maximal rate (G) involves a cost of $-C_G$, whereas the cost of guarding at rate γ is

TABLE 13.2. Payoff Configurations in an Irrigation Game with Guard[a]

Decision of TW	Event	Probability of event	Decision of TT S	Decision of TT $\neg S$
M	Guard detects	$\delta\gamma$	$R_G - C_M, -P_G$	$-C_M, 0$
	Some TW detects	$(1 - \delta\gamma)[1 - (1 - \alpha)(1 - \alpha\mu)^{N-1}]$	$R_M - C_M, -P_M$	$-C_M, 0$
	Nobody detects	$(1 - \delta\gamma)(1 - \alpha)(1 - \alpha\mu)^{N-1}$	$-L_M - C_M, B$	$-C_M, 0$
$\neg M$	Guard detects	$\delta\gamma$	$R_G, -P_G$	$0, 0$
	Some TW detects	$(1 - \delta\gamma)[1 - (1 - \alpha\mu)^{N-1}]$	$R_M, -P_M$	$0, 0$
	Nobody detects	$(1 - \delta\gamma)(1 - \alpha\mu)^{N-1}$	$-L_M, B$	$0, 0$

[a]The probabilities of the basic outcomes depend on the detection probability $\delta\gamma$ of the guard and on the probability $(1-\alpha\mu)^{N-1}$ that the other turnwaiters who are all monitoring at rate μ will not detect a stealing attempt.

given by $-\gamma \cdot C_G$. If the guard is a farmer himself or if he is paid according to the yield of the next harvest, he may well be harmed directly by an unde-tected stealing event. In such a case, the guard's loss due to undetected stealing will be denoted by $-L_G$. We assume that, up to the fixed cost of guarding $-\gamma \cdot C_G$, this is the only loss a guard may have as long as stealing remains undetected. In particular, the loss due to undetected stealing is as-sumed to be independent of his rate of guarding. If a stealing event is detect-ed, however, it may well matter whether the guard was at his place or not. If the guard himself was the one to detect it, he gets a positive "reward" which will be denoted by Q_G. If it was one of the turnwaiters who detected that stealing was going on, the guard's payoff (Q_M) will probably be different and may even be negative. In fact, the farmers might infer that the guard did not do his job properly and might sanction him.

Table 13.3 shows the guard's and the turntaker's payoff configurations together with their probabilities. The probabilities are calculated on basis of the assumption that all N turnwaiters monitor at the same rate μ. Notice that the payoff structure of the guard is essentially quite similar to that of the turnwaiters. However, the payoff parameters of the guard have usually quite a different interpretation than the corresponding payoff parameters of the turnwaiters.

A simple calculation shows that the best reply considerations can be rep-resented by

$$S \succ \neg S \quad \Leftrightarrow \quad (1-\delta\gamma)(1-\alpha\mu)^N > P_{rel} + \delta\gamma \frac{P_G - P_M}{B + P_M} ; \tag{13a}$$

$$M \succ \neg M \quad \Leftrightarrow \quad \sigma\left(1 - \delta\gamma\right)\left(1 - \alpha\mu\right)^{N-1} > C_{rel} \; ; \qquad (13b)$$

$$G \succ \neg G \quad \Leftrightarrow \quad \sigma > K_{rel}(\mu) \cdot \qquad (13c)$$

$K_{rel}(\mu)$, which is defined by

$$K_{rel}(\mu) := \frac{C_G}{\delta \cdot [Q_G - Q_M + (1 - \alpha\mu)^N \cdot (Q_M + L_G)]} , \qquad (14)$$

may be interpreted as the relative cost of guarding since it is the ratio between the absolute cost of guarding and the expected benefit of guarding in case of a stealing effort. For the rest of this chapter we shall assume that $Q_M + L_G$ is positive, which implies that the relative cost of guarding is positively correlated with the monitoring rate i of the turnwaiters. In particular, the relative cost of guarding attains its minimum, K_{min}, if the turnwaiters do not monitor at all:

$$K_{min} = \frac{C_G}{\sigma \cdot (Q_G + L_G)} . \qquad (15)$$

TABLE 13.3. Payoff Configurations in an Irrigation Game with Motivated Guard[a]

Decision of Guard	Event	Probability of event	Decision of *TT*	
			S	$\neg S$
	Guard detects	δ	$Q_G - C_G, -P_G$	$-C_G, 0$
G	Only *TW* detects	$(1 - \delta)(1 - (1 - \alpha\mu)^N)$	$Q_M - C_G, -P_M$	$-C_G, 0$
	Nobody detects	$(1 - \delta)(1 - \alpha\mu)^N$	$-L_G - C_G, B$	$-C_G, 0$
$\neg G$	Only *TW* detects	$1 - (1 - \alpha\mu)^N$	$Q_M, -P_M$	$0, 0$
	Nobody detects	$(1 - \alpha\mu)^N$	$-L_G, B$	$0, 0$

[a]The probabilities of the basic outcomes depend on the detection probability δ of the guard and on the probability $(1 - \alpha\mu)^N$ that none of the turnwaiters who are monitoring at rate μ will detect a stealing attempt.

The next theorem summarizes some basic aspects of the equilibrium structure of an irrigation game with a guard:

THEOREM 3: Stealing and Guarding Rate at Equilibrium

1. In an irrigation game with a motivated guard, the stealing rate is always positive at equilibrium.

2. At equilibrium, the guarding rate γ^* is restricted by the inequality $\delta\gamma^* < B/(B + P_G)$. In particular, the guarding rate is smaller than one at equilibrium if the guard is at least potentially efficient in the sense that

$$\delta > \frac{B}{B + P_G} \, . \tag{16}$$

3. If $K_{min} > 1$, the guarding rate is zero at equilibrium.

In contrast to irrigation games without guards, several symmetric Nash equilibrium points may coexist in an irrigation game with a motivated guard. In view of the large number of payoff parameters and their complex interaction, we do not attempt to give a complete equilibrium analysis for the most general irrigation game with a guard. In the next section, we rather ask whether more detailed results can be obtained if some reasonable restrictions are imposed on the payoff parameters. We consider two scenarios: (1) the use of integrated guards versus (2) disassociated guards.

4. Irrigation Systems with "Integrated" versus "Disassociated" Guards

In some natural settings, guard positions are filled by local farmers who themselves have a direct interest in how the system operates. In other words, some of an "integrated" guard's motivation is intrinsic to the operation of the system. In other settings, guard positions are filled by civil servants who are supposed to lack connections to the local setting and thus lack a personal or intrinsic interest in the operation of the system. Such guards are henceforth called "disassociated." In some cases, civil servants are regularly rotated for the express purpose of not allowing them to become interested in local affairs. The major justification for the rotation of civil servants is to prevent corruption, but rotation alone has not been a successful strategy for reducing corruption. Most self-organized irrigation systems rely on integrated guards, and most government-organized systems rely on disassociated guards.

In order to examine the difference that these two employment strategies may make in the capacity of guards to reduce the level of rule-breaking in a system, we will characterize two types of guards by different sets of restrictions on the payoff parameters of the system.[3] In the next sections, these two types of guards are characterized as follows:

An integrated guard: $\qquad P_G = P_M$ and $Q_G = Q_M$; $\qquad\qquad$ (17)

A disassociated guard: $\qquad L_G = 0$ and $Q_M = 0$. $\qquad\qquad\qquad$ (18)

The two equations in (17) incorporate the idea that rule enforcement by an integrated guard is intrinsically very similar to rule enforcement by the fellow farmers. Basically, (17) states that it does not make any difference for the turntaker or for the guard whether a stealing effort is detected by the guard or by one of the turnwaiters. In a system where this holds true, it is of minor importance for the participants which person detects a stealing event— the only thing that matters is whether a stealing effort is detected or not.[4]

The two restrictions in (18) reflect our assumption that a disassociated guard does not suffer losses due to undetected stealing and is not affected if a stealing effort is detected by a turnwaiter but not by himself. Disassociated guards are not farmers themselves, and their salary does not usually depend on the overall performance of the system. As a consequence, they do not suffer losses from undetected stealing. A disassociated guard may well have an incentive to detect a stealing effort himself (e.g., in order to obtain a gratification), but he is not per se interested in a reduction of the stealing rate. A disassociated guard may not even be aware of the monitoring activities of the irrigators. If a stealing event is detected by one of the turnwaiters but not by the guard, the incident will frequently not be reported to an external authority but rather will be regulated within the farmer community itself.

4.1. Irrigation Systems with Integrated Guards

If (17) holds true, the best reply structures for the three positions are "separable" and thus more transparent than in the general case:

$$S \succ \neg S \iff (1 - \delta\gamma) \cdot (1 - \alpha\mu)^N > P_{rel} \ ; \tag{19a}$$

$$M \succ \neg M \iff \sigma \cdot (1 - \delta\gamma) \cdot (1 - \alpha\mu)^{N-1} > C_{rel} \ ; \tag{19b}$$

$$G \succ \neg G \iff \sigma \cdot (1 - \alpha\mu)^N > K_{min} \cdot \tag{19c}$$

Let us for the rest of this section concentrate on those equilibrium points $(\delta^*, \mu^*, \gamma^*)$ where the stealing rate of the turntaker remains at an intermediate level. At such an equilibrium, the turntaker is indifferent between S and $\neg S$ or, equivalently,

$$(1 - \delta\gamma^*) \cdot (1 - \alpha\mu^*)^N = P_{rel} \cdot \tag{20}$$

For generic parameter combinations, (20) will not hold true if μ^* and γ^* both correspond to a pure strategy. Thus, there are five possible types of equilibrium with an intermediate stealing rate:

$(I_{\mu\gamma})$ $0<\sigma^*<1,\ 0<\mu^*<1,\ 0<\gamma^*<1$; $\qquad\qquad$ (21a)

$(I_{\mu0})$ $0<\sigma^*<1,\ 0<\mu^*<1,\ \gamma^*=0$; $\qquad\qquad$ (21b)

$(I_{\mu1})$ $0<\sigma^*<1,\ 0<\mu^*<1,\ \gamma^*=1$; $\qquad\qquad$ (21c)

$(I_{0\gamma})$ $0<\sigma^*<1,\ \mu^*=0,\ 0<\mu^*<1$; $\qquad\qquad$ (21d)

$(I_{1\gamma})$ $0<\sigma^*<1,\ \mu^*=1,\ 0<\mu^*<1$. $\qquad\qquad$ (21e)

The shape and location of parameter regimes $(I_{\mu0})$, $(I_{\mu1})$, $(I_{0\gamma})$, and $(I_{1\gamma})$ are illustrated in figure 13.3. Parameter regime $(I_{\mu\gamma})$ is a relatively small subset of the mixed-strategy regime (A) of the core game. A comparison of figure 13.3 with figure 13.1 shows that the introduction of an integrated guard considerably increases the parameter range of Nash equilibria with intermediate stealing rate. Thus, there are many parameter constellations for which the stealing rate can be reduced from the maximal to an intermediate level by the intro-

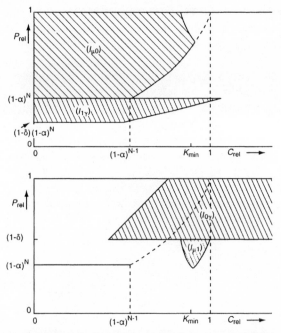

Fig. 13.3. Parameter regimes corresponding to equilibria with intermediate rate of stealing in an irrigation game with integrated guard. Regime $(I_{\mu\gamma})$ is not shown. See text for explanation.

duction of an integrated guard. For parameter combinations where the stealing rate is already intermediate in an irrigation game without guards, the introduction of an integrated guard is, however, not always efficient.

THEOREM 4: Effect of an Integrated Guard on the Stealing Rate

Consider a parameter combination in region (A) of the core game. Several equilibrium points of an irrigation game with integrated guard may coexist at such a parameter combination. The effect of an integrated guard on the stealing rate depends on the equilibrium regime:

$(I_{\mu\gamma})$: At the mixed-strategy equilibrium, the stealing rate will increase.
$(I_{\mu 0})$: If $\gamma^* = 0$, the stealing rate will not change.
$(I_{\mu 1})$: If $\gamma^* = 1$, the stealing rate will decrease.
$(I_{0\gamma})$: Here, the stealing rate will only decrease for particular parameter combinations.

4.2. Irrigation Systems with a Disassociated Guard

If (18) holds true, the relative cost of guarding is independent of the monitoring rate of the turnwaiters and the best reply structure of a disassociated guard can be characterized by

$$G \succ \neg G \Leftrightarrow \sigma > K_{min} .\tag{22}$$

Notice that $\neg G$ dominates G if $K_{min} > 1$ or, equivalently, if $C_G > \delta \cdot Q_G$. If this is the case, a disassociated guard does not guard at all, and—from the perspective of the irrigators—the game is equivalent to an irrigation game without guards. The condition $K_{min} > 1$ probably holds true in many natural systems with external guards because the detection rate and/or the rewards for detecting a stealing effort are often very small when compared to the costs of guarding. Since the case $K_{min} > 1$ has already been analyzed, we will from now on nevertheless assume

$$K_{min} < 1.\tag{23}$$

Again, there are five possible types of equilibrium with an intermediate stealing rate:

$$(D_{\mu\gamma}) \quad 0 < \sigma^* < 1, \ 0 < \mu^* < 1, \ 0 < \gamma^* < 1 ;\tag{24a}$$

$$(D_{\mu 0}) \quad 0 < \sigma^* < 1, \ 0 < \mu^* < 1, \ \gamma^* = 0 ;\tag{24b}$$

$(D_{\mu 1})$ $0 < \sigma^* < 1,\ 0 < \mu^* < 1,\ \gamma^* = 1$; (24c)

$(D_{0\gamma})$ $0 < \sigma^* < 1,\ \mu^* = 0,\ 0 < \gamma^* < 1$; (24d)

$(D_{1\gamma})$ $0 < \sigma^* < 1,\ \mu^* = 1,\ 0 < \gamma^* < 1$. (24e)

The shape and location of parameter regimes $(D_{\mu 0})$, $(D_{\mu 1})$, $(D_{0\gamma})$, and $(D_{1\gamma})$ is illustrated in figure 13.4. The parameter ψ is given by

$$\psi = \frac{(1-\delta) P_M}{(1-\delta) P_M + \delta P_G}, \qquad (25)$$

and it might be interpreted as the relative punishment effect of the turnwaiters. Qualitatively, the equilibrium regimes of an irrigation game with disassociated guard (fig. 13.4) are rather similar to the corresponding equilibrium regimes of an irrigation game with integrated guard (fig. 13.3). In both cases, the introduction of a guard considerably increases the parameter range of Nash equilibria with intermediate stealing rate. This is, however, only true if $K_{min} < 1$. Empirical studies (see sec. 5) indicate that this condition is much more often satisfied in systems with integrated guards than in comparable systems with disassociated guards.

Figure 13.4 illustrates that—for $K_{min} < 1$—the introduction of a disassociated guard may reduce the stealing rate from the maximal to an intermediate level. The situation is, however, not always improved by the introduction of a guard.

THEOREM 5: Effect of Disassociated Guard on Stealing Rate

Consider a parameter combination in region (A) of the core game. Several equilibrium points of an irrigation game with disassociated guard may coexist at such a parameter combination. For some combinations, the introduction of a disassociated guard may increase the stealing rate from an intermediate rate to the maximal level. Even if the stealing rate remains intermediate, it may be larger than the stealing rate at the corresponding equilibrium of the core game:

$(D_{\mu\gamma})$: At the mixed-strategy equilibrium, the stealing rate will increase if $P_G \leq P_M$.

$(D_{\mu 0})$: If $\gamma^* = 0$, the stealing rate will not change.

$(D_{\mu 1})$: If $\gamma^* = 1$, the stealing rate will decrease if and only if

$$P_G > P_M \left[1 + \frac{1 - (1-\delta)^{1/(N-1)}}{\delta \cdot (1-\delta)^{1/(N-1)}} \right]. \qquad (26)$$

$(D_{0\gamma})$: If $\mu^* = 0$, the stealing rate will decrease if $P_G > P_M$.

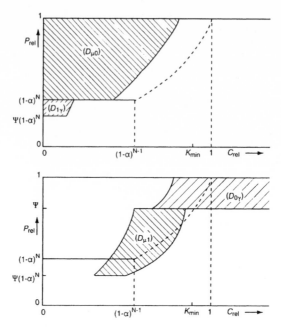

Fig. 13.4. Parameter regimes of equilibria with intermediate rate of stealing in an irrigation game with
disassociated guard. Regime $(D_{\mu Y})$ is not shown. See
text for explanation.

5. The Empirical Significance of Our Findings

What our formal results show is that enforcing rules allocating a highly valued good is difficult no matter whether attempted by a self-organized enterprise or by a government agency. If the temptation to steal is strong,
powerful counterincentives are needed to achieve an equilibrium where
stealing is not at maximal rates. The introduction of a guard—whether a
local farmer hired by a self-organized irrigation association or a civil servant hired by an external government agency—can lead to a reduction in
the stealing rate beyond that achievable in a system without any guards at
all. The introduction of a guard may, however, also lead to an increase in
the stealing rate above that achieved in a system without any guards at all.
This counterintuitive result may be explained by the fact that the introduction of a guard reduces the turnwaiters' incentives to monitor the turntaker's behavior.

   In all of our analyses the stealing rate at equilibrium is sensitively related to particular configurations of C_{rel}, K_{rel}, and P_{rel}. Thus, these are the key
variables that need to be related to empirical indicators. C_{rel} is the ratio be-

tween the absolute cost of monitoring and the maximal benefit of monitoring. Holding the other parameters constant C_{rel} rises when

- the rewards that a turnwaiter receives for successfully detecting a stealing event are reduced,
- the cost of monitoring by a turnwaiter increases,
- the losses that a turnwaiter suffers if stealing is not detected are reduced.

K_{rel} is the ratio between the absolute cost of guarding and the expected benefit of guarding if a stealing event has occurred. K_{rel} rises, holding other parameters constant when

- the rewards that a guard receives for successfully detecting a stealing event are reduced,
- the cost of guarding by the guard increases,
- the losses that the (internal) guard suffers if stealing is not detected are reduced.

While C_{rel} and K_{rel} represent the incentives facing the turntakers and turnwaiters, P_{rel} characterizes the incentives facing the turntaker. The relative cost of a detected stealing event to the turntaker rises when

- the punishment that is imposed upon a detected stealing event is increased,
- the benefit that could be obtained by stealing is reduced.

Tsebelis (1989, 1991) argues that much of the economics of crime literature has focused too narrowly on the incentives created by these variables for the potential rule breaker. His argument is partly based on the observation that, in two-person games, parameters associated with P_{rel} (particularly the punishment rate) have no influence on rule breaking rates at a mixed-strategy equilibrium. Our analysis shows that this result is certainly not generalizable to N-person games. In equilibrium regions where turntakers and turnwaiters are all playing a mixed strategy, the level of stealing is negatively correlated with P_{rel}. In addition, an increase in P_{rel} considerably expands the parameter domain for which an intermediate level of stealing is feasible at equilibrium.

We will pay primary attention to the incentives facing turntakers and guards since so little attention has been paid to these key variables and the presence of equilibria where stealing is less than maximal is strongly affected by them. These variables are the rewards assigned to a turntaker or a guard for successfully detecting stealing, the costs of monitoring, and the losses suffered if stealing is not detected. Here, we will mainly address the rewards to guards and turntakers in different systems. The costs of monitoring have been addressed in an earlier essay, and our findings will be briefly discussed below. We will not discuss at all the losses suffered if stealing is not detected since they are not easily subject to institutional change (but see Weissing and E. Ostrom 1991a, 245).

5.1. Rewards for Guards on Self-Organized Systems

As is documented in many descriptions by field workers, systematic differences in guarding rates are observed in self-organized systems versus government-managed irrigation systems throughout the world. It remains to be shown that the rewards are consistent with the restrictions imposed by our models. The rewards made available to turnwaiters and guards do appear to vary systematically in self-organized versus government-organized systems. In many of the self-organized systems in India, for example, positions called "common irrigators" or in the Telegu language of Southern India *neeruganti* (water men) or *neerukuttu* (water dividers or other variants such as neeranickams) are extensively used. Common irrigators are hired for a particular season of the year rather than for the entire year. Thus, part of the year, the system operates without guards at all. Employment frequently starts in October as the paddy rice crop is under way and continues for 60 to 90 days. Common irrigators are frequently paid in kind as the harvest is reaped. They collect their grain directly from each farmer based on the amount of land that a farmer has under irrigation (Wade 1988). Thus, a guard in such a system has to justify himself before others who are watching his performance in order to obtain his pay for the year.

In some villages in Southern India, the position of a common irrigator is an inherited position held by lower caste individuals. But currently in many villages, farmers are appointed to this position for one year and reappointed the next year if their work is considered efficient and impartial by the farmers they serve. In one village they are appointed by a village council a few days before their employment will begin.

> The council...sends the village crier out to announce that all who wish to be considered should assemble with the council on a certain evening. At the meeting the candidates' names are written down on a list; and at the same meeting or the following evening after less public discussions, the names of those appointed are announced. Since there are normally a few more candidates than positions, one means of control is not to reemploy a common irrigator whose work was thought unsatisfactory in the past. (Wade 1994, 83)

Two of the village elders make the specific work assignments so that the common irrigators do not work on an outlet connected to their own land.

In another South Indian village described by Sengupta (1991, 103–4), the common irrigators are appointed for a three- to four-month period each year. They work in pairs. One member of the team is a landless tenant and the other owns land served by the irrigation system.

> The Kandadevi villagers usually assemble in the evening in an open ground near the temple. Here a meeting is conducted and presided over by the important landowners in the ayacut. After appointing the *neer-paichys,* they are taken to the temple for oath-taking to remain impartial. With this vow they break a coconut. They are paid in cash at the rate of Rs. 10 per acres (Rs. 25 per hectare) per month by the cultivators. The *neerpaichys* themselves collect the money. (Sengupta 1991, 104)

In describing the system on another canal in South India, Sengupta mentions that the *neeranickams* on this system are paid at a relatively high rate and that they collect their own grain by going from door to door at the time of threshing. "If anyone is dissatisfied, the payments may be withheld. But no such thing has happened in the past" (Sengupta 1991, 167). In field settings, some of the reward given to guards for monitoring carefully and fairly is not assigned in each round as we do in the formal game. But given the impor-tance of these events to the individuals involved, one can think of guards earning imaginary "points" at each round that add up to full payment and a good tip at the end of the year for good performance or a tongue lashing and less than full payment for poor performance. The key point is that the re-wards assigned to integrated guards on self-organized systems are substantial and tightly connected to the work they do.

5.2. Rewards for Guards on Government-Managed Systems

The diversity of arrangements that are broadly grouped under the term *gov-ernment-managed system* is as great as the set included under the rubric of self-organized systems. Most government-managed irrigation systems em-ploy disassociated guards and are relatively large systems constructed by a regional or national agency. The position of a "guard" is usually held by a laborer who is directly supervised by an engineer. Both are employed in a relatively small operations and management (O&M) division of a large irri-gation department. Wade (1987) points out that in India, the O&M divisions of state-level irrigation departments are rarely any larger than 10 percent of the total personnel working in the agency. The professional staff do not have any particular identification with a local system, and they are transferred in and out of a particular posting every 15 to 20 months. Not only are O&M personnel thin on the ground, they may also not have adequate transportation. Consequently, the cost of guarding may be quite high.

Robert Wade points to three structural attributes that increase the problems of assuring rule conformance on the larger irrigation systems that employ external guards. The first problem is that "canal O&M is carried out by an organization whose primary function has been construction" (Wade 1987, 179). The second is "the lack of connection between the budget of the irrigation department and the collection of water rates" (Wade 1987, 179). The third is the lack of communication with other agencies that provide essential input to agricultural production. Consequently, water may be released at a time when the farmers have not yet been able to obtain key inputs to start the agricultural season.

Both the first and the second problems affect the incentives of those who are guards and the supervisors of the guards—particularly the latter. Engineers assigned to supervise the allocation of water do not gain professional or organizational rewards from this assignment. While the field laborers tend to have a stronger connection to the performance of the system simply because they are apt to live in the region and be affected generally by the productivity of local agriculture, they are supervised by a staff that is disinterested in and unaffected by what they are doing. Nor can the farmers affect the performance of the field guards or their supervisors by holding back irrigation fees. Since the fees go to a revenue department rather than to support the O&M division in most government-managed systems, whether the farmers pay their fees or do not pay them has no effect on whether the salary of a guard is paid or not. Contrast that with the problem of local guards who must go to each farmer to collect their year's payment in cash or in kind.

The descriptions of what happens on the field canals of large-scale, government-owned irrigation systems that are responsible for water distribution and rule enforcement in India and Sri Lanka are relatively bleak.[5] Either water distributors/guards are not to be found, or if present, they operate in a predatory rather than a protective role.[6]

These depressing empirical findings are consistent with our formal results. Where disassociated guards receive little rewards for their monitoring activities, and where the costs of monitoring are high, one should find low levels of monitoring activities. When monitoring rates are low, one should find higher rates of stealing. These perverse outcomes are, however, not the only equilibria that theoretically could be reached.

Are there not empirical settings where guards employed by external agencies are highly motivated to devote energy to monitoring and where stealing levels are relatively low? A positive response can be given to this question if one broadly interprets the concept of government-managed to include irrigation systems that are supervised by an external governmental agency, even though strong local supervision is also present. In other words, in irrigation systems that are closer to a federal structure, one should find some empirical examples where the parameter combinations are more conducive to good performance on the part of guards.

The empirical cases now come from Taiwan, where a combination of factors has apparently yielded some of the world's most technically efficient irrigation systems.[7] Since most of these systems are gravity-flow systems without much storage and serve rice-producing areas, the challenge of governing and managing these systems is substantial. Further, many of the rivers are short and steep, leading to unpredictable oscillations in the water supply. As Mike Moore points out, Taiwanese rice farmers face many incentives to steal water.

> [R]ice farmers, in Taiwan as elsewhere, face compelling temptations to steal water in excess of their needs. The reason lies in the unusual way in which rice yields respond to varying levels of water supply. A deficit supply produces immediate and relatively severe yield decreases. By contrast, over a wide range, an excess supply does not harm the plant at all. It makes a great deal of sense for rice farmers to seize any opportunity to appropriate more water than they are likely to need. They can store it in their flooded fields for several days, and feel that they have provided for their crop if for some reason canal supplies prove deficient. (Moore 1989, 1741)

And, Taiwanese farmers—even on the best regulated canal systems—do steal but at a low rate compared with other settings. It is not possible to achieve high levels of technical efficiency where water theft is common, as predictability is essential to efficient allocation of water (Levine 1980). Vandermeer (1971) found that stealing events were closely related to the opportunity to steal—partially the result of slack monitoring—rather than high benefit levels to the farmers.[8]

Government irrigation systems in Taiwan do not share the three structural attributes that Robert Wade relies upon, as elucidated above, to explain the dysfunctional performance of the irrigation departments in India. First, instead of the units responsible for O&M being a part of a large civil engineering hierarchy oriented to *construction*, O&M on Taiwanese irrigation systems is carried on by 14 large parastatal organizations called Irrigation Associations (IAs). Four-fifths of the irrigated lands in Taiwan are served by these IAs. The service areas of the IAs vary from 10,000 to 85,000 irrigated hectares (average = 30,000 hectares) (Moore 1989, 1741; in addition to Moore 1989, see also Bottrall 1985; Levine 1980; and Moore 1988). IAs have substantial autonomy even though they have been supervised by the

Provincial Water Conservancy Bureau since 1975. Construction of irrigation projects is undertaken by a totally separate national agency.

Second, instead of a lack of connection between the fees paid by farmers and the funds used to support O&M, fees paid by farmers account for one-third to three-quarters of the financing of the IAs.[9] Third, instead of being isolated from communication with other agencies and with the farmers, the staff of the Irrigation Associations are much more effectively linked to the local farmers as well as to national agencies. For each rotation area—covering 50 to 150 hectares and about the same number of farmers—an irrigation group chief is elected who supervises water distribution, conflict arrangement, and maintenance in this area (Moore 1989, 1742). These elected chiefs are usually local farmers. Nonpartisan elections for this purpose are separated from elections for other offices. The group chiefs closely monitor the jointly hired common irrigators who have full responsibility to distribute water and guard the system. Further, the staff themselves are more effectively monitored by the farmers.

> The IAs are overwhelmingly staffed by people who were born in the locality, have lived there all their lives, and in many cases farm there. Further, IA staff are not sharply differentiated from their members in terms of education or income levels. I have a strong overall impression that IA staff are so much part of local society that they can neither easily escape uncomfortable censure if they are conspicuously seen to be performing poorly at their work, nor ignore representations made to them by members in the context of regular and frequent social interactions. (Moore 1989, 1742)

Guarding rates and monitoring of the guards are also closely related to water conditions and the consequent temptation to steal. The number of staff and coverage of the canals is increased dramatically when drought conditions increase the value of water obtained by any means. During the severe drought of 1977, for example, guards were deployed on a 24-hour basis in the Yun-Lin system. Tents were set up for those watching the gates and control valves and "visits were made by the Chairman and Chief Engineer of the IA during the night to bring refreshments and encouragement [as well as surveillance] to these workers" (Levine 1978, 8).

A further aspect of the Taiwanese system that more closely links the monitoring behavior of the guards to the payoffs received by the guards is the way that the Provincial Water Conservancy Bureau interprets the speed with which members pay their dues. Contrary to the experience in many countries, fees have actually been collected from over 95 percent of the farmers served by IAs for many years.[10] Delays of a few days or more in paying fees are an institutionalized mechanism by which farmers can communicate dissatisfaction beyond the bounds of a local rotation area or IA.

During the two periods of the year when fees are due, most of the IAs' institutional machinery appears to be devoted to completing collections as expeditiously as possible. The same working station staff who provide farmers with irrigation services come to them to encourage them to part with their money if there is any sign of delayed payment. Each working station is required to make daily telephone reports to superiors about collections in their area....And the headquarters are obliged to report regularly to the Provincial Water Conservancy Bureau. At each level, delays in fee payment are taken as prima facie evidence of a problem which requires attention. (Moore 1989, 1743)

Moore goes on to point out that this signaling device fits into an overall system where staff at every level are subject to detailed annual job performance evaluations. "These evaluations result in gradings which directly or indirectly affect salary increments, promotions, and access to additional resources" (Moore 1989, 1743).[11]

These factors are part of the surrounding institutional environment in which a core irrigation game is embedded. The difference in the reward structure for a guard being in a large civil engineering agency versus working in an agency that specializes in distribution and maintenance problems can be substantial. Further, when the guard's budget is tied to the individuals served (and when higher budgetary authorities pay close attention to the speed and quantity of fee payment), the guard is far more motivated to provide fair and efficient water distribution/guarding activities. Also, guards working in their own community where high expectations of good performance exist, will find shirking on the job far less desirable than guards who are posted away from home or in places where their work is not observed at all.

The important point that we can learn from the Taiwanese systems is that when the incentives facing guards reward them sufficiently for fair and efficient water allocation services and punish them for slack performance, their level of activity rises, leading to a decrease in the stealing rate. What our models show, and what a brief review of the relevant empirical world also confirms, is that there are many parameter combinations that jointly affect monitoring and stealing rates and how they interact. Variations in these parameters make a major difference in the type of equilibria yielded by a game.

5.3. Costs of Monitoring

For most irrigation systems the cost of monitoring by a turnwaiter or guarding by a guard are affected by the same factors. In Weissing and E. Ostrom (1991a), we describe how many self-organized irrigation systems design their physical works and the rules they use to allocate water so as to reduce monitoring costs. Canals are designed so that the actions of the turntaker are noticeable at low cost to neighboring farmers. The rotation schedule itself brings those who are most directly affected by stealing activities to a similar physical location at about the same time. In contrast, many government-owned irrigation systems have designed physical works without any concern for how guards or farmers could observe activities at a low cost. Further, many of the guards hired by a government agency are given vast areas to cover, but are not even provided with a bicycle, let alone a motorized vehicle, to travel to the canals themselves. Often, irrigation guards in large, government-owned systems will be located in huts near the main control works, rather than out on the canals themselves. Thus, self-organized irrigation systems frequently have much lower monitoring costs than government-owned systems.

5.4. Punishment for Rule-Breaking on Self-Organized Systems

While the payoffs for turnwaiters and guards are somewhat more important in affecting the size and shape of equilibrium regions, the punishments assigned to turntakers who steal also have an important impact. Not much is written about the level of punishments meted out in self-organized systems.[12] Fortunately, David Guillet provides more detail about this aspect of some self-organized systems with integrated guards in a book about farmer-governed irrigation systems of the Colca valley in Peru. In this terraced, semiarid region, locally hired water distributors (*regidores*) are assigned the allocation and guarding duties.

General principles have long been established for dividing the year into agricultural seasons, each season into spatial rotations by terrace inlets, and, within seasons, for duly recognizing the water needs of different crops.[13] They do not follow a rigid rotation system, however, such as many systems around the world do. Decisions about the specific rotations to be followed are made at weekly meetings (reginas) attended by the farmers and the regidores. Reginas are held in locations overlooking the fields being discussed. "Farmers can see instantly which fields have been irrigated because they are dark and stand out clearly against the background of light colored dry fields" (Guillet 1992, 113). This enables the farmer to determine if the general spatial sequence, from head to tail, has been followed, whether "someone's field has been irrigated out of turn, or if someone who was advanced water at the last regina failed to irrigate" (Guillet 1992, 113).

Punishments are assigned to the water users as well as to the officials of the Irrigation Association responsible for the system. During the course of a recent year, 122 infractions were recorded in the minutes of the Irrigation Association. All but 21 of these involved water users who were members of the Irrigation Association. Theft of water was charged in one-fifth (19) of the cases against users, and some form of a general infraction involving minor types of water theft was charged in another 16 percent. Almost a third of the infractions involved failure to appear for corvée labor associated with the maintenance of the system. The other 33 infractions charged against users involved such actions as failure to use allocated water (9); water waste (6); offending an official (6); irrigating a nonallowed crop (4); and a variety of other minor offenses. During the same year, officials of the Irrigation Association were charged with 13 infractions involving such actions as missing a general meeting (8); improper scheduling (3); offending a user (1); and failing to produce receipts (1) (Guillet 1992, table 20).

The severity of punishments ranged from a simple fine to jail (Guillet 1992, table 21). While the latter only occurred in 3 instances, it is noteworthy that such a severe sanction was actually applied in 3 instances during a year's period. External guards on many large systems can rarely assess fines against misappropriation, let alone see someone go to jail. In 80 instances a simple fine was assessed, in 10 more instances a fine plus a water cutoff occurred. The fines ranged from "1,000 to 20,000 soles with proceeds going to the treasury of the Irrigation Association" (Guillet 1992, 208).

Guillet's study provides more detail, but many studies of self-organized systems do stress that the farmers are quite willing to impose modest to heavy sanctions on one another (see Tang 1992; E. Ostrom 1992). If those who were guarding or monitoring were not motivated to check on the behavior of turntakers, it would matter little whether punishments were in place or not. But in systems where effective monitoring occurs, being able to punish a rule breaker once discovered is an important aspect of keeping behavior within agreed-upon limits.

6. Conclusions and Implications

Among the striking aspects of all the games we have analyzed is the presence of multiple equilibria and the complex intertwining consequences of parameter changes. There are some strange feedback loops in games like these. For example, the introduction of a guard may not be efficient for the system as a whole, even if the guard does his job. In fact, the stealing rate may rise from an intermediate to the maximal level since the presence of the guard demotivates the turnwaiters to go on with their monitoring activities (see Weissing and E. Ostrom 1991a for other counterintuitive feedback effects).

Which types of equilibria are feasible in an irrigation game with guards depends on a complex combination of parameters that directly and indirectly impact on the incentives of the various participants. In each irrigation turn, these parameters are fixed. Between turns and, in particular, between seasons, however, the values of some parameters can be affected by the participants themselves or by outside agents.

One can imagine this process to be something like a "time-out" when the rules of the game are changed for future plays of the game (see E. Ostrom 1989, 1990). Thus, simply because an irrigation system is currently plagued with high levels of stealing (and thus unpredictability of water supply and low agricultural yields), this fate may not be permanent. The institutional arrangements in which any particular physical and social system is embedded can have subtle and interactive effects on the behavior of participants and the outcomes they obtain. In view of some counterintuitive feedback loops, it is also possible that a change in the rules affecting the parameters of the game may lead participants to change their patterns of interaction from those that were jointly beneficial to less desirable patterns.

The complexity of these strange loops has substantial implications for how best to "nest" a core irrigation game in an institutional structure. Is it possible to argue, for example, that farmer-governed and -managed irrigation systems (the upside-down hierarchies) will automatically achieve better equilibria than government-owned and -managed irrigation systems (the traditional pyramids)? The answer derived from our models is negative. There are perverse equilibria in all of the games we have studied. The empirical evidence is also negative. Not all farmer-governed and -managed irrigation systems achieve high levels of conformance to their water distribution rules. On the other hand, substantial evidence exists that a greater proportion of the self-organized and self-governed systems find mutually productive equilibria than those that rely on external enforcement agents (Tang 1991, 1992; Ascher and Healy 1990; Uphoff 1991). How might this happen?

Unfortunately, we cannot simply rely on trial-and-error learning or blind evolutionary processes to help explain the achievement of better equilibria in some institutional settings than in others. We know from evolutionary theory that blind, dynamic processes—even if they are adaptive in the short run— can easily lead to "not-so-nice" equilibria. Nor can one simply assume that individuals possess the skills needed to design the appropriate set of rules to achieve mutually productive equilibria. Without processes that cumulate and share knowledge so as to reduce the blind search aspect of learning, design may not be a more efficient process for finding better—as contrasted to worse—equilibria. Given the wide diversity of physical environments in the world, it is unlikely that just a few "pure type" institutions will provide the structure that induces more desirable equilibria. Thus, the repertoire of known institutional arrangements that might be called upon in crafting institutions in any one site may need to be quite large.

If one were to engage in the design of an irrigation game with a guard, for example, using the assumption (prevalent in the social science literature) that increasing the fines imposed on those that steal is the most important parameter to control in the design process, one might overlook the importance of motivating the guards (see Tsebelis 1989, 1991; Weissing and E. Ostrom 1991b). It certainly appears as if the importance of motivating the guards has been missed in the design and reform of most large-scale irrigation departments.[14]

A major disadvantage of establishing single, large-scale enterprises to govern and manage all irrigation systems in a region is the lack of learning from experience that can occur in such settings. Individual staff are shifted from one assignment to another before they become intimately familiar with any setting and can speculate in a skillful manner why perverse or beneficial results are achieved in that setting. They take with them the ideas with which they have been indoctrinated in school. Their performance ratings and promotion opportunities in one assignment may have little relevance for how well the irrigation systems in that area are helping farmers to increase agricultural output.

A major advantage of allowing considerable autonomy to design and reform rules in smaller, autonomous units nested in larger macropolitical regimes is that more information may be generated from experience. Many more independent experiments can be carried on. Whether this advantage is achieved, however, depends largely on whether each unit exists in physical and institutional isolation. Isolated self-organized entities may not learn much from one another. Some may do very well while others fail. On the other hand, if participants in these relatively autonomous units are nested in overarching regimes, considerable cumulation and sharing of the learning from experimental changes may be achieved. One important aspect of the Taiwanese case is the role of the Provincial Water Conservancy Bureau in pressuring the relatively autonomous Irrigation Associations to do well in terms of their speed of fee collection. The overarching bureau has stimulated a considerable amount of competition among the IAs. In some ways it serves to network the IAs together so that information on better practices and rules is shared. In other ways it is more like a traditional hierarchical supervisor monitoring the performance of lower units.

Consequently, an important factor in understanding how individuals might find the better equilibria that exist in multiequilibria games is how those games are nested in multiorganizational arrangements. As long as scholars focus on a market/hierarchy dichotomy, they do not examine the structure of multiorganizational arrangements in which individuals find themselves nested (Scharpf 1993). Our core game is certainly not a market. It can be nested in a wide diversity of multiorganizational arrangements. As one turns to the question of how individuals find the more attractive equilibria in multiequilibria games, the role of the multiorganizational system surrounding a core game increases in importance.

NOTES

1. This section draws on an earlier essay (Weissing and E. Ostrom 1991a) where we analyzed the models presented here in far more detail.

2. Strictly speaking, this is only true for generic parameter constellations. Some border cases in parameter space yield a continuum of symmetric Nash equilibrium points. Since they are of no relevance in real-world situations, we will henceforth neglect such border cases.

3. For this analysis, we limit ourselves to an analysis of two payoff scenarios. In addition to their incentives, guards hired by self-organized irrigation systems may differ from those hired by government systems in several respects, including their knowledge about the system and their norms of behavior.

4. Other parameter restrictions might also be used to describe an integrated guard. In natural settings, the rewards assigned to integrated guards are often closely connected to the overall performance of the system. As a consequence, integrated guards may be quite strongly affected by an undetected stealing event. In fact, their loss due to undetected stealing, $-L_G$, might be closely related to the turnwaiters' loss, $-L_M$. Other payoff parameters of the guard might be correlated with the corresponding parameters of the turnwaiters as well. We will, however, not impose more restrictions on the payoff parameters of an integrated guard. The weak condition stated here enables a rather complete analysis without sacrificing much generality.

5. The picture in most of Southeast Asia is consistent; see Ascher and Healy (1990) and Bottrall (1985).

6. In some cases, water distributors may be more dependent than predatory. In other words, powerful local farmers pressure distributors to allocate more water than their share or to look the other way while the farmer takes more water than allowed. Poorly paid, low-status individuals with no backing from their supervisors are greatly exposed in such circumstances to such pressure.

7. Levine (1980, 54) reported that the Tou Liu system (now the central core of the reconstituted Yun-Lin system) in Taiwan had a technical efficiency rating of over 90 percent, and that the majority of systems in Taiwan exceeded 60 percent. At the same time, systems in Malaysia were closer to 40 percent and in the Philippines below 25 percent.

8. Vandermeer found that the farmers who were most likely to steal were located at the head end of the canal. The benefit to a farmer of additional water at the head end is relatively low.

9. Since the IAs have substantial autonomy, budgetary support varies rather substantially. In recent years, some of the Irrigation Associations have received large subsidies, but even so, a substantial portion of their budget is still fees collected from the farmers they serve (see Moore 1988).

10. Leslie Small (1987) collected data from several countries and found that Korea was the only country with a higher level of fee collection than that of Taiwan. The Korean system is structurally relatively similar to that of Taiwan, and considerable emphasis is placed by central officials on fee collection. Small found that the Department of Irrigation in Nepal collected about 20 percent of the fees due. After

substantial change in the way fees were collected and assigned, fee collection in the Philippines had begun to approach 60 percent.

11. The complexity of this nested (federal) system is illustrated by the following description by Levine.

> The total area of the Yun-Lin Irrigation Association is 65,590 ha. It derives approximately 80% of its water supply from the Cho Shui River, with the remaining 20% coming from ground water and return flow drainage water. The Association has 4 management officers (regional), 43 working stations, 500 water groups, and 1,683 subwater subgroups (based on rotational areas)....there are over 30 identifiable systems; each system has its own fee schedule, some have more than one; each has a high degree of autonomy with respect to system operation, maintenance and improvement; each has specific water rights, usually based on historical development, which are respected in irrigation planning and operation, though they may be modified under emergency conditions. (Levine 1978, 3)

12. But, see E. Ostrom (1990, chap. 3) for some evidence from centuries-old, self-organized, Spanish irrigation systems.

13. Land and water use in this area have been tightly and stably associated for centuries. Guillet compares a map prepared in 1604 with a map of modern irrigation clusters and finds that 16 of the 19 clusters in use today were recognized by same name in 1604. The three that were not listed are all small sections fed by unreliable springs.

14. See Chambers (1988), Uphoff (1991), Sengupta (1991) for further discussion of this point.

REFERENCES

Ascher, William and Robert Healy, 1990: *Natural Resource Policymaking: A Framework for Developing Countries.* Durham, NC: Duke University Press.

Bottrall, Anthony, 1985: *Managing Large Irrigation Schemes: A Problem of Political Economy.* Occasional Paper 5. London: Overseas Development Institute.

Chambers, Robert, 1988: *Managing Canal Irrigation: Practical Analysis from South Asia.* Cambridge: Cambridge University Press.

de los Reyes, R. P., 1980: *47 Communal Gravity Systems: Organizational Profiles.* Quezon City: Institute of Philippine Culture.

Guillet, David, 1992: *Covering Ground: Communal Water Management and the State in the Peruvian Highlands.* Ann Arbor: University of Michigan Press.

Levine, Gilbert, 1978: *Irrigation Association Response to Severe Water Shortage— The Case of the Yun-Lin Irrigation Association, Taiwan.* Ithaca, NY: Cornell University. Manuscript.

Levine, Gilbert, 1980: The Relationship of Design, Operation and Management. In: E. Walter Coward (ed.), *Irrigation and Agricultural Development in Asia: Perspectives from the Social Sciences.* Ithaca, NY: Cornell University Press, 51–62.

Moore, Mike, 1988: Economic Growth and the Rise of Civil Society: Agriculture in

Taiwan and South Korea. In: Gilbert White (ed.), *Developmental States in East Asia.* London: Macmillan, 113–62.

Moore, Mike, 1989: The Fruits and Fallacies of Neoliberalism: The Case of Irrigation Policy. In: *World Development* 17, 1733–50.

Ostrom, Elinor, 1989: Microconstitutional Change in a Multiconstitutional Political System. In: *Rationality and Society* 1, 11–50.

Ostrom, Elinor, 1990: *Governing the Commons: The Evolution of Institutions for Collective Action.* New York: Cambridge University Press.

Ostrom, Elinor, 1992: *Crafting Institutions for Self-Governing Irrigation Systems.* San Francisco: Institute for Contemporary Studies Press.

Scharpf, Fritz W., 1993: Coordination in Hierarchies and Networks. In: Fritz W. Scharpf (ed.), *Games in Hierarchies and Networks: Analytical and Empirical Approaches to the Study of Governance Institutions.* Frankfurt am Main: Campus Verlag; Boulder, CO: Westview, 125–65.

Sengupta, Nirmal, 1991: *Managing Common Property: Irrigation in India and the Philippines.* London: Sage.

Small, Leslie, 1987: *Irrigation Service Fees in Asia.* ODI/IIMI Irrigation Management Network Paper 87/1c. London: Overseas Development Institute.

Tang, Shui Yan, 1991: Institutional Arrangements and the Management of Common-Pool Resources. In: *Public Administration Review* 51, 42–51.

Tang, Shui Yan, 1992: *Institutions and Collective Action: Self-Governance in Irrigation.* San Francisco: Institute for Contemporary Studies Press.

Tsebelis, George, 1989: The Abuse of Probability in Political Science: The Robinson Crusoe Fallacy. In: *American Political Science Review* 83, 77–92.

Tsebelis, George, 1991: The Effect of Fines on Regulated Industries: Game Theory vs. Decision Theory. In: *Journal of Theoretical Politics* 3, 81–101.

Uphoff, Norman, 1991: *Managing Irrigation: Analyzing and Improving the Performance of Bureaucracies.* London: Sage.

Vandermeer, Canute, 1971: Water Thievery in a Rice Irrigation System in Taiwan. In: *Annals of the Association of American Geographers* 61, 156–79.

Wade, Robert, 1987: Managing Water Managers: Deterring, Expropriation, or, Equity as a Control Mechanism. In: Wayne R. Jordan (ed.), *Water and Water Policy in World Food Supplies: Proceedings of the Conference held May 26–30, 1985.* College Station: Texas A&M University Press, 117–83.

Wade, Robert, 1994: *Village Republics: Economic Conditions for Collective Action in South India.* San Francisco: ICS Press.

Weissing, Franz and Elinor Ostrom, 1991a: Irrigation Institutions and the Games Irrigators Play: Rule Enforcement Without Guards. In: Reinhard Selten (ed.), *Game Equilibrium Models II: Methods, Morals, and Markets.* Berlin: Springer-Verlag, 188–262.

Weissing, Franz and Elinor Ostrom, 1991b: Crime and Punishment: Further Reflections on the Counterintuitive Results of Mixed Equilibria Games. In: *Journal of Theoretical Politics* 3, 343–50.

CHAPTER 14

Coping with Asymmetries in the Commons: Self-Governing Irrigation Systems Can Work

Elinor Ostrom and Roy Gardner

Common-pool resources are natural or manmade resources where exclusion is difficult and yield is subtractable (Gardner, E. Ostrom, and Walker 1990).[1] They share the first attribute with pure public goods; the second attribute, with pure private goods. Millions of common-pool resources exist in disparate natural settings ranging in scale from small inshore fisheries, irrigation systems, and pastures to the vast domains of the oceans and the biosphere.

The first attribute—difficulty of exclusion—stems from many factors, including the cost of parceling or fencing the resource and the cost of designing and enforcing property rights to exclude access to the resource. If exclusion is not accomplished by the design of appropriate institutional arrangements, free-riding related to the provision of the common-pool resource can be expected. After all, what rational actor would help to provide the maintenance of a resource system, if noncontributors can gain the benefits just as well as contributors? The extent to which a common-pool resource will be provided is a complicated problem, depending on how preferences are articulated, aggregated, and linked to the mobilization of resources.

The second attribute—subtractability—is the key to understanding the dynamics of how the "tragedy of the commons" can occur. The resource units (like acre-feet of water, tons of fish, or bundles of fodder) that one person appropriates from a common-pool resource are not available to others. Unless institutions change the incentives facing appropriators, one can expect substantial overappropriation. For example, those who fish from a lake derive all the benefit from catching additional fish. However, the depletion of

Originally published in *Journal of Economic Perspectives* 7, no. 4 (fall 1993): 93–112. Reprinted by permission of the American Economic Association and the authors.

Authors' note: The authors appreciate the support of the Ford Foundation (Grant No. 920–0701), the National Science Foundation (Grant No. SES–8921884), and the U.S. Agency for International Development for its support of the Decentralization: Finance and Management Project under Contract No. DHR–5446–Z–00–7033–00. The comments of John Ambler, Vincent Ostrom, James Walker, Franz Weissing, and the editors of *Journal of Economic Perspectives* on an earlier draft are deeply appreciated.

the fishery is a cost shared with other fishermen. The private gain is thus very likely to overbalance any single fisherman's share of the social loss. Or, to put it another way, no single fisherman can prevent depletion of the fishery by restricting his personal catch. The fishery is thus likely to be pushed to the brink of extinction unless institutions counteract these incentives.

In prevailing theories, the temptation to free-ride on the provision of key infrastructures is viewed as a major deterrent to successful economic development. David Freeman (1990, 115) concisely describes this problem in relation to irrigation.

> The logic of the individually rational utility seeker may not coincide with the logic of the community. If, for example, farmers individually observe that their leaky and misaligned water course requires improvement, they will not invest in corrective action on individually rational grounds. Assuming a sizable number of farmers, each will calculate as follows. If one farmer invests time, energy, and money required to improve the channel going through his or her own land and other farmers do not make comparable corrective investments in a coordinated fashion, then the payoff in improved water supply and control (the collective good) is negligible.
>
> However, if many farmers undertake the improvement effort on each of their sections, and one individually rational decision-maker does not do so, she or he will still enjoy a substantial share of the benefit provided by the work of others, at no personal cost. Therefore, the rational, calculating individual will choose to do nothing either way. The collective good will not automatically evolve, even though the individuals in question may possess full and accurate information about the potential benefits of improving the channel and may have the required know-how and resources to do so.

To make matters even more difficult, maintaining an irrigation system over the long term requires immediate and costly contributions of labor or fees, while benefits are hard to measure and dispersed over time and space. Whatever allocation rules that officials and/or farmers establish for an irrigation system, there is the temptation to cheat by taking more water than authorized, by taking water at an unauthorized time, or by contributing less inputs than required for provision of one's given water allocation. For example, rice farmers prefer to keep their rice paddies flooded continuously, since rice is intolerant to drying and highly tolerant to excess water. Extra water keeps weeds under control on those paddy fields that obtain the water but could be used more efficiently to yield a larger quantity of rice.

Empirical evidence from field and experimental settings amply demonstrates that, without effective institutions, common-pool resources will be underprovided and overused (Cordell 1978; E. Ostrom Gardner, and Walker 1994; Clark 1974; Larson and Bromley 1990). Substantial controversy exists, however, concerning how to remedy the problem. Many analysts presume that common-pool resource appropriators are trapped in a Hobbesian state of nature and cannot themselves create rules to counteract the perverse incentives they face. The logical consequence of this view is to recommend that an external authority—"the" government—take over the commons. Further, where technical knowledge and economies of scale are involved, it is presumed that this external force should be a large, central government. Central governments are seen as necessary agents of change to break the control of powerful individuals in rural areas who underinvest in collective action and obtain a disproportionate share of whatever is generated. Of course, recommending government action can lead to a wide range of policy interventions, from having the government impose a market to recommending that national governments manage common-pool resources themselves.

The theoretical presumption that an external, central government is necessary to supply and organize forms of collective action, such as providing irrigation works, has been reinforced by the colonial experience. During the colonial period in many parts of Asia and some parts of Africa, large-scale governmental bureaus were established to develop previously unirrigated areas. Such areas were opened for settlement and oriented toward the production of cash crops for export. The resulting centralization of governmental power over the supply of irrigation water has been continued, in most instances, by the governments that were created as colonial powers left the scene. In much of the developing world, irrigation development "has been highly distorted by the process of state concentration of investments and governance, and the concomitant demise of local rights and initiatives" (Barker et al. 1984, 26).

National governments in many parts of the developing world have come to be perceived as the "owners" of all water, as well as other natural resources (Sawyer 1992). From this viewpoint, national governments become the only agency that should or could invest in constructing and managing irrigation systems. This orientation toward the necessity of central authority is intensified by a second presumption that supplying irrigation requires considerable technical expertise, which is unlikely to be found locally. Together these lead to a belief that "scarce technical expertise is best located in a powerful state bureaucracy where it can be effectively dispensed" (Barker et al. 1984, 26). International aid agencies reinforce this predilection toward professional, central control over the supply of irrigation water by their inclination to work directly with the central ministries of the national government to whom aid funds are extended.

Considerable recent research has challenged the assumption that appropriators cannot set rules affecting the use of common-pool resources (Bardhan 1995). Considerable empirical evidence from field and experimental settings holds that appropriators frequently do constitute and enforce their own rules and that these rules work.[2] The initial research has focused primarily on groups where most of the relationships among participants are relatively symmetric with regard to assets, interests, and the physical situation and on situations where the problems they face are relatively simple to overcome. In such situations, appropriators do take "time out" from operational decisions to design rules that improve the joint yield that they can obtain (E. Ostrom 1990; E. Ostrom, Gardner, and Walker 1994; Berkes 1989; V. Ostrom, Feeny, and Picht 1993; Berkes et al. 1989; McCay and Acheson 1987; Wade 1988; Bromley et al. 1992). In Valençia, Spain, Hirano, Japan, and Törbel, Switzerland, for example, local appropriators have designed, monitored, and enforced their own rules to sustain intensive use of local common-pool resources for many centuries (Maass and Anderson 1986; McKean 1992; Netting 1981).

Now more difficult questions can be posed. These tougher questions center on whether individuals who differ substantially—in regard to their economic or political assets, their information, or their physical relationship—can craft rules that enhance joint output, distribute output equitably, or both (Johnson and Libecap 1982; Keohane, McGinnis, and E. Ostrom 1993). Of course, the answer to this question will likely be: "It depends." So the task of this research agenda is to develop a coherent understanding of the set of conditions that enhances or detracts from self-organizing capabilities when individuals differ substantially from one another.

For the sake of concreteness, this essay focuses on the asymmetry present on most irrigation systems between those who are physically near the source of water (the headenders) and those who are physically distant from it (the tailenders). This essay first explores the interaction between head-end and tail-end farmers, particularly their decisions about whether to devote resources to the upkeep of the irrigation system, and how bargaining between the parties can benefit all sides. Finally, we examine empirical evidence from a study of irrigation institutions in Nepal, and discuss the broader practical significance of our findings.

An Irrigation "State of Nature" Game

In large-scale, centrally constructed irrigation systems, the headenders and the tailenders are in very different positions. Narrowly selfish headenders would ignore the scarcity that they generate for those lower in the system. But if the headenders get most of the water, those at the tail end have even less reason to want to contribute to the continual maintenance of their system. All common-pool resources generate both appropriation and provision problems. In an irrigation common-pool resource, the appropriation problem concerns the allocation of water to agricultural production; the provision problem concerns the maintenance of the irrigation system. In addition, irrigation common-pool resources also have an asymmetry between headenders and tailenders, which increases the difficulty of providing irrigation systems over time.

We model the strategic interaction between headenders (player 1) and tailenders (player 2) of an irrigation system as follows. There is a temporary structure, a headworks, at the very beginning of the system, which brings water into the system. This structure has to be rebuilt annually. The total amount of water (W) brought into the system depends on how much labor headenders (L_1) and tailenders (L_2) provide. The decision to provide labor is taken simultaneously under the condition of complete information. Once water starts flowing, headenders, who get first crack at it, take the lion's share of it (75 percent), while tailenders get what is left (25 percent). The opportunity cost of providing a unit of labor is constant throughout the system and equal to 1.

Since the payoffs to each end of the system depend on what both ends do, the situation they face is a game of strategy. However, the incentives facing the two ends of the system are very different. The headenders have a physically advantageous position and get most of the water. The first order condition on headenders' labor says that the marginal product of labor equals its opportunity cost ($.75 \, \partial W/\partial L_1 = 1$). The first order condition on tailenders' labor says that the marginal product of their labor also equals opportunity cost ($.25 \, \partial W/\partial L_2 = 1$). Under the usual concavity assumptions, these first order conditions imply that headenders supply more labor than tailenders. We should observe a pattern over time in which headenders contribute more labor and get more water.[3]

This model is illustrated by the Thambesi irrigation system in Nepal, organized by farmers, where the Thambesi River provides a source of water that is easy to channel and requires little annual maintenance (unlike most farmer-organized systems). The headworks of the Thambesi system is a simple brush and stone diversion works that is easily adjusted each year to fit changes in the course of its source (Yoder 1986, 179). Routine maintenance carried out prior to the monsoon rains requires "only four to five hours of work with all the members participating" (180). As a result, it is possible for some of the farmers alone to keep this system going. "The members with holdings in the tail cannot force those with land above theirs to deliver water to them equally by not participating in maintenance and other system activities" (179).

Thambesi is one of the few farmer-organized systems where headenders have clearly established prior rights to water over those lower in the system.

Farmers at the head of each rotation unit "fill their fields with water first before those further down the secondary are able to take water" (Yoder 1986, 292). During the pre-monsoon seasons, farmers at the head of the system grow water-intensive rice. Consequently, no one lower in the system can grow an irrigated crop during this season. If the headenders were to grow wheat instead of rice, an area nearly ten times as large could be irrigated during the pre-monsoon season (313).[4] In this system, measured crop yields are correlated with distance from the headworks. Substantial land that could benefit from irrigation depends entirely on rainfall.

The game equilibrium with headenders contributing more than tailenders has undesirable properties, in the sense that production will be less than optimal and the system will be undermaintained. The tailenders work less for less water, and the whole system suffers. These considerations suggest that irrigators have good reason to leave the state of nature and reconstitute their system, crafting better rules to follow. Indeed, when the equilibrium is most seriously inefficient, the incentives to seek out a new system are greatest. The next section takes up the issue of bargaining over the rules of the game.

Bargaining over the Rules of the Game

If their rights to govern and manage their own irrigation system are recognized, or at least not interfered with, farmers on systems that require much more labor input than Thambesi can take a "time-out" between seasons and attempt to add or reform rules that improve their system outcome (Gardner and E. Ostrom 1991). The annual meetings of irrigators to decide on the rules affecting appropriation and provision activities can be modeled as a bargaining problem. If no agreement is reached in these annual meetings, the irrigators return to the equilibrium of the state of nature game discussed in the previous section.

The challenge to bargaining is to find an outcome that benefits both parties, relative to the prevailing state of nature equilibrium. After all, if either party is not improving itself, it will not accept the bargain.

We illustrate these ideas with a simple numerical example. The water production function is $W = 2(L_1^{0.5} + L_2^{0.5})$. The objective function for the entire system is to maximize water less the opportunity cost of labor, that is, maximize $W - L_1 - L_2$. The optimum is achieved when the marginal product of labor at each end of the system equals its opportunity cost, 1. Solving the resulting first order conditions, we find that each end of the system supplies 1 unit of labor, producing 4 units of water. By contrast, in the equilibrium in the state of nature, the headenders would supply .56 units of labor; the tailenders, .06 units; and only .2 units of water are produced.[5] Labor is seriously underprovided, and water is a fraction of what it should be. Of this flow of water, 75 percent (1.5

units) would go to the headenders; 25 percent (.5 units), to the tailenders.

A bargain in this situation would consist of a water-for-labor exchange. That is, the headenders would agree to work more, as would the tailenders. In exchange, the tailenders would get a lot more water. One possible bargaining solution would be as follows. The head end works .86 more; the tail end, .98 more. This gets them up to 1 unit of labor each, the optimal labor input. Extra water is proportional to extra labor, so the tail end gets $.94/(.44 + .94) = 68$ percent of the additional water produced. The water reaching the tail end rises from .5 to $.5 + .68(2) = 1.86$, while the head end comes out of this bargain with $1.5 + .32(2) = 2.14$ units of water. Everyone comes out ahead.[6]

This bargaining solution has an important empirical implication. Notice that the difference in the quantity of water allocated to the head end and the quantity allocated to the tail end has shrunk, due to the bargain struck. In the state of nature, the head-tail difference was $1.5 - .5 = 1$; in the bargain, the difference is only $2.14 - 1.86 = .28$. Although there is still a difference, it is considerably lower (36 percent). A reduction of the difference between head-tail water allocation of this magnitude should show up empirically. In the next section, we test the hypothesis that head-tail water differences are reduced in those systems where farmers are authorized to devise agreements between the two ends of the system.

Several factors in the field can affect the bargaining problem between headenders and tailenders. For instance, the presence of a permanent headworks, which reduces overall labor required, will favor headenders over tailenders in the event that negotiations break down. Headenders may even maintain the system all by themselves and take all the water they want. Such a production asymmetry reinforces the locational asymmetry. On the other hand, if there is real mutual dependence, and the productivity of labor of the tailenders affects the distributional advantage of the headenders, then the asymmetries tend to offset one another and the bargains struck are relatively symmetric.

In instances of symmetry, an entire family of rotation rules is used in practice that enables the irrigators to split the water and the labor about evenly. As examples, here are two rotation rules that suffice to transform the state of nature game into a game with a symmetric bargaining solution.

In *Rotation Rule A,* in odd-numbered years, water goes first to the headender, and in even-numbered years water goes first to the tailender. Both headenders and tailenders work side by side for any days devoted to maintaining the system. An example of a similar rule, based on season rather than year, is found on the Marpha farmer-organized system in the Mustang District of Nepal, where barley is planted in the winter and buckwheat is planted in the summer. Fields devoted to barley were watered from the top of the system to the bottom, while the ordering for buckwheat was reversed so that the tail-end fields received water first (Messerschmidt 1986).

In *Rotation Rule B,* all water in the system is allocated to the headenders for even days of the seasons and all to the tailenders for the odd days of the season. Headenders maintain the canal for one time period, and tailenders maintain the canal for an alternative, but equal, time period. This rule is illustrated by the Yampa Phant system described below.

Both of these rules result in an equal split of the rights of appropriation and the duties of provision. Whether written or unwritten, rotation rules of this sort are the common knowledge of all participants on a farmer-governed system. Further, farmers actively monitor each others' conformance to these rules (Weissing and E. Ostrom 1991, 1993).

Over time, these rules can easily be adapted to a situation where both water and labor are allocated in an unequal but proportional manner. For example, Rotation Rule A might be amended to say that in years not divisible by 3, water goes first to the headender, while in years divisible by 3, water goes first to the tailender. Headenders devote twice as much labor as tailenders on side-by-side workdays devoted to maintaining the system. The analogous version of Rotation Rule B would state that all water in the system is allocated to the headenders for days of the seasons not divisible by 3. All water in the system is allocated to the tailenders for days of the season divisible by 3. Headenders maintain the canal for twice the time period that tailenders do.

Of course, the rules as actually discovered in field settings are rarely this cut and dried. The rotations involved are often quite a bit more complicated. Rules that relate to routine maintenance frequently differ from those that relate to emergency repairs. Also, the proportionality may be multidimensional: water allocated is proportional to one variable, while labor provided is proportional to another. The set of variables upon which proportionality is based can include landholding, total labor in the household, labor relative to land held, physical capital, and votes held in a voting system. When rules are based on a clear principle of proportionality and all participants recognize that the rules enable them to reach better outcomes than feasible in the "state of nature" game, and all are prepared to punish rule breakers, more productive equilibria are reached and sustained over time. Although the details of any given irrigation system can appear bewildering, the strategic principles we have identified above are recurring regularities in data drawn from the field. It is to these data that we now turn.

Empirical Evidence

It is difficult to conduct empirical tests of these kinds of theoretical findings. One should expect to observe a wide diversity of outcomes regarding the level of net benefits achieved, their distribution between the head and tail portions of irrigation systems, their persistence over time, their efficiency, and their equity. The real world offers any number of potentially confounding variables, raising the risk that any seeming failure of the theory might be justified, after the fact, by introducing a new explanatory variable. Until recently, no large-scale data set with the appropriate variables existed that could be used for this purpose. However, the discovery of a large number of case descriptions of farmer- and government-operated irrigation systems in one country, Nepal, has led to the development of the Nepal Institutions and Irrigation Systems database, in which data for 127 systems has been coded (E. Ostrom, Benjamin, and Shivakoti 1992).

Nepal has an area of about 141,000 square kilometers, slightly larger than England. Its 18 million inhabitants are engaged largely in agriculture. Of the approximately 650,000 hectares of irrigated land, irrigation systems operated by farmers cover about 400,000 hectares, or 62 percent (Small, Adriano, and Martin 1986). The remaining irrigated land is served by a variety of agency-managed irrigation systems, many of which have been constructed since 1950 with extensive donor assistance. On some of the agency-managed systems, farmers have organized themselves and do participate in second-level, choice-of-rule games, but they are far more active in devising appropriation and provision rules on farmer-managed systems.

Irrigation occurs extensively in the hills (frequently quite steep), in the river-valleys (the terrain is more undulating), and in the flat and more fertile Terai, located in the southern part of the country, which has only been devoted to extensive agriculture since the successful eradication of malaria. In the hills, irrigation systems have several plateaus where it is very easy for farmers on the first plateau to get most of the water to their fields before any water goes on to the second or third plateau. Thus, one would expect that the problem of physical asymmetries would be easier to deal with in the Terai than in the hills.

In Nepal, farmer-managed irrigation systems achieve a high average level of agricultural productivity. Of the 127 systems in the Nepal database, we have productivity data for 108. The 86 farmer-managed irrigation systems average 6 metric tons per hectare a year; the 22 agency-managed irrigation systems only 5 metric tons per hectare, a statistically significant difference ($p = .05$). Farmer-managed irrigation systems tend also to achieve higher crop intensities.[7] A crop intensity of 100 percent means that all land in an irrigation system is put to full use for one season or partial use over multiple seasons amounting to the same crop coverage. Similarly, a crop intensity of 200 percent is full use for two seasons; 300 percent, full usage of all land for three seasons. Farmer-managed irrigation systems achieve a higher average crop intensity (247 percent) than do agency-managed irrigation systems (208 percent). Again, the difference is statistically significant.

The agricultural yields and crop intensities that farmers obtain depend on whether they can be assured of water during the winter and spring seasons, when water becomes progressively scarcer. A higher percentage of farmer-managed irrigation systems in Nepal are able to get adequate water to both the head and the tail of their systems across all three seasons, as shown in table 14.1. During the spring when water is normally very scarce, about 1 out of 4 farmer-managed irrigation systems is able to get adequate water to the tail of its system, while only 1 out of 12 agency-managed irrigation systems gets adequate water to the tail of its system. Even in the summer monsoon season, only about half of the agency-managed irrigation systems get adequate water to the tails, while almost 90 percent of the farmer-managed irrigation systems get water to the tails of their systems. It is pretty clear that most of the farmer-managed systems have engaged in substantial bargaining to get out of the state of nature game and achieve higher equilibria possible by adopting their own commonly understood rules.

TABLE 14.1. Water Adequacy[a] by Type of Governance as Arrangement and Season

Season of Year	Percentage of FMIS with Adequate Water at the Head	Percentage of FMIS with Adequate Water at the Tail	Percentage of AMIS with Adequate Water at the Head	Percentage of AMIS with Adequate Water at the Tail
Monsoon	97	88	92	46
Winter	48	38	42	13
Spring	35	24	25	8

[a]Water adequacy was measured on a four-point scale from adequate to nonexistent based on structured coding of field visits and case studies.

To further address why farmer-managed irrigation systems are more likely than agency-managed systems to distribute water more equitably between head and tail, we conduct a regression analysis. This shows how physical variables and type of governance structure combine to affect the difference in water availability achieved at the head and the tail of irrigation systems.

The dependent variable in our regression is called "water availability difference." This variable compares the availability of water at the head of the system minus the availability of water at the tail of the system, averaged across three seasons. We measured water availability on a scale with three possibilities: adequate, receiving a score of 2; limited, with a score of 1; and scarce or nonexistent, with a score of zero. Thus, an overall score of zero indicates that for all three seasons, the level of water adequacy was the same in the head and tail sections of the system. A score of .33 indicates that in one season, the head received adequate water and the tail received limited water or that the head received limited water and the tail received scarce water.[8]

The independent variables in the regression include the length of the canals in the system (measured in meters); the labor input (measured by the number of labor days devoted to regular maintenance each year, divided by the number of households served); four dummy variables, for the presence of permanent headworks, lining of the canals, whether the system is in the Terai, and whether the system is farmer managed, and a constant.

We have data on all of these variables for 76 of the irrigation systems.[9] The length and labor input variables are not significantly different from zero. However, the coefficients on headworks (.34) and whether the system is farmer managed (–.32) were both significant at the 95 percent level, while the coefficients on the lining of canals (–.14) and whether the system was in the Terai (–.10) are both significant at the 90 percent level. The regression also contained a constant term of .64, significant at the 95 percent level. However, the R^2 for the regression was only .28. At this point, we treat these results as initial and tentative. Field teams are collecting data from a larger set of systems, so additional analyses will be conducted on a larger data set in the future.

This preliminary analysis raises questions about how the physical characteristics of an irrigation system affect the capacity to distribute the gains from mutual cooperation equitably. The difference in water availability achieved at the head and the tail of these Nepali irrigation systems is significantly and negatively related to being in the Terai, presumably because the headender advantage is less marked in the plains than in the hills. The presence of permanent headworks—which are frequently considered as one of the hallmarks of a modern, well-operating, irrigation system—is positively related to an inequality between the water availability achieved at the head and the tail. Presumably, one reason is because permanent headworks increase the bargaining position of headenders, relative to tailenders. Lining of canals, on the other hand, preserves more water for tailenders and reduces water availability differences. Finally, the difference in water availability is significantly reduced at the tail of farmer-managed irrigation systems, as compared to that of agency-managed irrigation systems, presumably because farmer-managed systems are more likely to reach bargaining solutions about their own operational rules that more effectively take tailender interests into account.

The construction of a permanent headworks has frequently been funded by external sources, with farmers not required to repay the cost of this investment. This type of external "aid" substantially reduces the need for mobilizing labor (or other resources) to maintain the system each year, a reduction that has normally been interpreted in project plans entirely as a benefit. However, this claim of undiluted benefit is made too quickly; the construction of extensive infrastructure facilities without requiring that capital investment be paid back by beneficiaries has two adverse consequences. First, without a realistic requirement to pay back capital investments, farmers and host government officials are motivated to invest in rent-seeking activities and may overestimate previous annual costs to obtain external aid (Repetto 1986). Second, this form of aid can change the pattern of relationships among farmers within a system, reducing the recognition of mutual dependencies and patterns of reciprocity between headenders and tailenders that have long sustained the system. By denying the tailenders an opportunity to invest in the improvement of infrastructure, external assistance may prevent those who are most disadvantaged from being able to assert and defend rights to the flow of benefits (Ambler 1990). Let us offer an example of these distressing consequences as they occurred in an agency-managed irrigation system.

The Kamala Irrigation Project, located in the Terai, illustrates the problem of building sophisticated and expensive capital structures without paying attention to the design of institutions that will bring appropriation and provision decisions into close juxtaposition. Kamala was constructed during the 1970s by the Department of Irrigation (then called Department of Irrigation, Hydrology and Meteorology). It was originally designed to serve 25,000 hectares in a section of the Terai where farmers had practiced rain-fed agriculture but had not been previously organized to provide their own irrigation systems. A permanent, concrete headworks and a fully lined canal were constructed with the financial assistance of the Asian Development Bank. The system was completed during the 1983–84 agricultural year. Since then, no water fees have been imposed or collected. The system has never been able to supply irrigation water to all of the lands within its official service area.[10] The Kamala Project staff is financed from the general revenues of the central government, and few funds have been collected to provide the continued operation and maintenance of the system. Project personnel spend most of their time operating and maintaining the huge concrete headworks that divert water from the Kamala River to the main and branch canals and very little time maintaining the rest of the system. Few field canals have been constructed, and farmers have broken through the branch canals to obtain water.

Neither government nor farmers have undertaken responsibility for operating or maintaining the system below the headworks. The absence of organization at the water use level has generated a high level of conflict and led to an inequitable water distribution pattern. As described by a field research team (Laitos et al. 1986, 147):

> Water allocation is primarily first come, first served. Thus, farmers at the head . . . tend to get all the water they need, while farmers at the tail often receive inadequate and unreliable amounts of water. This situation has often led to conflict between head and tail farmers. Sometimes hundreds of farmers from the area near the middle village of Parshai will take spears and large sticks and go together to the head village of

Baramajhia to demand that water be released. At Baramajhia, farmers are often guarding their water with weapons. If water is released, Parshai farmers have had to maintain armed guards to assure that the . . . canal remains open.

Even with all of the investment made in physical works, this irrigation system is operating in a "state of nature" with regard to the establishment of rules to allocate water or provision responsibilities. The agricultural yields obtained in this system vary dramatically, depending on water availability, but crop intensity is generally below average: at the head, 180 percent; at the tail, 150 percent.

In marked contrast to the Kamala Irrigation Project is another agency-managed irrigation system, the Pithuwa Irrigation Project, also located in the Terai. While the Department of Irrigation invested in constructing and lining 16 branch canals, it did not attempt to build a permanent intake structure. The system was designed to serve about 600 hectares, but farmers have extended the system to serve 1,300 hectares of land by rotating the fields devoted to rice during the monsoon on an every-other-year basis. The location of many of the larger landowners near the tail of the system (which is also near to the east-west highway and thus low cost transportation of produce to markets) is a fortuitous circumstance that counterbalanced the lack of attention by the government to the design of complementary allocation and resource mobilization rules. Even with many absentee landowners and sharecropping arrangements, this system has managed considerable self-organization.

While there are many effective farmer-managed irrigation systems in its district, the area served by the Pithuwa system depended on rain-fed agriculture and was not organized prior to canal construction. In the early days of project operation, water distribution was based on a "might is right" principle, and there were many conflicts and feuds similar to those that still exist on the Kamala Project. The origin of the current high level of farmer participation at Pithuwa is interesting as it evolved from organization on one branch at the tail of the system to the organization of the entire system (Laitos et al. 1986, 126–27).

[O]ne prominent farmer took the initiative to organize the other farmers on Branch 14 into a committee, which formulated rules for water allocation and distribution along Branch 14. With farmer participation in committee activities, conflicts over water sharing along the branch canal decreased in a short time. Other branches started to follow the example set by the farmers of Branch 14. Eventually, all of the branch farmers created branch committees for water allocation and distribution....Once the branch canal committees were working satisfactorily, a federation of the branch canal committees created a general assembly of farmers and a main canal committee.

From this initial organization, a two-tier system now covers the entire system and governs many aspects related to the operation of the overall system, as well as individually crafted rules to operate each of the 16 branches. The main canal committee is responsible for allocating water among the 16 branch canals. During the monsoon season when water is abundant, all branches receive water. Given the crop rotation, half the farmers plant rice, while the other half plant vegetables, seeds, and fibers.[11] This crop rotation, which is tantamount to a water rotation, allows a doubling of the irrigated area. When water is scarce, however, the "committee arranges a rotation system. They then allocate water to the tail outlets first for a set number of days and then to the head outlets" (Laitos et al. 1986, 130). Each branch committee determines its own allocation rules and the rules differ substantially from branch to branch.

> In branches 1 and 2, four hours of water per *bigha* (0.66 hectare) are allocated, whereas in branches 3 and 16, two hours per *bigha* are allocated. The time for allocation is based on the nature of the soil, the size of the fields, the volume of water available, and the frequency of watering required for the crop. On some branches, daytime water is allocated for transplantation, and nighttime water is allocated for fields that are transplanted. Each committee has adopted rules that suit their soil, crops, and the availability of water in the branch canal.

Further, adequate representation of the head and tail sections of each branch is assured by a rule that if the chairman of a branch committee comes from the head portion of that branch, the secretary must come from the tail and vice versa (Giri and Aryal 1989, 15).

Given the strength of the branch and system committees, the Department of Irrigation has gradually turned over the maintenance and operation of this system to the farmers. Repairing the intake structure each spring before the monsoon season is a huge task, implemented with the assistance of a government bulldozer and a budget to purchase the fuel for the bulldozer. The responsibility for cleaning the branch canals has been assigned to each branch canal committee using several methods. Some branches "contract out" the cleaning of their canals based on a competitive bidding system, whereby the lowest bidder is awarded the contract for a year. The funds to pay for this maintenance are raised by the farmers through an assessment imposed by the farmers' committee based on the size of a farmer's holding. On other branches, the farmers themselves clean the canals, and the rules for labor mobilization are determined by that branch.[12]

Agricultural practices on this system are considered to be among the best of the agency-managed irrigation systems. The average crop intensity achieved at the tail end of the system (228 percent) is slightly higher than at the head end (221 percent), which is explained by the location of larger farms at the tail of the system and the strong incentives created by the proximity of the tail to an all-weather road. Soils and other factors are similar throughout the system. Even though this is formally an agency-managed irrigation system, the farmers have nearly the same local authority as in a farmer-managed irrigation system.

On farmer-managed irrigation systems, the diversity of appropriation and provision rules adopted is substantial and closely related to the types of labor requirements needed to keep the systems operational. The owner-operators of the Yampa Phant irrigation system, a very old 40-hectare system located in the hills, for example, do not need to invest in massive resource mobilization during the spring to build or repair their headworks, as they and a neighboring system have built a permanent storage structure to retain water from a perennial spring. They are, however, concerned about the daily upkeep of their 12 outlets during the monsoon rains and have devised a labor rotation system during the period of peak labor demand (Laitos et al. 1986, 97).

> During the summer paddy season, maintenance responsibility is rotated among the 12 outlets daily. One laborer per day per outlet is required. After 12 days, responsibility shifts back to the farmers served by the first outlet. Within each field channel served by an outlet, farmers also rotate the responsibility for main system maintenance. Each farmer takes a turn inspecting the main canal and making necessary repairs. Everyone participates during an emergency.

During periods of water abundance, water is available on demand. During the winter season of water scarcity, the upper six outlets receive water for one 24-hour period, and the lower six outlets receive water for the next 24-hour period. The farmers of Yampa Phant have thus devised a set of rules that is remarkably like stylized Rule B noted earlier. These farmers average 7.75 metric tons of grain per hectare per year. Most farmers at both ends of the system grow three crops every year, so the difference between head and tail end production is negligible.

The farmers of the Kerabari irrigation system, in the Terai, face a different type of resource mobilization problem and have adapted a different set of rules. Constructed by farmers during the 1970s, this farmer-managed irrigation system draws water from a stream (the Khadam Khola) that carries a large quantity of sediment from the foothills during the monsoon rains. Even though the main canal built by the farmers is considered by outsiders to be "quite an engineering feat," several attempts by the farmers and by the government to build a permanent intake structure have been washed away (Laitos et al. 1986, 217). Thus, despite past government efforts to help the farmers of Kerabari economize on the annual effort required by a temporary headworks, the farmers must continue to deal with floods and washouts. During the spring of 1985, for example, 150 farmers had to work for 15 days to repair the main canal (Laitos et al. 1986, 219). Two branch canals serve the upper Khadam and lower Khadam farmers. All farmers on this system own their own land, and land is relatively equally distributed.

When this farmer-managed irrigation system was first organized, there was one committee for both subsystems, but the "lower Khadam farmers felt that the upper Khadam farmers were less active and enthusiastic about system maintenance and operation. They divided the committee into the upper and lower Khadam committees but agreed to have one common chairman [who owns land in both systems] for both committees" (Laitos et al. 1986, 22). When water is abundant, each farmer withdraws whenever desired. During the spring when water is scarce, decisions about cropping patterns are made within the two branch committees, and rotation systems are devised that ensure that water is adequate for the cropping patterns that have been jointly established.

All households owning less than two *bighas* (1.32 hectares) of land—about two-thirds of the owner-operators—send one laborer for each day of maintenance decided upon by the joint committee. Those owning more than two *bighas* of land contribute one laborer for each two *bighas* owned. Given that many farmers own much less than one *bigha* of land, this rule places a heavier burden on small landowners than on large landowners.[13] The joint committee has mobilized cash from the farmers to line the canal and has sought external assistance to try to address the headworks issue. At least 90 percent of the fields located in both the upper and lower sections are planted during each of the three growing seasons. Farmers have adopted high yield varieties and good agricultural practices and produce about 9.1 metric tons per hectare per year—well above average.

Implications

Asymmetries among participants facing common-pool resource provision and appropriation problems can present substantial barriers to overcoming the disincentives of the "state of nature" game between head-end and tail-end farmers. However, these asymmetries are frequently overcome in settings where farmers are made aware of their mutual dependencies; after all, head-enders may need the resources provided by tailenders when it comes to maintaining the system over time. Moreover, such bargaining over new rules can only work if the players have some assurance that the efforts they make to devise and enforce new appropriation and provision rules will not be undercut by external authorities. In a monetized and self-financed system, for example, tailenders are unwilling to contribute their water fees unless they obtain sufficient, reliable water so that the increased yield obtained by taking water is greater than the fees assessed on them.[14]

What is striking from an examination of the farmers in Nepal, the Philippines, and Indonesia and in other developing countries where they have been effectively allowed to self-organize, is the diversity of rules that results from the tough bargaining that farmers engage in during their annual meetings (Coward 1980; Geertz 1980; Hunt 1989; Korten and Siy 1988; E. Ostrom 1992; Siy 1982; Tang 1992). As would be expected from our theoretical analysis, not all of these bargaining efforts achieve rules that enhance efficiency and/or equity.[15] But as long as mutual dependencies are clear to all participants, and they expect to relate to one another for a long time into the future, farmers in the developing world demonstrate substantial capabilities to craft rules that lead to higher yields and to a reduction in the asymmetry of results—and to enforce those rules as well. Many government agencies have wanted to impose the type of cropping patterns that exist on the Pithuwa and Kerabari systems, or the monetary fees that are collected by the Pithuwa farmers, but have not been able to gain sufficient cooperation from farmers to implement such policies.

Much of the emphasis in the development literature has been on the importance of physical technology to improve irrigation and agricultural performance, rather than institutions. There is little question that appropriately designed modern irrigation works can enhance the agricultural yield and efficiency of many farmer-organized systems. But interventions designed by outsiders that ignore the potential disruption of the mutual dependencies and reciprocal relationships among farmers may cause more harm than good. For example, aid in the form of grants (or loans that are never paid back by those who directly benefit from them) used to invest in the physical capital of irrigation systems may remove any need by headenders to recognize the needs of tailenders for water. In the ensuing state of nature battle over water, if no one pays fees or contributes labor to maintain the system, then the supply of irrigation may decline, even in the presence of outside aid.

Belief in the capability of external agencies to solve common-pool resource problems should be balanced with a recognition that external agencies can sometimes be disruptive. There is much more to learn about the capabilities and limits of various institutional arrangements to cope with diverse problems, but self-governing common-pool resource communities have demonstrated their ability to perform at very high levels of efficiency and equity.

NOTES

1. Our definition of common-pool resources focuses on the structure of the *resource,* while Seabright's (1993) essay focuses on the *property regime* related to a resource

(or other common property). Most farmer-organized irrigation systems fit Seabright's conception of a common-property regime.

2. This research is closely related to the more general question of how institutions evolve and perform (Calvert 1996; Knight 1992; V. Ostrom 1991; Milgrom, North, and Weingast 1990).

3. Corner solutions are also possible, but we still have $L_1 > L_2$.

4. Yoder's (1986) research also demonstrates that where the farmers own shares of the water system and can sell these shares, the incentives faced by headenders are entirely different. In such a property-rights regime, headenders can capture part of the capitalized value of the benefits of allocating more water to those lower in the system. Thus, a change in the institutional arrangements related to a physical environment dramatically changes the incentives of participants and the efficiency and equity of outcomes.

5. In the state of nature, the first order condition for headenders is $.75\partial W/\partial L_1 = 1$; for tailenders, $.25\partial W/\partial L_2 = 1$. Since $\partial W/\partial L_1 = L_1^{-0.5}$, we get $L_1 = (.75)^2$. Similarly, we get $L_2 = (.25)^2$.

6. There are many solutions to this bargaining problem. Well-known solutions, such as those of Nash and Kalai-Smorodinsky, yield predictions qualitatively similar to this one.

7. It should be noted that farmer-managed systems are on average smaller than agency-managed irrigation systems, but the size of the system is not significantly related to agricultural productivity when we control for the type of governance and for other physical attributes controlled by governance, like the presence of permanent headworks and at least partial lining of the irrigation channels (Lam, Lee, and E. Ostrom 1997). Some farmer systems as large as 15,000 farmers have sustained themselves for very long periods of time. Scale is, however, an important fact affecting the cost of organizing these (or agency-operated) systems.

8. For the 118 systems for which we have data, the difference score ranges from −.66 to 1.66. The regression presented in the text is based on data for 76 systems for which we had data on all variables in the regression equation. A parallel analysis using multinomial probit estimates yields parallel findings concerning the direction and significance of permanent headworks and type of system, but the negative relationship between terrain and the difference score does not reach statistical significance.

9. This subset of 76 systems does not appear to differ on any relevant variable from the larger data set.

10. This could be due to an overoptimistic estimate of the service area in the first place, a problem that happens frequently when a positive benefit-cost ratio is required in the project design phase.

11. Crop rotations such as this require the full cooperation of farmers since they are forgoing choice about what to plant on the land they own. Once established, they are easy to monitor.

12. Farmers are allowed to register for more irrigation water than the land they have. This requires such farmers to invest more in maintenance but also to obtain more water in proportion to their contribution. "Water rights earned through this are

saleable. Any farmer who has excess water rights in a particular year can sell his water to potential buyers" (Giri and Aryal 1989, 43).

13. Donald Curtis (1991, 76) provides a fascinating account of the change in rules for the Bansbote system on the Sewar River as it evolved from a system originally organized by a Zamindar, a local land-owning tax farmer. Under the Zamindar, each area of this system was required to send a fixed number of laborers whenever the system needed maintenance. When a farmer's committee took charge of the system, they changed the labor mobilization rule to each household sending one laborer (so households owning land in different areas were relieved of their double or triple obligations). Next, the small landowners prevailed on the large landowners to provide labor at the rate of one laborer per *bigha.* "Since in local terms this is still a fairly large unit of land the pressure is now on to have the rules changed again to require labour contributions on the basis of a smaller unit of land. But this move has not yet succeeded."

14. Svendsen (1992) analyzes the effects of imposing a water fee on users by the National Irrigation Agency of the Philippines, which was in turn assigned primarily to that agency rather than the general treasury. Since the fees were to cover the full costs of the project, one effect was the reduction of the subsidy given by the national government to the irrigation sector. A second major effect is a significant increase in the service areas of government systems and in the equity of water distribution. The evidence from Svendsen's study strongly indicates that the imposition of a user fee improved the situation of the more disadvantaged who could refuse to pay user fees unless they obtained adequate water for their investment.

15. The survival rate of farmer-governed systems is obviously not as large as that of government-owned systems that can draw on coerced taxation (and donor aid) to survive. It is, thus, not too surprising that farmer-run systems are more efficient than government systems given selection pressures to eliminate the less efficient.

REFERENCES

Ambler, John S., "The Influence of Farmer Water Rights on the Design of Water-Proportioning Devices." In Yoder, Robert, and Juanita Thurston, eds., *Design Issues in Farm-Managed Irrigation Systems.* Colombo, Sri Lanka: International Irrigation Management Institute, 1990, 37–52.

Bardhan, Pranab, "Rational Fools and Cooperation in a Poor Hydraulic Economy." In Basu, K., et al., eds., *Development, Welfare and Ethics: A Festschrift for Amartya Sen.* Oxford, England: Clarendon Press, 1995.

Barker, Randolph, E., Walter Coward, Jr., Gilbert Levine, and Leslie E. Small, *Irrigation Development in Asia: Past Trends and Future Directions.* Ithaca: Cornell University Press, 1984.

Berkes, Fikret, ed., *Common Property Resources: Ecology and Community-Based Sustainable Development.* London: Belhaven Press, 1989.

Berkes, Fikret, David Feeny, Bonnie J. McCay, and James M. Acheson, "The Benefits of the Commons." *Nature,* July 1989, 340, 91–93.

Bromley, Daniel W., David Feeny, Margaret McKean, Pauline Peters, Jere Gilles, Ronald Oakerson, C. Ford Runge, and James Thomson, eds., *Making the Commons Work: Theory, Practice, and Policy.* San Francisco: Institute for Contemporary Studies Press, 1992.

Calvert, Randall L., "Rational Actors, Equilibrium, and Social Institutions." In Knight, Jack, and Itai Sened, eds., *Explaining Social Institutions.* Ann Arbor: University of Michigan Press, 1996.

Clark, Colin W., "The Economics of Over-exploitation." *Science,* August 1ᶜ /4, 181:17, 630–34.

Cordell, John C., "Carrying Capacity Analysis of Fixed Territorial Fishing." *Ethnology,* January 1978, 17, 1–24.

Coward, E. Walter, Jr., ed., *Irrigation and Agricultural Development in Asia: Perspectives from the Social Sciences.* Ithaca: Cornell University Press, 1980.

Curtis, Donald, *Beyond Government: Organisations for Common Benefit.* London: Macmillan, 1991.

Freeman, David M., "Designing Local Irrigation Organization for Linking Water Demand with Supply." In Sampath, Rajan K., and Robert A. Young, eds., *The Social Economies, and Institutional Issues on Third World Management.* Boulder: Westview Press, 1991, 111–40.

Gardner, Roy, and Elinor Ostrom, "Rules and Games." *Public Choice,* May 1991, 70, 121–49.

Gardner, Roy, Elinor Ostrom, and James Walker, "The Nature of Common-Pool Resource Problems." *Rationality and Society,* July 1990, 2, 335–58.

Geertz, Clifford, *Negara: The Theatre State in Nineteenth-Century Bali.* Princeton: Princeton University Press, 1980.

Giri, Khadka, and Murari Aryal, *Turnover Process of Agency Managed Irrigation Systems in Nepal.* Kathmandu, Nepal: Department of Irrigation, Irrigation Management Project, 1989.

Hunt, Robert C., "Appropriate Social Organization? Water User Associations in Bureaucratic Canal Irrigation Systems." *Human Organization,* spring 1989, 48:1, 79–90.

Johnson, Ronald N., and Gary D. Libecap, "Contracting Problems and Regulation: The Case of the Fishery." *American Economic Review,* December 1982, 72:3, 1005–22.

Keohane, Robert O., Michael McGinnis, and Elinor Ostrom, *Linking Local and Global Commons. Proceedings of a conference held at the Harvard Center for International Affairs.* Cambridge, April 23–25, 1992. Bloomington: Indiana University, Workshop in Political Theory and Policy Analysis, 1993.

Knight, Jack, *Institutions and Social Conflict.* New York: Cambridge University Press, 1992.

Korten, Frances, and Robert Siy, Jr., eds., *Transforming a Bureaucracy: The Experience of the Philippine National Irrigation Administration.* West Hartford: Kumarian Press, 1988.

Laitos, Robby et al., "Rapid Appraisal of Nepal Irrigation Systems." Water Management Synthesis Report No. 43. Fort Collins: Colorado State University, 1986.

Lam, Wai Fung, Myungsuk Lee, and Elinor Ostrom, "The Institutional Analysis and Development Framework: Application to Irrigation Policy in Nepal." In Brinkerhoff, Derick, ed., *Policy Studies and Developing Nations: An Institutional and Implementation Focus,* vol. 5. Greenwich, CT: JAI Press, 1997, 53–85.

Larson, Bruce A., and Daniel W. Bromley, "Property Rights, Externalities, and Resource Degradation: Locating the Tragedy." *Journal of Development Economics,* October 1990, 33:2, 235–62.

Maass, Arthur, and Raymond L. Anderson, . . . *and the Desert Shall Rejoice: Conflict, Growth, and Justice in Arid Environments.* Malabar: R. E. Krieger, 1986.

McCay, Bonnie J., and James M. Acheson, *The Question of the Commons: The Culture and Ecology of Communal Resources.* Tucson: University of Arizona Press, 1987.

McKean, Margaret A., "Management of Traditional Common Lands (*Iriaichi*) in Japan." In Bromley, Daniel W., ed., *Making the Commons Work: Theory, Practice, and Policy.* San Francisco: Institute for Contemporary Studies Press, 1992, 63–98.

Messerschmidt, Donald A., "People and Resources in Nepal: Customary Resource Management Systems of the Upper Kali Gandaki." In National Research Council, *Proceedings of a Conference on Common Property Resource Management.* Washington, D.C.: National Academy Press, 1986, 455–80.

Milgrom, Paul R., Douglass C. North, and Barry R. Weingast, "The Role of Institutions in the Revival of Trade: The Law Merchant, Private Judges, and the Champagne Fairs," *Economics and Politics,* March 1990, 2:1, 1–23.

Netting, Robert McC., *Balancing on an Alp.* New York: Cambridge University Press, 1981.

Ostrom, Elinor, *Governing the Commons: The Evolution of Institutions for Collective Action.* New York: Cambridge University Press, 1990.

Ostrom, Elinor, *Crafting Institutions for Self-Governing Irrigation Systems.* San Francisco: Institute for Contemporary Studies Press, 1992.

Ostrom, Elinor, Paul Benjamin, and Ganesh Shivakoti, *Institutions, Incentives, and Irrigation in Nepal.* Vol. 1. Bloomington: Indiana University, Workshop in Political Theory and Policy Analysis, 1992.

Ostrom, Elinor, Roy Gardner, and James Walker, *Rules, Games, and Common-Pool Resources.* Ann Arbor: University of Michigan Press, 1994.

Ostrom, Vincent, *The Meaning of American Federalism: Constituting a Self-Governing Society.* San Francisco: Institute for Contemporary Studies Press, 1991.

Ostrom, Vincent, David Feeny, and Hartmut Picht, eds., *Rethinking Institutional Analysis and Development: Issues, Alternatives and Choices.* 2nd ed. San Francisco: Institute for Contemporary Studies Press, 1993.

Repetto, Robert, *Skimming the Water: Rent Seeking and the Performance of Public Irrigation Systems.* Research Report No. 41. Washington, D.C.: World Resources Institute, 1986.

Sawyer, Amos, *The Emergence of Autocracy in Liberia: Tragedy and Challenge.* San Francisco: Institute for Contemporary Studies Press, 1992.

Seabright, Paul, "Managing Local Commons: Theoretical Issues in Incentive Design," *Journal of Economic Perspectives,* Fall 1993, 7:4, 113–34.

Siy, Robert Y., Jr., *Community Resource Management: Lessons from the Zanjera.* Quezon City: University of Philippines Press, 1982.

Small, Leslie, Marietta Adriano, and Edward D. Martin, *Regional Study on Irrigation Service Fees: Final Report.* Colombo, Sri Lanka: International Irrigation Management Institute, 1986.

Svendsen, Mark, "Assessing Effects of Policy Change on Philippine Irrigation Performance," Working Papers on Irrigation Performance 2. Washington, D.C.: International Food Policy Research Institute, December 1992.

Tang, Shui Yan, *Institutions and Collective Action: Self-Government in Irrigation.* San Francisco: Institute for Contemporary Studies Press, 1992.

Wade, Robert, *Village Republics: Economic Conditions for Collective Action in South India.* Cambridge: Cambridge University Press, 1988.

Weissing, Franz J., and Elinor Ostrom, "Irrigation Institutions and the Games Irrigators Play: Rule Enforcement without Guards." In Selten, Reinhard, ed., *Game Equilibrium Models II: Methods, Morals, and Markets.* Berlin: Springer-Verlag, 1991, 188–262.

Weissing, Franz J., and Elinor Ostrom, "Irrigation Institutions and the Games Irrigators Play: Rule Enforcement on Governmentand Farmer-Managed Systems." In Scharpf, Fritz W., ed., *Games in Hierarchies and Networks: Analytical and Empirical Approaches to the Study of Governance Institutions.* Frankfurt: Campus Verlag; Boulder: Westview Press, 1993, 387–428. (Chapter 13 in this volume.)

Yoder, Robert D. "Performance of Farmer-Managed Irrigation Systems in the Hills of Nepal," Ph. D. dissertation, Cornell University, 1986.

Part V
Continuing Challenges
for Research and Policy

Introduction to Part V

The two readings in part V provide general overviews of a large number of laboratory experiments that have investigated the ability of groups of experimental subjects to achieve a collectively desirable outcome. The relatively controlled environment of a laboratory experiment allows researchers to isolate the effects of particular factors on the ability of groups of individuals to cooperate for their common good. In effect, these laboratory settings can be seen as experiments in self-governance: under what conditions can a group of individuals brought together on a relatively random basis learn to cooperate to achieve a common goal? From this perspective, an experimental laboratory is a reflection in microcosm of the dilemmas faced by groups of humans throughout history and throughout all realms of social life.

Cooperation in Laboratory Experiments

In "Neither Markets nor States: Linking Transformation Processes in Collective Action Arenas" (chap. 15), Elinor Ostrom and James Walker survey a long series of research reports prepared by Workshop scholars and their collaborators, especially related to the behavior of experimental subjects in public goods settings. They compare experimental treatments that vary the size of the group, the magnitude of the group's endowments, the similarity or heterogeneity of interests among participants, and especially the amount of information available to the experimental subjects as they make their decisions.

In a typical baseline laboratory setup, subjects have no opportunity to communicate or make any kinds of deals with other participants. The authors' review of a large body of evidence highlights the ability of groups of experimental subjects to achieve better outcomes simply by being allowed to talk with each other, even if only briefly at the beginning of the play of the game. This result would be considered surprising by many game theorists because rational actors should not consider costless communication, so-called cheap talk, to be effective when each individual can obtain a higher payoff (at least in the short run) by ignoring these signals. Yet, these experiments provide rigorous and replicable evidence that even cheap talk does seem to help in social dilemma situations. Laboratory subjects can do even better if they are given the opportunity to make arrangements to sanction the behavior of others. Since this experimental treatment comes the closest to reflecting the basic structure found in field settings, it should not be surprising that these results sound so similar to the field research reported elsewhere.

These experimental results address the central concern of politics: the nature of governance. E. Ostrom, Walker, and Gardner (1992) use a series of laboratory experiments to investigate the ability of experimental subjects to devise their own means of governance. (The results of this and related essays are summarized in chap. 16.) One might think that it should come as no surprise to U.S. political scientists that self-governance is possible; after all, that is the whole basis for the U.S. system of constitutional order. Yet, what was surprising is how easily a modicum of self-governance could be introduced into the admittedly artificial confines of a laboratory experiment. Even in these stark settings, where long-term considerations were absent, experimental subjects demonstrated an ability, even an eagerness, to form their own system of governance, to monitor each other's behavior, and to enforce their agreement by punishing transgressors. What was remarkable was that all this was so easy to achieve simply by allowing the subjects a limited opportunity to communicate and to sanction each other. If self-governance is possible in such a starkly limited environment, then it is certainly relevant to more consequential interactions in the real world. These experimental results remind us of the crucial importance of self-governance.

Rethinking the Nature of Rational Choice

Although often dismissed as merely a technical tool, game theory can force scholars to confront foundational issues of the nature of human cognition and common understandings. One important outgrowth of this use of game theory to understand cognition is the alternative version of game rationality presented in the conclusion of E. Ostrom, Gardner, and Walker (1994). In an effort to explain the results of their laboratory experiments, they introduce a "measure-for-measure" behavioral strategy in which individuals react in a measured way to the rule violations of other participants. (This strategy was also influenced by field research demonstrating the importance of graduated sanctions as a contributor to successful management of common-pool resources; see E. Ostrom 1990.)

Several game theorists have struggled to develop more realistic models of the cognitive processes of real strategic players as an alternative to the "hyperrationality" characterizing the standard approach to game theory (Binmore 1990; Gardner 1995; McGinnis 1992; Rubinstein 1998; Samuelson 1997; Scharpf 1997). This remains very much a work in progress, but it is likely that future game models will become much more sophisticated and realistic in their depictions of individual cognition and institutional arrangements.

In her presidential address to the American Political Science Association, "A Behavioral Approach to the Rational Choice Theory of Collective Action" (chap. 16), Elinor Ostrom uses the results of experimental research to suggest a new conceptualization of the nature of individual actors. While doing so, she revisits many of the conceptual issues covered in part I of this volume. In what she describes as a "second generation model" of rational choice, Ostrom assumes that most individuals are predisposed to use reciprocity norms in many circumstances and to pay careful attention to the behavior of other individuals in all circumstances. Thus, their interactions unfold in different directions depending on the nature of these initial expectations and the intermediate play of the game. Nonetheless, certain regularities of behavior emerge from years of experimental research, regularities that do not always comport with the expectations of rational choice theorists.

Much remains to be done to formalize this "second generation" behavioral model of rational choice and to fully test its implications in diverse institutional settings, but even at this preliminary stage it is clear that the type of individual actors posited in this speech sounds very much like the farmers, fishers, and other groups of individuals who have been found to have the capacity to organize themselves to achieve common goals in real-life situations. If we can find even a glimmer of a capacity for self-governance among anonymous individuals randomly thrown together in the confines of a computer laboratory, then we should hardly be surprised when communities of individuals who have come to know each other well are able to cooperate on tangible and important problems. Unfortunately, many social scientists continue to be puzzled by the human capacity for collective self-governance in the absence of governmental direction or coordination. Familiarity with the research reported in this book should suffice to dispel that sense of puzzlement.

REFERENCES

Binmore, Kenneth G. 1990. *Essays on the Foundations of Game Theory.* Cambridge, MA: Blackwell.

Gardner, Roy. 1995. *Games for Business and Economics.* New York: Wiley.

McGinnis, Michael D. 1992. "Bridging or Broadening the Gap? A Comment on Wagner's 'Rationality and Misperception in Deterrence Theory'." *Journal of Theoretical Politics* 4 (4): 443–57.

Ostrom, Elinor. 1990. *Governing the Commons: The Evolution of Institutions for Collective Action.* New York: Cambridge University Press.

Ostrom, Elinor, Roy Gardner, and James Walker, with Arun Agrawal, William Blomquist, Edella Schlager, and Shui Yan Tang. 1994. *Rules, Games, and Common-Pool Resources.* Ann Arbor: University of Michigan Press.

Ostrom, Elinor, James Walker, and Roy Gardner. 1992. "Covenants with and without a Sword: Self-Governance Is Possible." *American Political Science Review* 86: 404–17.

Rubinstein, Ariel. 1998. *Modeling Bounded Rationality.* Cambridge, MA: MIT Press.

Samuelson, Larry. 1997. *Evolutionary Games and Equilibrium Selection.* Cambridge, MA: MIT Press.

Scharpf, Fritz Wilhelm. 1997. *Games Real Actors Play: Actor-Centered Institutionalism in Policy Research.* Boulder, CO: Westview Press.

CHAPTER 15

Neither Markets nor States: Linking Transformation Processes in Collective Action Arenas

Elinor Ostrom and James Walker

1. Collective Action Problems

Since the foundational work of Mancur Olson (1965), the concept of collective action has gained prominence in all of the social sciences. Given the structure of an initial situation, collective action problems occur when individuals, as part of a group, select strategies generating outcomes that are suboptimal from the perspective of the group. In the most commonly examined case, individuals are involved in a collective action game where the unique Nash equilibrium for a single iteration of the game yields less than an optimal outcome for all involved. If such a game is finitely repeated and individuals share complete information about the structure of the situation, the predicted outcome for each iteration is the Nash equilibrium of the constituent game. If uncertainty exists about the number of iterations, or if iterations are infinite, possible equilibria explode in number. Among the predicted equilibria are strategies yielding the deficient Nash equilibria, optimal outcomes, and virtually everything in between. The *problem* of collective action is finding a way to avoid deficient outcomes and to move closer to optimal outcomes. Those who find a way to coordinate strategies receive a "cooperation dividend" equal to the difference between the payoffs at a deficient outcome and the more efficient outcome.

Public choice theorists have focused primarily on those collective action problems related to public goods, common-pool resources, and club goods. All three of these types of goods potentially involve deficient equilibria in

Originally published in Dennis C. Mueller, ed., *Perspectives on Public Choice: A Handbook* (Cambridge: Cambridge University Press, 1997), 35–72. Copyright © 1997, Cambridge University Press. Reprinted with the permission of Cambridge University Press and the authors.

Authors' note: The authors are appreciative of the support they have received from the National Science Foundation (Grants Nos. SBR–9319835 and SES–8820897) and the useful comments received from Vincent Ostrom and Audun Sandberg.

provision or consumption activities. Pure public goods have been considered the paradigm case for the necessity of the state. The state is viewed as the vehicle necessary to select and provide the "proper" outcome as opposed to that which would occur without the state. On the other hand, the problems of providing public goods, regardless of scope, are not so obviously solved by a state (McGinnis and E. Ostrom 1996). If the purpose of an analysis is to demonstrate how market institutions fail to provide optimal levels for a wide variety of goods and services, demonstrating market weakness or failure is relatively easy. Demonstrating the failure of one type of institution is not, however, equivalent to establishing the superiority of a second type of institution.[1] An institutional arrangement with a monopoly on the legitimate use of force—the state—has itself been shown to suffer substantial failures and weaknesses as a provider of public goods in many countries of the world (see Bates 1981; V. Ostrom 1984, 1991, 1993; Wunsch and Olowu 1990; Sawyer 1992).

The choice that citizens face is not between an imperfect market, on the one hand, and an all-powerful, all-knowing, and public-interest-seeking institution on the other. The choice is, rather, from among an array of institutions—all of which are subject to weaknesses and failures. Various forms of associations and networks of relationships are successfully used to solve aspects of collective action problems. These include families and clans, neighborhood associations, communal organizations, trade associations, buyers and producers' cooperatives, local voluntary associations and clubs, special districts, international regimes, public-service industries, arbitration and mediation associations, and charitable organizations. Some institutional arrangements, such as gangs, criminal associations, and cartels, solve collective action problems for some participants by harming others. Predatory states may solve some collective action problems for those who are in power but do so by diminishing productivity and benefits for others.[2]

No doubt exists about the importance of governments—at supranational, national, and subnational levels—in coping with aspects of collective action. Given recent events in Russia and Eastern Europe, however, it is particularly important in the post-1989 era to explore how a wide diversity of institutions that are neither markets nor states operate to enhance the joint benefits that individuals achieve in collective action situations. Many of these institutions are constituted by participants in a self-governing process rather than imposed by external authorities. The creation of these institutions is itself a collective action problem. Understanding how individuals solve different types of collective action problems is of substantial analytical and normative importance.

To understand how institutions that are neither markets nor states evolve and cope with collective action problems, we need to unpack larger and more complex problems into a series of transformations that occur between the provision of *any* good and its consumption. For each transformation process, we need to understand the kind of behavior that individuals adopt. Research conducted in the field and in experimental settings provides substantial evidence about the capacity of those affected to develop institutions that are neither markets nor states. After discussing key theoretical concepts, we provide an overview from the field and from laboratory experiments of how individuals cope with diverse collective action problems.

1.1. Transformation Processes

Many processes transform inputs into outputs that jointly affect some set of individuals. Drawing on Plott and Meyer (1975) and our own previous research (E. Ostrom, Schroeder, and Wynne 1993; E. Ostrom, Gardner, and Walker 1994), we define and discuss the following eight transformation processes: provision, production, distribution, appropriation, use, organization, monitoring, and sanctioning.[3]

The linkage among these processes may be accomplished within diverse institutional arrangements, including (1) strictly private market or governmental enterprises, (2) independent actors engaged in quid pro quo transactions or more loosely defined reciprocal relations, (3) independent actors who lack communication or contracting capabilities, or (4) some combination of the above. The particular attributes of a collective action problem depend upon the transformation functions involved and the institutional rules that define available strategies. Considerable confusion has been generated by conflating these processes and/or ignoring the importance of alternative institutional arrangements. Treating the provision, production, and consumption of a public good, for example, as if these were a single process or independent of institutional rules is often misleading.

We emphasize that each of these transformation processes is involved in all efforts to produce and consume goods, regardless of the attributes of the good to be produced or the institutional context. In our discussion, however, we highlight each process in its relation to market and/or nonmarket processes. Our goal is to emphasize how each is naturally linked to the diverse attributes of particular goods and/or the institutional setting in which those goods are produced.

Transformations Normally Associated with a Market Context

First consider four of the processes that are normally associated with markets.

Production:	Combining private and/or public inputs to yield one or more outputs.
Distribution:	Making outputs available.

Appropriation: Taking, harvesting, or receiving one or more outputs.

Use: Transforming one or more outputs into final consumption and waste (or through another production process into other outputs).

When institutional conditions allow production and distribution to be handled by one entity—the supplier—and appropriation and use to be handled by another—the appropriator—the key linking transaction occurs at the point of appropriation. When quid pro quo market exchanges link suppliers and appropriators (who are buyers in this context), anyone who appropriates a unit of a good is liable under an institutional constraint to pay for it or be penalized if caught stealing (Plott and Meyer 1975).

Transformations Associated with a Nonmarket Context

If appropriators are not liable to pay for goods and services they receive, then suppliers cannot exclude potential beneficiaries. Thus, the dimension of exclusion and nonexclusion relates to the type of liability an individual faces who appropriates goods that are made available by a supplier (or nature) (Plott and Meyer 1975). When exclusion is complete, a potential consumer is fully liable for any goods or services appropriated (whether or not the individual exercised choice in appropriating benefits). When exclusion is entirely incomplete, a supplier does not have the physical or legal means to withhold goods from individuals who receive a positive or negative benefit from their receipt. Clearly, goods and services vary substantially in the difficulty of excluding potential beneficiaries once the good is made available. This variation results both from the particular attributes of the good involved and the legal structure in which the transformation processes occur.

1.2. Public Goods

In the case of *pure public goods,* there exists a shared outcome of production and distribution (whose value may be either positive, zero, or negative to recipients). Exclusion does not occur, and consumption is nonsubtractable. When *all* individuals are affected by the provision of a public good—without crowding or subtracting from its availability—the public good is global. The availability of knowledge and gravity are examples of global public goods. Most public goods, however, have a scale of effects that is less than global. Local public goods include a subset of individuals, often as small as a household (sharing the lighting and heating in the household) or a neighborhood (sharing the security provided by the surveillance efforts of neighbors and a local police department).

Turning once again to transformation processes, the difficulties of exclusion lead to a fundamental problem to be solved in relation to public goods—that of provision (V. Ostrom, Tiebout, and Warren 1961).

Provision: The articulation of demand, the arranging for production, and the supplying of funding/inputs necessary for production.

In quid pro quo market exchanges, all aspects of provision are accomplished as a by-product of a series of transactions. Buyers articulate their demands by selecting and purchasing those items they prefer. When beneficiaries are not excluded from appropriating, however, potential appropriators are not motivated to pay for the positive value they receive from the goods provided. If the value of the public good is negative, individual appropriators are not motivated to pay for the collective action required to reduce the negative public good. Underprovision (overprovision) occurs, implying an inefficient allocation of resources. This behavior, often referred to as free riding, leads to a fundamental proposition concerning the provision of public goods:

P1   Without some form of coordination or organization to enable individuals to agree upon, monitor, and sanction the contributions of individuals to the provision of a public good with positive (negative) value, the good will be underprovided (overprovided).

When the institutional setting is extremely sparse and yields equilibria that imply suboptimal outcomes, there is no theoretical challenge to this proposition. And we know of no evidence that contradicts it.[4] There is, however, considerable debate in the literature on two issues: the extent of suboptimality that results and the form of coordination or institutional setting that can overcome the suboptimal provision of public goods.

1.3. Common-pool Resources

Common-pool resources (CPRs) are goods for which subtractability in units appropriated from and restricting access to the resource or facility is a nontrivial institutional problem. For CPRs, either appropriation or consumption (or both) occurs *within* a resource or facility. If the resource is a fishing grounds, one must fish in the grounds even though the tons of fish captured can then be bought and sold as private goods. If the CPR is a commonly provided and maintained irrigation system, water appropriated becomes a private good, but the facility itself must be provided and maintained. Further, allocation mechanisms for appropriating water must be crafted by those who govern the use of this facility.

Confusion often exists about the meanings of nonexclusion and subtractability as they apply to CPRs. One difficulty is that theorists are not always clear about what "the" good that was being accessed or appropriated is. Clarity is introduced by recognizing that jointly accessed or appropriated goods involve both a facility and a flow of services (or use units) from that facility (Blomquist and E. Ostrom 1985; E. Ostrom, Gardner, and Walker 1994). The facility, the flow of services to one individual, and the flow of services to other individuals may all potentially affect an individual's benefit stream.[5]

In the case of a market-produced private good, attention is generally focused on the attribute of the good that creates a benefit stream, that is, the flow of goods or services that one individual appropriates and uses. Appropriating and consuming these goods is not generally undertaken in a shared manner. The decoupling of production, appropriation and use, and the liability function that links appropriation eventually to production are basic reasons that private goods are not plagued with the panoply of collective action problems discussed in this essay.

Finding the right mix of incentive-compatible institutional arrangements to provide the optimal facility and an acceptable rule system for defining appropriation rights is a difficult problem of collective action. Further, creating institutions to facilitate exclusion can range from the relatively simple to the impossible. The problem of exclusion must be solved by those who wish to utilize a large array of CPRs efficiently over the long run (E. Ostrom, Gardner, and Walker 1994). Such resources range from the global commons—generating joint benefits with reference to the stratosphere and biodiversity—to moderately sized fisheries, grazing lands, forests, irrigation systems, oil pools, and groundwater basins. At the other extreme, the users of many jointly used facilities, such as theaters, golf clubs, and toll roads, solve problems of exclusion at relatively low cost given the existence of particular legal rights and appropriate technologies. These goods have been called *club goods* (Buchanan 1965; Cornes and Sandler 1986).

The *initial provision* of natural resources that have the incentive structure of a CPR may be nonproblematic. For example, nature has already provided fisheries, lakes, oceans, groundwater basins, and oil pools. The key problems for such collective action situations are the maintenance of the resource and the *appropriation* of the benefits. Frequently, individual users acting independently face temptations to overharvest from a CPR leading to the well-known "tragedy of the commons" (G. Hardin 1968). In other cases, individual users acting independently generate technological externalities for others or fight over the best sites located within a CPR. If no agreement exists about how to regulate the production technology or the quantity, timing, and location of harvesting a CPR, little incentive exists for individuals to take into account their adverse effects on each other or the resource. The problems associated with diverse externalities of appropriation lead to a second fundamental proposition concerning the independent appropriation of CPRs.

> P2   Without some form of coordination or organization to enable individuals to agree upon, monitor, and sanction the patterns of appropriation by individuals from a CPR, the resource will be overused.

As with the first proposition, we know of no contradictory theoretical or empirical challenge. There are many open questions, however, about how coordination or organization can evolve and about whether one should impose market or state institutions on those who use CPRs.

1.4. Crafting Institutions

To minimize the degree of suboptimal provision and appropriation, appropriators of collective goods may find recourse to the final three transformation processes—organization, monitoring, and sanctioning. These processes, which are related to rule orderings, are defined as

> Organization: Devising rules and agreements about membership, distribution of obligations, and benefits, and about how future collective choice decisions will be made.
>
> Monitoring: Measuring inputs, outputs, and conformance to rules and agreements.
>
> Sanctioning: Allocating inducements or punishments related to conformance or lack thereof to agreements.

Farmers who jointly use an irrigation system, for example, must organize a variety of provision activities primarily related to maintenance or watch their system deteriorate toward uselessness (Hilton 1992). Organizing the provision side of an irrigation CPR involves deciding upon how many days a year should be devoted to routine maintenance, how work will be allocated to individual farmers, how emergency repairs should be handled, who is responsible for repairing broken embankments caused by grazing animals, and how new control gates and regulatory devices are to be installed and paid for. Appropriation activities are closely linked to these provision activities. How much water is available for distribution is dependent upon whether a system is kept in good repair. The level of conflict over water distribution is apt to be higher on a poorly maintained system than on a well-maintained system. Organizing the appropriation side of an irrigation CPR involves deciding upon the method of water allocation and the formulae for allocating water during the different seasons of the year (E. Ostrom 1992).

When individuals craft their own institutions, some arrangements are recorded on paper. Many however, are left unrecorded as problem-solving individuals try to do a better job in the future than was done in the past. In a democratic society, problem-solving individuals craft their own institutions all the time. Individuals also participate in less fluid decision-making arrangements (elections to select representatives, for example). Elected representatives may then engage in open, good-faith attempts to solve a wide diversity of problems brought to them by their constituents. It is also possible in a formal governance system where individuals are elected for patterns to emerge that are not strictly problem solving. Incentives exist to create mechanisms whereby one set of individuals dominates others. Attempts to dominate and avoid being dominated increase the complexity of the processes by which new institutions are crafted.

Many activities of self-organization do not occur in formal decision-making arenas. Consequently, external public officials frequently presume that individuals in local settings cannot coordinate activities and thus face the suboptimalities predicted by the first and second fundamental propositions. The working rules that individuals develop are invisible to outsiders unless substantial time and effort are devoted to ascertain their presence and structure.[6]

In legal systems where individuals have considerable autonomy to self-organize, they may develop rules of their own. These might be called customary rules, contracts, communal organizations, or associations. They are often referred to as "informal" by those who perceive governments as the source of all rules. Many of these locally devised rules would, however, be considered as binding if challenged in a court of law. Thus, calling them informal confuses these self-organized but legal rules with other self-organized but illegal rules. What is important for the analysis of collective action problems, however, is to recognize that individuals can consciously decide to adopt their own rules that either replace or complement the rules governing an initial collective action situation.

Once one recognizes that those involved in collective action may shift out of a current "game" to a deeper-level game, the necessity of using multiple levels of analysis becomes apparent. All rules are nested in another set of rules that, if enforced, defines how the first set of rules can be changed. Changes in the rules used to order action at one level occur within a currently "fixed" set of rules at a deeper level. Changes in deeper-level rules usually are more difficult and more costly to accomplish, thus increasing the stability of mutual expectations among individuals interacting according to a set of rules.

It is useful to distinguish three levels of rules that cumulatively affect the actions taken and the outcomes obtained in any setting (Kiser and E. Ostrom 1982).

1. *operational rules* directly affecting day-to-day decisions made by the participants in any setting;
2. *collective choice rules* affecting operational activities and outcomes through their effects in determining who is eligible and the specific rules to be used in changing operational rules; and
3. *constitutional choice rules* affecting operational activities and their outcomes in determining who is eligible and the rules to be used in crafting the set of collective choice rules that in turn affect the sets of operational rules.

At each level of analysis there may be one or more arenas in which decisions are made. Policymaking regarding the rules that will be used to regulate operational-level actions is usually carried out in one or more collective choice arenas and enforced at an operational level as well. Dilemmas are not limited to an operational level of analysis. They frequently occur at the collective choice and constitutional levels.

The art of crafting institutions can be viewed as one of creating coordinated strategies for players in multilevel games. Two types of coordinated strategies enable participants to extricate themselves from collective action dilemmas: one exists when individuals agree upon a joint strategy within a set of preexisting rules; another when an effort is made to change the rules themselves by moving to a collective choice or constitutional choice arena. The possibility of switching arenas is frequently ignored in current analyses of collective action problems.

There are two reasons why the possibility of changing the rules of a game have been ignored. The first is a methodological position that eliminates analysis of structural change while examining the effects of one structure on outcomes; in other words, the existing conditions that one uses to specify a problem for analysis are not to be changed in the process of analysis. It is easy to overcome this limit by overtly taking a long-term perspective. It is with a long-term perspective that the given constraints of a particular physical facility are changed into variables that can be changed and thus analyzed. A similar approach can be taken with rules. The second is the assumption that rules are a public good. Agreement on better rules affects all individuals in the group whether they participate in the reform effort or not. The temptation to free ride in the effort to craft new rules may be offset by the strong interest that most individuals have in ensuring that their own interests are taken into account in any set of new rules. Further, the group might be "privileged" in the sense that one or a very small group of individuals might expect such a high return from provision that they pay the full cost themselves (Olson 1965).

It is generally assumed that the higher the order of a collective action dilemma, the more difficult it is to solve. An even higher-order dilemma than a rule change is the dilemma involving monitoring and sanctioning to enforce a set of rules (Bendor and Mookherjee 1987; Bianco and Bates 1990). Even the best rules are rarely self-enforcing. It is usually the case that once most appropriators follow these rules, there are strong temptations for some appropriators to break them. If most farmers take only the legal amount of water from an irrigation system so that the system operates predictably, each farmer will be tempted from time to time to take more than a legal amount of water because his or her crops need water. Once some farmers take more than their allotment of water, others are tempted to do the same, and the agreed-upon rules rapidly crumble. Monitoring each other's activities and imposing sanctions are costly activities. Unless the rewards received by the individual who monitors and sanctions someone else are high enough, each potential monitor faces a situation in which not monitoring and not sanctioning may be the individually preferred strategy even though everyone would be better off if that strategy were not chosen (Weissing and E. Ostrom 1991, 1993). Designing monitoring and sanctioning arrangements that sustain themselves over time is a difficult task involved in transforming a collective action dilemma.

2. Evidence from Field Settings

Most of the emphasis in the public choice tradition has been on predicting behavior *within* the structure of a game, rather than on the processes of organizing new games and on self-monitoring and sanctioning activities.[7] Consequently, the literature from field settings concerned with organizing, monitoring, and sanctioning activities has been largely written by scholars in the fields of anthropology, sociology, and history and by specialists who focus on one sector or one region. During the mid-1980s, an interdisciplinary panel was established by the National Academy of Sciences to examine common-property institutions and their performance in managing smaller-scale natural resources. Since the publication of the Panel's report (National Research Council 1986), a flurry of book-length publications has documented the extraordinary diversity of self-organized institutions operating at constitutional, collective choice, and operational levels, as well as situations in which the participants undertake most of their own organization, monitoring, and sanctioning activities.[8]

The most important findings regarding the emergence and consequences of self-organized institutional arrangements in field settings include the following.

1 . Overuse, conflict, and potential destruction of *valuable* natural resources are likely to occur when users act independently due to lack of meaningful collective action or the inability to make credible commitments (Christy and Scott 1965; Anderson 1977; Libecap and Wiggins 1984).

2. Conditional on existing collective choice rules, overuse, conflict, and destruction can be substantially reduced through the crafting, implementation, and sustainment of alternative norms, rules, and property-rights systems (Schlager 1990; Tang 1992; Schlager and E. Ostrom 1993).
3. Locally selected systems of norms, rules, and property rights that are not recognized by external authorities may collapse if their legitimacy is challenged or if large exogenous economic or physical shocks undermine their foundations (Alexander 1982; Davis 1984; Cordell and McKean 1992).
4. Regulation by state authorities is effective in some settings but is frequently less effective than regulation by local users, especially in relation to smaller-scale systems where user groups are stable and interact frequently in regard to the use of resources (Thomson 1977; Feeny 1988; Gadgil and Iyer 1989; Lam, Lee, and E. Ostrom 1997).
5. Efforts to establish marketable property rights to natural resource systems have substantially increased efficiency in some cases (e.g., Jessup and Peluso 1986; Blomquist 1992) and encountered difficulties of implementation in others (e.g., Wilson 1982; Copes 1986; Townsend 1986; Townsend and Wilson 1987).

2.1. Design Principles and Robust Institutions

In addition to knowing that various institutions can reduce the externalities involved in collective action arenas, we are also beginning to understand the design principles of robust institutions. *Robust institutions* are found in systems that have survived for long periods of time. Such institutions survive when operational rules are adapted in relation to a set of collective choice and constitutional choice rules (Shepsle 1989). There appear to be several key design principles found in robust institutions (these are summarized in table 15.1). Evidence from the field suggests that *fragile institutions* tend to be characterized by only some of these design principles, though *failed institutions* are characterized by only a few (if any) of these principles. In addition, initial analysis from irrigation systems governed by farmers suggests that systems characterized by most of these principles are associated with higher agriculture yields and crop intensities—provided the physical characteristics of the systems are controlled for (Lam, Lee, and E. Ostrom 1997).[9]

2.2. Factors Affecting Institutional Change

Not only is there substantial variety in the institutions crafted to facilitate efficient use of resources across types of resources, but it seems that neighbor-

TABLE 15.1. Design Principles Illustrated by Long-Enduring CPR Institutions

1. *Clearly defined boundaries*
Individuals or households have rights to withdraw resource units from the CPR, and the boundaries of the CPR itself are clearly defined.

2. *Congruence between appropriation and provision rules and local conditions*
Appropriation rules restricting time, place, technology, and/or quantity of resource units are related to local conditions and to provision rules requiring labor, materials, and/or money.

3. *Collective choice arrangements*
Most individuals affected by operational rules can participate in modifying operational rules.

4. *Monitoring*
Monitors, who actively audit CPR conditions and appropriator behavior, are accountable to the appropriators and/or are the appropriators themselves.

5. *Graduated sanctions*
Appropriators who violate operational rules are likely to receive graduated sanctions (depending on the seriousness and context of the offense) from other appropriators, from officials accountable to these appropriators, or from both.

6. *Conflict resolution mechanisms*
Appropriators and their officials have rapid access to low-cost, local arenas to resolve conflict among appropriators or between appropriators and officials.

7. *Minimal recognition of rights to organize*
The rights of appropriators to devise their own institutions are not challenged by external governmental authorities.

For CPRs that are part of larger systems

8. *Nested enterprises*
Appropriation, provision, monitoring, enforcement, conflict resolution, and governance activities are organized in multiple layers of nested enterprises.

Source: E. Ostrom (1990, 90)

ing systems that appear to face similar situations frequently adopt different institutional solutions. The variety of rules selected by individuals facing similar circumstances raises the question of whether institutional change is an evolutionary process in which more efficient institutions are selected over time.[10] If such an argument is correct, then to explain institutional change, one would have to analyze the specific relationships between variables characterizing the resource, the community of individuals involved, and the rules for making and changing rules.

The existing theoretical and empirical research on CPR situations enables one to specify several important variables that appear to be conducive to the development of institutions that increase efficiency.[11]

1. Accurate information—regarding the condition of the resource, provision costs, and consequences of appropriation—is technologically and economically feasible to obtain.
2. Users are relatively homogeneous in regard to asset structure, infor-

mation, and preferences.

3. Users share a common understanding about the potential benefits and risks associated with the continuance of the status quo as opposed to alternative (feasible) options.
4. Users share generalized norms of reciprocity and trust that can be used as initial social capital.
5. The user group sharing the resource is relatively stable.
6. Users' discounting of future resource use is sufficiently low.
7. Users have the autonomy to craft operational rules that are supported by, and potentially enforced by, external authorities.
8. Users use collective choice rules that fall between the extremes of unanimity and control by a few, avoiding high transaction or high deprivation costs.
9. Users can develop relatively accurate and low-cost monitoring and sanctioning arrangements.

Many of these variables are in turn affected by the larger regime in which users are embedded. If the larger regime facilitates local self-organization by providing accurate information about natural resource systems, provides arenas in which participants can engage in discovery and conflict-resolution processes, and provides mechanisms that back up local monitoring and sanctioning efforts, the probability of participants adapting more effective norms and rules over time is higher than in regimes that either ignore resource problems entirely or, at the other extreme, presume that all decisions about governance and management must be made by central authorities.

The extensive evidence from field settings poses an initial challenge to our current theoretical understanding of the role of either states or markets in solving collective action problems. Many of the institutions that are designed by those directly involved in appropriating from a CPR use elements frequently associated with markets (such as auctions or transferable rights to the use units generated by a CPR). But on close examination, these institutions are not fully market institutions since entry and exit into the "market" are strictly controlled and a group of individuals jointly owns or manages the resource or facility itself. Furthermore, many of these self-organized institutions use elements frequently associated with states (such as the election of officials who prepare budgets, make rules, monitor conformance to rules, and sanction those who break these rules). But, they are not fully state institutions either, since they exist in a polycentric system involving many public organizations all with varying levels of autonomy and organized in a nested manner at many scales. In some instances, the self-organized institutions are not even recognized by state officials. In others, these self-organized institutions may come close to being outlaw organizations, since their activities are contrary to what is written in formal legislation or administrative regulations.

3. Public Goods and CPRs: Experimental Evidence

Turning from the field to the laboratory allows a closer took at the strategic nature and behavior in collective action situations. While research in field settings is essential for understanding the effect of institutions in diverse physical, social, and economic settings, it is exactly the diversity of these settings that makes it difficult to understand which of many variables account for differences in observed behavior. Laboratory experiments are particularly useful for testing well-specified models of behavior where one variable at a time can be changed and control is therefore enhanced. In parameterizing experiments, the experimenter controls (1) the number of participants, (2) the positions they may hold, (3) the specific actions they can take, (4) the outcomes they can effect, (5) how actions are linked to outcomes, (6) the information they obtain, and (7) the potential payoffs. That is, the experimenter creates the institutions that structure the decision situation facing the subjects.

A key step in linking a theoretical model to laboratory conditions is inducing incentives. That is, the laboratory situation must lead the subjects to perceive and act on payoffs that have the same properties as the payoffs in the linked theoretical model. Value over alternative outcomes is most commonly induced using a cash reward structure. As discussed by Smith (1982), the following conditions constitute a set of sufficient conditions for inducing values.

Nonsatiation: Subjects' utility must be monotone-increasing in payoffs. This guarantees that given alternatives that are identical except in payoff space, a subject will always choose the alternative with the greater reward.

Saliency: For rewards to be motivationally relevant, or *salient,* their level must be directly tied to participants' decisions.

Dominance: An important element of the economic environment is the subjects' utility functions, which contain many arguments in addition to the rewards associated with payoffs. To induce value successfully, the cash reward structure must dominate the subjective benefits and costs of an action derived from these other utility function arguments.

Privacy: Subjects may not be autonomous own-reward maximizers. To the extent allowed by the underlying theoretical model under investigation, subjects' payoffs should be made private.

TABLE 15.2. Illustration of VCM Decision Problem

Round 1 currently in progress

Your endowment of tokens in each round: 50; Group size: 10
Total group endowment of tokens in each round: 500
Each token retained in your private account earns: $0.01

Examples of Possible Earnings from the Group Account		
Tokens in Group Account (from the entire group)	Total Group Earnings	Your 10% Share of Group Earnings
0	$ 0.000	$ 0.000
31	$ 0.930	$ 0.093
63	$ 1.890	$ 0.189
94	$ 2.820	$ 0.282
125	$ 3.750	$ 0.375
156	$ 4.680	$ 0.468
188	$ 5.640	$ 0.564
219	$ 6.570	$ 0.657
250	$ 7.500	$ 0.750
281	$ 8.430	$ 0.843
313	$ 9.390	$ 0.939
344	$ 10.320	$ 1.032
375	$ 11.250	$ 1.125
406	$ 12.180	$ 1.218
438	$ 13.140	$ 1.314
469	$ 14.070	$ 1.407
500	$ 15.000	$ 1.500

-HELP→ review instructions.
-LAB→ view the earnings from the Group Account for any possible value of "Tokens in Group Account."
How many tokens do you wish to place in the Group Account?

While a laboratory situation can be designed to match a theoretical model closely, it cannot be expected to parallel far more complex naturally occurring settings to which the theoretical model may also be applied. Nevertheless, to the extent that a logically consistent model is put forward as a useful explanation of naturally occurring phenomena, laboratory experimental methods offer the advantage of control, observation, and measurement.

3.1. Decision Settings: Stark Institutions

Our discussion of laboratory studies applied to public goods and CPR decision situations begins with two "stark" institutional settings referred to as the "Voluntary Contributions Mechanism" in the case of public goods and the "CPR Appropriation Game" in the case of CPRs. The term *stark* is used to emphasize that the institutional settings of these decision situations are null in regard to subjects' ability to coordinate decisions overtly or create an alternative institutional situation. This discussion draws substantially on the work found in Isaac, Walker, and Williams (1994, hereafter IWW) and E. Ostrom, Gardner, and Walker (1994, hereafter OGW).[12]

The Voluntary Contributions Mechanism (VCM)

In the VCM decision setting, N subjects participate in a series of decision rounds. Each participant is endowed with z tokens that are to be divided between a "private account" and a "group account." Tokens cannot be carried across rounds. The subject is informed that for each token she places in the private account she earns p cents with certainty. The subject is also informed that earnings from the group account are dependent upon the decisions of all group members. For a given round, let X represent the sum of tokens placed in the group account by all individuals in the group. Earnings from the group account are dependent upon the preassigned earnings function $G(X)$. Each individual receives earnings from the group account regardless of whether she allocates tokens to that account—thus the publicness of the group account. For simplicity, each individual is symmetric with respect to her earnings from the group account. Each earns an equal amount from the group account of $[G(X)]/N$ cents. Table 15.2 illustrates the type of information subjects receive for a given parameterization of the game. Prior to the start of each decision round, each individual knows the number of remaining rounds and the groups' aggregate token endowment for prior rounds. The decisions for each round are binding, and rewards are based on the sum of earnings from all rounds. During each round, subjects can view their personal token allocations, earnings, and total tokens placed in the group account for all previous rounds.

The CPR Appropriation Game

Contrast the VCM public goods provision game with the CPR appropriation game. In the appropriation game, subjects are instructed that in each decision round they are endowed with a given number of tokens that they can invest in two markets. Market 1 is described as an investment opportunity in which each token yields a fixed (constant) rate of output and each unit of output yields a fixed (constant) return. Market 2 (the CPR) is described as a market that yields a rate of output per token dependent upon the total number of tokens invested by the entire group. Subjects receive a level of output from market 2 that is equivalent to the percentage of total group tokens they invest. Subjects furthermore know that each unit of output from market 2 yields a fixed (constant) rate of return. Table 15.3 illustrates the type of information subjects see in a given parameterization of the game. Subjects know the total number of decision makers in the group, total group tokens, and that endowments are identical.

TABLE 15.3. Illustration of CPR Decision Problem

Units produced and cash return from investments in market 2 commodity 2
value per unit=$0.01

Tokens invested by Group	Units of Commodity 2 Produced	Total Group Return	Average Return per Token	Additional Return per Token
8	168	$ 1.68	$ 0.21	$ 0.21
16	304	$ 3.04	$ 0.19	$ 0.17
24	408	$ 4.08	$ 0.17	$ 0.13
32	480	$ 4.80	$ 0.15	$ 0.09
40	520	$ 5.20	$ 0.13	$ 0.05
48	528	$ 5.28	$ 0.11	$ 0.01
56	504	$ 5.04	$ 0.09	$-0.03
64	448	$ 4.48	$ 0.07	$-0.07
72	360	$ 3.60	$ 0.05	$-0.11
80	240	$ 2.40	$ 0.03	$-0.15

The table shown above displays information on investments in market 2 at various levels of group investment. Your return from market 2 depends on what percentage of the total group investment is made to you. Market 1 returns you one unit of commodity 1 for each token you invest in market 1. Each unit of commodity 1 for each token you invest in market 1. Each unit of commodity 1 pays you $0.05.

3.2. Behavior in Stark Settings

The discussion below focuses on several regularities from an extensive program of experimental investigations related to VCM and the CPR appropriation game. Results are reported as summary observations with emphasis placed on the relationship between behavior and game theoretic predictions for noncooperative games of complete information.

Theory and Behavior in VCM

The strategic nature of the VCM decision problem is straightforward. Each participant's decision to allocate a marginal token to the group account costs that individual p cents. For appropriate initializations, however, allocations to the group account yield a positive gain in group surplus of $[G'(\cdot) - p]$. In the VCM mechanism, the marginal per capita return from the group account (MPCR) is defined as the ratio of "$" benefits to costs for moving a single token from the individual to the group account, or $[G'(\cdot)/N]/p$. In the experiments that we focus upon here, $p = \$.01$ and G' is a constant greater than $.01$ so that the Pareto optimum (defined simply as the outcome that maximizes group earnings) is for each individual to place all tokens in the group account. On the other hand, the single-period dominant strategy is for each individual to place zero tokens in the group account. The "social dilemma" follows strategically because p and G $(\cdot)$ are chosen so that the MPCR < 1. For finitely repeated play, the outcome of *zero* allocations to the group account is also the unique, backward induction,

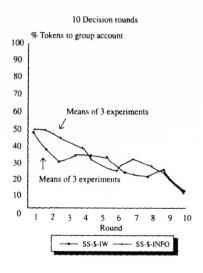

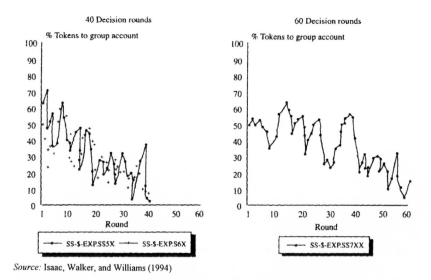

Source: Isaac, Walker, and Williams (1994)

Fig. 15.1. Experiments with additional payoff information and a larger number of decision rounds

complete information Nash equilibrium. In summary, for all parameterizations in which MPCR < 1, complete information noncooperative game theory yields the same prediction—zero allocations to the group account. In actuality, behavior is affected by the specific parameterizations of the setting.

Figure 15.1 portrays the type of behavior that is "typical" for VCM experiments with group size of $N = 10$ and MPCR = .30. Shown are allocations to the group account as a percentage of optimum across decision rounds. In the upper panel, average decisions are shown for a series of experiments in which subjects received (did not receive) additional information on the consequences of alternative individual and group decisions. In summary, the additional information explained to subjects how an individual and how the group could maximize and/or minimize their experimental earnings. Note the lack of any strong effect of this treatment condition. In the lower panels, results are shown from several experiments using highly experienced VCM subjects and a larger number of decision rounds.

For groups of size 4, 10, 40, and 100, experiments reported by IWW yield the following conclusions.

1. Depending upon specific parameterizations, replicable behavior is observed where allocations are very near the predicted outcome of zero allocations to the group account or are significantly above zero allocations to the group account.
2. Allocations to the group account are either unaffected by MPCR or are inversely related to MPCR.
3. Holding MPCR constant, allocations to the group account are either unaffected by group size or are *positively* related to group size.
4. Increasing group size in conjunction with a *sufficient* decrease in MPCR leads to lower allocations to the group account.
5. There tends to be some decay (but generally incomplete) to the predicted outcome of zero allocations to the group account.
6. Even with a richer information set regarding the implications of alternative allocation decisions, highly experienced subject groups continue to follow a pattern of behavior generally inconsistent with the predictions of the complete information Nash model.
7. Inconsistent with models of learning, the rate of decay of allocations to the group account is inversely related to the number of decision rounds.

Theory and Behavior in the CPR.
In the experimental investigations reported by OGW, the CPR is operationalized with eight appropriators ($n = 8$) and quadratic production functions F (Σx_i) for market 2, where

$$F\left(\sum x_i\right) = a \sum x_i - b\left(\sum x_i\right)^2$$

with $F'(0) = a > w$ and $F'(ne) = a - 2bne < 0$, where w is the per token return from market 1, n is the number of subjects, and e is individual token endowments. Their analysis uses the parameter specifications summarized in table 15.4. Subjects are endowed each round with either 10 or 25 tokens depending upon design conditions. With these payoff parameters, a group investment of 36 tokens yields the optimal level of investment. The complete information noncooperative Nash equilibrium is for each subject to invest 8 tokens in market 2 (regardless of the endowment condition)—for a total group investment in market 2 of 64 tokens.[13]

Much of the discussion of experimental results for the CPR game focuses on what is termed *Maximum Net Yield* from the CPR. This measure captures the degree of optimal yield earned from the CPR. Specifically, net yield is the return from market 2 minus the opportunity costs of tokens invested in market 2 divided by the optimal return from market 2 minus the opportunity costs of tokens invested in market 2 at the optimum. For the CPR game, opportunity costs equal the potential return that could have been earned by investing the tokens in market 1. Dissipation of yield from the CPR is known in the resource literature as "rent" dissipation. Note, as with the symmetric Nash equilibrium, optimal net yield is invariant to the level of subjects' endowments across the two designs OGW examine. Thus, even though the range for subject investment decisions is increased with an increase in subjects' endowments, the equilibrium and optimal levels of investment are not altered. At the Nash equilibrium, subjects earn 39 percent of maximum net yield from the CPR.

Figure 15.2 presents data from a series of CPR experiments in which individual token endowments were either 10 or 25. Displayed is average net

TABLE 15.4. Experimental Design Baseline: Parameters for a Given Decision Period

Experiment Type:	Low Endowment	High Endowment
Number of subjects	8	8
Individual token endowment	10	25
Production function: Mkt.2[a]	$23(\sum x_i) - .25(\sum x_i)^2$	$23(\sum x_i) - .25(\sum x_i)^2$
Market 2 return/unit of output	\$.01	\$.01
Market 1 return/unit of output	\$.05	\$.05
Earnings/subject at group maximum[b]	\$.91	\$.83
Earnings/subject at Nash equilibrium	\$.66	\$.70
Earnings/subject at zero rent	\$.50	\$.63

[a] $\sum x_i =$ the total number of tokens invested by the group in market 2. The production function shows the number of units of output produced in market 2 for each level of tokens invested in market 2.

[b] In the high-endowment design, subjects were paid in cash one-half of their "computer" earnings. Amounts shown are potential cash payoffs.

Average net yield as a percentage of maximum
10 - Token parameterization

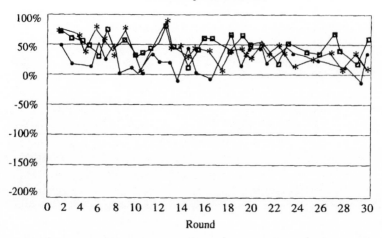

25 - Token parameterization

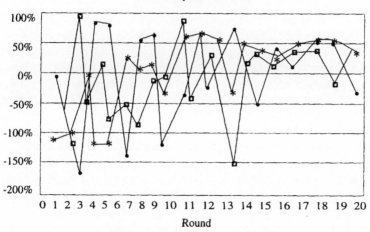

Source: E. Ostrom, Gardner, and Walker (1994)

Fig. 15.2. Individual baseline experiments

yield from the CPR as a percentage of optimum across decision rounds. The data in these panels are typical of the "baseline" experiments reported by OGW, leading to the following summary observations.

1. Subjects make investment allocations to market 2 (the CPR) well above optimum, leading to significant "rent" dissipation.
2. Investments in market 2 are characterized by a "pulsing" pattern in which investments are increased, leading to a reduction in yield, at which time investors tend to reduce their investments in market 2 and yields increase. This pattern reoccurs across decision rounds within an experiment, with a tendency for the variation across rounds to diminish as the experiment continues.
3. Investment behavior is affected by token endowments. Yields as a percentage of optimum are less in 25-token experiments than in 10-token experiments.
4. The Nash equilibrium is the best predictor of *aggregate* outcomes for low-endowment experiments. In the high-endowment setting, aggregate behavior is far from Nash in early rounds but approaches Nash in later rounds. However, at the individual decision level, there is virtually no evidence that behavior converges to the Nash equilibrium.

In an alternative design, Walker and Gardner (1992, hereafter WG) alter the CPR decision setting to add "probabilistic destruction" as one consequence of overinvestment in the CPR. In one of their designs, destruction implies the experiment ends—there is no further opportunity for earning money. A safe appropriation zone is created by allowing for a minimal level of appropriation for which the probability of destruction is zero. For market 2 investments above the minimum "threshold" level, the probability of destruction increases monotonically. In the design investigated by WG, the threshold appropriation level is set at a point that generates near optimum investments—100 percent net yield. This threshold becomes one equilibrium for the game. There is, however, another subgame perfect equilibrium in which overinvestment occurs, yielding an expected return of less than 25 percent of optimum. Interestingly, in their stark institutional setting with no overt possibilities for coordination, WG find that subjects do not tend to coordinate on the best equilibrium. In every experimental session the CPR is destroyed and, in most cases, rather quickly. The consequence of this destruction is a significant loss in yield from the resource.

VCM and CPR Behavior: Puzzles
The results summarized above offer several puzzles from the perspective of noncooperative game theory. In the VCM environment, especially for some

parameterizations that are inconsequential from the perspective of the theory, there is just "too much" cooperative play. What aspects of the VCM game are fostering the degree of cooperation exhibited, and what explains the interaction between the level of cooperation, MPCR, and group size? In the CPR environment, game theory does a pretty good job of explaining aggregate behavior (especially with repetition of the game). On the other hand, game theory predicts no endowment effect, but the data clearly show one. Finally, the subjects used in both of these decision settings are drawn from the same sampling population. What might explain the tendency for what appears to be greater cooperative play in the VCM game in comparison to the CPR game?

In their efforts to explain discrepancies between the theory and the data, IWW consider learning and/or the failure of backward induction. Their experiments with richer information sets and longer game horizons suggest, however, that there is more going on than just learning the consequences of alternative strategies. Complementing this research, Laury, Walker, and Williams (1995) investigate the VCM decision setting in an operationalization in which (1) group members' identity is anonymous and (2) subjects' decisions and earnings are anonymous to other group members and the experimenters. Their results, however, parallel those reported by IWW. Anonymity (or the lack of it) does not appear to be a significant variable that can be used to explain behavior in the VCM decision setting.

It is interesting to consider the "knife-edged" prediction of the theory for the VCM game. If the MPCR > 1, a 100 percent allocation of all tokens to the group good is the single-period dominant strategy. If the MPCR < 1, zero allocations of tokens to the group good is the single-period dominant strategy. On the other hand, in experimental payoff space, if the MPCR is near 1, the payoff consequences of different strategies are minimal from an individual's perspective but can be great from the group's perspective. The greater the MPCR, the greater the gains from some form of group (subgroup) cooperation. Further, holding MPCR constant, in any public goods setting, increasing group size increases the gains from the group achieving some form of tacit cooperation.

In an attempt to model MPCR and group size overtly as an artifact of the game that may affect behavior, several modeling approaches have been suggested. Ledyard (1995) proposes an equilibrium model in which individuals "get some satisfaction (a warm glow) from participating in a group that implicitly and successfully cooperates." In modeling a "warm glow," Ledyard's assumption is related to the work of Andreoni (1989). Individuals are distinguished by types, based upon the strength of their "warm glow" preferences.[14] Under certain assumptions on the population distribution of preferences, Ledyard finds that (1) there can be deviations from complete free riding even in a single-shot game and (2) individuals will be more likely to deviate from complete free riding in large groups.

Miller and Andreoni (1991) present a model based on the adaptive behavior of replicator dynamics. Their approach is consistent with the findings of IWW that the percentage of tokens allocated to the group account appears to be directly related to both group size and MPCR. On the other hand, while the replicator dynamic approach also predicts decay toward complete free riding, it does not explain behavior in which the rate of decay is inversely related to the number of replications of the decision setting.

IWW propose a "forward" looking model of behavior in which players allocate tokens toward the group good in an effort to solicit cooperation. The approach is composed of three principal components: (1) the assumption that individual i believes her decisions have signaling content to others, (2) a benchmark earnings level for measuring the success of signaling, and (3) the formulation of a subjective probability function for evaluating the likelihood of success. In this model, ceteris paribus, the likelihood that a signal (allocations to the group account) succeeds increases with both group size and MPCR, and rate of decay in group allocations is inversely related to number of replications of the game.

One explanation for the discrepancy between theory and data in both VCM and CPR games may be that the theory accounts for the behavior of many individuals but not *all*. The play of these "nontheoretical" players prevents convergence to a theoretical equilibrium, and the impact of such players can be decision setting and institution specific. Contrast, for example, the low-endowment CPR game with the high-endowment game. Assume that in the CPR game, some players are trying implicitly to reach a cooperative solution to the game. Noncooperative players reap (at least short term) benefits from taking advantage of such players by investing more heavily in the CPR. In the low-endowment game, however, there are clear constraints on an *individual's* ability to unilaterally defeat (take advantage of) such an attempt. In the high-endowment parameterization, an individual's strategic "leverage" is increased. One or two players can invest sufficiently in the CPR to take full advantage of attempts at cooperation, yielding an aggregate result that is near the deficient game equilibrium. Now consider the VCM game. Especially for large groups and/or high MPCR parameterizations, an individual has very limited strategic leverage to take advantage of cooperative attempts. It is the strategic decision space of the VCM game that differs fundamentally from that of the CPR game. For example, if in the VCM game, players could either contribute toward provision or appropriate tokens that have been contributed by others, the strategy space of the game would be altered in such a way that a noncooperative player could more readily offset attempts at cooperation by others.

Understanding the linkages and discrepancies between the data and theory for these stark institutional settings remains an important goal for future research in collective action decision settings. But results from field settings suggest that equally important is behavior in settings that are institutionally richer than the stark settings discussed above. Considerable experimental research has been undertaken using more complex institutions. We summarize part of this research to give the reader an idea of the types of institutional settings that may significantly alter individual decisions in collective action arenas.

3.3. Behavior in Richer Institutional Settings

Face-to-Face Communication
The effect of communication in collective action situations is open to considerable debate. Words alone are viewed by many as frail constraints when individuals make private, repetitive decisions between short-term, profit-maximizing strategies and strategies negotiated by a verbal agreement.[15] The inability to make enforceable agreements is at the core of the distinction between cooperative and noncooperative theories for

> [T]he decisive question is whether the players can make enforceable agreements, and it makes little difference whether they are allowed to talk to each other. Even if they are free to talk and to negotiate a agreement, this fact will be of no real help if the agreement has little chance of being kept. An ability to negotiate agreements is useful only if the rules of the game make such agreements binding and enforceable. (Harsanyi and Selten 1988, 3)[16]

Thus, much of contemporary, noncooperative game theory treats the ability to communicate as inessential and unlikely to change results unless the individuals involved can call on external agents to enforce agreements.[17]

Studies of repetitive collective action situations in field settings, however, show that individuals in many settings adopt cooperative strategies that enhance their joint payoffs without the presence of external enforcers. Many situational factors appear to affect the capacity to arrive at and maintain agreed-upon play. The ability to communicate appears to be a necessary but not a sufficient condition. Experimental research on face-to-face communication has shown this mechanism to be a powerful tool for enhancing efficiency. As Dawes states, "The salutary effects of communication on cooperation are ubiquitous" (1980, 185).[18] Hypotheses forwarded to explain why communication increases the selection of cooperative strategies identify a process that communication is posited to facilitate (1) offering and extracting promises, (2) changing the expectations of others' behavior, (3) changing the payoff structure, (4) reenforcing of prior normative orientations, and (5) developing of a group identity. Experimental examination of communication has demonstrated the independent effect of all five of these processes, but they also appear to reenforce one another in an interactive manner.[19] Prior research that relied on signals exchanged via computer terminals rather than face-to-face communication has not had the same impact on behavior. Sell and Wilson (1991, 1992), whose experimental design allowed participants in a public good experiment to signal a promise to cooperate via their terminals, found much less sustained cooperation than reported for face-to-face communication.

A deeper examination of the role of communication in facilitating the selection and retention of efficient strategies is thus of considerable theoretical (as well as policy) interest.[20] The summary below focuses primarily on the findings from a series of experiments in which face-to-face communication was operationalized (without the presence of external enforcement) in the CPR appropriation environment of OGW.[21] The role of communication and its success in fostering outcomes more in line with social optimality are investigated in settings in which (1) the communication mechanism is provided as a costless one-shot opportunity, (2) the communication mechanism is provided as a costless opportunity on a repeated basis, and (3) the subjects face a dilemma of having to provide the communication mechanism in a voluntary contribution decision environment.

The procedure for operationalizing costless face-to-face communication in the laboratory generally follows a protocol similar to the following announcement used by OGW for a one-shot opportunity to communicate.

> Some participants in experiments like this have found it useful to have the opportunity to discuss the decision problem you face. You will be given ten minutes to hold such a discussion. You may discuss anything you wish during your ten-minute discussion period, with the following restrictions. (1) You are not allowed to discuss side payments. (2) You are not allowed to make physical threats. (3) You are not allowed to see the private information on anyone's monitor.

If the experimental design calls for repeated opportunities to communicate, the subjects are told that after each decision round they will return to a common area and have further opportunities to discuss the decision problem.

The results of face-to-face communication reported by OGW can be summarized as follows.

1. Subjects in repeated, high-endowment CPR games, with one and only one opportunity to communicate, obtain an average percentage of net yield above that obtained in baseline experiments in the same decision rounds without communication (55 percent compared to 21 percent).

2. Subjects in repeated, high-endowment CPR games, with *repeated opportunities* to communicate, obtain an average percentage of net yield that is substantially above that obtained in baseline experiments without communication (73 percent compared to 21 percent). In low-endowment games, the average net yield is 99 percent as compared to 34 percent.
3. Repeated communication opportunities in high-endowment games lead to higher joint outcomes (73 percent) than one-shot communication (55 percent), as well as lower defection rates (13 percent compared to 25 percent).
4. In no experiment where one or more subjects deviated from an agreed-upon joint strategy did the other subjects then follow a grim trigger strategy of substantially increasing their investments in the CPR.

The OGW experiments summarized above focused on situations in which individuals were homogeneous in decision attributes. It is possible that the strong efficiency-enhancing properties of face-to-face communication are dependent upon homogeneities in decision attributes. In fact, the literature provides several arguments that point to heterogeneity as a serious deterrent to cooperation (R. Hardin 1982; Johnson and Libecap 1982; Libecap and Wiggins 1984; Wiggins and Libecap 1985, 1987; Isaac and Walker 1988a; E. Ostrom 1990; Kanbur 1992; Hackett 1992). For example, Kanbur argues, "theory and evidence would seem to suggest that cooperative agreements are more likely to come about in groups that are homogeneous in the relevant economic dimension, and they are more likely to break down as heterogeneity along this dimension increases" (1992, 21–2).

The task of agreeing to and sustaining agreements is more difficult for heterogeneous individuals because of the distributional conflict associated with alternative sharing rules. In heterogeneous settings, different sharing rules generally produce different distributions of earnings across individuals. While all individuals may be made better off by cooperating, some benefit more than others, depending upon the sharing rule chosen. Consequently, individuals may fail to cooperate on the adoption of a sharing rule because they cannot agree upon what would constitute a fair distribution of benefits produced by cooperating.[22]

Hackett, Schlager, and Walker (1994, hereafter HSW), building on the experimental research of OGW, examine the CPR decision setting with heterogeneous players. Heterogeneity is introduced by varying the input endowments of the subjects. Heterogeneities in endowments imply that alternative rules adopted to reduce overappropriation from the CPR will have differential effects on earnings across subjects. HSW find that, even with heterogeneity, face-to-face communication remains a very effective institution for increasing efficiency. Investigating two designs, one in which heterogeneities in endowments were assigned randomly and one in which they were assigned through an auction mechanism, both treatment conditions led to significant increases in net yield over baseline (no-communication) conditions. With noncommunication, HSW report a level of net yield relatively close to that predicted by the Nash equilibrium for their designs (48.9 percent). With communication, overall yield increases to over 94 percent on average in both designs.

The opportunity to communicate in a CPR dilemma situation can be viewed as enabling individuals to coordinate strategies to solve the first order CPR dilemma. In the field, providing such an institution for communication would require costly investments on the part of group members. E. Ostrom and Walker (1991) report results from a set of experiments in which the opportunity to communicate became a costly public good that subjects had to first provide through voluntary (anonymous) contributions. The provision mechanism imposed on the right to communicate placed subjects in a second order public good dilemma (with a provision point). Second order dilemma games exist whenever individuals must expend resources to provide a mechanism that may alter the strategic nature of a first order dilemma game (Oliver 1980; Taylor 1987).[23] The results reported in E. Ostrom and Walker (1991) can be summarized as follows.

> The provision problem players faced in the costly communication experiments was not trivial and did in fact create a barrier. In all experiments with costly communication, the problem of providing the institution for communication diminished the success of either: (a) having the ability to develop a coordinated strategy and/or (b) dealing with players who cheated on a previous agreement. On the other hand, all groups succeeded to some degree in providing the communication mechanism and in dealing with the CPR dilemma. On average, efficiency in these groups increased from approximately 42 percent to 80 percent.

In general, these results are consistent with research on public goods reported by Isaac and Walker (1988a, 1991). They found face-to-face communication as a very effective mechanism in facilitating increased efficiency in the VCM decision setting. Further, increasing the complexity of the environment by making communication a costly second order public good reduced the success of face-to-face communication, but even with this reduction, the institution remained a successful mechanism for improving allocative efficiency.

Sanctioning

As we have found in both field and laboratory settings, nonbinding, face-to-face communication can be a very effective institution for facilitating cooperation. In some instances, such as those in which the high endowments of players lead to very strong temptations to defect on a verbal agreement, the nonbinding character of the institution can be problematic. Further, no well-specified theoretical model explains the diverse types of sanctioning behavior observed in field settings. In many instances, theory predicts that individuals will not self-monitor or impose sanctions on others that are costly on the sanctioner. Understanding internally imposed monitoring and sanctioning behavior more thoroughly provides a key to understanding endogenous solutions to collective action dilemmas. OGW report the results from several experiments in which a sanctioning institution was exogenously imposed on decision makers and several face-to-face communication experiments in which the opportunity to use a sanctioning institution was an endogenous decision for the group of decision makers.

The sanctioning mechanism investigated by OGW required that each subject incur a cost (a fee) to sanction another. OGW found in experiments with an imposed sanctioning institution and no communication.

1. Significantly more sanctioning occurs than predicted by subgame perfection, and the frequency is inversely related to cost.
2. Sanctioning is primarily focused on heavy market 2 (CPR) investors. However, there is a nontrivial amount of sanctioning that can be classified as error, lagged punishment, or "blind" revenge.
3. Average net yield increases from 21 percent with no sanctioning to 37 percent with sanctioning. When the costs of fees and fines are subtracted from average net yield, however, net yield drops to 9 percent.

Thus, subjects overused the sanctioning mechanism, and sanctioning without communication reduced net yield. In experiments where communication and sanctioning were combined, on the other hand, the results are quite positive. OGW report that:

1. With an imposed sanctioning mechanism and a single opportunity to communicate, subjects achieve an average net yield of 85 percent. When the costs of fees and fines are subtracted, average net yield is still 67 percent. These represent substantial gains over baseline where net yield averaged 21 percent.
2. With the right to choose a sanctioning mechanism and a single opportunity to communicate, subjects who adopt a sanctioning mecha-

nism achieve an average net yield of 93 percent. When the costs of fees and fines are subtracted, average net yield is still 90 percent. In addition, the defection rate from agreements is only 4 percent.

3. With the right to choose a sanctioning mechanism and a single opportunity to communicate, subjects who do not adopt a sanctioning mechanism achieve an average net yield of only 56 percent. In addition, the defection rate from agreements is 42 percent.

Thus, subjects who use the opportunity to communicate and agree upon a joint strategy and choose their own sanctioning mechanism achieve close to optimal results based entirely on the promises they make, their own efforts to monitor, and their own investments in sanctioning. This is especially impressive in the high-endowment environment, where defection by a few subjects is very disruptive.

Public Good Settings with Provision Points

One useful method for organizing collective action experiments is to partition experiments along two conditions: (1) settings with a unique Nash equilibrium yielding suboptimal provision and (2) settings based on more complex parameterizations or with contribution-facilitating mechanisms that yield multiple equilibria.

One direct way of changing the public goods decision setting is to investigate provision where the public goods are discrete (provision point or step function public goods). Such experimental situations have naturally occurring counterparts in decision settings in which a minimum level of provision support is necessary for productive services (a bridge, for example). For illustration, consider the VCM setting with an MPCR = .30 and $N = 4$. Isaac, Schmidtz, and Walker (1989, hereafter ISW) examined this setting but with the following change. If allocations to the group account did not meet a specified minimum threshold, there was no provision of the public good and all allocations were lost (that is, had a zero value). ISW examined several designs in which they varied the minimum threshold. This type of decision situation creates an "assurance problem." Zero contributions to the group good is no longer a dominant strategy or the unique Nash strategy. Players have an incentive to contribute to the public good if they have some expectation (assurance) that others will contribute. On the other hand, if others will provide the public good, the individual has an incentive to free ride on their contributions. ISW found that

1. In designs with provision points that require relatively low levels of contributions, numerous experimental groups were able, in early decision rounds, to overcome the assurance problem and provide the public good.

2. In experiments with higher provision points and in later decision rounds of most experiments, free-riding behavior tended to increase with resulting low levels of efficiency.

These results are similar to results from a closely related provision point setting discussed by van de Kragt, Orbell, and Dawes (1983). In this study (where subjects made a binary decision to contribute or not to contribute to a group good), groups met the provision point in less than 35 percent of the decision trials. In an interesting modification to the standard decision setting, Dorsey (1992) examines the VCM decision setting with provision point settings and "real time" revision of individual decisions. More specifically, Dorsey allows subjects a continuous information update of "others' " contributions and the opportunity to revise one's own contribution. In provision point designs with revisions that can *only increase*, allocations to the group good significantly increase.

In anticipation that the specific rules of the contribution mechanism might significantly affect decision behavior, several studies have examined the provision point decision setting using an alternative contribution mechanism. For example, Dawes, Orbell, and van de Kragt (1984), Isaac, Schmidtz, and Walker (1989), and Bagnoli and McKee (1991) investigate several versions of what is commonly referred to as the "payback" mechanism.[24] Contributions are made toward the provision of the public good. If they do not meet the specified minimum, all are returned to the players making the contributions. As one might expect, this simple change can significantly affect decision incentives, equilibria, and observed behavior. Certainly the risks involved in making contributions are reduced. On the other hand, there is still an incentive to free ride if others will provide the public good.

The variation in findings of the three studies cited above is quite interesting. In one-shot decisions (with no value for contributions above the provision point and binary decisions to contribute or not to contribute), Dawes, Orbell, and van de Kragt found no significant effects on levels of contributions when comparing provision point experiments with and without the payback mechanism. On the other hand (in a decision setting in which contributions above the provision point have a positive value and subjects make nonbinary choices), ISW found that using the payback mechanism substantially increased efficiency in settings with higher provision points and to a lesser extent in the low provision point setting. ISW still observed significant problems in low and medium provision point settings (especially later decision periods) due to what they refer to as "cheap" riding. Significant numbers of subjects strategically attempted to provide a smaller share of the public good than their counterparts, in some cases leading to a failure to meet the provision point. Finally (in a decision setting in which contributions above the provision point have no value and subjects make nonbinary decisions), Bagnoli and McKee found very strong support for the cooperative facilitating features of the payback institution. In their experiments, the public good was provided in 85 of 98 possible cases, and there was very little loss in efficiency due to overinvestments.

4. Summary Comments and Conclusions

Collective action problems offer a unique and challenging setting for social scientists interested in the linkages between theory, institutions, and behavior. Evidence from field and experimental studies provides support for the following two fundamental propositions.

P1   Without some form of coordination or organization to enable individuals to agree upon, monitor, and sanction contributions to the provision of a public good, the good is *underprovided.*

P2   Without some form of coordination or organization to enable individuals to agree upon, monitor, and sanction the patterns of appropriation from a CPR, the resource is *overused.*

Evidence from field and experimental studies, on the other hand, challenges many currently accepted theoretical understandings of the likelihood of those directly involved in smaller-scale public good or CPR settings to develop their own forms of coordination or organization.

Field studies have generated a wealth of examples of individuals organizing themselves and changing the structure of their own settings to reduce the incentives to underprovide public goods or CPRs and to overappropriate from CPRs. Many of these endogenously designed institutions do not easily fit with prior conceptions of the state or the market. But it is also obvious from field studies that self-organization is not universal and frequently fails. Consequently, evidence from laboratory experiments that more precisely identify the specific variables that are associated with particular behavioral tendencies is crucial for further theoretical development. Below we provide a brief summary of what we have learned in the lab about provision, appropriation, organization, and sanctioning.

4.1. Findings Related to Provision

1.   In stark institutional settings where subjects cannot communicate, the theoretical prediction of suboptimal provision is supported.

a.   In repeated decision rounds, initial efficiencies were greater than predicted, generally with some decay over time.

  b. In repeated decision rounds, the path of decline toward the Nash equilibrium often showed a distinct pulsing pattern.
2. Ceteris paribus, when the marginal value of contributing to a public good was higher, efficiencies were higher.
3. Increasing the number of individuals—holding marginal return from the public good constant—generally had no impact or *increased* efficiency.
4. Face-to-face communication increased efficiency—even under circumstances where individuals were not symmetric in regard to asset endowments.
5. Establishing a payback mechanism (whereby those who contributed to a good when a sufficient number of others did not contribute were assured a rebate of their contribution) led to an increase in efficiency.
6. A provision point (or, lumpy production function) was often more conducive to the efficient provision of public goods than a smooth production function.

4.2. Findings Related to Appropriation

1. In stark institutional settings where subjects could not communicate, the theoretical prediction of overappropriation (or, where relevant, destruction) was supported.
    a. In repeated decision rounds, initial inefficiencies were often greater than predicted by the Nash equilibrium for a one-shot game.
    b. In repeated decision rounds, the path of movement toward the Nash equilibrium showed a distinct pulsing pattern.
2. Increasing asset endowments used for appropriation reduced efficiencies, even when the predicted Nash equilibrium was unchanged.
3. Face-to-face communication increased efficiencies.

4.3. Findings Related to Organization, Monitoring, and Sanctioning

1. When given an opportunity, subjects spent time and effort designing agreements to share the costs and benefits of providing and appropriating under mutually developed constraints.
2. When the only sanctioning mechanisms available for punishing those who did not follow agreements were verbal chastisement or switching to deficient outcomes:
    a. small deviations by a few subjects were usually met by small deviations by other subjects and a return to behavior consistent

with the agreement, but

b.   large deviations by a few subjects tended to lead to the unraveling of agreements.

3.   When given an opportunity, subjects overutilized a costly exogenous sanctioning system.

4.   When given an opportunity, subjects who chose to impose a sanctioning system upon themselves were able to earn higher efficiencies in repeated decision rounds.

Findings from experimental settings complement those from field settings and help to identify relevant variables associated with the endogenous development of efficiency-enhancing institutions. While the two fundamental propositions concerning suboptimal provision and overappropriation in stark institutional settings are supported in both field and experimental studies, many other commonly accepted propositions about the helplessness of individuals to change the structure of their own situations are not.

When turning to policy prescriptions, the importance of the complex mixture of variables that exist in distinct settings is substantial. The particular features of a natural setting that might effectively be used by participants in selecting rules cannot be included in general models. The likelihood is small that any set of uniform rules for all natural settings within a large territory, such as developing marketable rights or imposing state regulations, will produce optimal results. This is unfortunately the case, whether or not the particular rules can be shown to generate optimal rules in sparse theoretical or experimental settings. Theoretical and empirical research, though, can be used to help inform those who are close to particular natural resource systems as well as those in larger, overarching agencies about the principles that may be used to improve performance. It is important to keep these differences in mind when making policy prescriptions. Slogans such as "privatization" or "regulation" may mask important underlying principles rather than provide useful guides for reform.

The findings summarized above are also important because they challenge the propensity to use an extremely limited set of simple models to characterize behavior in highly complex settings. All theory involves necessary simplification. One of the important lessons from game theory, however, is that a very small difference in the structure of a game can make an immense difference in the predicted behavior and outcomes. Empirical research conducted in experimental laboratories reinforces this lesson, as does systematic field research conducted in multiple sites. Presuming that the institutional world of relevance to the study of public choice is limited to either the market or the state brackets together an immense variety of rich institutional arrangements that should not be lumped together. Our conceptual language loses considerable analytical bite when highly disparate phenomena are viewed as if they were essentially equivalent in oversimplified models of the world.

NOTES

1. This is especially true when the model of the second kind of institution—the state—is based on a set of assumptions about the behavior of officials who are "economic eunuchs," to use James Buchanan's description of Pigovian policymakers (Buchanan 1973).

2. See Sandler (1992) and Schmidtz (1991) for recent work on related issues.

3. We do not present these eight as the exhaustive list of transformation processes of relevance to understanding public economies but rather as a core set that has been useful in organizing our own work. Each process can be further unpacked. Also, we do not focus here on the epistemic orders that are closely related to the economic and political orders we do discuss (see V. Ostrom 1997).

4. The experimental evidence presented later in this essay suggests that even in some sparse settings the degree of suboptimality that is theoretically predicted may be overly "pessimistic."

5. Eitan Berglas (1981, 390) stresses that to understand the equilibrium conditions of the private provision of some club goods, "one has to distinguish between the manufacture of swimming pools and the services that swimming pools provide" (see also Berglas 1976). One might even wonder why these are not strictly private goods in the first place since exclusion is easy and consumption is rival. The collective aspect is the joint facility that must be utilized with others; deciding upon the appropriate size of membership and the rules of allocation is, however, a collective choice problem.

6. It is not always the case, however, that participants explain their actions to outsiders in the same way they explain them to fellow participants. Consequently, learning about the working rules used in a particular CPR may be very difficult. Further, rule following or conforming actions are not as predictable as biological or physical behavior explained by physical laws. Rules are formulated in human language. As such, rules share the problems of lack of clarity, misunderstanding, and change that typify any language-based phenomenon. Words are "symbols that name, and thus, stand for classes of things and relationships" (V. Ostrom 1980, 312). Words are always simplifications of the phenomena to which they refer (V. Ostrom 1991).

7. The obvious exception is the work of James Buchanan and many colleagues who have been associated with Buchanan throughout the years. See in particular Buchanan and Tullock (1962) and Brennan and Buchanan (1985), as well as literature reviewed by Mueller.

8. See McCay and Acheson (1987); Fortmann and Bruce (1988); Wade (1988); Berkes (1989); Pinkerton (1989); Sengupta (1991); Blomquist (1992); Bromley et al. (1992); Dasgupta and Mäler (1992); Tang (1992); Thomson (1992); Netting (1993); and V. Ostrom, Feeny, and Picht (1993). See also the influential article by Feeny et al. (1990) and recent important works on property rights (Libecap 1989; Eggertsson

1990; Bromley 1991) that examine common-property institutions.

9. For a theoretical discussion of why these design principles work in practice see Coward (1979); Siy (1982); Ruddle (1988); E. Ostrom (1990); E. Ostrom and Gardner (1993); and E. Ostrom, Gardner, and Walker (1994).

10. See Gardner, E. Ostrom, and Walker (1993) for further discussion of this topic.

11. See E. Ostrom (1990); McKean (1992); and Schlager, Blomquist, and Tang (1994) for further discussion.

12. For other related studies, see for example Marwell and Ames (1979, 1980, 1981); Isaac, Walker, and Thomas (1984); Kim and Walker (1984); Isaac, McCue, and Plott (1985); Isaac and Walker (1988b, 1993); Andreoni (1989, 1993); Brookshire, Coursey, and Redington (1989); Dorsey (1992); Asch, Gigliotti, and Polito (1993); Chan et al. (1993); Fisher et al. (1993); Palfrey and Prisbry (1993); Sefton and Steinberg (1993); Ledyard (1995); and Laury, Walker, and Williams (1995).

13. See E. Ostrom, Gardner, and Walker (1994) for details of the derivation of this game equilibrium.

14. Also see Palfrey and Rosenthal (1988) where they model "uncontrolled preferences" that derive from "acts of social cooperation or contribution, the utility of altruism, or social duty."

15. See E. Ostrom, Walker, and Gardner (1992) for further discussion of this topic.

16. Harsanyi and Selten (1988, 3) add that in real life, "agreements may be enforced externally by courts of law, government agencies, or pressure from public opinion; they may be enforced internally by the fact that the players are simply unwilling to violate agreements on moral grounds and know that this is the case." To model self-commitment using noncooperative game theory, the ability to break the commitment is removed by trimming the branches that emanate from a self-commitment move to remove any alternative contrary to that which has been committed. In a lab setting, this would mean changing the structure of the alternatives made available to subjects after an agreement (which was not done).

17. Self-commitment is also possible, but whether the agreement is backed by external agents or self-commitment, the essential condition is that all branches of the game tree are removed that correspond to moves violating the agreement that has been made (Harsanyi and Selten 1988, 4). In the lab, this would mean that the experimenters would reprogram the experiment so that no more than the agreed upon number of tokens could be invested in the CPR. This condition was never imposed in the laboratory experiments. In the field, this would mean that some action was taken to remove the feasibility of certain types of activities—an almost impossible task.

18. Among the studies showing a positive effect of the capacity to communicate are Jerdee and Rosen (1974); Caldwell (1976); Dawes, McTavish, and Shaklee (1977); Edney and Harper (1978); Braver and Wilson (1984, 1986); Dawes, Orbell, and van de Kragt (1984); Kramer and Brewer (1986); van de Kragt et al. (1986); Bornstein and Rapoport (1988); Isaac and Walker (1988a, 1991); Bornstein et al. (1989); Orbell, Dawes, and van de Kragt (1990); Orbell, van de Kragt, and Dawes (1991); E. Ostrom and Walker (1991); and Hackett, Schlager, and Walker (1994).

19. Orbell, van de Kragt, and Dawes (1988) summarize the findings from 10 years

of research on one-shot public good experiments by stressing both the independent and interdependent nature of the posited factors explaining why communication has such a powerful effect on rates of cooperation.

20. See Banks and Calvert (1992a,b) for an important discussion of the theoretical significance of communication in incomplete information games.

21. See E. Ostrom and Walker (1991) and Isaac and Walker (1991) for a more detailed discussion of the role of communication and the experimental evidence summarized here.

22. Hackett (1992) suggests that heterogeneous resource endowments can lead to disagreement over the supply and implementation of rules that allocate access to CPRs.

23. Yamagishi (1986, 1988) examines the imposition of sanctioning to change the structure of a simple public good dilemma situation.

24. Palfrey and Rosenthal (1984) discuss strategic equilibria in provision point games with and without the payback mechanism; see Bagnoli, Shaul, and McKee (1992) for further results.

REFERENCES

Alexander, Paul. 1982: *Sri Lankan Fishermen: Rural Capitalism and Peasant Society*. Canberra: Australian National University.

Anderson, Lee G. 1977: *The Economics of Fisheries Management*. Baltimore, Md: Johns Hopkins University Press.

Andreoni, James. 1989: Giving with impure altruism: Applications to charity and Ricardian equivalence. *Journal of Political Economy*, 97, 1447–58.

Andreoni, James. 1993: An experimental test of the public-goods crowding-out hypothesis. *American Economic Review*, 83(5) (December), 1317–27.

Asch, Peter, Gigliotti, Gary A., and Polito, James A. 1993: Free riding with discrete and continuous public goods: Some experimental evidence. *Public Choice*, 77 (2) (October), 293–305.

Bagnoli, Mark and McKee, Michael. 1991: Voluntary contribution games: Efficient private provision of public goods. *Economic Inquiry*, 29(2) (April), 351–66.

Bagnoli, Mark, Shaul, Ben-David, and McKee, Michael. 1992: Voluntary provision of public goods, the multiple unit case. *Journal of Public Economics*, 47, 85–106.

Banks, Jeffrey S. and Calvert, Randall L. 1992a: A battle-of-the-sexes game with incomplete information. *Games and Economic Behavior*, 4, 1–26.

Banks, Jeffrey S. and Calvert, Randall L. 1992b: Communication and efficiency in coordination games. Working paper. University of Rochester, Department of Economics and Department of Political Science.

Bates, Robert H. 1981: *Markets and States in Tropical Africa: The Political Basis of Agricultural Policies*. Berkeley: University of California Press.

Bendor, Jonathan and Mookherjee, Dilip. 1987: Institutional structure and the logic of ongoing collective action. *American Political Science Review*, 81(1) (March),

129–54.

Berglas, Eitan. 1976: On the theory of clubs. *American Economic Review,* 66(2) (May), 116–21.

Berglas, Eitan. 1981: The market provision of club goods once again. *Journal of Public Economics,* 15(3) (June), 389–93.

Berkes, Fikret (ed). 1989: *Common Property Resources: Ecology and Community-Based Sustainable Development.* London: Belhaven.

Bianco, William T. and Bates, Robert H. 1990: Cooperation by design: Leadership, structure, and collective dilemmas. *American Political Science Review,* 84 (March), 133–47.

Blomquist, William. 1992: *Dividing the Waters: Governing Groundwater in Southern California.* San Francisco, Calif.: Institute for Contemporary Studies Press.

Blomquist, William and Ostrom, Elinor. 1985: Institutional capacity and the resolution of a commons dilemma. *Policy Studies Review,* 5(2) (November), 383–93.

Bornstein, Gary and Rapoport, Amnon. 1988: Intergroup competition for the provision of step-level public goods: Effects of preplay communication. *European Journal of Social Psychology,* 18, 125–42.

Bornstein, Gary, Rapoport, Amnon, Kerpel, Lucia, and Katz, Tani. 1989: Within- and between-group communication in intergroup competition for public goods. *Journal of Experimental Social Psychology,* 25, 422–36.

Braver, Sanford L. and Wilson, L. A. 1984: A laboratory study of social contracts as a solution to public goods problems: Surviving on the lifeboat. Presented at the Western Social Science Association meeting, San Diego, California.

Braver, Sanford L. and Wilson, L. A. 1986: Choices in social dilemmas: Effects of communication within subgroups. *Journal of Conflict Resolution,* 30(1) (March), 51–62.

Brennan, Geoffrey and Buchanan, James. 1985: *The Reason of Rules.* Cambridge: Cambridge University Press.

Bromley, Daniel W. 1991: *Environment and Economy: Property Rights and Public Policy.* Oxford: Basil Blackwell.

Bromley, Daniel W., et al. (eds). 1992: *Making the Commons Work: Theory, Practice, and Policy.* San Francisco, Calif.: Institute for Contemporary Studies Press.

Brookshire, David, Coursey, Don, and Redington, Douglas. 1989: Special interests and the voluntary provision of public goods. Typescript.

Buchanan, James M. 1965: An economic theory of clubs. *Economica,* 32(125) (February), 1–14.

Buchanan, James M. 1973: The Coase theorem and the theory of the state. *Natural Resources Journal,* 13, 579–94.

Buchanan, James M. and Tullock, Gordon. 1962: *The Calculus of Consent.* Ann Arbor: University of Michigan Press.

Caldwell, Michael D. 1976: Communication and sex effects in a five-person prisoners' dilemma game. *Journal of Personality and Social Psychology,* 33(3), 273–80.

Chan, Kenneth, Mestelman, Stuart, Moir, Rob, and Muller, Andrew. 1993: The voluntary provision of public goods under varying endowment distributions. Working paper. Hamilton, Ontario: McMaster University, Department of Economics.

Christy, Francis T., Jr. and Scott, Anthony D. 1965: *The Common Wealth in Ocean Fisheries*. Baltimore, Md.: Johns Hopkins University Press.

Copes, Parzival. 1986: A critical review of the individual quota as a device in fisheries management. *Land Economics,* 62(3) (August), 278–91.

Cordell, John C. and McKean, Margaret A. 1992: Sea tenure in Bahia, Brazil. In Daniel W. Bromley et al. (eds.), *Making the Commons Work: Theory, Practice, and Policy*. San Francisco, Calif.: Institute for Contemporary Studies Press, 183–205.

Cornes, Richard and Sandler, Todd. 1986: *The Theory of Externalities, Public Goods, and Club Goods*. Cambridge: Cambridge University Press.

Coward, E. Walter, Jr. 1979: Principles of social organization in an indigenous irrigation system. *Human Organization,* 38(1) (spring), 28–36.

Dasgupta, Partha and Mäler, Karl Göran. 1992: *The Economics of Transnational Commons*. Oxford: Clarendon Press.

Davis, Anthony. 1984: Property rights and access management in the small boat fishery: A case study from Southwest Nova Scotia. In Cynthia Lamson and Arthur J. Hanson (eds.), *Atlantic Fisheries and Coastal Communities: Fisheries Decision-Making Case Studies*. Halifax: Dalhousie Ocean Studies Programme, 133–64.

Dawes, Robyn M. 1980: Social dilemmas. *Annual Review of Psychology,* 31, 169–93.

Dawes, Robyn M., McTavish, Jeanne, and Shaklee, Harriet. 1977: Behavior, communication, and assumptions about other people's behavior in a commons dilemma situation. *Journal of Personality and Social Psychology,* 35(1) (January), 1–11.

Dawes, Robyn M., Orbell, John M., and van de Kragt, Alphons. 1984: Normative constraint and incentive compatible design. University of Oregon, Department of Psychology. Typescript.

Dorsey, Robert E. 1992: The voluntary contributions mechanism with real time revisions. *Public Choice,* 73(3) (April), 261–82.

Edney, Julian J. and Harper, Christopher S. 1978: The commons dilemma: A review of contributions from psychology. *Environmental Management,* 2(6), 491–507.

Eggertsson, Thráinn. 1990: *Economic Behavior and Institutions*. New York: Cambridge University Press.

Feeny, David. 1988: Agricultural expansion and forest depletion in Thailand, 1900–1975. In John F. Richards and Richard P. Tucker (eds.), *World Deforestation in the Twentieth Century*. Durham, N.C.: Duke University Press, 112–43.

Feeny, David, Berkes, Fikret, McCay, Bonnie J., and Acheson, James M. 1990: The tragedy of the commons: Twenty-two years later. *Human Ecology,* 18(1), 1–19.

Fisher, Joseph, Isaac, R. Mark, Schatzberg, Jeffrey W., and Walker, James M. 1993: Heterogenous demand for public goods: Effects on the voluntary contributions mechanism. Working paper. University of Arizona and Indiana University, Department of Economics.

Fortmann, Louise and Bruce, John W. (eds.). 1988: *Whose Trees? Proprietary Dimensions of Forestry*. Boulder, Colo.: Westview Press.

Gadgil, Madhav and Iyer, Prema. 1989: On the diversification of common-property resource use by Indian society. In Fikret Berkes (ed.), *Common Property Resources, Ecology and Community-Based Sustainable Development.* London: Belhaven Press, 240–72.

Gardner, Roy, Ostrom, Elinor, and Walker, James M. 1993: Crafting institutions in common-pool resource situations. Paper presented at the Annual Meetings of the Association for Evolutionary Economics (AFEE), Boston, Massachusetts, January 3–5, 1994.

Hackett, Steven. 1992: Heterogeneity and the provision of governance for common-pool resources. *Journal of Theoretical Politics,* 4(3), 325–42.

Hackett, Steven, Schlager, Edella, and Walker, James M. 1994: The role of communication in resolving commons dilemmas: Experimental evidence with heterogeneous appropriators. *Journal of Environmental Economics and Management,* 27, 99–126.

Hardin, Garrett. 1968: The tragedy of the commons. *Science,* 162, 1,243–48.

Hardin, Russell. 1982: *Collective Action.* Baltimore: Johns Hopkins University Press.

Harsanyi, John C. and Selten, Reinhard. 1988: *A General Theory of Equilibrium Selection in Games.* Cambridge, Mass.: MIT Press.

Hilton, Rita. 1992: Institutional incentives for resource mobilization: An analysis of irrigation schemes in Nepal. *Journal of Theoretical Politics,* 4(3), 283–308.

Isaac, R. Mark, McCue, Kenneth, and Plott, Charles R. 1985: Public goods provision in an experimental environment. *Journal of Public Economics,* 26, 51–74.

Isaac, R. Mark, Schmidtz, David, and Walker, James M. 1989: The assurance problem in a laboratory market. *Public Choice,* 62(3), 217–36.

Isaac, R. Mark and Walker, James M. 1988a: Communication and free-riding behavior: The voluntary contribution mechanism. *Economic Inquiry,* 26(4) (October), 585–608.

Isaac, R. Mark and Walker, James M. 1988b: Group size effects in public goods provision: The voluntary contributions mechanism. *Quarterly Journal of Economics,* 103 (February), 179–99.

Isaac, R. Mark and Walker, James M. 1991: Costly communication: An experiment in a nested public goods problem. In Thomas R. Palfrey (ed.), *Laboratory Research in Political Economy.* Ann Arbor: University of Michigan Press, 269–86.

Isaac, R. Mark and Walker, James M. 1993: Nash as an organizing principle in the voluntary provision of public goods: Experimental evidence. Working paper. Bloomington: Indiana University.

Isaac, R. Mark, Walker, James M., and Thomas, Susan. 1984: Divergent evidence on free riding: An experimental examination of some possible explanations. *Public Choice,* 43(2), 113–49.

Isaac, R. Mark, Walker, James M., and Williams, Arlington. 1994: Group size and the voluntary provision of public goods: Experimental evidence utilizing large groups. *Journal of Public Economics,* 54(1) (May), 1–36.

Jerdee, Thomas H. and Rosen, Benson. 1974: Effects of opportunity to communicate and visibility of individual decisions on behavior in the common interest. *Journal of Applied Psychology,* 59(6), 712–16.

Jessup, Timothy C. and Peluso, Nancy Lee. 1986: Minor forest products as common property resources in East Kalimantan, Indonesia. In National Research Council, *Proceedings of the Conference on Common Property Management.* Washington, D.C.: National Academy Press, 501–31.

Johnson, Ronald and Libecap, Gary D. 1982: Contracting problems and regulation: The case of the fishery. *American Economic Review,* 72(5) (December), 1005–23.

Kanbur, S. M. Ravi. 1992: Heterogeneity, distribution, and cooperation in common property resource management. Background Paper for the World Development Report. Washington, D.C.: World Bank.

Kim, Oliver and Walker, Mark. 1984: The free rider problem: Experimental evidence. *Public Choice,* 43(1), 3–24.

Kiser, Larry L. and Ostrom, Elinor. 1982: The three worlds of action: A metatheoretical synthesis of institutional approaches. In Elinor Ostrom (ed.), *Strategies of Political Inquiry.* Beverly Hills, Calif.: Sage, 179–222. (Chapter 2 in this volume.)

Kramer, R. M. and Brewer, Marilyn M. 1986: Social group identity and the emergence of cooperation in resource conservation dilemmas. In Henk A. Wilke, David M. Messick, and Christel G. Rutte (eds.), *Experimental Social Dilemmas.* Frankfurt am Main: Lang, 205–34.

Lam, Wai Fung, Lee, Myungsuk, and Ostrom, Elinor. 1997: The institutional analysis and development framework: Application to irrigation policy in Nepal. In Derick W. Brinkerhoff (ed.), *Policy Studies and Developing Nations: An Institutional and Implementation Focus,* Vol.5. Greenwich, Conn.: JAI Press, 53–85.

Laury, Susan, Walker, James M., and Williams, Arlington. 1995: Anonymity and the voluntary provision of public goods. *Journal of Economic Behavior and Organization,* 27(3) (August), 365–80.

Ledyard, John. 1995: Is there a problem with public goods provision? In John Kagel and Alvin Roth (eds.), *The Handbook of Experimental Economics.* Princeton, N.J.: Princeton University Press, 111–94.

Libecap, Gary D. 1989: Distributional issues in contracting for property rights. *Journal of Institutional and Theoretical Economics,* 145, 6–24.

Libecap, Gary D. and Wiggins, Steven N. 1984: Contractual responses to the common pool: Prorationing of crude oil production. *American Economic Review,* 74, 87–98.

Marwell, Gerald and Ames, Ruth E. 1979: Experiments on the provision of public goods I: Resources, interest, group size, and the free rider problem. *American Journal of Sociology,* 84, 1335–60.

Marwell, Gerald and Ames, Ruth E. 1980: Experiments on the provision of public goods II: Provision points, stakes, experience and the free rider problem. *American Journal of Sociology,* 85, 926–37.

Marwell, Gerald and Ames, Ruth E. 1981: Economists free ride: Does anyone else? *Journal of Public Economics,* 15, 295–310.

McCay, Bonnie J. and Acheson, James M. 1987: *The Question of the Commons: The Culture and Ecology of Communal Resources.* Tucson: University of Arizona Press.

McGinnis, Michael and Ostrom, Elinor. 1996: Design principles for local and global commons. In Oran R. Young (ed.), *The International Political Economy and International Institutions*, Vol. II. Cheltanham, United Kingdom: Edward Elgar Publishing, 465–93.

McKean, Margaret A. 1992: Success on the commons: A comparative examination of institutions for common property resource management. *Journal of Theoretical Politics*, 4(3), 247–82.

Miller, John and Andreoni, James. 1991: A coevolutionary model of free riding behavior: Replicator dynamics as an explanation of the experimental results. *Economics Letters*, 36, 9–15.

National Research Council. 1986: *Proceedings of the Conference on Common Property Resource Management*. Washington, D.C.: National Academy Press.

Netting, Robert McC. 1993: *Smallholders, Householders: Farm Families and the Ecology of Intensive, Sustainable Agriculture*. Stanford, Calif.: Stanford University Press.

Oliver, Pamela. 1980: Rewards and punishments as selective incentives for collective action: Theoretical investigations. *American Journal of Sociology*, 85, 356–75.

Olson, Mancur. 1965: *The Logic of Collective Action: Public Goods and the Theory of Groups*. Cambridge, Mass.: Harvard University Press.

Orbell, John M., Dawes, Robyn M., and van de Kragt, Alphons. 1990: The limits of multilateral promising. *Ethics*, 100(4) (April), 616–27.

Orbell, John M., van de Kragt, Alphons, and Dawes, Robyn M. 1988: Explaining discussion-induced cooperation. *Journal of Personality and Social Psychology*, 54(5), 811–19.

Orbell, John M., van de Kragt, Alphons, and Dawes, Robyn M. 1991: Covenants without the sword: The role of promises in social dilemma circumstances. In Kenneth J. Kofford and Jeffrey B. Miller (eds.), *Social Norms and Economic Institutions*. Ann Arbor: University of Michigan Press, 117–34.

Ostrom, Elinor. 1990: *Governing the Commons: The Evolution of Institutions for Collective Action*. New York: Cambridge University Press.

Ostrom, Elinor. 1992: *Crafting Institutions for Self-Governing Irrigation Systems*. San Francisco, Calif.: Institute for Contemporary Studies Press.

Ostrom, Elinor and Gardner, Roy. 1993: Coping with asymmetries in the commons: Self-governing irrigation systems can work. *Journal of Economic Perspectives* 7 (4) (fall), 93–112. (Chapter 14 in this volume.)

Ostrom, Elinor, Gardner, Roy, and Walker, James M. 1994: *Rules, Games, and Common-Pool Resources*. Ann Arbor: University of Michigan Press.

Ostrom, Elinor, Schroeder, Larry, and Wynne, Susan. 1993: *Institutional Incentives and Sustainable Development: Infrastructure Policies in Perspective*. Boulder, Colo.: Westview Press.

Ostrom, Elinor and Walker, James M. 1991: Communication in a commons: Cooperation without external enforcement. In Thomas R. Palfrey (ed.), *Laboratory Research in Political Economy*. Ann Arbor: University of Michigan Press, 287–322.

Ostrom, Elinor, Walker, James M., and Gardner, Roy. 1992. Covenants with and without a sword: Self-governance is possible. *American Political Science Review*, 86(2) (June), 404–17.

Ostrom, Vincent. 1980: Artisanship and artifact. *Public Administration Review*, 40(4) (July-August), 309–17. (Reprinted in Michael D. McGinnis, ed., *Polycentric Governance and Development* [Ann Arbor: University of Michigan Press, 1999].)

Ostrom, Vincent. 1984: Why governments fail: An inquiry into the use of instruments of evil to do good. In James M. Buchanan and Robert Tollison (eds.), *The Theory of Public Choice—II*. Ann Arbor: University of Michigan Press, 422–35.

Ostrom, Vincent. 1991: *The Meaning of American Federalism: Constituting a Self-Governing Society*. San Francisco, Calif.: Institute for Contemporary Studies Press.

Ostrom, Vincent. 1993: Cryptoimperialism, predatory states, and self-governance. In Vincent Ostrom, David Feeny, and Hartmut Picht (eds.), *Rethinking Institutional Analysis and Development: Issues, Alternatives, and Choices*. San Francisco, Calif.: Institute for Contemporary Studies Press, 43–68. (Reprinted in Michael D. McGinnis, ed., *Polycentric Governance and Development* [Ann Arbor: University of Michigan Press, 1999].)

Ostrom, Vincent. 1997. *The Meaning of Democracy and the Vulnerability of Democracies: A Response to Tocqueville's Challenge*. Ann Arbor: University of Michigan Press.

Ostrom, Vincent, David Feeny, and Hartmut Picht (eds). 1993: *Rethinking Institutional Analysis and Development: Issues, Alternatives, and Choices*. 2d ed. San Francisco, Calif.: Institute for Contemporary Studies Press.

Ostrom, Vincent, Tiebout, Charles, and Warren, Robert. 1961: The organization of government in metropolitan areas: A theoretical inquiry. *American Political Science Review*, 55 (December), 831–42. (Reprinted in Michael D. McGinnis, ed., *Polycentricity and Local Public Economies* [Ann Arbor: University of Michigan Press, 1999].)

Palfrey, Thomas R. and Prisby, Jeffrey E. 1993: Anomalous behavior in linear public goods experiments: How much and why? Social Science Working Paper no. 833. Pasadena, Calif.: California Institute of Technology.

Palfrey, Thomas R. and Rosenthal, Howard. 1984: Participation and the provision of discrete public goods: A strategic analysis. *Journal of Public Economics*, 24(2) (July), 171–93.

Palfrey, Thomas R. and Rosenthal, Howard. 1988: Private incentives in social dilemmas. *Journal of Public Economics*, 35, 309–32.

Pinkerton, Evelyn (ed). 1989: *Co-operative Management of Local Fisheries: New Directions for Improved Management and Community Development*. Vancouver: University of British Columbia Press.

Plott, Charles R. and Meyer, Robert A. 1975: The technology of public goods, externalities, and the exclusion principle. In Edwin S. Mills (ed.), *Economic Analysis of Environmental Problems*. New York: Columbia University Press, 65–94.

Ruddle, Kenneth. 1988: Social principles underlying traditional inshore fishery management systems in the Pacific basin. *Marine Resource Economics*, 5, 351–63.

Sandler, Todd. 1992: *Collective Action: Theory and Applications*. Ann Arbor: University of Michigan Press.

Sawyer, Amos. 1992: *The Emergence of Autocracy in Liberia: Tragedy and Challenge*. San Francisco, Calif.: Institute for Contemporary Studies Press.

Schlager, Edella. 1990: Model specification and policy analysis: The governance of coastal fisheries. Ph.D. diss., Indiana University, Department of Political Science.

Schlager, Edella, William Blomquist, and Shui Yan Tang. 1994: Mobile flows, storage, and self-organized institutions for governing common-pool resources. *Land Economics*, 70(3) (August), 294–317. (Reprinted in Michael D. McGinnis, ed., *Polycentric Governance and Development* [Ann Arbor: University of Michigan Press, 1999].)

Schlager, Edella and Ostrom, Elinor. 1993: Property-rights regimes and coastal fisheries: An empirical analysis. In Randy Simmons and Terry Anderson (eds.), *The Political Economy of Customs and Culture: Informal Solutions to the Commons Problem*. Lanham, Md.: Rowman & Littlefield, 13–41. (Reprinted in Michael McGinnis, ed., *Polycentric Governance and Development* [Ann Arbor: University of Michigan Press, 1999].)

Schmidtz, David. 1991: *The Limits of Government: An Essay on the Public Goods Argument*. Boulder, Colo.: Westview Press.

Sefton, Martin and Steinberg, Richard. 1993: Reward structures in public good experiments. Indiana University-Purdue University Indianapolis, Department of Economics.

Sell, Jane and Wilson, Rick. 1991: Levels of information and contributions to public goods. *Social Forces,* 70(1) (September), 107–24.

Sell, Jane and Wilson, Rick. 1992: Liar, liar, pants on fire: Cheap talk and signalling in repeated public goods settings. Working paper. Rice University, Department of Political Science.

Sengupta, Nirmal. 1991: *Managing Common Property: Irrigation in India and the Philippines*. London and New Delhi: Sage.

Shepsle, Kenneth A. 1989: Studying institutions: Some lessons from the rational choice approach. *Journal of Theoretical Politics*, 1, 131–49.

Siy, Robert Y., Jr. 1982: *Community Resource Management: Lessons from the Zanjera*. Quezon City, Philippines: University of the Philippines Press.

Smith, Vernon. 1982: Microeconomic systems as an experimental science. *American Economic Review,* 72 (December), 923–55.

Tang, Shui Yan. 1992: *Institutions and Collective Action: Self-Governance in Irrigation*. San Francisco, Calif.: Institute for Contemporary Studies Press.

Taylor, Michael. 1987: *The Possibility of Cooperation*. New York: Cambridge University Press.

Thomson, James T. 1977: Ecological deterioration: Local-level rule making and enforcement problems in Niger. In Michael H. Glantz (ed.), *Desertification: Environmental Degradation in and around Arid Lands*. Boulder, Colo.: Westview Press, 57–79.

Thomson, James T. 1992: *A Framework for Analyzing Institutional Incentives in Community Forestry*. Rome, Italy: Food and Agriculture Organization of the

United Nations, Forestry Department, Via delle Terme di Caracalla.

Townsend, Ralph E. 1986: A critique of models of the American lobster fishery. *Journal of Environmental Economics and Management,* 13, 277–91.

Townsend, Ralph E. and Wilson, James. 1987: An economic view of the "tragedy of the commons." In Bonnie J. McCay and James M. Acheson (eds.), *The Question of the Commons: The Culture and Ecology of Communal Resources.* Tucson: University of Arizona Press, 311–26.

van de Kragt, Alphons, Dawes, Robyn M., Orbell, John M., Braver, S. R., and Wilson, L. A. 1986: Doing well and doing good as ways of resolving social dilemmas. In Henk A. M. Wilke, Dave Messick, and Christel Rutte (eds.), *Experimental Social Dilemmas.* Frankfurt am Main: Verlag Peter Lang, 177–204.

van de Kragt, Alphons, Orbell, John M., and Dawes, Robyn M. 1983: The minimal contributing set as a solution to public goods problems. *American Political Science Review,* 77 (March), 112–22.

Wade, Robert. 1988: *Village Republics: Economic Conditions for Collective Action in South India.* New York: Cambridge University Press.

Walker, James M. and Gardner, Roy. 1992: Probabilistic destruction of common-pool resources: Experimental evidence. *Economic Journal,* 102(414) (September), 1,149–61.

Weissing, Franz J. and Ostrom, Elinor. 1991: Irrigation institutions and the games irrigators play: Rule enforcement without guards. In Reinhard Selten (ed.), *Game Equilibrium Models II: Methods, Morals, and Markets.* Berlin: Springer-Verlag, 188–262.

Weissing, Franz J. and Ostrom, Elinor. 1993: Irrigation institutions and the games irrigators play: Rule enforcement on government- and farmer-managed systems. In Fritz W. Scharpf (ed.), *Games in Hierarchies and Networks: Analytical and Empirical Approaches to the Study of Governance Institutions.* Frankfurt am Main: Campus Verlag; Boulder, Colo.: Westview Press, 387–428. (Chapter 13 in this volume.)

Wiggins, Steven N. and Libecap, Gary D. 1985: Oil field unitization: Contractual failure in the presence of imperfect information. *American Economic Review,* 75, 368–85.

Wiggins, Steven N. and Libecap, Gary D. 1987: Firm heterogeneities and cartelization efforts in domestic crude oil. *Journal of Law, Economics, and Organization,* 3, 1–25.

Wilson, James A. 1982: The economical management of multispecies fisheries. *Land Economics,* 58(4) (November), 417–34.

Wunsch, James S. and Olowu, Dele (eds.). 1990: *The Failure of the Centralized State: Institutions and Self-Governance in Africa.* Boulder, Colo.: Westview Press.

Yamagishi, Toshio. 1986: The provision of a sanctioning system as a public good. *Journal of Personality and Social Psychology,* 51(1), 110–16.

Yamagishi, Toshio. 1988: Seriousness of social dilemmas and the provision of a sanctioning system. *Social Psychology Quarterly,* 51(1), 32–42.

CHAPTER 16

A Behavioral Approach to the Rational Choice Theory of Collective Action

Elinor Ostrom

Let me start with a provocative statement. You would not be reading this essay if it were not for some of our ancestors learning how to undertake collective action to solve social dilemmas. Successive generations have added to the stock of everyday knowledge about how to instill productive norms of behavior in their children and to craft rules to support collective action that produces public goods and avoids "tragedies of the commons."[1] What our ancestors and contemporaries have learned about engaging in collective action for mutual defense, child rearing, and survival is not, however, understood or explained by the extant theory of collective action.

Yet, the theory of collective action is *the* central subject of political science. It is the core for the justification of the state. Collective-action problems pervade international relations, face legislators when devising public budgets, permeate public bureaucracies, and are at the core of explanations of voting, interest group formation, and citizen control of governments in a democracy. If political scientists do not have an empirically grounded theory of collective action, then we are hand-waving at our central questions. I am afraid that we do a lot of hand-waving.

The lessons of effective collective action are not simple—as is obvious

Originally published in *American Political Science Review* 92, no. 1 (March 1998), 1–22. Reprinted by permission of the American Political Science Association and the author. This essay was presented as the Presidential Address to the American Political Science Association, 1997.

Author's note: The author gratefully acknowledges the support of the National Science Foundation (Grants Nos. SBR–9319835 and SBR–9521918), the Ford Foundation, the Bradley Foundation, and the MacArthur Foundation. My heartiest thanks go to James Alt, Jose Apesteguia, Patrick Brandt, Kathryn Firmin-Sellers, Roy Gardner, Derek Kauneckis, Fabrice Lehoucq, Margaret Levi, Thomas Lyon, Tony Matejczyk, Mike McGinnis, Trudi Miller, John Orbell, Vincent Ostrom, Eric Rasmusen, David Schmidt, Sujai Shivakumar, Vernon Smith, Catherine Tucker, George Varughese, Jimmy Walker, John Williams, Rick Wilson, Toshio Yamagishi, and Xin Zhang for their comments on earlier drafts and to Patty Dalecki for all her excellent editorial and moral support.

from human history and the immense tragedies that humans have endured, as well as the successes we have realized. As global relationships become even more intricately intertwined and complex, however, our survival becomes more dependent on empirically grounded scientific understanding. We have not yet developed a *behavioral theory of collective action* based on models of the individual consistent with empirical evidence about how individuals make decisions in social-dilemma situations. A behavioral commitment to theory grounded in empirical inquiry is essential if we are to understand such basic questions as why face-to-face communication so consistently enhances cooperation in social dilemmas or how structural variables facilitate or impede effective collective action.

Social dilemmas occur whenever individuals in interdependent situations face choices in which the maximization of short-term self-interest yields outcomes leaving all participants worse off than feasible alternatives. In a public-good dilemma, for example, all those who would benefit from the provision of a public good—such as pollution control, radio broadcasts, or weather forecasting—find it costly to contribute and would prefer others to pay for the good instead. If everyone follows the equilibrium strategy, then the good is not provided or is underprovided. Yet, everyone would be better off if everyone were to contribute.

Social dilemmas are found in all aspects of life, leading to momentous decisions affecting war and peace as well as the mundane relationships of keeping promises in everyday life. Social dilemmas are called by many names, including the public-good or collective-good problem (Olson 1965; P. Samuelson 1954), shirking (Alchian and Demsetz 1972), the free-rider problem (Edney 1979; Grossman and Hart 1980), moral hazard (Holmstrom 1982), the credible commitment dilemma (Williams, Collins, and Lichbach 1997), generalized social exchange (Ekeh 1974; Emerson 1972a,b; Yamagishi and Cook 1993), the tragedy of the commons (G. Hardin 1968), and exchanges of threats and violent confrontations (Boulding 1963). The prisoners' dilemma has become the best-known social dilemma in contemporary scholarship. Among the types of individuals who are posited to face these kinds of situations are politicians (Geddes 1994), international negotiators (Sandler 1992; Snidal 1985), legislators (Shepsle and Weingast 1984), managers (Miller 1992), workers (Leibenstein 1976), long-distance traders (Greif, Milgrom, and Weingast 1994), ministers (Bullock and Baden 1977), oligopolists (Cornes, Mason, and Sandler 1986), labor union organizers (Messick 1973), revolutionaries (Lichbach 1995), homeowners (Boudreaux and Holcombe 1989), even cheerleaders (Hardy and Latané 1988), and, of course, all of us—whenever we consider trusting others to cooperate with us on long-term joint endeavors.

In prehistoric times, simple survival was dependent both on the aggressive pursuit of self-interest and on collective action to achieve cooperation in defense, food acquisition, and child rearing. Reciprocity among close kin was used to solve social dilemmas, leading to a higher survival rate for those individuals who lived in families and used reciprocity within the family (Hamilton 1964). As human beings began to settle in communities and engage in agriculture and long-distance trade, forms of reciprocity with individuals other than close kin were essential to achieve mutual protection, to gain the benefits of long-distance trading, to build common facilities, and to preserve common-pool resources.[2] Evolutionary psychologists have produced substantial evidence that human beings have evolved the capacity—similar to that of learning a language—to learn reciprocity norms and general social rules that enhance returns from collective action (Cosmides and Tooby 1992). At the same time, cognitive scientists have also shown that our genetic inheritance does not give us the capabilities to do unbiased, complex, and full analyses without substantial acquired knowledge and practice as well as reliable feedback from the relevant environment. Trial-and-error methods are used to learn individual skills as well as rules and procedures that increase the joint returns individuals may obtain through specialization, coordination, and exchange. All long-enduring political philosophies have recognized human nature to be complex mixtures of the pursuit of self-interest combined with the capability of acquiring internal norms of behavior and following enforced rules when understood and perceived to be legitimate. Our evolutionary heritage has hardwired us to be boundedly self-seeking at the same time that we are capable of learning heuristics and norms, such as reciprocity, that help achieve successful collective action.

One of the most powerful theories used in contemporary social sciences—rational choice theory—helps us understand humans as self-interested, short-term maximizers. Models of complete rationality have been highly successful in predicting marginal behavior in competitive situations in which selective pressures screen out those who do not maximize external values, such as profits in a competitive market or the probability of electoral success in party competition. Thin models of rational choice have been unsuccessful in explaining or predicting behavior in one-shot or finitely repeated social dilemmas in which the theoretical prediction is that no one will cooperate. In indefinitely (or infinitely) repeated social dilemmas, standard rational choice models predict a multitude of equilibria ranging from the very best to the very worst of available outcomes without any hypothesized process for how individuals might achieve more productive outcomes and avert disasters.[3] Substantial evidence from experiments demonstrates that cooperation levels for most one-shot or finitely repeated social dilemmas far exceed the predicted levels and are systematically affected by variables that play no theoretical role in affecting outcomes. Field research also shows that individuals systematically engage in collective action to provide local public goods or manage common-pool resources without an external authority to offer inducements or impose sanctions. Simply assuming that individuals use long-range thinking "to achieve the goal of establishing and/or maintaining continued mutual cooperation" (Pruitt and Kimmel 1977, 375) is not a sufficient theory either. It does not explain why some groups fail to obtain joint outcomes easily available to them or why initial cooperation can break down.

We now have enough scholarship from multiple disciplines to expand the range of rational choice models we use. For at least five reasons, we need to formulate a behavioral theory of boundedly rational and moral behavior.

First, behavior in social dilemmas is affected by many structural variables, including size of group, heterogeneity of participants, their dependence on the benefits received, their discount rates, the type and predictability of transformation processes involved, the nesting of organizational levels, monitoring techniques, and the information available to participants.[4] In theories that predict either zero or 100 percent cooperation in one-shot or finitely repeated dilemmas, structural variables do not affect levels of cooperation at all. A coherent explanation of the relationship among structural variables and the likelihood of individuals solving social dilemmas depends on developing a behavioral theory of rational choice. This will allow scholars who stress structural explanations of human behavior and those who stress individual choice to find common ground, rather than continue the futile debate over whether structural variables or individual attributes are the most important.

Second, scholars in all the social and some biological sciences have active research programs focusing on how groups of individuals achieve collective action. An empirically supported theoretical framework for the analysis of social dilemmas would integrate and link their efforts. Essential to the development of such a framework is a conception of human behavior that views complete rationality as one member of a family of rationality models rather than the only way to model human behavior. Competitive institutions operate as a scaffolding structure so that individuals who fail to learn how to maximize some external value are no longer in the competitive game (Alchian 1950; Clark 1995; Satz and Ferejohn 1994). If all institutions involved strong competition, then the thin model of rationality used to explain behavior in competitive markets would be more useful. Models of human behavior based on theories consistent with our evolutionary and adaptive heritage need to join the ranks of theoretical tools used in the social and biological sciences.

Third, sufficient work by cognitive scientists, evolutionary theorists, game theorists, and social scientists in all disciplines (Axelrod 1984; Boyd and Richerson 1988, 1992; Cook and Levi 1990; Güth and Kliemt 1995; Sethi and Somanathan 1996; Simon 1985, 1997) on the use of heuristics and norms of behavior, such as reciprocity, has already been undertaken. It is now possible to continue this development toward a firmer behavioral foundation for the study of collective action to overcome social dilemmas.

Fourth, much of our current public policy analysis—particularly since Garrett Hardin's (1968) evocative paper, "The Tragedy of the Commons"— is based on an assumption that rational individuals are helplessly trapped in social dilemmas from which they cannot extract themselves without induce- ment or sanctions applied from the outside. Many policies based on this as- sumption have been subject to major failure and have exacerbated the very problems they were intended to ameliorate (Arnold and Campbell 1986; Baland and Platteau 1996; Morrow and Hull 1996). Policies based on the assumptions that individuals can learn how to devise well-tailored rules and cooperate conditionally when they participate in the design of institutions affecting them are more successful in the field (Berkes 1989; Bromley et al. 1992; Ellickson 1991; Feeny et al. 1990; McCay and Acheson 1987; McKean and E. Ostrom 1995; Pinkerton 1989; Yoder 1994).

Fifth, the image of citizens we provide in our textbooks affects the long- term viability of democratic regimes. Introductory textbooks that presume rational citizens will be passive consumers of political life—the masses—and focus primarily on the role of politicians and officials at a national level—the elite—do not inform future citizens of a democratic polity of the actions they need to know about and can undertake. While many political scientists claim to eschew teaching the normative foundations of a democratic polity, they actually introduce a norm of cynicism and distrust without providing a vision of how citizens could do anything to challenge corruption, rent seeking,[5] or poorly designed policies.

The remainder of this essay is divided into six sections. In the first I briefly review the theoretical predictions of currently accepted rational choice theory related to social dilemmas. The next will summarize the chal- lenge to the sole reliance on a complete model of rationality presented by extensive experimental research. Then I examine two major empirical find- ings that begin to show how individuals achieve results that are "better than rational" (Cosmides and Tooby 1994) by building conditions in which reci- procity, reputation, and trust can help to overcome the strong temptations of short-run self-interest. The following section raises the possibility of devel- oping second-generation models of rationality, and the next develops an ini- tial theoretical scenario. I conclude by examining the implications of placing reciprocity, reputation, and trust at the core of an empirically tested, behav- ioral theory of collective action.

Theoretical Predictions for Social Dilemmas

The term *social dilemma* refers to a large number of situations in which indi- viduals make independent choices in an interdependent situation (Dawes

1975, 1980; R. Hardin 1971). In all *N*-person social dilemmas, a set of participants has a choice of contributing (*C*) or not contributing (~*C*) to a joint benefit. While I represent this as an either-or choice in figure 16.1, it frequently is a choice of how much to contribute rather than whether to contribute or not.

If everyone contributes, they get a net positive benefit (*G*). Everyone faces a temptation (*T*) to shift from the set of contributors to the set of those who do not contribute. The theoretical prediction is that everyone will shift and that no one will contribute. If this happens, then the outcome will be at the intercept. The difference between the predicted outcome and everyone contributing is *G–X*. Since the less-valued payoff is at a Nash equilibrium, no one is independently motivated to change his or her choice, given the choices of other participants. These situations are dilemmas because at least one outcome exists that yields greater advantage for all participants. Thus, a Pareto-superior alternative exists, but rational participants making isolated choices are not predicted to realize this outcome. A conflict is posed between

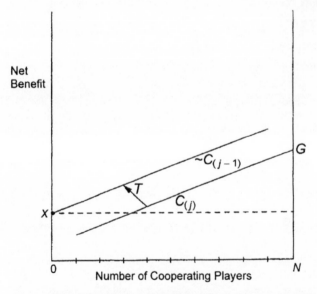

Note: *N* players choose between cooperating (*C*) or not cooperating (~*C*). When *j* individuals cooperate, their payoffs are always lower than the *j* –1 individuals who do not cooperate. The predicted outcome is that no one will cooperate and all players will receive *X* benefits. The temptation (*T*) not to cooperate is the increase in benefit any cooperator would receive for switching to not cooperating. If all cooperate, they all receive *G – X* more benefits than if all do not cooperate.

Fig. 16.1. *N*-person social dilemma

individual and group rationality. The problem of collective action raised by social dilemmas is to find a way to avoid Pareto-inferior equilibria and to move closer to the optimum. Those who find ways to coordinate strategies in some fashion receive a "cooperators' dividend" equal to the difference between the predicted outcome and the outcome achieved.

Many models of social dilemmas exist in the literature (see Schelling 1978; Lichbach 1996 for reviews of alternative formalizations). In all models, a set of individuals is involved in a game in which a strategy leading to a Nash equilibrium for a single iteration of the game yields less than an optimal outcome for all involved. The equilibrium is thus Pareto inferior. The optimal outcome could be achieved if those involved "cooperated" by selecting strategies other than those prescribed by an equilibrium solution to a noncooperative game (Harsanyi and Selten 1988). Besides these assumptions regarding the structure of payoffs in the one-shot version of the game, other assumptions are made in almost all formal models of social dilemmas. (1) All participants have common knowledge of the exogenously fixed structure of the situation and of the payoffs to be received by all individuals under all combinations of strategies. (2) Decisions about strategies are made independently, often simultaneously. (3) In a symmetric game, all participants have available the same strategies. (4) No external actor (or central authority) is present to enforce agreements among participants about their choices.

When such a game is finitely repeated, participants are assumed to solve the game through backward induction. Assumptions about the particular payoff functions differ. Rather than describe the Nash equilibrium and Pareto-efficient outcome for all models considered in this article, aggregate behavior consistent with the Nash equilibrium will be described as zero cooperation, while behavior consistent with the efficient outcome will be described as 100 percent cooperation.

The grim predictions evoked considerable empirical challenges as well as important theoretical breakthroughs. The predictions ran counter to so many everyday experiences that some scholars turned to survey and field studies to examine the level of voluntary contributions to public goods (see Lichbach 1995 for a summary). Others turned to the experimental lab and confirmed much higher than predicted levels of cooperation in one-shot experiments. Game theorists were challenged to rethink their own firm conclusions and to pose new models of when cooperation might emerge (see Benoit and Krishna 1985).

The introduction of two kinds of uncertainty into repeated games—about the number of repetitions and about the types of players participating in a social dilemma—has led to more optimistic predictions. When individuals, modeled as fully rational actors with low discount rates, interact in a repeated social dilemma whose end point is determined stochastically, it is now theoretically well established that it is *possible* for them to achieve optimal or near optimal outcomes and avoid the dominant strategies of one-shot and finitely repeated games that yield suboptimal outcomes (Fudenberg and Maskin 1986). This is possible when players achieve self-enforcing equilibria by committing themselves to punish noncooperators sufficiently to deter noncooperation. Kreps et al. (1982) introduced a second kind of uncertainty related to whether all the players use complete rationality as their guide to action. The probability of the presence of an "irrational" player, who reciprocates cooperation with cooperation, is used as the grounds for a completely rational player to adopt the strategy of cooperating early in a sequence of games and switching to noncooperation at the end. Once either of these two forms of uncertainty is introduced, the number of possible equilibria explode in number (Abreau 1988). Everything is predicted: optimal outcomes, the Pareto-inferior Nash equilibria, and everything in between.

To generate predictions other than noncooperation, theorists using standard rational choice theory have found it necessary to assume real uncertainty about the duration of a situation or to assume that some players are "irrational" in their willingness to reciprocate cooperation with cooperation. To assume that if some players *irrationally* choose reciprocity, then others can *rationally* choose reciprocity is a convoluted explanation—to say the least—of the growing evidence that reciprocity is a core norm used by many individuals in social-dilemma situations.

The Lack of a General Fit

In all the social sciences, experiments have been conducted on various types of social dilemmas for several decades. While some scholars question the value of laboratory experiments for testing the predictions of major theories in the social sciences, this method has many advantages. First, one can design experiments that test multiple predictions from the same theory under controlled conditions. Second, replication is feasible. Third, researchers can challenge whether a particular design adequately captures the theoretically posited variables and conduct further experiments to ascertain how changes in a design affect outcomes. The evidence discussed below is based on multiple studies by diverse research teams. Fourth, experimental methods are particularly relevant for studying human choice under diverse institutional arrangements. Subjects in experimental studies draw on the modes of analysis and values they have learned throughout their lives to respond to diverse incentive structures. Experiments thus allow one to test precisely whether individuals behave within a variety of institutional settings as predicted by theory (Plott 1979; Smith 1982).

In this section, I will summarize four consistently replicated findings that directly challenge the general fit between behavior observed in social-dilemma experiments and the predictions of noncooperative game theory using complete rationality and complete information for one-shot and finitely repeated social dilemmas. I focus first on the fit between theory and behavior, because the theoretical predictions are unambiguous and have influenced so much thinking across the social sciences. Experiments on market behavior do fit the predictions closely (see Davis and Holt 1993 for an overview). If one-shot and finitely repeated social-dilemma experiments were to support strongly the predictions of noncooperative game theory, then we would have a grounded theory with close affinities to a vast body of economic theory for which there is strong empirical support. We would need to turn immediately to the problem of indefinitely repeated situations for which noncooperative game theory faces an embarrassment of too many equilibria. As it turns out, we have a different story to tell. The four general findings are as follows.

1. High levels of initial cooperation are found in most types of social dilemmas, but the levels are consistently less than optimal.
2. Behavior is not consistent with backward induction in finitely repeated social dilemmas.
3. Nash equilibrium strategies are not good predictors at the individual level.
4. Individuals do not learn Nash equilibrium strategies in repeated social dilemmas.

High but Suboptimal Levels of Initial Cooperation

Most experimental studies of social dilemmas with the structure of a public-goods provision problem have found levels of cooperative actions in one-shot games, or in the first rounds of a repeated game, that are significantly above the predicted level of zero.[6] "In a wide variety of treatment conditions, participants rather persistently contributed 40 to 60 percent of their token endowments to the [public good], far in excess of the 0 percent contribution rate consistent with a Nash equilibrium" (Davis and Holt 1993, 325). Yet, once an experiment is repeated, cooperation levels in public-good experiments tend to decline. The individual variation across experiment sessions can be very great.[7] While many have focused on the unexpectedly high rates of cooperation, it is important to note that in sparse institutional settings with no feedback about individual contributions, cooperation levels never reach the optimum. Thus, the prediction of zero levels of cooperation can be rejected, but cooperation at a suboptimal level is consistently observed in sparse institutional settings.

Behavior in Social Dilemmas Inconsistent with Backward Induction

In all finitely repeated experiments, players are predicted to look ahead to the last period and determine what they would do in that period. In the last period, there is no future interaction; the prediction is that they will not cooperate in that round. Since that choice would be determined at the beginning of an experiment, the players are presumed to look at the second-to-last period and ask themselves what they would do there. Given that they definitely would not cooperate in the last period, it is assumed that they also would not cooperate in the second-to-last period. This logic would then extend backward to the first round (Luce and Raiffa 1957, 98–99).

While backward induction is still the dominant method used in solving finitely repeated games, it has been challenged on theoretical grounds (Binmore 1997; R. Hardin 1997). Furthermore, as discussed above, uncertainty about whether others use norms like tit-for-tat rather than follow the recommendations of a Nash equilibrium may make it rational for a player to signal a willingness to cooperate in the early rounds of an iterated game and then defect at the end (Kreps et al. 1982). What is clearly the case from experimental evidence is that players do not use backward induction in their decision-making plans in an experimental laboratory. Amnon Rapoport (1997, 122) concludes from a review of several experiments focusing on resource dilemmas that "subjects are not involved in or capable of backward induction."[8]

Nash Equilibrium Strategies Do Not Predict Individual Behavior in Social Dilemmas

From the above discussion, it is obvious that individuals in social dilemmas tend not to use the predicted Nash equilibrium strategy, even though this is a good predictor at both an individual and group level in other types of situations. While outcomes frequently approach Nash equilibria at an aggregate level, the variance of individual actions around the mean is extremely large. When groups of eight subjects made appropriation decisions in repeated common-pool resource experiments of 20 to 30 rounds, the unique symmetric Nash equilibrium strategy was never played (Walker, Gardner, and E. Ostrom 1990). Nor did individuals use Nash equilibrium strategies in repeated public-good experiments (Dudley 1993; Isaac and Walker 1991, 1993). In a recent set of 13 experiments involving seven players making 10 rounds of decisions without communication or any other institutional structure, Walker et al. (1997) did not observe a single individual choice of a symmetric Nash equilibrium strategy in the 910 opportunities available to subjects. Chan et al. (1996, 58) also found little evidence to support the use of Nash equilibria when they examined the effect of heterogeneity of income on outcomes: "It is clear that the outcomes of the laboratory sessions reported here cannot be characterized as Nash equilibria outcomes."

Individuals Do Not Learn Nash Equilibrium Strategies in Social Dilemmas

In repeated experiments without communication or other facilitating institutional conditions, levels of cooperation fall (rise) toward the Nash equilibrium in public-good (common-pool resource) experiments. Some scholars have speculated that it just takes some time and experience for individuals to learn Nash equilibrium strategies (Ledyard 1995). But this does not appear to be the case. In all repeated experiments, there is considerable pulsing as subjects obtain outcomes that vary substantially with short spurts of increasing and decreasing levels of cooperation, while the general trend is toward an aggregate that is consistent with a Nash equilibrium (Isaac, McCue, and Plott 1985; E. Ostrom, Gardner, and Walker 1994).[9] Furthermore, there is substantial variation in the strategies followed by diverse participants within the same game (Dudley 1993; Isaac and Walker 1988b; E. Ostrom, Gardner, and Walker 1994).

It appears that subjects learn something other than Nash strategies in finitely repeated experiments. Isaac, Walker, and Williams (1994) compare the rate of decay when experienced subjects are explicitly told that an experiment will last 10, 40, or 60 rounds. The rate of decay of cooperative actions is inversely related to the number of decision rounds. Instead of learning the noncooperative strategy, subjects appear to be learning how to cooperate at a moderate level for even longer periods. Cooperation rates approach zero only in the last few periods, whenever these occur.

Two Internal Ways out of Social Dilemmas

The combined effect of these four frequently replicated, general findings represents a strong rejection of the predictions derived from a complete model of rationality. Two more general findings are also contrary to the predictions of currently accepted models. At the same time, they also begin to show how individuals are able to obtain results that are substantially "better than rational" (Cosmides and Tooby 1994), at least as *rational* has been defined in currently accepted models. The first is that simple, cheap talk allows individuals an opportunity to make conditional promises to one another and potentially to build trust that others will reciprocate. The second is the capacity to solve second-order social dilemmas that change the structure of the first-order dilemma.

Communication and Collective Action

In noncooperative game theory, players are assumed to be unable to make enforceable agreements.[10] Thus, communication is viewed as cheap talk (Farrell 1987). In a social dilemma, self-interested players are expected to use communication to try to convince others to cooperate and promise cooperative action but then to choose the Nash equilibrium strategy when they make their private decision (Barry and Hardin 1982, 381; Farrell and Rabin 1996, 113).[11] Or, as Gary Miller (1992, 25) expresses it: "It is obvious that simple communication is not sufficient to escape the dilemma."[12]

From this theoretical perspective, face-to-face communication should make no difference in the outcomes achieved in social dilemmas. Yet, consistent, strong, and replicable findings are that substantial increases in the levels of cooperation are achieved when individuals are allowed to communicate face to face.[13] This holds true across all types of social dilemmas studied in laboratory settings and in both one-shot and finitely repeated experiments. In a meta-analysis of more than 100 experiments involving more than 5,000 subjects conducted by economists, political scientists, sociologists, and social psychologists, Sally (1995) finds that opportunities for face-to-face communication in one-shot experiments significantly raise the cooperation rate, on average, by more than 45 percentage points. When subjects are allowed to talk before each decision round in repeated experiments, they achieve 40 percentage points more on average than in repeated games without communication. No other variable has as strong and consistent an effect on results as face-to-face communication. Communication even has a robust and positive effect on cooperation levels when individuals are not provided with feedback on group decisions after every round (Cason and Khan 1996).

The efficacy of communication is related to the capability to talk face to face. Sell and Wilson (1991, 1992), for example, developed a public-good experiment in which subjects could signal promises to cooperate via their computer terminals. There was much less cooperation than in the face-to-face experiments using the same design (Isaac and Walker 1988a, 1991). Rocco and Warglien (1995) replicated all aspects of prior common-pool resource experiments, including the efficacy of face-to-face communication.[14] They found, however, that subjects who had to rely on computerized communication did not achieve the same increase in efficiency as did those who were able to communicate face to face.[15] Palfrey and Rosenthal (1988) report that no significant difference occurred in a provision point public-good experiment in which subjects could send a computerized message stating whether they intended to contribute.

The reasons offered by those doing experimental research for why communication facilitates cooperation include (1) transferring information from those who can figure out an optimal strategy to those who do not fully understand what strategy would be optimal, (2) exchanging mutual commitment, (3) increasing trust and thus affecting expectations of others' behavior, (4) adding additional values to the subjective payoff structure, (5) reinforcement of prior normative values, and (6) developing a group identity (Davis and Holt 1993; Orbell, Dawes, and van de Kragt 1990; Orbell, van de Kragt, and Dawes 1988; E. Ostrom and Walker 1997). Carefully crafted experiments demonstrate that the effect of communication is not primarily due to the first reason. When information about the individual strategy that produces an optimal joint outcome is clearly presented to subjects who are not able to communicate, the information makes little difference in outcomes achieved (Isaac, McCue, and Plott 1985; Moir 1995).

Consequently, exchanging mutual commitment, increasing trust, creating and reinforcing norms, and developing a group identity appear to be the most important processes that make communication efficacious. Subjects in experiments do try to extract mutual commitment from one another to follow the strategy they have identified as leading to their best joint outcomes. They frequently go around the group and ask each person to promise the others that he or she will follow the joint strategy. Discussion sessions frequently end with such comments as: "Now remember, everyone, that we all do much better if we all follow X strategy" (see transcripts in E. Ostrom, Gardner, and Walker 1994). In repeated experiments, subjects use communication opportunities to lash out verbally at unknown individuals who did not follow mutually agreed strategies, using such evocative terms as *scumbuckets* and *finks*. Orbell, van de Kragt, and Dawes (1988) summarize the findings from 10 years of research on one-shot public-good experiments by stressing how many mutually reinforcing processes are evoked when communication is allowed.[16] Without increasing mutual trust in the promises that are exchanged, however, expectations of the behavior of others will not change. Given the very substantial difference in outcomes, communication is most likely to affect individual trust that others will keep to their commitments. As discussed below, the relationships among trust, conditional commitments, and a reputation for being trustworthy are key links in a second-generation theory of boundedly rational and moral behavior.

As stakes increase and it is difficult to monitor individual contributions, communication becomes less efficacious, however. E. Ostrom, Gardner, and Walker (1994) found that subjects achieved close to fully optimal results when each subject had relatively low endowments and was allowed opportunities for face-to-face communication. When endowments were substantially increased—increasing the temptation to cheat on prior agreements—subjects achieved far more in communication experiments as contrasted to non-communication experiments but less than in small-stake situations. Failures to achieve collective action in field settings in which communication has been feasible point out that communication alone is not a sufficient mechanism to assure successful collective action under all conditions.

Innovation and Collective Action

Changing the rules of a game or using scarce resources to punish those who do not cooperate or keep agreements are usually not considered viable options for participants in social dilemmas, since these actions create public goods. Participants face a second-order social dilemma (of equal or greater difficulty) in any effort to use costly sanctions or change the structure of a game (Oliver 1980). The predicted outcome of any effort to solve a second-order dilemma is failure.

Yet, participants in many field settings and experiments do exactly this. Extensive research on how individuals have governed and managed common-pool resources has documented the incredible diversity of rules designed and enforced by participants themselves to change the structure of underlying social-dilemma situations (Blomquist 1992; Bromley et al. 1992; Lam 1998; McKean 1992; E. Ostrom 1990; Schlager 1990; Schlager and E. Ostrom 1993; Tang 1992). The particular rules adopted by participants vary radically to reflect local circumstances and the cultural repertoire of acceptable and known rules used generally in a region. Nevertheless, general design principles characterize successfully self-organized, sustainable, local, regional, and international regimes (E. Ostrom 1990). Most robust and longlasting common-pool regimes involve clear mechanisms for monitoring rule conformance and graduated sanctions for enforcing compliance. Thus, few self-organized regimes rely entirely on communication to sustain cooperation in situations that generate strong temptations to break mutual commitments. Monitors—who may be participants themselves—do not use strong sanctions for individuals who rarely break rules. Modest sanctions indicate to rule breakers that their lack of conformance has been observed by others. By paying a modest fine, they rejoin the community in good standing and learn that rule infractions are observed and sanctioned. Repeated rule breakers are severely sanctioned and eventually excluded from the group. Rules meeting these design principles reinforce contingent commitments and enhance the trust participants have that others are also keeping their commitments.

In field settings, innovation in rules usually occurs in a continuous trial-and-error process until a rule system is evolved that participants consider yields substantial net benefits. Given the complexity of the physical world that individuals frequently confront, they are rarely ever able to "get the rules right" on the first or second try (E. Ostrom 1990). In highly unpredictable environments, a long period of trial and error is needed before individuals can find rules that generate substantial positive net returns over a sufficiently long time horizon. Nonviolent conflict may be a regular feature of successful institutions when arenas exist to process conflict cases regularly and, at times, to innovate new rules to cope with conflict more effectively (V. Ostrom 1987; V. Ostrom, Feeny, and Picht 1993).

In addition to the extensive field research on changes that participants make in the structure of situations they face, subjects in a large number of experiments have also solved second-order social dilemmas and consequently moved the outcomes in their first-order dilemmas closer to optimal levels (Dawes, Orbell, and van de Kragt 1986; Messick and Brewer 1983; Rutte and Wilke 1984; Sato 1987; van de Kragt, Orbell, and Dawes 1983; Yamagishi 1992). Toshio Yamagishi (1986), for example, conducted experiments with subjects who had earlier completed a questionnaire including items from a scale measuring trust. Subjects who ranked higher on the trust scale consistently contributed about 20 percent more to collective goods than those who ranked lower. When given an opportunity to contribute to a substantial "punishment fund" to be used to fine the individual who contributed the least to their joint outcomes, however, low-trusting individuals contributed significantly more to the punishment fund and also achieved the highest level of cooperation. In the last rounds of this experiment, they were contributing 90 percent of their resources to the joint fund. These results, which have now been replicated with North American subjects (Yamagishi 1988ab), show that individuals who are initially the least trusting are willing to contribute to sanctioning systems and then respond more to a change in the structure of the game than those who are initially more trusting.

E. Ostrom, Walker, and Gardner (1992) also examined the willingness of subjects to pay a "fee" in order to "fine" another subject. Instead of the predicted zero use of sanctions, individuals paid fees to fine others at a level significantly above zero.[17] When sanctioning was combined with a single opportunity to communicate or a chance to discuss and vote on the creation of their own sanctioning system, outcomes improved dramatically. With only a single opportunity to communicate, subjects were able to obtain an average of 85 percent of the optimal level of investments (67 percent with the costs of sanctioning subtracted). Those subjects who met face to face and agreed by majority vote on their own sanctioning system achieved 93 percent of optimal yield. The level of defections was only 4 percent, so that the costs of the sanctioning system were low, and net benefits were at a 90 percent level (E. Ostrom, Walker, and Gardner 1992).

Messick and his colleagues have undertaken a series of experiments designed to examine the willingness of subjects to act collectively to change institutional structures when facing common-pool resource dilemmas (see Messick et al. 1983; Samuelson et al. 1984; C. Samuelson and Messick

1986). In particular, they have repeatedly given subjects the opportunity to relinquish their individual decisions concerning withdrawals from the common resource to a leader who is given the authority to decide for the group. They have found that "people want to change the rules and bring about structural change when they observe that the common resource is being depleted" (C. Samuelson and Messick 1995, 147). Yet, simply having an unequal distribution of outcomes is not a sufficient inducement to affect the decision whether to change institutional structure.

What do these experiments tell us? They complement the evidence from field settings and show that individuals temporarily caught in a social-dilemma structure are likely to invest resources to innovate and change the structure itself in order to improve joint outcomes. They also strengthen the earlier evidence that the currently accepted, noncooperative game-theoretical explanation relying on a particular model of the individual does not adequately predict behavior in one-shot and finitely repeated social dilemmas. Cooperative game theory does not provide a better explanation. Since both cooperative and noncooperative game theory predict extreme values, neither provides explanations for the conditions that tend to enhance or detract from cooperation levels.

The really big puzzle in the social sciences is the development of a consistent theory to explain why cooperation levels vary so much and why specific configurations of situational conditions increase or decrease cooperation in first- or second-level dilemmas. This question is important not only for our scientific understanding but also for the design of institutions to facilitate individuals achieving higher levels of productive outcomes in social dilemmas. Many structural variables affect the particular innovations chosen and the sustainability and distributional consequences of these institutional changes (Knight 1992). A coherent theory of institutional change is not within reach, however, with a theory of individual choice that predicts no innovation will occur. We need a second-generation theory of boundedly rational, innovative, and normative behavior.

Toward Second-Generation Models of Rationality

First-generation models of rational choice are powerful engines of prediction when strong competition eliminates players who do not aggressively maximize immediate external values. While incorrectly confused with a general theory of human behavior, complete rationality models will continue to be used productively by social scientists, including the author. But the thin model of rationality needs to be viewed, as Selten (1975) points out, as the limiting case of bounded or incomplete rationality. Consistent with all models of rational choice is a general theory of human behavior that views all humans as complex, fallible learners who seek to do as well as they can given the constraints that they face and who are able to learn heuristics, norms, rules, and how to craft rules to improve achieved outcomes.

Learning Heuristics, Norms, and Rules

Because individuals are boundedly rational, they do not calculate a complete set of strategies for every situation they face. Few situations in life generate information about all potential actions that one can take, all outcomes that can be obtained, and all strategies that others can take. In a model of complete rationality, one simply assumes this level of information. In field situations, individuals tend to use heuristics—rules of thumb—that they have learned over time regarding responses that tend to give them good outcomes in particular kinds of situations. They bring these heuristics with them when they participate in laboratory experiments. In frequently encountered, repetitive situations, individuals learn better and better heuristics that are tailored to particular situations. With repetition, sufficiently large stakes, and strong competition, individuals may learn heuristics that approach best-response strategies.

In addition to learning instrumental heuristics, individuals also learn to adopt and use norms and rules. By *norms* I mean that the individual attaches an internal valuation—positive or negative—to taking particular types of action. Crawford and E. Ostrom (1995) refer to this internal valuation as a delta parameter that is added to or subtracted from the objective costs of an action.[18] Andreoni (1989) models individuals who gain a "warm glow" when they contribute resources that help others more than they help themselves in the short term. Knack (1992) refers to negative internal valuations as "duty."[19] Many norms are learned from interactions with others in diverse communities about the behavior that is expected in particular types of situations (Coleman 1987). The change in preferences represents the internalization of particular moral lessons from life (or from the training provided by one's elders and peers).[20] The strength of the commitment (Sen 1977) made by an individual to take particular types of future actions (telling the truth, keeping promises) is reflected in the size of the delta parameter. After experiencing repeated benefits from other people's cooperative actions, an individual may resolve that he or she should always initiate cooperative actions in the future.[21] Or, after many experiences of being the "sucker" in such experiences, an individual may resolve never to be the first to cooperate.

Since norms are learned in a social milieu, they vary substantially across cultures, across individuals within any one culture, within individuals across different types of situations they face, and across time within any particular situation. The behavioral implications of assuming that individuals acquire norms do not vary substantially from the assumption that individuals learn to use heuristics. One may think of norms as heuristics that individuals adopt from a moral perspective, in that these are the kinds of actions they wish to follow in living their life. Once some members of a population acquire norms of behavior, they affect the expectations of others.

By *rules* I mean that a group of individuals has developed shared understandings that certain actions in particular situations must, must not, or may be undertaken and that sanctions will be taken against those who do not conform. The distinction between internalized but widely shared norms for what are appropriate actions in broad types of situations and rules that are self-consciously adopted for use in particular situations is at times difficult to draw when doing fieldwork. Analytically, individuals can be thought of as learning norms of behavior that are general and fit a wide diversity of particular situations. Rules are artifacts related to particular actions in specific situations (V. Ostrom 1980, 1997). Rules are created in private associations as well as in more formalized public institutions, where they carry the additional legal weight of being enforced legal enactments.[22] Rules can enhance reciprocity by making mutual commitments clear and overt. Alternatively, rules can assign authority to act so that benefits and costs are distributed inequitably and thereby destroy reliance on positive norms.

Reciprocity: An Especially Important Class of Norms

That humans rapidly learn and effectively use heuristics, norms, and rules is consistent with the lessons learned from evolutionary psychology (see Barkow, Cosmides, and Tooby 1992), evolutionary game theory (see Güth and Kliemt 1996; Hirshleifer and Rasmusen 1989),[23] biology (Trivers 1971), and bounded rationality (Selten 1990, 1991; Selten, Mitzkewitz, and Uhlich 1997; Simon 1985). Humans appear to have evolved specialized cognitive modules for diverse tasks, including making sense out of what is seen (Marr 1982), inferring rules of grammar by being exposed to adult speakers of a particular language (Pinker 1994), and increasing their long-term returns from interactions in social dilemmas (Cosmides and Tooby 1992). Humans dealt with social dilemmas related to rearing and protecting offspring, acquiring food, and trusting one another to perform future promised action millennia before such oral commitments could be enforced by external authorities (de Waal 1996). Substantial evidence has been accumulated (and reviewed in Cosmides and Tooby 1992) that humans inherit a strong capacity to learn reciprocity norms and social rules that enhance the opportunities to gain benefits from coping with a multitude of social dilemmas.

Reciprocity refers to a family of strategies that can be used in social dilemmas involving (1) an effort to identify who else is involved, (2) an assessment of the likelihood that others are conditional cooperators, (3) a decision to cooperate initially with others if others are trusted to be conditional cooperators, (4) a refusal to cooperate with those who do not reciprocate, and (5) punishment of those who betray trust. All reciprocity norms share the common ingredients that individuals tend to react to the positive actions of others with positive responses and the negative actions of others with negative responses. Reciprocity is a basic norm taught in all societies (see Becker 1990; Blau 1964; Gouldner 1960; Homans 1961; Oakerson 1993; V. Ostrom 1997; Thibaut and Kelley 1959).

By far the most famous reciprocal strategy—tit-for-tat—has been the subject of considerable study from an evolutionary perspective. In simulations, pairs of individuals are sampled from a population, and they then interact with one another repeatedly in a prisoners' dilemma game. Individuals are each modeled as if they had inherited a strategy that included the fixed maxims of always cooperate, always defect, or the reciprocating strategy of tit-for-tat (cooperate first and then do whatever the others did in the last round). Axelrod and Hamilton (1981) and Axelrod (1984) have shown that when individuals are grouped so that they are more likely to interact with one another than with the general population, and when the expected number of repetitions is sufficiently large, reciprocating strategies such as tit-for-tat can successfully invade populations composed of individuals following an all-defect strategy. The size of the population in which interactions are occurring may need to be relatively small for reciprocating strategies to survive potential errors of players (Bendor and Mookherjee 1987; but see Boyd and Richerson 1988, 1992; Hirshleifer and Rasmusen 1989; Yamagishi and Takahashi 1994).

The reciprocity norms posited to help individuals gain larger cooperators' dividends depend upon the willingness of participants to use retribution to some degree. In tit-for-tat, for example, an individual must be willing to "punish" a player who defected in the last round by defecting in the current round. In grim trigger, an individual must be willing to cooperate initially but then "punish" everyone for the rest of the game if any defection is noticed in the current round.[24]

Human beings do not inherit particular reciprocity norms via a biological process. The argument is more subtle. Individuals inherit an acute sensitivity for learning norms that increase their own long-term benefits when confronting social dilemmas with others who have learned and value similar norms. The process of growing up in any culture provides thousands of incidents (learning trials) whereby parents, siblings, friends, and teachers provide the specific content of the type of mutual expectations prevalent in that culture. As Mueller (1986) points out, the first dilemmas that humans encounter are as children. Parents reward and punish them until cooperation is a learned response. In the contemporary setting, corporate managers strive for a trustworthy corporate reputation by continuously reiterating and rewarding the use of key principles or norms by corporate employees (Kreps 1990).

Since particular reciprocity norms are *learned,* not everyone learns to use the same norms in all situations. Some individuals learn norms of behavior that are not so "nice." Clever and unscrupulous individuals may learn how to lure others into dilemma situations and then defect on them. It is possible to gain substantial resources by such means, but one has to hide intentions and actions, to keep moving, or to gain access to power over others. In any group composed only of individuals who follow reciprocity norms, skills in detecting and punishing cheaters could be lost. If this happens, it will be subject to invasion and substantial initial losses by clever outsiders or local deviants who can take advantage of the situation. Being too trusting can be dangerous. The presence of some untrustworthy participants hones the skills of those who follow reciprocity norms.

Thus, individuals vary substantially in the probability that they will use particular norms, in how structural variables affect their level of trust and willingness to reciprocate cooperation in a particular situation, and in how they develop their own reputation. Some individuals use reciprocity only in situations in which there is close monitoring and strong retribution is likely. Others will only cooperate in dilemmas when they have publicly committed themselves to an agreement and have assurances from others that their trust will be returned. Others find it easier to build an external reputation by building their own personal identity as someone who always trusts others until proven wrong. If this trust proves to be misplaced, then they stop cooperating and either exit the situation or enter a punishment phase. As Hoffman, McCabe, and Smith (1996a, 23–24) express it:

> A one-shot game in the laboratory is part of a life-long sequence, not an isolated experience that calls for behavior that deviates sharply from one's reputational norm. Thus we should expect subjects to rely upon reciprocity norms in experimental settings unless they discover in the process of participating in a particular experiment that reciprocity is punished and other behaviors are rewarded. In such cases they abandon their instincts and attempt other strategies that better serve their interests.

In any population of individuals, one is likely to find some who use one of three reciprocity norms when they confront a repeated social dilemma.[25]

1. Always cooperate first; stop cooperating if others do not reciprocate; punish noncooperators if feasible.
2. Cooperate immediately only if one judges others to be trustworthy;

stop cooperating if others do not reciprocate; punish noncooperators if feasible.

3. Once cooperation is established by others, cooperate oneself; stop cooperating if others do not reciprocate; punish noncooperators if feasible.

In addition, one may find at least three other norms.

4. Never cooperate.
5. Mimic (1) or (2) but stop cooperating if one can successfully free ride on others.
6. Always cooperate (an extremely rare norm in all cultures).

The proportion of individuals who follow each type of norm will vary from one subpopulation to another and from one situation to another.[26] Whether reciprocity is advantageous to individuals depends sensitively on the proportion of other individuals who use reciprocity and on an individual's capacity to judge the likely frequency of reciprocators in any particular situation and over time. When there are many others who use a form of reciprocity that always cooperates first, then even in one-shot situations cooperation may lead to higher returns when diverse situations are evaluated together. Boundedly rational individuals would expect other boundedly rational individuals to follow a *diversity* of heuristics, norms, and strategies rather than expect to find others who adopt a single strategy—except in those repeated situations in which institutional selection processes sort out those who do not search out optimal strategies. Investment in detection of other individuals' intentions and actions improves one's own outcomes. One does not have to assume that others are "irrational" in order for it to be rational to use reciprocity (Kreps et al. 1982).

Evidence of the Use of Reciprocity in Experimental Settings

Laboratory experiments provide evidence that a substantial proportion of individuals use reciprocity norms even in the very short-term environments of an experiment (McCabe, Rassenti, and Smith 1996). Some evidence comes from experiments on ultimatum games. In such games, two players are asked to divide a fixed sum of money. The first player suggests a division to the second, who then decides to accept or reject the offer. If the offer is accepted, then the funds are divided as proposed. If it is rejected, then both players receive zero. The predicted equilibrium is that the first player will offer a minimal unit to the second player, who will then accept anything more than zero. This prediction has repeatedly been falsified, starting with the work of Güth, Schmittberger, and Schwarze (1982; see Frey and Bohnet 1996; Güth and Tietz 1990; Roth 1995; Samuelson, Gale, and Binmore 1995).[27] Subjects assigned to the first position tend to offer substantially more than the minimum unit. They frequently offer the "fair" division of splitting the sum. Second movers tend to reject offers that are quite small. The acceptance level for offers tends to cluster around different values in diverse cultures (Roth et al. 1991). Given that the refusal to accept the funds offered contradicts a basic tenet in the complete model of rationality, these findings have represented a major challenge to the model's empirical validity in this setting.

Several hypotheses have been offered to explain these findings, including a "punishment hypothesis" and a "learning hypothesis."

> The punishment hypothesis is in essence a reciprocity argument. In contrast to adaptive learning, punishment attributes a motive to the second mover's rejection of an unequal division asserting that it is done to punish the first mover for unfair treatment. This propensity toward negative reciprocity is the linchpin of the argument. Given this propensity, first movers should tend to shy away from the perfect equilibrium offer out of fear of winding up with nothing. (Abbink et al. 1996, 6)

Abbink and his colleagues designed an experiment in which the prediction of the learning and punishment hypotheses is clearly different and found strong support for the punishment hypothesis. "We found that second movers were three times more likely to reject the unequal split when doing so punished the first mover...than when doing so rewarded the first mover" (Abbink et al. 1996, 15–16). Consequently, second movers do appear to punish first movers who propose unfair divisions.

Two additional findings from one-shot social dilemmas provide further evidence of the behavioral propensities of subjects. First, those who intend to cooperate in a particular one-shot social dilemma also expect cooperation to be returned by others at a much higher rate than those who intend to defect (Dawes, McTavish, and Shaklee 1977; Dawes, Orbell, and van de Kragt 1986). As Orbell and Dawes (1991, 519) summarize their own work: "One of our most consistent findings throughout these studies—a finding replicated by others' work—is that cooperators expect significantly more cooperation than do defectors." Second, when there is a choice whether to participate in a social dilemma, those who intend to cooperate exhibit a greater willingness to enter such transactions (Orbell and Dawes 1993). Given these two tendencies, reciprocators are likely to be more optimistic about finding others following the same norm and disproportionately enter more voluntary social dilemmas than nonreciprocators. Given both propensities, the feedback from such voluntary activities will generate confirmatory evidence that they have adopted a norm that serves them well over the long run.

Thus, while individuals vary in their propensity to use reciprocity, the evidence from experiments shows that a substantial proportion of the population drawn on by social science experiments has sufficient trust that others are reciprocators to cooperate with them even in one-shot, no-communication experiments. Furthermore, a substantial proportion of the population is also willing to punish noncooperators (or individuals who do not make fair offers) at a cost to themselves. Norms are learned from prior experience (socialization) and are affected by situational variables yielding systematic differences among experimental designs. The level of trust and resulting levels of cooperation can be increased by (1) providing subjects with an opportunity to see one another (Frey and Bohnet 1996; Orbell and Dawes 1991), (2) allowing subjects to choose whether to enter or exit a social-dilemma game (Orbell and Dawes 1991, 1993; Orbell, Schwartz-Shea, and Simmons 1984; Schuessler 1989; Yamagishi 1988c; Yamagishi and Hayashi 1996), (3) sharing the costs equally if a minimal set voluntarily contributes to a public good (Dawes, Orbell, and van de Kragt 1986), (4) providing opportunities for distinct punishments of those who are not reciprocators (Abbink et al. 1996; McCabe, Rassenti, and Smith 1996), and, as discussed above, (5) providing opportunities for face-to-face communication.

The Core Relationships: Reciprocity, Reputation, and Trust

When many individuals use reciprocity, there is an incentive to acquire a *reputation* for keeping promises and performing actions with short-term costs but long-term net benefits (Keohane 1984; Kreps 1990; Milgrom, North, and Weingast 1990; Miller 1992). Thus, trustworthy individuals who trust others with a reputation for being trustworthy (and try to avoid those who have a reputation for being untrustworthy) can engage in mutually productive social exchanges, even though they are dilemmas, so long as they can limit their interactions primarily to those with a reputation for keeping promises. A reputation for being trustworthy, or for using retribution against those who do not keep their agreements or keep up their fair share, becomes a valuable asset. In an evolutionary context, it increases fitness in an environment in which others use reciprocity norms. Similarly, developing *trust* in an environment in which others are trustworthy is also an asset (Braithwaite and Levi 1998; Fukuyama 1995; Gambetta 1988; Putnam 1993). Trust is the expectation of one person about the actions of others that affects the first person's choice, when an action must be taken before the actions of others are known (Dasgupta 1997, 5). In the context of a social dilemma, trust affects whether an individual is willing to initiate cooperation in the expectation that it will be reciprocated. Boundedly rational individuals enter situations with an initial probability of using reciprocity based on their own prior training and experience.

Thus, at the core of a behavioral explanation are the links between the trust that individuals have in others, the investment others make in trustworthy reputations, and the probability that participants will use reciprocity norms (see fig. 16.2). This mutually reinforcing core is affected by structural variables as well as the past experiences of participants. In the initial round of a repeated dilemma, individuals do or do not initiate cooperative behavior based on their own norms, how much trust they have that others are reciprocators (based on any information they glean about one another), and how structural variables affect their own and their expectation of others' behavior.

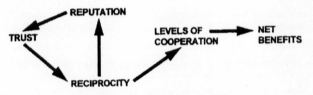

Fig. 16.2. The core relationships

If initial levels of cooperation are moderately high, then individuals may learn to trust one another, and more may adopt reciprocity norms. When more individuals use reciprocity norms, gaining a reputation for being trustworthy is a better investment. Thus, levels of trust, reciprocity, and reputations for being trustworthy are positively reinforcing. This also means that a decrease in any one of these can lead to a downward spiral. Instead of explaining levels of cooperation directly, this approach leads one to link structural variables to an inner triangle of trust, reciprocity, and reputation as these, in turn, affect levels of cooperation and net benefits.

Communication and the Core Relationships

With these core relationships, one can begin to explain why repeated face-to-face communication substantially changes the structure of a situation (see discussion in E. Ostrom, Gardner, and Walker 1994, 199). With a repeated chance to see and talk with others, a participant can assess whether he or she trusts others sufficiently to try to reach a simple contingent agreement regarding the level of joint effort and its allocation. In a contingent agreement, individuals agree to contribute X resources to a common effort so long as at least Y others also contribute. Contingent agreements do not need to include all those who benefit. The benefit to be obtained from the contribution of Y

proportion of those affected may be so substantial that some individuals are willing to contribute so long as Y proportion of others also agree and perform.

Communication allows individuals to increase (or decrease) their trust in the reliability of others.[28] When successful, individuals change their expectations from the initial probability that others use reciprocity norms to a higher probability that others will reciprocate trust and cooperation. When individuals are symmetric in assets and payoffs, the simplest agreement is to share a contribution level equally that closely approximates the optimum joint outcome. When individuals are not symmetric, finding an agreement is more difficult, but various fairness norms can be used to reduce the time and effort needed to achieve an agreement (see Hackett, Dudley, and Walker 1995; Hackett, Schlager, and Walker 1994).

Contingent agreements may deal with punishment of those who do not cooperate (Levi 1988). How to punish noncooperative players, keep one's own reputation, and sustain any initial cooperation that has occurred in N-person settings is more difficult than in two-person settings.[29] In an N-person, uncertain situation, it is difficult to interpret from results that are less than expected whether one person cheated a lot, several people cheated a little, someone made a mistake, or everyone cooperated and an exogenous random variable reduced the expected outcome. If there is no communication, then the problem is even worse. Without communication and an agreement on a sharing formula, individuals can try to signal a willingness to cooperate through their actions, but no one has agreed to any particular contribution. Thus, no one's reputation (external or internal) is at stake.

Once a verbal agreement in an N-person setting is reached, that becomes the focal point for further action within the context of a particular ongoing group. If everyone keeps to the agreement, then no further reaction is needed by someone who is a reciprocator. If the agreement is not kept, however, then an individual following a reciprocity norm—without any prior agreement regarding selective sanctions for nonconformance—needs to punish those who did not keep their commitment. A frequently posited punishment is the grim trigger, whereby a participant plays the Nash equilibrium strategy forever upon detecting any cheating. Subjects in repeated experiments frequently discuss the use of a grim trigger to punish mild defections but reject the idea because it would punish everyone—not just the cheater(s) (E. Ostrom, Gardner, and Walker 1994). A much less drastic punishment strategy is the measured reaction. "In a measured reaction, a player reacts mildly (if at all) to a small deviation from an agreement. Defections trigger mild reactions instead of harsh punishments. If defections continue over time, the measured response slowly moves from the point of agreement toward the Nash equilibrium" (199–200).

For several reasons, this makes sense as the initial "punishment" phase in an N-person setting with a minimal institutional structure and no feedback concerning individual contributions. If only a small deviation occurs, then the cooperation of most participants is already generating positive returns. By keeping one's own reaction close to the agreement, one keeps up one's own reputation for cooperation, keeps cooperation levels higher, and makes it easier to restore full conformance. Using something like a grim trigger immediately leads to the unraveling of the agreement and the loss of substantial benefits over time. To supplement the measured reaction, effort is expended on determining who is breaking the agreement, on using verbal rebukes to try to get that individual back in line, and on avoiding future interactions with that individual.[30]

Thus, understanding how trust, reciprocity, and reputation feed one another (or their lack, which generates a cascade of negative effects) helps to explain why repeated, face-to-face communication has such a major effect. Coming to an initial agreement and making personal promises to one another places at risk an individual's own identity as one who keeps one's word, increases trust, and makes reciprocity an even more beneficial strategy. Tongue-lashing can be partially substituted in a small group for monetary losses and, when backed by measured responses, can keep many groups at high levels of cooperation. Meeting only once can greatly increase trust, but if some individuals do not cooperate immediately, the group never has a further opportunity to hash out these problems. Any evidence of lower levels of cooperation undermines the trust established in the first meeting, and there is no further opportunity to build trust or use verbal sanctioning. It is also clearer now why sending anonymous, computerized messages is not as effective as face-to-face communication. Individuals judge one another's trustworthiness by watching facial expressions and hearing the way something is said. It is hard to establish trust in a group of strangers who will make decisions independently and privately without seeing and talking with one another.

Illustrative Theoretical Scenarios

I have tried to show the need for the development of second-generation models of rationality in order to begin a coherent synthesis of what we know from empirical research on social dilemmas. Rather than try to develop a new formal model, I have stayed at the theoretical level to identify the attributes of human behavior that should be included in future formal models. The individual attributes that are particularly important in explaining behavior in social dilemmas include the expectations individuals have about others' behavior (trust), the norms individuals learn from socialization and life's experiences (reciprocity), and the identities individuals create that project their intentions and norms (reputation). Trust, reciprocity, and reputation can be included in formal models of individual behavior (see the works cited by Boyd and Richerson 1988; Güth and Yaari 1992; Nowak and Sigmund 1993).

In this section, I construct theoretical scenarios of how exogenous variables combine to affect endogenous structural variables that link to the core set of relationships shown in figure 16.2. It is not possible to relate all structural variables in one large causal model, given the number of important variables and the fact that many depend for their effect on the values of other variables. It is possible, however, to produce coherent, cumulative, theoretical scenarios that start with relatively simple baseline models. One can then begin the systematic exploration of what happens as one variable is changed. Let me illustrate what I mean by theoretical scenarios.

Let us start with a scenario that should be conducive to cooperation—a small group of 10 farmers who own farms of approximately the same size. These farmers share the use of a creek for irrigation that runs by their relatively flat properties. They face the problem each year of organizing one collective workday to clear out the fallen trees and brush from the prior winter. All 10 expect to continue farming into the indefinite future. Let us assume that the creek delivers a better water supply directly in response to how many days of work are completed. All farmers have productive opportunities for their labor that return more at the margin than the return they would receive from their own input into this effort. Thus, free riding and hoping that the others contribute labor is objectively attractive. The value to each farmer, however, of participation in a successful collective effort to clear the creek is greater than the costs of participating.

Now let us examine how some structural variables affect the likelihood of collective action (see fig. 16.3). Because they are a small group, it would be easy for them to engage in face-to-face communication. Since their interests and resources are relatively symmetric, arriving at a fair, contingent agreement regarding how to share the work should not be too difficult. One simple agreement that is easy to monitor is that they all work on the same day, but each is responsible for clearing the part of the creek going through his or her property. Conformance to such an agreement would be easy to verify. While engaged in discussions, they can reinforce the importance of everyone participating in the workday. In face-to-face meetings, they can also gossip about anyone who failed to participate in the past, urge them to change their ways, and threaten to stop all labor contributions if they do not "shape up." Given the small size of the group, its symmetry, and the relatively low cost of providing the public good, combined with the relatively long time horizon, we can predict with some confidence that a large proportion of individuals facing such a situation will find a way to cooperate and overcome the dilemma. Not only does the evidence from experimental research support that prediction, but also substantial evidence from the field is consistent with this explanation (see E. Ostrom 1998).

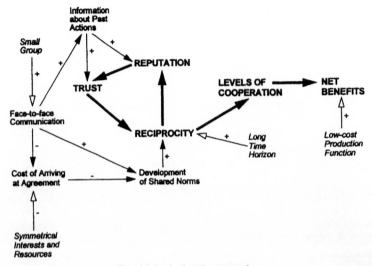

Fig. 16.3. A simple scenario

This is a rough but coherent causal theory that uses structural variables (small size, symmetry of assets and resources, long time horizon, and a low-cost production function) to predict with high probability that participants can themselves solve this social dilemma. Changes in any of the structural variables of this relatively easy scenario affect that prediction. Even a small change may suffice to reverse the predicted outcome. For example, assume that another local farmer buys five parcels of land with the plan to farm them for a long time. Now there are only six farmers, but one of them holds half the relevant assets. If that farmer shares the norm that it is fair to share work allocated to a collective benefit in the same ratio as the benefits are allocated, then the increased heterogeneity will not be a difficult problem to overcome. They would agree—as farmers around the world have frequently agreed (see Lam 1998; Tang 1992)—to share the work in proportion to the amount of land they own. If the new farmer uses a different concept of fairness, then the smaller group may face a more challenging problem than the larger group due to its increased heterogeneity.

Now, assume that the five parcels of land are bought by a local developer to hold for future use as a suburban housing development. The time horizon of one of the six actors—the developer—is extremely short with regard to investments in irrigation. From the developer's perspective, he is not a "free rider," as he sees no benefit to clearing out the creek. Thus, such a change actually produces several: A decrease in the N of the group, an introduction of an asymmetry of interests and resources, and the presence of one participant with half the resources but a short time horizon and no interest in the joint benefit. This illustrates how changes in one structural variable can lead to a cascade of changes in the others and thus how difficult it is to make simple bivariate hypotheses about the effect of one variable on the level of cooperation. In particular, this smaller group is much less likely to cooperate than the larger group of 10 symmetric farmers, exactly the reverse of the standard view of the effect of group size.

Implications

The implications of developing second-generation models of empirically grounded, boundedly rational, and moral decision making are substantial. Puzzling research questions can now be addressed more systematically. New research questions will open up. We need to expand the type of research methods regularly used in political science. We need to increase the level of understanding among those engaged in formal theory, experimental research, and field research across the social and biological sciences. The foundations of policy analysis need rethinking. And civic education can be based on empirically validated theories of collective action empowering citizens to use the "science and art of association" (Tocqueville [1835 and 1840] 1945) to help sustain democratic polities in the twenty-first century.

Implications for Research

What the research on social dilemmas demonstrates is a world of *possibility* rather than of *necessity*. We are neither trapped in inexorable tragedies nor free of moral responsibility for creating and sustaining incentives that facilitate our own achievement of mutually productive outcomes. We cannot adopt the smug presumption of those earlier group theorists who thought groups would always form whenever a joint benefit would be obtained. We can expect many groups to fail to achieve mutually productive benefits due to their lack of trust in one another or to the lack of arenas for low-cost communication, institutional innovation, and the creation of monitoring and sanctioning rules (V. Ostrom 1997). Nor can we simply rest assured that only one type of institution exists for all social dilemmas, such as a competitive market, in which individuals pursuing their own preferences are led to produce mutually productive outcomes. While new institutions often facilitate collective action, the key problems are to design new rules, motivate participants to conform to rules once they are devised, and find and appropriately punish those who cheat. Without individuals viewing rules as appropriate mechanisms to enhance reciprocal relationships, no police force and court system on earth can monitor and enforce all the needed rules on its own. Nor would most of us want to live in a society in which police were really the thin blue line enforcing all rules.

While I am proposing a further development of second-generation theories of rational choice, theories based on complete but thin rationality will continue to play an important role in our understanding of human behavior. The clear and unambiguous predictions stemming from complete rational choice theories will continue to serve as a critical benchmark in conducting empirical studies and for measuring the success or failure of any other explanation offered for observed behavior. A key research question will continue to be: What is the difference between the predicted equilibrium of a complete rationality theory and observed behavior? Furthermore, game theorists are already exploring ways of including reputation, reciprocity, and various norms of behavior in game-theoretic models (see Abbink et al. 1996; Güth 1995; Kreps 1990; Palfrey and Rosenthal 1988; Rabin 1994; Selten 1990, 1991). Thus, bounded and complete rationality models may become more complementary in the next decade than appears to be the case today.

For political scientists interested in diverse institutional arrangements, complete rational choice theories provide well-developed methods for analyzing the vulnerability of institutions to the strategies devised by talented, analytically sophisticated, short-term hedonists (Brennan and Buchanan 1985). Any serious institutional analysis should include an effort to understand how institutions—including ways of organizing legislative procedures, formulas used to calculate electoral weights and minimal winning coalitions, and international agreements on global environmental problems—are vulnerable to manipulation by calculating, amoral participants.[31] In addition to the individuals who have learned norms of reciprocity in any population, others exist who may try to subvert the process so as to obtain very substantial returns for themselves while ignoring the interests of others. One should always know the consequences of letting such individuals operate in any particular institutional setting.

The most immediate research questions that need to be addressed using second-generation models of human behavior relate to the effects of structural variables on the likelihood of organizing for successful modes of collective action. It will not be possible to relate all structural variables in one large causal theory, given that they are so numerous and that many depend for their effect on the values of other variables. What is possible, however, is the development of coherent, cumulative, theoretical scenarios that start with relatively simple baseline models and then proceed to change one variable at a time, as briefly illustrated above. From such scenarios, one can proceed to formal models and empirical testing in field and laboratory settings. The kind of *theory* that emerges from such an enterprise does not lead to the global bivariate (or even multivariate) predictions that have been the ideal to which many scholars have aspired. Marwell and Oliver (1993) have constructed such a series of theoretical scenarios for social dilemmas involving large numbers of heterogeneous participants in collective action. They have come to a similar conclusion about the nature of the theoretical and empirical enterprise: "This is not to say that general theoretical predictions are impossible using our perspective, only that they cannot be simple and global. Instead, the predictions that we can validly generate must be complex, interactive, and conditional" (25).

As political scientists, we need to recognize that political systems are complexly organized and that we will rarely be able to state that one variable is always positively or negatively related to a dependent variable. One can do comparative statics, but one must know the value of the other variables and not simply assume that they vary around the average.

The effort to develop second-generation models of boundedly rational and moral behavior will open up a variety of *new* questions to be pursued that are of major importance to all social scientists and many biologists interested in human behavior. Among these questions are: How do individuals gain trust in other individuals? How is trust affected by diverse institutional arrangements? What verbal and visual clues are used in evaluating others' behavior? How do individuals gain common understanding so as to craft and follow self-organized arrangements (V. Ostrom 1990)? John Orbell (personal communication) posits a series of intriguing questions: "Why do people join together in these games in the first place? How do we select partners in these games? How do our strategies for selecting individual partners differ from our strategies for adding or removing individuals from groups?"

An important set of questions is related to how institutions enhance or restrict the building of mutual trust, reciprocity, and reputations. A recent set of studies on tax compliance raises important questions about the trust heuristics used by citizens and their reactions to governmental efforts to monitor compliance (see Scholz 1998). Too much monitoring may have the counterintuitive result that individuals feel they are *not* trusted and thus become less trustworthy (Frey 1993). Bruno Frey (1997) questions whether some formal institutional arrangements, such as social insurance and paying people to contribute effort, reduce the likelihood that individuals continue to place a positive intrinsic value on actions taken mainly because of internal norms. Rather, they may assume that formal organizations are charged with the responsibility of taking care of joint needs and that reciprocity is no longer needed (see also Taylor 1987).

Since all rules legitimate the use of sanctions against those who do not comply, rules can be used to assign benefits primarily to a dominant coalition. Those who are, thus, excluded have no motivation to cooperate except in order to avoid sanctions. Using first-generation models, that is what one expects in any case. Using second-generation models, one is concerned with how constitutional and collective-choice rules affect the distribution of benefits and the likelihood of reciprocal cooperation. While much research has been conducted on long-term successful self-organized institutions, less has been documented about institutions that never quite got going or failed after years of success. More effort needs to be made to find reliable archival information concerning these failed attempts and why they failed.

It may be surprising that I have relied so extensively on experimental research. I do so for several reasons. As theory becomes an ever more important core of our discipline, experimental studies will join the ranks of basic empirical research methods for political scientists. As an avid field researcher for the past 35 years, I know the importance and difficulty of testing theory in field settings—particularly when variables function interactively. Large-scale field studies will continue to be an important source of empirical data, but frequently they are a very expensive and inefficient method for addressing how institutional incentives combine to affect individual behavior and outcomes. We can advance much faster and more coherently when we examine hypotheses about contested elements among diverse models or theories of a coherent framework. Careful experimental research designs frequently help sort out competing hypotheses more effectively than does trying to find the precise combination of variables in the field. By adding experimental methods to the battery of field methods already used extensively, the political science of the twenty-first century will advance more rapidly in acquiring well-grounded theories of human behavior and of the effect of diverse institutional arrangements on behavior. Laboratory research will still need to be complemented by sound field studies to meet the criteria of external validity.

Implications for Policy

Using a broader theory of rationality leads to potentially different views of the state. If one sees individuals as helpless, then the state is the essential external authority that must solve social dilemmas for everyone. If, however, one assumes individuals can draw on heuristics and norms to solve some problems and create new structural arrangements to solve others, then the image of what a national government might do is somewhat different. There is a very considerable role for large-scale governments, including national defense, monetary policy, foreign policy, global trade policy, moderate redistribution, keeping internal peace when some groups organize to prey on others, provision of accurate information and of arenas for resolving conflicts with national implications, and other large-scale activities. But national governments are too small to govern the global commons and too big to handle smaller scale problems.

To achieve a complex, multitiered governance system is quite difficult. Many types of questions are raised. How do different kinds of institutions support or undermine norms of reciprocity both within hierarchies (Miller 1992) and among members of groups facing collective action problems (Frohlich and Oppenheimer 1970; Galjart 1992)? Field studies find that monitoring and graduated sanctions are close to universal in all robust common-pool resource institutions (E. Ostrom 1990). This tells us that without some external support of such institutions, it is unlikely that reciprocity *alone* completely solves the more challenging common-pool resource problems. Note that sanctions are graduated rather than initially severe. Our current theory of crime—based on a strict expected value theory—does not explain this. If people can learn reciprocity as the fundamental norm for organizing their lives, and if they agree to a set of rules contingent upon others following these rules, then graduated sanctions do something more than deter rule infractions.

Reciprocity norms can have a dark side. If punishment consists of escalating retribution, then groups who overcome social dilemmas may be limited to very tight circles of kin and friends, who cooperate only with one another, embedded in a matrix of hostile relationships with outsiders (R. Hardin 1995). This pattern can escalate into feuds, raids, and overt warfare (Boyd and Richerson 1992; Chagnon 1988; Elster 1985; Kollock 1993). Or tight circles of individuals who trust one another may discriminate against anyone of a different color, religion, or ethnicity. A focus on the return of favors for favors can also be the foundation for corruption. It is in everyone else's interest that some social dilemmas are *not* resolved, such as those involved in monopolies and cartel formation, those that countervene basic moral standards and legal relationships, and those that restrict the opportunities of an open society and an expanding economy. Policies that provide alternative opportunities for those caught in dysfunctional networks are as important as those that stimulate and encourage positive networks (Dasgupta 1997).

Implications for Civic Education

Human history teaches us that autocratic governments often wage war on their own citizens as well as on those of other jurisdictions. Democracies are characterized by the processing of conflict among individuals and groups without resorting to massive killings. Democracies are, however, themselves fragile institutions that are vulnerable to manipulation if citizens and officials are not vigilant (V. Ostrom 1997). For those who wish the twenty-first century to be one of peace, we need to translate our research findings on collective action into materials written for high school and undergraduate students. All too many of our textbooks focus exclusively on leaders and, worse, only national-level leaders. Students completing an introductory course in U.S.

government, or political science more generally, will not learn that they play an essential role in sustaining democracy. Citizen participation is presented as contacting leaders, organizing interest groups and parties, and voting. That citizens need additional skills and knowledge to resolve the social dilemmas they face is left unaddressed. Their moral decisions are not discussed. We are producing generations of cynical citizens with little trust in one another, much less in their governments. Given the central role of trust in solving social dilemmas, we may be creating the very conditions that undermine our own democratic ways of life. It is ordinary persons and citizens who craft and sustain the workability of the institutions of everyday life. We owe an obligation to the next generation to carry forward the best of our knowledge about how individuals solve the multiplicity of social dilemmas—large and small—that they face.

NOTES

1. The term *tragedy of the commons* refers to the problem that common-pool resources, such as oceans, lakes, forests, irrigation systems, and grazing lands, can easily be overused or destroyed if property rights to these resources are not well defined (see G. Hardin 1968).

2. The term *reciprocal altruism* is used by biologists and evolutionary theorists to refer to strategies of conditional cooperation with *nonkin* that produce higher benefits for the individuals who follow these strategies if they interact primarily with others who are reciprocators (Trivers 1971). Since these strategies benefit the individual using them in the long term, I prefer the term *reciprocity*.

3. See Farrell and Maskin (1989) for a different approach to this problem.

4. This is only a short list of the more important variables found to affect behavior within social dilemmas (for summary overviews of this literature, see Goetze and Orbell 1988; Ledyard 1995; Lichbach 1996; E. Ostrom 1990, 1998; E. Ostrom, Gardner, and Walker 1994; Sally 1995; Schroeder 1995).

5. The term *rent seeking* refers to nonproductive activities directed toward creating opportunities for profits higher than would be obtained in an open, competitive market.

6. See Isaac, McCue, and Plott (1985); Kim and Walker (1984); Marwell and Ames (1979, 1980, 1981); Orbell and Dawes (1991, 1993); Schneider and Pommerehne (1981). An important exception to this general finding is that when subjects are presented with an experimental protocol with an opportunity to invest tokens in a common-pool resource (the equivalent of harvesting from a common pool), they tend to *overinvest* substantially in the initial rounds (see E. Ostrom, Gardner, and Walker 1994; and comparison of public goods and common-pool resource experiments in Goetze 1994; and E. Ostrom and Walker 1997). Ledyard (1995) considers common-pool resource dilemmas to have the same underlying structure as public-good dilemmas, but behavior in common-pool resource experiments without communication is

consistently different from public-good experiments without communication. With repetition, outcomes in common-pool resource experiments approach the Nash equilibrium from below rather than from above, as is typical in public good experiments.

7. In a series of eight experiments with different treatments conducted by Isaac, Walker, and Thomas (1984), in which the uniform theoretical prediction was zero contributions, contribution rates varied from nearly 0 percent to around 75 percent of the resources available to participants.

8. Subjects in Centipede games also do not use backward induction (see McKelvey and Palfrey 1992).

9. The pulsing cannot be explained using a complete model of rationality, but it can be explained as the result of a heuristic used by subjects to raise or lower their investments depending upon the average return achieved on the most recent round (see E. Ostrom, Gardner, and Walker 1994).

10. In cooperative game theory, in contrast, it is assumed that players can communicate and make *enforceable agreements* (Harsanyi and Setten 1988, 3).

11. In social-dilemma experiments, subjects make anonymous decisions and are paid privately. The role of cheap talk in coordination experiments is different since there is no dominant strategy. In this case, preplay communication may help players coordinate on one of the possible equilibria (see Cooper, DeJong, and Forsythe 1992).

12. As Aumann (1974) cogently points out, the players are faced with the problem that whatever they agree upon has to be self-enforcing. That has led Aumann and most game theorists to focus entirely on Nash equilibria, which, once reached, are self-enforcing. In coordination games, cheap talk can be highly efficacious.

13. See E. Ostrom, Gardner, and Walker (1994) for extensive citations to studies showing a positive effect of the capacity to communicate. Dawes, McTavish, and Shaklee (1977); Frey and Bohnet (1996); Hackett, Schlager, and Walker (1994); Isaac and Walker (1988a, 1991); Orbell, Dawes, and van de Kragt (1990); Orbell, van de Kragt, and Dawes (1988, 1991); E. Ostrom, Gardner, and Walker (1994); Sally (1995).

14. Moir (1995) also replicated these findings with face-to-face communication.

15. Social psychologists have found that groups who perform tasks using electronic media do much better if they have had an opportunity to work face to face prior to the use of electronic communication only (Hollingshead, McGrath, and O'Connor 1993).

16. See also Banks and Calvert (1992a,b) for a discussion of communication in incomplete information games.

17. Furthermore, they invested more when the fine was lower or when it was more efficacious, and they tended to direct their fines to those who had invested the most on prior rounds. Given the cost of the sanctioning mechanism, subjects tended to overuse it and to end up with a less efficient outcome after sanctioning costs were subtracted from their earnings. This finding is consistent with the Boyd and Richerson (1992) result that moralistic strategies may result in negative net outcomes.

18. When constructing formal models, one can include overt delta parameters in the model (see Crawford and Ostrom 1995; Palfrey and Rosenthal 1988). Alternatively, one can assume that these internal delta parameters lead individuals to enter new situations with differing probabilities that they will follow norms such as reci-

procity. These probabilities not only vary across individuals but also increase or decrease as a function of the specific structural parameters of the situation and, in repeated experiments, the patterns of behavior and outcomes achieved in that situation vary over time.

19. The change in valuations that an individual may attach to an action-outcome linkage may be generated strictly internally or may be triggered by external observation and, thus, a concern with how others will evaluate the normative appropriateness of actions.

20. Gouldner (1960, 171) considers norms of reciprocity to be universal and as important in most cultures as incest taboos, even though the "concrete formulations may vary with time and place."

21. See Selten (1986) for a discussion of his own and John Harsanyi's (1977) conception of "rule utilitarianism" as contrasted to "act utilitarianism."

22. Crawford and E. Ostrom (1995) discuss these issues in greater depth. See also Piaget ([1932] 1969).

23. The evolutionary approach has been strongly influenced by the work of Robert Axelrod (see, in particular, Axelrod 1984, 1986; Axelrod and Hamilton 1981; and Axelrod and Keohane 1985).

24. The grim trigger has been used repeatedly as a support for cooperative outcomes in infinitely (or indefinitely) repeated games (Fudenberg and Maskin 1986). In games in which substantial joint benefits are to be gained over the long term from mutual cooperation, the threat of the grim trigger is thought to be sufficient to encourage everyone to cooperate. A small error on the part of one player or exogenous noise in the payoff function, however, makes this strategy a dangerous one to use in larger groups, where the cooperators' dividend may also be substantial.

25. This is not the complete list of all types of reciprocity norms, but it captures the vast majority.

26. The proportion of individuals who follow the sixth norm—cooperate always—will be minuscule or nonexistent. Individuals following the first norm will be those, along with those following the sixth norm, who cooperate in the first few rounds of a finitely repeated experimental social dilemma without prior communication. Individuals following the second norm will cooperate (immediately) in experiments if they have an opportunity to judge the intentions and trustworthiness of the other participants and expect most of the others to be trustworthy. Those following the third norm will cooperate (after one or a few rounds) in experiments in which others cooperate.

27. The results obtained by Hoffman, McCabe, and Smith (1996b) related to dictator games under varying conditions of social distance are also quite consistent with the behavioral approach of this article.

28. Frank, Gilovich, and Regan (1993) found, for example, that the capacity of subjects to predict whether others would play cooperatively was significantly better than chance after a face-to-face group discussion. Kikuchi, Watanabe, and Yamagishi (1996) found that high trusters predicted other players' trustworthiness significantly better than did low trusters.

29. In a two-person situation of complete certainty, individuals can easily follow

the famous tit-for-tat (or tit-for-tat or exit) strategy even without communication. When a substantial proportion of individuals in a population follows this norm, and they can identify with whom they have interacted in the past (to either refuse future interactions or to punish prior uncooperative actions), and when discount rates are sufficiently low, tit-for-tat has been shown to be a highly successful strategy, yielding higher payoffs than are available to those using other strategies (Axelrod 1984, 1986). With communication, it is even easier.

30. In a series of 18 common-pool resource experiments, each involving eight subjects in finitely repeated communication experiments, E. Ostrom, Gardner, and Walker (1994, 215) found that subjects kept to their agreements or used measured responses in two-thirds of the experiments. In these experiments, joint yields averaged 89% of optimum. In the six experiments in which some players deviated substantially from agreements and measured responses did not bring them back to the agreement, cooperation levels were substantially less, and yields averaged 43% of optimum (which is still far above zero levels of cooperation).

31. Consequently, research on the effect of institutional arrangements on strategies and outcomes continues to be crucial to future developments. See Agrawal 1999; Alt and Shepsle 1990; Bates 1989; Dasgupta 1993; Eggertsson 1990; Gibson 1999; Levi 1997; V. Ostrom 1997; V. Ostrom, Feeny, and Picht 1993; Scharpf 1997.

REFERENCES

Abbink, Klaus, Gary E. Bolton, Abdolkarim Sadrieh, and Fang Fang Tang. 1996. "Adaptive Learning versus Punishment in Ultimatum Bargaining." Discussion paper no. B–381. Rheinische Friedrich-Wilhelms-Universität Bonn. Typescript.

Abreau, Dilip. 1988. "On the Theory of Infinitely Repeated Games with Discounting." *Econometrica* 80(4):383–96.

Agrawal, Arun. 1999. *Greener Pastures: Politics, Markets, and Community among a Migrant Pastoral People.* Durham, NC: Duke University Press.

Alchian, Armen A. 1950. "Uncertainty, Evolution, and Economic Theory." *Journal of Political Economy* 58(3):211–21.

Alchian, Armen A., and Harold Demsetz. 1972. "Production, Information Costs, and Economic Organization." *American Economic Review* 62(December):777–95.

Alt, James E., and Kenneth A. Shepsle, eds. 1990. *Perspectives on Positive Political Economy.* New York: Cambridge University Press.

Andreoni, James. 1989. "Giving with Impure Altruism: Applications to Charity and Ricardian Equivalence." *Journal of Political Economy* 97(December):1, 447–51, 458.

Arnold, J. E. M., and J. Gabriel Campbell. 1986. "Collective Management of Hill Forests in Nepal: The Community Forestry Development Project." In *Proceedings of the Conference on Common Property Resource Management, National Research Council.* Washington, DC: National Academy Press. 425–54.

Aumann, Robert J. 1974. "Subjectivity and Correlation in Randomized Strategies." *Journal of Mathematical Economics* 1(March):6796.

Axelrod, Robert. 1984. *The Evolution of Cooperation.* New York: Basic Books.

Axelrod, Robert. 1986. "An Evolutionary Approach to Norms." *American Political Science Review* 80(December):1095–111.

Axelrod, Robert, and William D. Hamilton. 1981. "The Evolution of Cooperation." *Science* 211(March):1390–96.

Axelrod, Robert, and Robert O. Keohane. 1985. "Achieving Cooperation under Anarchy: Strategies and Institutions." *World Politics* 38(October):226–54.

Baland, Jean-Marie, and Jean-Philippe Platteau. 1996. *Halting Degradation of Natural Resources. Is There a Role for Rural Communities?* Oxford: Clarendon Press.

Banks, Jeffrey S., and Randall L. Calvert. 1992a. "A Battle-of-the-Sexes Game with Incomplete Information." *Games and Economic Behavior* 4(July):347–72.

Banks, Jeffrey S., and Randall L. Calvert. 1992b. "Communication and Efficiency in Coordination Games." Working paper. Department of Economics and Department of Political Science, University of Rochester, New York. Typescript.

Barkow, Jerome H., Leda Cosmides, and John Tooby, eds. 1992. *The Adapted Mind. Evolutionary Psychology and the Generation of Culture.* Oxford: Oxford University Press.

Barry, Brian, and Russell Hardin. 1982. *Rational Man and Irrational Society? An Introduction and Source Book.* Beverly Hills, CA: Sage.

Bates, Robert H. 1989. *Beyond the Miracle of the Market: The Political Economy of Agrarian Development in Kenya.* New York: Cambridge University Press.

Becker, Lawrence C. 1990. *Reciprocity.* Chicago: University of Chicago Press.

Bendor, Jonathan, and Dilip Mookherjee. 1987. "Institutional Structure and the Logic of Ongoing Collective Action." *American Political Science Review* 81 (March):129–54.

Benoit, Jean-Pierre, and Vijay Krishna. 1985. "Finitely Repeated Games." *Econometrica* 53(July):905–22.

Berkes, Fikret, ed. 1989. *Common Property Resources: Ecology and Community-Based Sustainable Development.* London: Belhaven.

Binmore, Kenneth. 1997. "Rationality and Backward Induction." *Journal of Economic Methodology* 4:23–41.

Blau, Peter M. 1964. *Exchange of Power in Social Life.* New York: Wiley.

Blomquist, William. 1992. *Dividing the Waters: Governing Groundwater in Southern California.* San Francisco, CA: Institute for Contemporary Studies Press.

Boudreaux, Donald J., and Randall G. Holcombe. 1989. "Government by Contract." *Public Finance Quarterly* 17(July):264–80.

Boulding, Kenneth E. 1963. "Towards a Pure Theory of Threat Systems." *American Economic Review* 53(May):424–34.

Boyd, Robert, and Peter J. Richerson. 1988. "The Evolution of Reciprocity in Sizable Groups." *Journal of Theoretical Biology* 132(June):337–56.

Boyd, Robert, and Peter J. Richerson. 1992. "Punishment Allows the Evolution of Cooperation (or Anything Else) in Sizable Groups." *Ethology and Sociobiology* 13(May):171–95.

Braithwaite, Valerie, and Margaret Levi, eds. 1998. *Trust and Governance.* New

York: Russell Sage Foundation.

Brennan, Geoffrey, and James Buchanan. 1985. *The Reason of Rules.* Cambridge: Cambridge University Press.

Bromley, Daniel W., David Feeny, Margaret McKean, Pauline Peters, Jere Gilles, Ronald Oakerson, C. Ford Runge, and James Thomson, eds. 1992. *Making the Commons Work: Theory, Practice, and Policy.* San Francisco, CA: Institute for Contemporary Studies Press.

Bullock, Kari, and John Baden. 1977. "Communes and the Logic of the Commons." In *Managing the Commons,* ed. Garrett Hardin and John Baden. San Francisco, CA: Freeman. 182–99.

Cason, Timothy N., and Feisal U. Khan. 1996. "A Laboratory Study of Voluntary Public Goods Provision with Imperfect Monitoring and Communication." Working paper. Department of Economics, University of Southern California, Los Angeles.

Chagnon, N. A. 1988. "Life Histories, Blood Revenge, and Warfare in a Tribal Population." *Science* 239(February):985–92.

Chan, Kenneth, Stuart Mestelman, Rob Moir, and Andrew Muller. 1996. "The Voluntary Provision of Public Goods under Varying Endowments." *Canadian Journal of Economics* 29(1):54–69.

Clark, Andy. 1995. "Economic Reason: The Interplay of Individual Learning and External Structure." Working paper. Department of Philosophy, Washington University in St. Louis.

Coleman, James S. 1987. "Norms as Social Capital." In *Economic Imperialism: The Economic Approach Applied Outside the Field of Economics,* ed. Gerard Radnitzky and Peter Bernholz. New York: Paragon House. 133–55.

Cook, Karen S., and Margaret Levi. 1990. *The Limits of Rationality.* Chicago: University of Chicago Press.

Cooper, Russell, Douglas V. DeJong, and Robert Forsythe. 1992. "Communication in Coordination Games." *Quarterly Journal of Economics* 107(2):739–71.

Cornes, Richard, C. F. Mason, and Todd Sandler. 1986. "The Commons and the Optimal Number of Firms." *Quarterly Journal of Economics* 101(August):641–46.

Cosmides, Leda, and John Tooby. 1992. "Cognitive Adaptations for Social Exchange." In *The Adapted Mind. Evolutionary Psychology and the Generation of Culture,* ed. Jerome H. Barkow, Leda Cosmides, and John Tooby. New York: Oxford University Press. 163–228.

Cosmides, Leda, and John Tooby. 1994. "Better than Rational: Evolutionary Psychology and the Invisible Hand." *American Economic Review* 84(May):327–32.

Crawford, Sue E. S., and Elinor Ostrom. 1995. "A Grammar of Institutions." *American Political Science Review* 89(September): 582–600. (Chapter 4 in this volume.)

Dasgupta, Partha S. 1993. *An Inquiry into Well-Being and Destitution.* Oxford: Clarendon Press.

Dasgupta, Partha S. 1997. "Economic Development and the Idea of Social Capital." Working paper. Faculty of Economics, University of Cambridge.

Davis, Douglas D., and Charles A. Holt. 1993. *Experimental Economics.* Princeton, NJ: Princeton University Press.

Dawes, Robyn M. 1975. "Formal Models of Dilemmas in Social Decision Making." In *Human Judgment and Decision Processes: Formal and Mathematical Approaches,* ed. Martin F. Kaplan and Steven Schwartz. New York: Academic Press. 87–108.

Dawes, Robyn M. 1980. "Social Dilemmas." *Annual Review of Psychology* 31:169–93.

Dawes, Robyn M., Jeanne McTavish, and Harriet Shaklee. 1977. "Behavior, Communication, and Assumptions about Other People's Behavior in a Commons Dilemma Situation." *Journal of Personality and Social Psychology* 35(1):1–11.

Dawes, Robyn M., John M. Orbell, and Alphons van de Kragt. 1986. "Organizing Groups for Collective Action." *American Political Science Review* 80 (December):1171–85.

de Waal, Frans. 1996. *Good Natured: The Origins of Right and Wrong in Humans and Other Animals.* Cambridge, MA: Harvard University Press.

Dudley, Dean. 1993. "Essays on Individual Behavior in Social Dilemma Environments: An Experimental Analysis." Ph.D. diss., Indiana University.

Edney, Julian. 1979. "Freeriders en Route to Disaster." *Psychology Today* 13 (December):80–102.

Eggertsson, Thráinn. 1990. *Economic Behavior and Institutions.* New York: Cambridge University Press.

Ekeh, P. P. 1974. *Social Exchange Theory: The Two Traditions.* Cambridge, MA: Harvard University Press.

Ellickson, Robert C. 1991. *Order without Law: How Neighbors Settle Disputes.* Cambridge, MA: Harvard University Press.

Elster, Jon. 1985. *Sour Grapes: Studies in the Subversion of Rationality.* Cambridge: Cambridge University Press.

Emerson, Richard. 1972a. "Exchange Theory, Part 1: A Psychological Basis for Social Exchange." In *Sociological Theories in Progress,* ed. Joseph Berger, Morris Zelditch, and Bo Anderson. Vol. 2. Boston: Houghton Mifflin. 38–57.

Emerson, Richard. 1972b. "Exchange Theory, Part 11: Exchange Relations and Networks." In *Sociological Theories in Progress,* ed. Joseph Berger, Morris Zelditch, and Bo Anderson. Vol. 2. Boston: Houghton Mifflin. 58–87.

Farrell, Joseph. 1987. "Cheap Talk, Coordination, and Entry." *Rand Journal of Economics* 18(spring):34–9.

Farrell, Joseph, and Eric Maskin. 1989. "Renegotiation in Repeated Games." *Games and Economic Behavior* 1(December):327–60.

Farrell, Joseph, and Matthew Rabin. 1996. "Cheap Talk." *Journal of Economic Perspectives* 10(summer):103–18.

Feeny, David, Fikret Berkes, Bonnie J. McCay, and James M. Acheson. 1990. "The Tragedy of the Commons: Twenty-Two Years Later." *Human Ecology* 18(1):1–19.

Frank, Robert H., Thomas Gilovich, and Dennis T. Regan. 1993. "The Evolution of One-Shot Cooperation: An Experiment." *Ethology and Sociobiology* 14 (July):247–56.

Frey, Bruno S. 1993. "Does Monitoring Increase Work Effort? The Rivalry with

Trust and Loyalty." *Economic Inquiry* 31(October): 663–70.

Frey, Bruno S. 1997. *Not Just for the Money: An Economic Theory of Personal Motivation.* Cheltenham, UK: Edward Elgar.

Frey, Bruno S., and Iris Bohnet. 1996. "Cooperation, Communication and Communitarianism: An Experimental Approach." *Journal of Political Philosophy* 4 (4):322–36.

Frohlich, Norman, and Joe Oppenheimer. 1970. "I Get by with a Little Help from My Friends." *World Politics* 23(October):104–20.

Fudenberg, Drew, and Eric Maskin. 1986. "The Folk Theorem in Repeated Games with Discounting or with Incomplete Information." *Econometrica* 54(3):533–54.

Fukuyama, Francis. 1995. *Trust: The Social Virtues and the Creation of Prosperity.* New York: Free Press.

Galjart, Bruno. 1992. "Cooperation as Pooling: A Rational Choice Perspective." *Sociologia Ruralis,* 32(4):389–407.

Gambetta, Diego, ed. 1988. *Trust: Making and Breaking Cooperative Relations.* Oxford: Basil Blackwell.

Geddes, Barbara. 1994. *Politician's Dilemma: Building State Capacity in Latin America.* Berkeley: University of California Press.

Gibson, Clark. 1999. *Politicians and Poachers: The Political Economy of Wildlife Policy in Africa.* Cambridge: Cambridge University Press.

Goetze, David. 1994. "Comparing Prisoner's Dilemma, Commons Dilemma, and Public Goods Provision Designs in Laboratory Experiments." *Journal of Conflict Resolution* 38(March):56–86.

Goetze, David, and John Orbell. 1988. "Understanding and Cooperation in Social Dilemmas." *Public Choice* 57(June):275–9.

Gouldner, Alvin W. 1960. "The Norm of Reciprocity: A Preliminary Statement." *American Sociological Review* 25(April):161–78.

Greif, Avner, Paul Milgrom, and Barry R. Weingast. 1994. "Coordination, Commitment, and Enforcement: The Case of the Merchant Guild." *Journal of Political Economy* 102(August):745–76.

Grossman, Sanford J., and Oliver D. Hart. 1980. "Takeover Bids, the Free-Rider Problem, and the Theory of the Corporation." *Bell Journal of Economics* 11 (spring):42–64.

Güth, Werner. 1995. "An Evolutionary Approach to Explaining Cooperative Behavior by Reciprocal Incentives." *International Journal of Game Theory* 24 (4):323–44.

Güth, Werner, and Hartmut Kliemt. 1995. "Competition or Cooperation. On the Evolutionary Economics of Trust, Exploitation and Moral Attitudes." Working paper. Humboldt University, Berlin.

Güth, Werner, and Hartmut Kliemt. 1996. "Towards a Completely Indirect Evolutionary Approach—a Note." Discussion Paper 82. Economics Faculty, Humboldt University, Berlin.

Güth, Werner, Rolf Schmittberger, and Bernd Schwarze. 1982. "An Experimental Analysis of Ultimatum Bargaining." *Journal of Economic Behavior and Or-*

ganization 3(December):367–88.

Güth, Werner, and Reinhard Tietz. 1990. "Ultimatum Bargaining Behavior. A Survey and Comparison of Experimental Results." *Journal of Economic Psychology* 11(September):417–49.

Güth, Werner, and M. Yaari. 1992. "An Evolutionary Approach to Explaining Reciprocal Behavior in a Simple Strategic Game." In *Explaining Process and Change. Approaches to Evolutionary Economics,* ed. Ulrich Witt. Ann Arbor: University of Michigan Press. 23–34.

Hackett, Steven, Dean Dudley, and James Walker. 1995. "Heterogeneities, Information and Conflict Resolution: Experimental Evidence on Sharing Contracts." In *Local Commons and Global Interdependence: Heterogeneity and Cooperation in Two Domains,* ed. Robert O. Keohane and Elinor Ostrom. London: Sage. 93–124.

Hackett, Steven, Edella Schlager, and James Walker. 1994. "The Role of Communication in Resolving Commons Dilemmas: Experimental Evidence with Heterogeneous Appropriators." *Journal of Environmental Economics and Management* 27 (September):99–126.

Hamilton, W. D. 1964. "The Genetical Evolution of Social Behavior." *Journal of Theoretical Biology* 7(July): 1–52.

Hardin, Garrett. 1968. "The Tragedy of the Commons." *Science* 162 (December):1243–48.

Hardin, Russell. 1971. "Collective Action as an Agreeable n-Prisoners' Dilemma." *Science* 16(September-October):472–81.

Hardin, Russell. 1995. *One for All: The Logic of Group Conflict.* Princeton, NJ: Princeton University Press.

Hardin, Russell. 1997. "Economic Theories of the State." In *Perspectives on Public Choice: A Handbook,* ed. Dennis C. Mueller. Cambridge: Cambridge University Press. 21–34.

Hardy, Charles J., and Bibb Latané. 1988. "Social Loafing in Cheerleaders: Effects of Team Membership and Competition." *Journal of Sport and Exercise Psychology* 10(March):109–14.

Harsanyi, John. 1977. "Rule Utilitarianism and Decision Theory." *Erkenntnis* 11 (May):25–53.

Harsanyi, John C., and Reinhard Selten. 1988. *A General Theory of Equilibrium Selection in Games.* Cambridge, MA: MIT Press.

Hirshleifer, David, and Eric Rasmusen. 1989. "Cooperation in a Repeated Prisoner's Dilemma with Ostracism." *Journal of Economic Behavior and Organization* 12 (August):87–106.

Hoffman, Elizabeth, Kevin McCabe, and Vernon Smith. 1996a. "Behavioral Foundations of Reciprocity: Experimental Economics and Evolutionary Psychology." Working paper. Department of Economics, University of Arizona, Tucson.

Hoffman, Elizabeth, Kevin McCabe, and Vernon Smith. 1996b. "Social Distance and Other-Regarding Behavior in Dictator Games." *American Economic Review* 86 (June):653–60.

Hollingshead, Andrea B., Joseph E. McGrath, and Kathleen M. O'Connor. 1993.

"Group Task Performance and Communication Technology: A Longitudinal Study of Computer-Mediated versus Face-to-Face Work Groups." *Small Group Research* 24(August):307–33.

Holmstrom, Bengt. 1982. "Moral Hazard in Teams." *Bell Journal of Economics* 13 (autumn):324–40.

Homans, George C. 1961. *Social Behavior: Its Elementary Forms.* New York: Harcourt, Brace, & World.

Isaac, R. Mark, Kenneth McCue, and Charles R. Plott. 1985. "Public Goods Provision in an Experimental Environment." *Journal of Public Economics* 26 (February):51–74.

Isaac, R. Mark, and James Walker. 1988a. "Communication and Free-Riding Behavior: The Voluntary Contribution Mechanism." *Economic Inquiry* 26 (October):585–608.

Isaac, R. Mark, and James Walker. 1988b. "Group Size Effects in Public Goods Provision: The Voluntary Contributions Mechanism." *Quarterly Journal of Economics* 103(February): 179–99.

Isaac, R. Mark, and James Walker. 1991. "Costly Communication: An Experiment in a Nested Public Goods Problem." In *Laboratory Research in Political Economy,* ed. Thomas R. Palfrey. Ann Arbor: University of Michigan Press. 269–86.

Isaac, R. Mark, and James Walker. 1993. "Nash as an Organizing Principle in the Voluntary Provision of Public Goods: Experimental Evidence." Working paper. Indiana University, Bloomington.

Isaac, R. Mark, James Walker, and Susan Thomas. 1984. "Divergent Evidence on Free Riding: An Experimental Examination of Some Possible Explanations." *Public Choice* 43(2):113–49.

Isaac, R. Mark, James Walker, and Arlington W. Williams. 1994. "Group Size and the Voluntary Provision of Public Goods: Experimental Evidence Utilizing Large Groups." *Journal of Public Economics* 54(May):1–36.

Keohane, Robert O. 1984. *After Hegemony.* Princeton, NJ: Princeton University Press.

Kikuchi, Masako, Yoriko Watanabe, and Toshio Yamagishi. 1996. "Accuracy in the Prediction of Others' Trustworthiness and General Trust: An Experimental Study." *Japanese Journal of Experimental Social Psychology* 37(1):23–36.

Kim, Oliver, and Mark Walker. 1984. "The Free Rider Problem: Experimental Evidence." *Public Choice* 43(1):3–24.

Knack, Stephen. 1992. "Civic Norms, Social Sanctions, and Voter Turnout." *Rationality and Society* 4(April):133–56.

Knight, Jack. 1992. *Institutions and Social Conflict.* Cambridge: Cambridge University Press.

Kollock, Peter. 1993. "An Eye for an Eye Leaves Everyone Blind: Cooperation and Accounting Systems." *American Sociological Review* 58(6):768–86.

Kreps, David M. 1990. "Corporate Culture and Economic Theory." In *Perspectives on Positive Political Economy,* ed. James E. Alt and Kenneth A. Shepsle. New York: Cambridge University Press. 90–143.

Kreps, David M., Paul Milgrom, John Roberts, and Robert Wilson. 1982. "Rational

Cooperation in the Finitely Repeated Prisoner's Dilemma." *Journal of Economic Theory* 27(August): 245–52.

Lam, Wai Fung. 1998. *Governing Irrigation Systems in Nepal: Institutions, Infrastructure, and Collective Action.* Oakland, CA: Institute for Contemporary Studies Press.

Ledyard, John. 1995. "Public Goods: A Survey of Experimental Research." In *The Handbook of Experimental Economics,* ed. J. Kagel and Alvin Roth. Princeton, NJ: Princeton University Press. 111–94.

Leibenstein, Harvey. 1976. *Beyond Economic Man.* Cambridge, MA: Harvard University Press.

Levi, Margaret. 1988. *Of Rule and Revenue.* Berkeley: University of California Press.

Levi, Margaret. 1997. *Consent, Dissent, and Patriotism.* New York: Cambridge University Press.

Lichbach, Mark Irving. 1995. *The Rebel's Dilemma.* Ann Arbor: University of Michigan Press.

Lichbach, Mark Irving. 1996. *The Cooperator's Dilemma.* Ann Arbor: University of Michigan Press.

Luce, R. Duncan, and Howard Raiffa. 1957. *Games and Decisions: Introduction and Critical Survey.* New York: Wiley.

Marr, David. 1982. *Vision: A Computational Investigation into the Human Representation and Processing of Visual Information.* San Francisco, CA: W. H. Freeman.

Marwell, Gerald, and Ruth E. Ames. 1979. "Experiments on the Provision of Public Goods 1: Resources, Interest, Group Size, and the Free Rider Problem." *American Journal of Sociology* 84(May):1335–60.

Marwell, Gerald, and Ruth E. Ames. 1980. "Experiments on the Provision of Public Goods II: Provision Points, Stakes, Experience and the Free Rider Problem." *American Journal of Sociology* 85(January):926–37.

Marwell, Gerald, and Ruth E. Ames. 1981. "Economists Free Ride: Does Anyone Else?" *Journal of Public Economics* 15(November):295–310.

Marwell, Gerald, and Pamela Oliver. 1993. *The Critical Mass in Collective Action: A Micro-Social Theory.* New York: Cambridge University Press.

McCabe, Kevin, Stephen Rassenti, and Vernon Smith. 1996. "Game Theory and Reciprocity in Some Extensive Form Bargaining Games." Working paper. Economic Science Laboratory, University of Arizona, Tucson.

McCay, Bonnie J., and James M. Acheson. 1987. *The Question of the Commons: The Culture and Ecology of Communal Resources.* Tucson: University of Arizona Press.

McKean, Margaret. 1992. "Success on the Commons: A Comparative Examination of Institutions for Common Property Resource Management." *Journal of Theoretical Politics* 4(July):247–82.

McKean, Margaret, and Elinor Ostrom. 1995. "Common Property Regimes in the Forest: Just a Relic from the Past?" *Unasylva* 46(January):3–15.

McKelvey, Richard D., and Thomas Palfrey. 1992. "An Experimental Study of the

Centipede Game." *Econometrica* 60(July):803–36.

Messick, David M. 1973. "To Join or Not to Join: An Approach to the Unionization Decision." *Organizational Behavior and Human Performance* 10(August):146–56.

Messick, David M., and Marilyn B. Brewer. 1983. "Solving Social Dilemmas: A Review." In *Annual Review of Personality and Social Psychology,* ed. L. Wheeler and P. Shaver. Beverly Hills, CA: Sage. 11–44.

Messick, David M., H. A. M. Wilke, Marilyn B. Brewer, R. M. Kramer, P. E. Zemke, and Layton Lui. 1983. "Individual Adaptations and Structural Change as Solutions to Social Dilemmas." *Journal of Personality and Social Psychology* 44 (February):294–309.

Milgrom, Paul R., Douglass C. North, and Barry R. Weingast. 1990. "The Role of Institutions in the Revival of Trade: The Law Merchant, Private Judges, and the Champagne Fairs." *Economics and Politics* 2(March):1–23.

Miller, Gary. 1992. *Managerial Dilemmas. The Political Economy of Hierarchy.* New York: Cambridge University Press.

Moir, Rob. 1995. "The Effects of Costly Monitoring and Sanctioning upon Common Property Resource Appropriation." Working paper. Department of Economics, University of New Brunswick, Saint John.

Morrow, Christopher E., and Rebecca Watts Hull. 1996. "Donor-Initiated Common Pool Resource Institutions: The Case of the Yanesha Forestry Cooperative." *World Development* 24(10):1641–57.

Mueller, Dennis. 1986. "Rational Egoism versus Adaptive Egoism as Fundamental Postulate for a Descriptive Theory of Human Behavior." *Public Choice* 51 (1):3–23.

Nowak, Martin A., and Karl Sigmund. 1993. "A Strategy of Win-Stay, Lose-Shift that Outperforms Tit-for-Tat in the Prisoner's Dilemma Game." *Nature* 364 (July):56–58.

Oakerson, Ronald J. 1993. "Reciprocity: A Bottom-up View of Political Development." In *Rethinking Institutional Analysis and Development: Issues, Alternatives, and Choices,* ed. Vincent Ostrom, David Feeny, and Hartmut Picht. San Francisco, CA: Institute for Contemporary Studies Press. 141–58.

Oliver, Pamela. 1980. "Rewards and Punishments as Selective Incentives for Collective Action: Theoretical Investigations." *American Journal of Sociology* 85 (May):1356–75.

Olson, Mancur. 1965. *The Logic of Collective Action: Public Goods and the Theory of Groups.* Cambridge, MA: Harvard University Press.

Orbell, John M., and Robyn M. Dawes. 1991. "A 'Cognitive Miser' Theory of Cooperators' Advantage." *American Political Science Review* 85(June):515–28.

Orbell, John M., and Robyn M. Dawes. 1993. "Social Welfare, Cooperators' Advantage, and the Option of Not Playing the Game." *American Sociological Review* 58(December):787–800.

Orbell, John M., Robyn M. Dawes, and Alphons van de Kragt. 1990. "The Limits of Multilateral Promising." *Ethics* 100(April):616–27.

Orbell, John M., Peregrine Schwartz-Shea, and Randy Simmons. 1984. "Do Cooperators Exit More Readily than Defectors?" *American Political Science Review* 78(March):147–62.

Orbell, John M., Alphons van de Kragt, and Robyn M. Dawes. 1988. "Explaining Discussion-Induced Cooperation." *Journal of Personality and Social Psychology* 54(5):811–19.

Ostrom, Elinor. 1990. *Governing the Commons: The Evolution of Institutions for Collective Action.* New York: Cambridge University Press.

Ostrom, Elinor. 1998. "Self-Governance of Common-Pool Resources." In *The New Palgrave Dictionary of Economics and the Law,* vol. 3, ed. Peter Newman, 424–33. London: Macmillan Press.

Ostrom, Elinor, Roy Gardner, and James Walker. 1994. *Rules, Games, and Common-Pool Resources.* Ann Arbor: University of Michigan Press.

Ostrom, Elinor, and James Walker. 1997. "Neither Markets nor States: Linking Transformation Processes in Collective Action Arenas." In *Perspectives on Public Choice: A Handbook,* ed. Dennis C. Mueller. Cambridge: Cambridge University Press. 35–72. (Chapter 15 in this volume.)

Ostrom, Elinor, James Walker, and Roy Gardner. 1992. "Covenants with and without a Sword: Self-Governance Is Possible." *American Political Science Review* 86 (June):404–17.

Ostrom, Vincent. 1980. "Artisanship and Artifact." *Public Administration Review* 40 (July-August):309–17. (Reprinted in Michael D. McGinnis, ed. *Polycentric Governance and Development* [Ann Arbor: University of Michigan Press, 1999].)

Ostrom, Vincent. 1987. *The Political Theory of a Compound Republic: Designing the American Experiment.* 2d rev. ed. San Francisco, CA: Institute for Contemporary Studies Press.

Ostrom, Vincent. 1990. "Problems of Cognition as a Challenge to Policy Analysts and Democratic Societies." *Journal of Theoretical Politics* 2(3):243–62. (Reprinted in Michael D. McGinnis, ed. *Polycentric Governance and Development* [Ann Arbor: University of Michigan Press, 1999].)

Ostrom, Vincent. 1997. *The Meaning of Democracy and the Vulnerability of Democracies: A Response to Tocqueville's Challenge.* Ann Arbor: University of Michigan Press.

Ostrom, Vincent, David Feeny, and Hartmut Picht, eds. 1993. *Rethinking Institutional Analysis and Development: Issues, Alternatives, and Choices.* San Francisco, CA: Institute for Contemporary Studies Press.

Palfrey, Thomas R., and Howard Rosenthal. 1988. "Private Incentives in Social Dilemmas." *Journal of Public Economics* 35(April):309–32.

Piaget, Jean. [1932] 1969. *The Moral Judgment of the Child.* New York: Free Press.

Pinker, Steven. 1994. *The Language Instinct.* New York: W. Morrow.

Pinkerton, Evelyn, ed. 1989. *Co-operative Management of Local Fisheries: New Directions for Improved Management and Community Development.* Vancouver: University of British Columbia Press.

Plott, Charles R. 1979. "The Application of Laboratory Experimental Methods to Public Choice." In *Collective Decision Making: Applications from Public*

Choice Theory, ed. Clifford S. Russell. Baltimore, MD: Johns Hopkins University Press. 137–60.

Pruitt, D. G., and M. J. Kimmel. 1977. "Twenty Years of Experimental Gaming: Critique, Synthesis, and Suggestions for the Future." *Annual Review of Psychology* 28:363–92.

Putnam, Robert D., with Robert Leonardi and Raffaella Nanetti. 1993. *Making Democracy Work: Civic Traditions in Modern Italy*. Princeton, NJ: Princeton University Press.

Rabin, Matthew. 1994. "Incorporating Behavioral Assumptions into Game Theory." In *Problems of Coordination in Economic Activity*, ed. J. Friedman. Norwell, MA: Kluwer Academic Press.

Rapoport, Amnon. 1997. "Order of Play in Strategically Equivalent Games in Extensive Form." *International Journal of Game Theory* 26(1):113–36.

Rocco, Elena, and Massimo Warglien. 1995. "Computer Mediated Communication and the Emergence of 'Electronic Opportunism.' " Working paper RCC 13659. Universita degli Studi di Venezia.

Roth, Alvin E. 1995. "Bargaining Experiments." In *Handbook of Experimental Economics*, ed. John Kagel and Alvin E. Roth. Princeton, NJ: Princeton University Press.

Roth, Alvin E., Vesna Prasnikar, Masahiro Okuno-Fujiwara, and Shmuel Zamir. 1991. "Bargaining and Market Behavior in Jerusalem, Ljubljana, Pittsburgh, and Tokyo: An Experimental Study." *American Economic Review* 81 (December):1068–95.

Rutte, Christel G., and H. A. M. Wilke. 1984. "Social Dilemmas and Leadership." *European Journal of Social Psychology* 14(January–March):105–21.

Sally, David. 1995. "Conservation and Cooperation in Social Dilemmas. A Meta-Analysis of Experiments from 1958 to 1992." *Rationality and Society* 7 (January):58–92.

Samuelson, Charles D., and David M. Messick. 1986. "Alternative Structural Solutions to Resource Dilemmas." *Organizational Behavior and Human Decision Processes* 37(February):139–55.

Samuelson, Charles D., and David M. Messick. 1995. "When Do People Want to Change the Rules for Allocating Shared Resources." In *Social Dilemmas. Perspectives on Individuals and Groups*, ed. David A. Schroeder. Westport, CT: Praeger. 143–62.

Samuelson, Charles D., David M. Messick, Christel G. Rutte, and H. A. M. Wilke. 1984. "Individual and Structural Solutions to Resource Dilemmas in Two Cultures." *Journal of Personality and Social Psychology* 47(July):94–104.

Samuelson, Larry, John Gale, and Kenneth Binmore. 1995. "Learning to be Imperfect: The Ultimatum Game." *Games and Economic Behavior* 8(January):56–90.

Samuelson, P. A. 1954. "The Pure Theory of Public Expenditure." *Review of Economics and Statistics* 36(November):387–9.

Sandler, Todd. 1992. *Collective Action: Theory and Applications*. Ann Arbor: University of Michigan Press.

Sato, Kaori. 1987. "Distribution of the Cost of Maintaining Common Property Resources." *Journal of Experimental Social Psychology* 23(January):19–31.

Satz, Debra, and John Ferejohn. 1994. "Rational Choice and Social Theory." *Journal of Philosophy* 91(February):71–82.

Scharpf, Fritz W. 1997. *Games Real Actors Play: Actor Centered Institutionalism in Policy Research.* Boulder, CO: Westview Press.

Schelling, Thomas C. 1978. *Micromotives & Macrobehavior.* New York: W. W. Norton.

Schlager, Edella. 1990. "Model Specification and Policy Analysis: The Governance of Coastal Fisheries." Ph.D. diss., Indiana University.

Schlager, Edella, and Elinor Ostrom. 1993. "Property-Rights Regimes and Coastal Fisheries: An Empirical Analysis." In *The Political Economy of Customs and Culture: Informal Solutions to the Commons Problem,* ed. Randy Simmons and Terry Anderson. Lanham, MD: Rowman & Littlefield. 13–41. (Reprinted in Michael D. McGinnis, ed. *Polycentric Governance and Development* [Ann Arbor: University of Michigan Press, 1999].)

Schneider, Friedrich, and Werner W. Pommerehne. 1981. "Free Riding and Collective Action: An Experiment in Public Microeconomics." *Quarterly Journal of Economics* 96(November):689–704.

Scholz, John T. 1998. "Trust, Taxes, and Compliance." In *Trust and Governance,* ed. Valerie Braithwaite and Margaret Levi. New York: Russell Sage Foundation.

Schroeder, David A., ed. 1995. *Social Dilemmas. Perspectives on Individuals and Groups.* Westport, CT: Praeger.

Schuessler, Rudolph. 1989. "Exit Threats and Cooperation under Anonymity." *Journal of Conflict Resolution* 33(December):728–49.

Sell, Jane, and Rick Wilson. 1991. "Levels of Information and Contributions to Public Goods." *Social Forces* 70(September):107–24.

Sell, Jane, and Rick Wilson. 1992. "Liar, Liar, Pants on Fire: Cheap Talk and Signalling in Repeated Public Goods Settings." Working paper. Department of Political Science, Rice University.

Selten, Reinhard. 1975. "Reexamination of the Perfectness Concept for Equilibrium Points in Extensive Games." *International Journal of Game Theory* 4(1):25–55.

Selten, Reinhard. 1986. "Institutional Utilitarianism." In *Guidance, Control, and Evaluation in the Public Sector,* ed. Franz-Xaver Kaufmann, Giandomenico Majone, and Vincent Ostrom. New York: de Gruyter. 251–63.

Selten, Reinhard. 1990. "Bounded Rationality." *Journal of Institutional and Theoretical Economics* 146(December):649–58.

Selten, Reinhard. 1991. "Evolution, Learning, and Economic Behavior." *Games and Economic Behavior* 3(February):3–24.

Selten, Reinhard, Michael Mitzkewitz, and Gerald R. Uhlich. 1997. "Duopoly Strategies Programmed by Experienced Players." *Econometrica* 65(May):517–55.

Sen, Amartya K. 1977. "Rational Fools: A Critique of the Behavioral Foundations of Economic Theory." *Philosophy and Public Affairs* 6(Summer):317–44.

Sethi, Rajiv, and E. Somanathan. 1996. "The Evolution of Social Norms in Common Property Resource Use." *American Economic Review* 86(September):766–88.

Shepsle, Kenneth A., and Barry R. Weingast. 1984. "Legislative Politics and Budget

Outcomes." In *Federal Budget Policy in the 1980's,* ed. Gregory Mills and John Palmer. Washington, DC: Urban Institute Press. 343–67.

Simon, Herbert A. 1985. "Human Nature in Politics: The Dialogue of Psychology with Political Science." *American Political Science Review* 79(June):293–304.

Simon, Herbert A. 1997. *Models of Bounded Rationality: Empirically Grounded Economic Reason.* Cambridge, MA: MIT Press.

Smith, Vernon. 1982. "Microeconomic Systems as an Experimental Science." *American Economic Review* 72(December):923–55.

Snidal, Duncan. 1985. "Coordination versus Prisoner's Dilemma: Implications for International Cooperation and Regimes." *American Political Science Review* 79 (December):923–42.

Tang, Shui Yan. 1992. *Institutions and Collective Action: Self-Governance in Irrigation.* San Francisco, CA: Institute for Contemporary Studies Press.

Taylor, Michael. 1987. *The Possibility of Cooperation.* New York: Cambridge University Press.

Thibaut, J. W., and H. H. Kelley. 1959. *The Social Psychology of Groups.* New York: Wiley.

Tocqueville, Alex de. [1835 and 1840] 1945. *Democracy in America.* 2 vols. Ed. Phillips Bradley. New York: Alfred A. Knopf.

Trivers, Robert L. 1971. "The Evolution of Reciprocal Altruism." *Quarterly Review of Biology* 46(March):35–57.

van de Kragt, Alphons, John M. Orbell, and Robyn M. Dawes. 1983. "The Minimal Contributing Set as a Solution to Public Goods Problems." *American Political Science Review* 77(March):112–22.

Walker, James, Roy Gardner, Andrew Herr, and Elinor Ostrom. 1997. "Voting on Allocation Rules in a Commons: Predictive Theories and Experimental Results." Presented at the 1997 annual meeting of the Western Political Science Association, Tucson, Arizona, March 13–15.

Walker, James, Roy Gardner, and Elinor Ostrom. 1990. "Rent Dissipation in a Limited-Access Common-Pool Resource: Experimental Evidence." *Journal of Environmental Economics and Management* 19(November):203–11.

Williams, John T., Brian Collins, and Mark I. Lichbach. 1997. "The Origins of Credible Commitment to the Market." Presented at the 1995 annual meeting of the American Political Science Association, Chicago, Illinois.

Yamagishi, Toshio. 1986. "The Provision of a Sanctioning System as a Public Good." *Journal of Personality and Social Psychology* 51(1):110–6.

Yamagishi, Toshio. 1988a. "Exit from the Group as an Individualistic Solution to the Free Rider Problem in the United States and Japan." *Journal of Experimental Social Psychology* 24(6):530–42.

Yamagishi, Toshio. 1988b. "The Provision of a Sanctioning System in the United States and Japan." *Social Psychology Quarterly* 51(3):265–71.

Yamagishi, Toshio. 1988c. "Seriousness of Social Dilemmas and the Provision of a Sanctioning System." *Social Psychology Quarterly* 51(1):32–42.

Yamagishi, Toshio. 1992. "Group Size and the Provision of a Sanctioning System in a Social Dilemma." In *Social Dilemmas: Theoretical Issues and Research Find-

ings, ed. W. B. G. Liebrand, David M. Messick, and H. A. M. Wilke. Oxford, England: Pergamon Press. 267–87.

Yamagishi, Toshio, and Karen S. Cook. 1993. "Generalized Exchange and Social Dilemmas." *Social Psychological Quarterly* 56(4):235–48.

Yamagishi, Toshio, and Nahoko Hayashi. 1996. "Selective Play: Social Embeddedness of Social Dilemmas." In *Frontiers in Social Dilemmas Research,* ed. W. B. G. Liebrand and David M. Messick. Berlin: Springer-Verlag.

Yamagishi, Toshio, and Nobuyuki Takahashi. 1994. "Evolution of Norms without Metanorms." In *Social Dilemmas and Cooperation,* ed. Ulrich Schutz, Wulf Albers, and Ulrich Mueller, Berlin: Springer-Verlag. 311–26.

Yoder, Robert. 1994. *Locally Managed Irrigation Systems.* Colombo, Sri Lanka: International Irrigation Management Institute.

Suggested Further Readings

The best place to begin is with *Rules, Games, and Common-Pool Resources* (University of Michigan Press, 1994), written by Elinor Ostrom, Roy Gardner, and James Walker with the assistance of four Workshop colleagues. This is a truly unique book that fully integrates field research, formal models, and laboratory experiments, all focused on the management of common-pool resources. In short, this book it exemplifies the Workshop's multifaceted approach to institutional analysis. Field research is the primary topic of Elinor Ostrom's award-winning book *Governing the Commons: The Evolution of Institutions for Collective Action* (Cambridge University Press, 1990), but this book also includes a brief introduction to the use of Prisoner's Dilemma games and other formal representations of social dilemmas.

As readers of this volume are well aware, the Workshop approach to institutional analysis draws inspiration from a wide variety of sources. For influential overviews of the closely related fields of research known as public choice, social choice, new institutional economics, and constitutional economics, readers should consult (respectively) Dennis C. Mueller, *Public Choice II* (Cambridge University Press, 1989), William H. Riker, *Liberalism against Populism: A Confrontation between the Theory of Democracy and the Theory of Social Choice* (W. H. Freeman, 1982), Thráinn Eggertsson, *Economic Behavior and Institutions* (Cambridge University Press, 1990), and James M. Buchanan and Gordon Tullock, *The Calculus of Consent: Logical Foundations of Constitutional Democracy* (University of Michigan Press, 1962). *Perspectives on Public Choice: A Handbook* (Cambridge University Press, 1997), edited by Dennis C. Mueller, includes excellent overviews contributed by most of the major players in the public choice tradition.

Three Workshop-affiliated scholars have written textbooks on game theory or environmental economics: Roy Gardner, *Games for Business and Economics* (Wiley, 1995), Eric Rasmusen, *Games and Information: An Introduction to Game Theory*, 2d ed. (Blackwell, 1994), and Steven Hackett, *Environmental and Natural Resources Economics: Theory, Policy, and the Sustainable Society* (M. E. Sharpe, 1998). John Williams and Kenneth Bickers are writing a general textbook on policy analysis, *Public Policy Analysis: A*

Political Economy Approach (Houghton-Mufflin, forthcoming), that incorporates the Workshop perspective. Textbooks focusing on particular policy areas include *Local Government in the United States* (Institute for Contemporary Studies [ICS] Press, 1988), by Vincent Ostrom, Robert Bish, and Elinor Ostrom; and *Institutional Incentives and Sustainable Development: Infrastructure Policies in Perspective* (Westview Press, 1993), by Elinor Ostrom, Larry Schroeder, and Susan Wynne.

Workshop-affiliated scholars have also contributed to several edited volumes in which game models have been applied to a diverse array of empirical settings, including *Games in Hierarchies and Networks: Analytical and Empirical Approaches to the Study of Governance Institutions* (Westview Press, 1993), edited by Fritz W. Scharpf; *Social Dilemmas and Cooperation* (Springer-Verlag, 1994), edited by Ulrich Schulz, Wulf Albers, and Ulrich Mueller; *Understanding Strategic Interaction: Essays in Honor of Reinhard Selten* (Springer-Verlag, 1997), edited by Wulf Albers, Werner Güth, Peter Hammerstein, Benny Moldovanu, and Eric van Damme; and a four-volume set edited by Reinhard Selten, *Game Equilibrium Models I-IV* (Springer-Verlag, 1991).

Empirical applications and laboratory experiments have played important roles in the development of the research programs surveyed in this volume. Examples of such applications are collected in *Laboratory Research in Political Economy* (University of Michigan Press, 1991), edited by Thomas R. Palfrey; and *Empirical Studies in Institutional Change* (Cambridge University Press, 1996), edited by Lee J. Alston, Thráinn Eggertsson, and Douglass C. North. Calvin Jillson and Rick K. Wilson, *Congressional Dynamics: Structure, Coordination, and Choice in the First American Congress, 1774–1789* (Stanford University Press, 1994), shows how the findings of experimental research on voting systems can inform our understanding of important historical processes.

Earlier versions of essays included in this volume can be found in several edited volumes that also include contributions by scholars from other research institutes. The contributors to Franz-Xaver Kaufmann, Giandomenico Majone, and Vincent Ostrom, eds. *Guidance, Control, and Evaluation in the Public Sector* (Walter de Gruyter, 1986), suggest several related approaches toward resolution of the conceptual difficulties involved in modeling large-scale systems. A shorter version of this report on the results of the Bielefeld interdisciplinary project was later published as *The Public Sector—Challenge for Coordination and Learning* (Walter de Gruyter, 1991), edited by Franz-Xaver Kaufmann. The broader implications of Workshop research for development issues are explored in *Rethinking Institutional Analysis and Development,* 2d ed. (ICS Press, 1993), edited by Vincent Ostrom, David Feeny, and Hartmut Picht. This book includes a mixture of microlevel and macrolevel applications.

The current volume was prepared in conjunction with two other volumes of previously published reports on research by Workshop-affiliated scholars. *Polycentric Governance and Development* (University of Michigan Press, 1999) focuses on the management of common-pool resources by self-governing communities throughout the world. This book includes early statements of the Workshop perspective, dating back before game theory had made much of an impact on the study of institutions. This volume complements the current one very well, because most of the models and experiments developed here are based on a generic representation of the problems associated with managing common-pool resources.

It remains to be seen whether simple game models can be equally successful in helping us to understand the complexities of public service provision in metropolitan areas, which is the primary focus of *Polycentricity and Local Public Economies* (University of Michigan Press, 1999). This work includes classic essays on the nature of polycentric order and a series of research reports comparing the performances of large and small police agencies in selected metropolitan areas of the United States.

Charlotte Hess maintains an extensive bibliography on game theory at http://www.indiana.edu/~workshop/wsl/gamebib.html. Bibliographies on related topics are also available at http://www.indiana.edu/~workshop/wsl/bibs. html. Four volumes of *Common-Pool Resources and Collective Action: A Bibliography* have been published by the Workshop in Political Theory and Policy Analysis. Volumes 1 (1989) and 2 (1992) were edited by Fenton Martin and volumes 3 (1996) and 4 (1998) by Charlotte Hess. All four volumes are available in a single file on CD-ROM. Finally, readers are encouraged to check the Workshop's home page (http://www.indiana.edu/~workshop) for recent updates and for links to other reference and teaching materials.

Contributors

Arun Agrawal carries out research on the politics of environment, development, and decentralization. After completing his dissertation at Duke University, he taught for three years at the University of Florida, Gainesville, and is currently assistant professor at Yale University in the political science department. He has published articles in a number of journals, including *Development and Change, Human Ecology, Human Organization, Journal of Theoretical Politics,* and *World Development.* His first book, on migrant shepherds, is entitled *Greener Pastures: Politics, Markets, and Community among a Migrant Pastoral People.* He is currently about to complete his second book, *Community-in-Conservation: Forest Councils of Kumaon in Comparative Perspective.*

Sue E. S. Crawford attended Workshop courses and worked with Elinor Ostrom as a research assistant from 1989 through 1995 during her graduate studies in political science at Indiana University. She currently teaches political science and policy courses at Creighton University and conducts research on the policy activities of clergy. She has published work on institutions and citizen participation, on the politics of urban government institutions, on the policy agendas of clergy, and on teaching policy and U.S. government courses.

Roy Gardner is Chancellor's Professor of Economics and West European Studies at Indiana University. Dr. Gardner specializes in the theory of games and economic behavior. He has applied game theory to such diverse topics as class struggle in the USSR, the processes of veto and purge, spoils systems, elections, draft resistance, budget processes, and corruption. A recent major focus of his research has been human dimensions of global environmental change. In addition to working with pure game theory applications, he has also been part of teams studying game playing behavior in controlled laboratory situations. His research has been supported by the national science foundations of the United States, France, and Germany.

527

Clark C. Gibson is assistant professor of political science at Indiana University. He focuses on environmental politics, especially the governance of natural resources at the local and national levels. He is the author of *Politicians and Poachers: The Political Economy of Wildlife Policy in Africa* and is co-editor (with Elinor Ostrom and Margaret McKean) of *People and Forests: Communities, Institutions, and Governance*.

Steven C. Hackett is associate professor and coordinator of the economics program at Humboldt State University in Arcata, California. Prior to coming to Humboldt State, he was a member of the department of economics at Indiana University from 1989 to 1994, during which time he was affiliated with the Workshop. He has published over 15 articles in various scholarly journals and edited volumes and has been the recipient of a National Science Foundation grant. He has recently published a textbook entitled *Environmental and Natural Resources Economics: Theory, Policy, and the Sustainable Society*.

Roberta Herzberg is an associate professor of political science and administrative director of the Institute of Political Economy at Utah State University. She was a research associate at the Workshop in Political Theory and Policy Analysis and a member of the political science faculty at Indiana University from 1982 to 1992. She specializes in public policy, analytic modeling, and U.S. politics. She currently serves as a commissioner for the Utah Health Policy Commission and on the Utah Graduate Medical Education Council. Her current work develops a new model of risk assignment in health care insurance to outline a new approach to managing health care cost, access, and quality.

Larry L. Kiser is professor of economics and associate vice provost of university studies at Eastern Washington University. He spent a two-year postdoctoral fellowship (1978–80) at the Workshop in Political Theory and Policy Analysis and returned for a sabbatical year in 1983–84. He has published work derived from his Workshop experience in the *Policy Studies Journal, Urban Affairs Quarterly, Constitutional Political Economy*, and the edited volume *Strategies of Political Inquiry*. His current administrative responsibility for interdisciplinary programs also grew directly from his experience at the Workshop.

Thomas P. Lyon received his B.S.E. from Princeton University in 1981 and his Ph.D. in engineering-economic systems from Stanford University in 1989. During the 1995–96 academic year he was visiting associate professor at the University of Chicago, and in 1997 he held the Pisa Chair in the

Economics and Management of Innovation at the Scuola Sant' Anna, in Pisa, Italy. His research interests are in the economics of regulation, industrial organization, contracts, and technological innovation. His varied work experience includes positions with Benjamin Schlesinger and Associates, a management consultancy; the U.S. Department of Energy; the National Audubon Society; and Public Service Electric and Gas Company. He has served as a consultant to a variety of organizations, including Air Products Corporation, AT&T, and Enron Corporation. Professor Lyon teaches managerial economics, game theory, and business and public policy at the Indiana University Kelley School of Business and has lectured on business strategy and the economics of innovation at various universities in Europe.

Stuart A. Marks has been an ethnographer and ecologist while working for academia, government, and the private sector as tenured professor, consultant, and director. Currently, he is senior scientist for research and community development for Safari Club International and a research professor of anthropology at the University of North Carolina–Chapel Hill. Besides his long-term research in Africa, he is the author of *Southern Hunting in Black and White: Nature, History, and Ritual in a Carolina Community.*

Michael D. McGinnis is associate professor of political science and associate director of publications at the Workshop in Political Theory and Policy Analysis. After receiving a B.S. degree in mathematics from Ohio State University and a Ph.D. degree in political science from the University of Minnesota, he joined the faculty at Indiana University in 1985. He became a Workshop research associate in 1990 and a co-associate director in 1997. His initial research on arms rivalries and international conflict was published in several articles and in a forthcoming book, *Compound Dilemmas*, co-authored with John Williams. His current research interests concern institutional arrangements for the provision of humanitarian aid to communities displaced by conflict and famine.

Michael Moore is associate professor of resource and environmental economics in the School of Natural Resources and Environment at the University of Michigan. His research interests include economic analysis of water law and policy in the U.S. West and the study of collective action mechanisms. He is presently investigating consumption of green electricity as a noncooperative approach to private provision of environmental public goods.

Elinor Ostrom is Arthur F. Bentley Professor of Political Science, co-director of the Workshop in Political Theory and Policy Analysis, and co-director of the Center for the Study of Institutions, Population, and Environmental

Change, all at Indiana University in Bloomington. She received her Ph.D. degree from the University of Southern California in 1965 and began teaching at Indiana University the same year. She has served as chair of the political science department and as president of four professional associations: the American Political Science Association, International Association for the Study of Common Property, Midwest Political Science Association, and Public Choice Society. She has served as a consultant and member of advisory boards for several local and national organizations. She is a fellow of the American Academy of Arts and Sciences and received the Frank E. Seidman Distinguished Award in political economy in 1997. She was a cofounder and coeditor of *Journal of Theoretical Politics.* Her best-known book is *Governing the Commons,* and she has authored, coauthored, and edited numerous other books as well as many journal articles and book chapters.

Vincent Ostrom is Arthur F. Bentley Professor Emeritus of Political Science and co-director of the Workshop in Political Theory and Policy Analysis. He received his Ph.D. degree from the University of California at Los Angeles in 1950. After teaching at the Universities of Wyoming and Oregon and at UCLA, he joined the political science faculty at Indiana University in 1964. Throughout his career he has kept active in resource policy, serving, for example, as consultant for the Wyoming Legislative Interim Committee, the Alaska Constitutional Convention, the Territory of Hawaii, and the National Water Commission. He was editor-in-chief for *Public Administration Review* (1963–66) and president of the Public Choice Society (1967–69). He is the author of several books and numerous articles and book chapters. Among his principal works are *The Intellectual Crisis in American Public Administration, The Political Theory of a Compound Republic, The Meaning of American Federalism,* and *The Meaning of Democracy and the Vulnerability of Democracies.*

Eric Rasmusen received his B.A. and M.A. in economics from Yale University in 1980 and his Ph.D. from M.I.T. in 1984, after attending University High School in Urbana, Illinois. He has taught at UCLA's Anderson School and Indiana University's Kelley School of Business and has visited at Yale Law School and the University of Chicago Graduate School of Business. He is best known for his book, *Games and Information,* now going into its third edition, but his recent research has turned toward law and economics.

James Walker is a research associate of the Workshop in Political Theory and Policy Analysis and professor of economics at Indiana University. His principal research focus is the use of experimental methods in the investigation of individual and group behavior related to the voluntary provision of public goods and the use of common-pool resources. Recent publications include *Rules, Games, and Common-Pool Resources* (with E. Ostrom and R. Gardner); "Group Size and Voluntary Provision of Public Goods: Experimental Evidence Utilizing Large Groups" (with Mark Isaac and Arlington Williams) (*Journal of Public Economies*); and "Appropriation Externalities in the Commons: Repetition, Time Dependence, and Group Size" (with A. Herr and R. Gardner) (*Games and Economic Behavior*).

Franz J. Weissing studied mathematics and biology at the University of Bielefeld (Germany), where he obtained his Ph.D. As a student of Nobel laureate Reinhard Selten, he centered his research around evolutionary game theory. In particular, he has investigated the question of whether "evolutionarily stable strategies" do indeed characterize the outcome of natural selection. From 1987 to 1989, he was Reinhard Selten's research assistant at the Centre for Interdisciplinare Research (Bielefeld). During the research year on the subject "Game Theory in the Behavioral Sciences," he and Elinor Ostrom started their collaboration on the game theoretical analysis of rule following and rule enforcement, taking irrigation institutions as their main example. Since 1989, he has been an associate professor in theoretical biology at the University of Gronigen, the Netherlands.

John T. Williams is professor of the Department of Political Science at Indiana University. He is also a member of the Workshop faculty. After receiving degrees in political science from the University of North Texas, he received his Ph.D. from the University of Minnesota. His research has focused on defense and economic policy. He also does ample research on political methodology, particularly time series analysis. Recent work has focused on policy credibility, distributive politics in a local setting, and time series methods for event count data. John Williams and Mike McGinnis have a forthcoming book, titled *Compound Dilemmas,* about the rivalry between the Soviet Union and United States. He is also an author of a forthcoming text called *Public Policy: A Political Economy Approach,* with Kenneth Bickers.

Rick K. Wilson has been an associate of the Workshop since 1977, when he entered graduate school at Indiana University. He received his Ph.D. in 1982. From 1982 to 1983, he was a postdoctoral fellow at Washington University. In 1983, he joined the faculty of Rice University, where he is currently professor of political science. He was in residence at the Workshop for the academic year 1989–90 while on sabbatical. During 1996–98, he served as program officer of political science at the National Science Foundation. He has published many journal articles and is coauthor (with Cal Jillson) of *Congressional Dynamics: Structure, Coordination and Choice in the First American Congress, 1774–1789.*

Index

INSTITUTIONAL ANALYSIS

Editors, Michael D. McGinnis and Elinor Ostrom